Subcellular Biochemistry

Volume 16

Intracellular Transfer of Lipid Molecules

SUBCELLULAR BIOCHEMISTRY

Recent Volumes in This Series:

Volumes 5–11	Edited by Donald B. Roodyn
Volume 12	**Immunological Aspects** Edited by J. R. Harris
Volume 13	**Fluorescence Studies on Biological Membranes** Edited by H. J. Hilderson
Volume 14	**Artificial and Reconstituted Membrane Systems** Edited by J. R. Harris and A.-H. Etémadi
Volume 15	**Virally Infected Cells** Edited by J. R. Harris
Volume 16	**Intracellular Transfer of Lipid Molecules** Edited by H. J. Hilderson

Subcellular Biochemistry

Volume 16

Intracellular Transfer of Lipid Molecules

Edited by

H. J. Hilderson

University of Antwerp
Antwerp, Belgium

SPRINGER SCIENCE+BUSINESS MEDIA, LLC

The Library of Congress cataloged the first volume of this title as follows:

Sub-cellular biochemistry.

London, New York, Plenum Press.
v. illus. 23 cm. quarterly.
Began with Sept. 1971 issue. Cf. New serial titles.
1. Cytochemistry – Periodicals. 2. Cell organelles – Periodicals.
QH611.S84 574.8′76 73-643479

ISBN 978-1-4899-1623-5 ISBN 978-1-4899-1621-1 (eBook)
DOI 10.1007/978-1-4899-1621-1

This series is a continuation of the journal *Sub-Cellular Biochemistry*, Volumes 1 to 4 of which were published quarterly from 1972 to 1975

Originally published by Plenum Press, New York in 1990
Softcover reprint of the hardcover 1st edition 1990

Contributors

Akira Abe Department of Biochemistry, Cancer Research Institute, Sapporo Medical College, Sapporo 060, Japan

J. F. Billheimer E. I. du Pont de Nemours & Co., Medical Products Department, Experimental Station, Wilmington, Delaware 19880-0400

Rhoderick E. Brown The Hormel Institute, University of Minnesota, Austin, Minnesota 55912

Richard C. Crain Department of Molecular and Cell Biology, The University of Connecticut, Storrs, Connecticut 06268

G. Daum Institut für Biochemie und Lebensmittelchemie, Technische Universität Graz, A-8010 Graz, Austria

M. De Wolf RUCA-Laboratory for Human Biochemistry, University of Antwerp, B2020 Antwerp, Belgium

W. Dierick RUCA-Laboratory for Human Biochemistry, University of Antwerp, B2020 Antwerp, Belgium

Ulf Eriksson Ludwig Institut for Cancer Research, Stockholm Branch, S-104 01 Stockholm, Sweden

George M. Helmkamp, Jr. Department of Biochemistry and Molecular Biology, School of Medicine, University of Kansas Medical Center, Kansas City, Kansas 66103-8410

H. J. Hilderson RUCA-Laboratory for Human Biochemistry, University of Antwerp, B2020 Antwerp, Belgium

Jean-Claude Kader Laboratoire de Physiologie Cellulaire, Unité de Recherches Associée au CNRS 1180, Université Pierre et Marie Curie, 75005 Paris, France

A. Lagrou RUCA-Laboratory for Human Biochemistry, University of Antwerp, B2020 Antwerp, Belgium

Manju Mukherjea Department of Biochemistry, University of Calcutta, Calcutta-700019, India

F. Paltauf Institut für Biochemie und Lebensmittelchemie, Technische Universität Graz, A-8010 Graz, Austria

René J. A. Paulussen Department of Biochemistry, University of Nijmegen, 6500 HB Nijmegen, The Netherlands

M. P. Reinhart USDA–ARS–ERRC, Wyndmoor, Pennsylvania 19118

Frank Roerink Department of Biochemistry, Cancer Research Institute, Sapporo Medical College, Sapporo 060, Japan

Terukatsu Sasaki Department of Biochemistry, Cancer Research Institute, Sapporo Medical College, Sapporo 060, Japan

P. Somerharju Department of Medical Chemistry, University of Helsinki, Helsinki 17, Finland

Friedrich Spener Department of Biochemistry, University of Münster, D-4400 Münster, Federal Republic of Germany

G. Van Dessel UIA–Laboratory for Pathological Biochemistry, University of Antwerp, B2610 Antwerp, Belgium

P. A. van Paridon Center for Biomembranes and Lipid Enzymology, University of Utrecht, 3508 TB Utrecht, The Netherlands

Jacques H. Veerkamp Department of Biochemistry, University of Nijmegen, 6500 HB Nijmegen, The Netherlands

K. W. A. Wirtz Center for Biomembranes and Lipid Enzymology, University of Utrecht, 3508 TB Utrecht, The Netherlands

Preface

Corpora non agunt nisi fixata. This old saying of Ehrlich's describing the physiological role of receptors and their ligands might be paraphrased into *Corpora non ambulant nisi fixata* when considering lipid transport between and within cells. Volume 16 of *Subcellular Biochemistry* is intended to bring the reader up to date with this young field. Indeed, lipid transfer proteins have only recently become the subject of a more systematic study. In this book the current status and the emerging trends are discussed. Chapters cover protein-mediated transfer of fatty acids, phospholipids, phosphatidylinositol, glycolipids, dolichol, retinoids, and cholesterol in animal, plant, yeast, and other eukaryotic cells. Details are included of the study of lipid transport proteins by means of fluorescent phospholipid analogues and of the lipid transfer proteins as probes of membrane structure and function, as well as spontaneous lipid transfer as it occurs between biological membranes. Some of the chapters should be read in conjunction with Volume 13 of this series, devoted to fluorescence studies on biological membranes, in particular Chapter 2 (Somerharju *et al.*) concentrating on studies in which fluorescent phospholipid analogues have been used. Chapter 10 (Billheimer and Reinhart), dealing with cholesterol trafficking, should be compared with Chapter 12 of Volume 13 (Van Blitterswijk), pointing to the existence of a preferential association of cholesterol with sphingomyelin, which drags cholesterol to the plasma membrane.

In one chapter (Chapter 8: Van Dessel *et al.*) the authors warn of the danger of underestimating the complexity of subcellular methodology. In addition, more attention should be paid in the literature to a correct use of subcellular terminology (e.g., the term *microsomes* is too frequently misused as a synonym for *endoplasmic reticulum*).

A more systematic discussion of each chapter is given by Friedrich Spener in his introductory chapter. In general, most authors recognize that much is already understood about protein-mediated lipid transfer *in vitro* but that it is frustratingly difficult to define any *in vivo* function for lipid transfer proteins. Obviously, lipid transport processes observed *in vitro* do not necessarily reflect the cellular situation. The most important function assigned up to now to lipid transfer proteins seems to be the facilitation of transport in order to maintain a nonrandom distribution of lipids in biological membranes. Another interesting feature is the ability of these proteins to modulate metabolic pathways by storage of substrates, by sequestration of substrates and/or effectors, or by protection against inhibitors. In this way they can be considered as being involved in cell compartmentalization. Finally, lipid transfer proteins of well-defined specificity, which can be readily purified in large amounts, have become very useful for the study of membrane structure and function (Chapter 3: Crain).

As scientists sometimes tend to obfuscate and confuse, I am most grateful to all of the contributors for doing just the opposite. I also want to express my thanks to Dr. G. Van Dessel for his advice during the preparation of this book.

Herwig Hilderson

Antwerp, Belgium

Contents

Chapter 1

Nonenzymatic Proteins Mediating Intracellular Lipid Transport and Metabolism: Current Status and Emerging Trends

Friedrich Spener and Manju Mukherjea

Chapter 2

Application of Fluorescent Phospholipid Analogues to Studies on Phospholipid Transfer Proteins

P. J. Somerharju, P. A. van Paridon, and K. W. A. Wirtz

Chapter 3

Phospholipid Transfer Proteins as Probes of Membrane Structure and Function

Richard C. Crain

Chapter 4

Intracellular Transfer of Phospholipids, Galactolipids, and Fatty Acids in Plant Cells

Jean-Claude Kader

Chapter 5

Glycolipid Transfer Protein in Animal Cells

Terukatsu Sasaki, Akira Abe, and Frank Roerink

Chapter 6

Transport and Metabolism of Phosphatidylinositol in Eukaryotic Cells

George M. Helmkamp, Jr.

Chapter 7

Intracellular Fatty-Acid-Binding Proteins: Characteristics and Function

René J. A. Paulussen and Jacques H. Veerkamp

Chapter 8

Intracellular and Extracellular Flow of Dolichol

G. Van Dessel, M. De Wolf, H. J. Hilderson, A. Lagrou, and W. Dierick

Chapter 9

Phospholipid Transfer in Microorganisms

Friedrich Paltauf and G. Daum

Chapter 10

Intracellular Trafficking of Sterols

J. T. Billheimer and M. P. Reinhart

Chapter 1

Nonenzymatic Proteins Mediating Intracellular Lipid Transport and Metabolism

Current Status and Emerging Trends

Friedrich Spener and Manju Mukherjea

1. INTRODUCTION

Lipids fascinate cell biologists and biochemists because they can have profound effects on cell function. Encoded in these simple molecules is the ability to form macroscopic, two-dimensional membrane systems spontaneously. In addition to functioning as physical and chemical barriers separating aqueous compartments, membranes are involved in many regulatory processes such as secretion, transport, endocytosis, and signal transduction. The interaction between lipids and

Abbreviations used in this chapter: SCP, sterol carrier protein; CoA, coenzyme A; ACTH, adrenocorticotropic hormone; PLTP, phospholipid transfer protein; PC, phosphatidylcholine; PE, phosphatidylethanolamine; PI, phosphatidylinositol; PS, phosphatidylserine; PG, phosphatidylglycerol; LTP, lipid transfer protein; nsLTP, nonspecific lipid transfer protein; GLTP, glycolipid transfer protein; FABP, fatty acid-binding protein; ACBP, acyl-CoA-binding protein; DBI, diazepam-binding inhibitor; GABA, γ-aminobutyric acid; RBP, retinol-binding protein; CRABP, cellular retinoic acid-binding protein.

Friedrich Spener Department of Biochemistry, University of Münster, D-4400 Münster, Federal Republic of Germany. **Manju Mukherjea** Department of Biochemistry, University of Calcutta, Calcutta-700019, India.

proteins is essential to such membrane activities. Broadly speaking, the metabolic turnover of membrane lipids and proteins encompasses synthesis, sorting, and degradation. Lipids serve as one of the major sources of energy, both directly and also potentially when stored in adipose tissues. They also act as thermal insulators in the subcutaneous tissues and around certain organs, and membrane lipids along myelinated nerves serve as electrical insulators, allowing rapid propagation of waves of depolarization. Some lipids act as biological modulators and signal transducers (e.g., pheromones, prostaglandins, thromboxanes, leukotriens, steroids, platelet-activating factor, and phosphatidylinositol and derivatives) and as the vehicles for carrying fat-soluble vitamins. Some other lipids, which are particularly enriched when present in certain microorganisms and in the spermatozoa and brain of higher animals, exhibit antibacterial, antifungal, and antitumor activities.

1.1. Intracellular Location of Lipids

The various organelles in a eukaryotic cell differ widely in both protein and lipid composition. Although prokaryotic cells contain only a plasma membrane, eukaryotes have evolved an elaborate network of membrane-limited organelles in addition to the plasma membrane. Of the membrane phospholipids, phosphatidylcholine (PC) and phosphatidylethanolamine (PE) are the most common constituents. PC is enriched in the endoplasmic reticulum, whereas the highest concentration of PE occurs in the mitochondria. Phosphatidylinositol (PI) and its phosphorylated derivatives are found in the plasma membrane. Cardiolipin, interestingly, is located exclusively on the inner mitochondrial membrane. The highest levels of sphingomyelin and cholesterol occur in the plasma membrane, with a significant concentration of the former in both Golgi and lysosomal membranes and an almost complete absence in the endoplasmic reticulum and mitochondria. Triglycerides are stored in special globuli or assembled for export with apolipoproteins. These examples indicate that specific sorting and transport processes are necessary to ensure the correct ultimate location of these molecules within the cell. Moreover, all these lipids are highly insoluble in the aqueous media of biological systems. Thus, specific mechanisms must exist for transport, and failure to transport a given lipid to its correct destination could result in malfunction of cellular processes.

1.2. Cellular Uptake of Lipids and Their Transfer through the Cell

Lipids cross the boundary of cells either as complexes with proteins or as single molecules. Several models have been proposed to explain the mechanism of their uptake into mammalian cells; these include (1) receptor-mediated endocytosis via lipoproteins; (2) passive diffusion through the lipid bilayer of plasma membrane, by which lipids from the aqueous phase can enter the outer leaflet and spontaneously move over to the inner half; (3) facilitated diffusion; and (4) membrane-associated protein receptors that facilitate the dissociation of a lipid-protein complex, which is followed by passive diffusion of the lipid.

The pathways by which water-insoluble compounds move between different compartments within cells are gradually being better understood. The enzymes responsible for lipid biosynthesis are spread mainly over the endoplasmic reticulum and Golgi apparatus, although other organelle membranes are involved to some extent. Thus, a major problem in the cell biology of lipids is to understand how newly synthesized lipids are sorted and how these molecules are translocated or targeted to various destinations inside the cell. The same consideration applies to imported lipids. Potential mechanisms which direct lipid traffic in cells are (1) spontaneous lipid transfer between membranes, (2) intra-organelle transport of lipids by vesicle budding and fusion as well as transport by lateral diffusion between organelles connected by transient junctions, and (3) carrier protein-mediated lipid transport.

Spontaneous transfer of lipids is discussed in Chapter 11 of this volume, whereas concepts of vesicle budding and fusion as well as membrane flow can be found elsewhere (Farquhar, 1985; Scow and Blanchette-Mackie, 1985). Most chapters in this volume are devoted to the nature and function of nonenzymatic proteins that are believed to mediate intracellular transfer and metabolism of different lipid classes. In this introductory chapter we provide a synopsis covering these proteins.

2. CARRIER PROTEIN-MEDIATED LIPID TRANSFER

The release of lipid molecules from the plasma membrane into the aqueous cytosol may require the help of carrier molecules. Considerable evidence indicates that lipids are translocated between intracellular compartments complexed with proteins that have variable specificities for binding to hydrophobic ligands or to membranes. These proteins do not exhibit enzymatic activity in the usual sense, but they may exert a permissive or enhancing effect on the entry of these ligands into reactions catalyzed by enzymes.

2.1. Sterol Carrier Proteins

Sterols are prominent membrane components of eukaryotic cells but are absent among prokaryotes. In specialized eukaryotic cells cholesterol is also needed in free or esterified form for incorporation into nascent lipoproteins prior to secretion, for storage in oil droplets, or as a precursor for bile acids and hormones. Thus, from their site of synthesis (the endoplasmic reticulum), sterols have to travel to their intracellular targets according to tissue need. Spontaneous and vesicular transfer are both considered potential mechanisms; however, a number of sterol carrier proteins (SCPs) are reported to be involved in the transfer of sterols. These proteins were initially isolated because of their possible role in cholesterol synthesis from squalene.

2.1.1. Types and Functions

Proteins have been isolated from rat liver cytosol, namely, squalene and sterol carrier protein, SCP_1, and SCP_2. It is clear nowadays that SCP and hepatic-type fatty acid-binding protein are identical and do not bind cholesterol (Bass, 1988). SCP_1 is a 47-kDa protein (also known as supernatant protein factor) and increases the conversion of squalene to lanosterol by increasing squalene epoxidase activity in intact microsomes (Ono and Bloch, 1975).

SCP_2 increases the synthesis of cholesterol from lanosterol, the esterification of cholesterol by acyl coenzyme A (acyl-CoA):cholesterol acyltransferase (Gavey *et al.*, 1981), and the oxidation of cholesterol by the side-chain cleavage enzyme or 7-hydroxycholesterol oxidase (Chanderbhan *et al.*, 1982). Such increases in enzyme activity can be achieved *in vitro* by the transfer of sterol from liposomes to the organelles that contain the enzymes (Noland *et al.*, 1980). Attempts have been made to extend *in vitro* observations to the intracellular transport of cholesterol *in vivo*. In adrenocortical cells treated with adrenocorticotropic hormone (ACTH), a threefold increase in SCP_2 synthesis has been demonstrated (Trzeciak *et al.*, 1987). Moreover, fusion of adrenocortical cells with liposomes containing anti-SCP_2-antibody decreased ACTH-stimulated steroidogenesis by 45%, indicating the involvement of SCP_2 in steroidogenic cells.

Besides these SCPs, a triglyceride and cholesterol ester transfer protein has been partially purified from liver (Wetterau and Zilversmit, 1986). It has been suggested that this protein also plays a role in the intestine for cholesterol ester incorporation into nascent lipoproteins.

Oxygenated derivatives of cholesterol suppress cholesterol biosynthesis in mammalian cells. This regulation appears to involve a specific, high-affinity binding of the oxysterols to a cytosolic oxysteroid-binding protein (Taylor and Kandutsch, 1985).

2.1.2. Properties of SCP_2

Of the proteins initially isolated for their ability to transfer cholesterol, the most extensively studied is SCP_2. This protein was purified from the livers of various mammals and from rat ovary. It is a basic, heat-stable, 14.0-kDa protein and may be synthesized from a higher-molecular-mass precursor protein (Trzeciak *et al.*, 1987). Its concentration is highest in the cytosol of the liver

(0.78 μg/mg) and intestine (0.46 μg/mg), with the supernatant of all other tissues having about 0.1 μg of SCP_2/mg of protein. Several investigators have shown, by using immunohistochemical staining, that a peroxisomal protein reacted with anti-SCP_2-antibody (Tsuneoka *et al.*, 1988; Keller *et al.*, 1989). This observation was strengthened by the fact that the C terminus of SCP_2 contains a peroxisomal targeting signal (Morris *et al.*, 1988). At least 50% of SCP_2 immunoreactivity has been estimated to be peroxisomal (Tsuneoka *et al.*, 1988), although this organelle does not play a major role in cholesterol synthesis or metabolism. It is now recognized that SCP_2 is present in many subcellular organelles (Van Noort *et al.*, 1988). According to Van Amerongen *et al.* (1989), only 19% of adrenal SCP_2 was cytosolic, whereas two-thirds of liver SCP_2 was in the supernatant, despite the similarity of the total SCP_2 content in rat liver and adrenal gland. This indicates that analysis of only the supernatant fraction can be misleading for the estimation of total cellular SCP_2 content.

SCP_2 probably does not act as a classic binding protein for cholesterol, but SCP_2-mediated transfer of sterols between membranes may serve (1) to promote membrane fusion, (2) to increase the off-rate of sterols from a membrane, (3) to carry bound sterol across the aqueous milieu, and (4) to facilitate the close association of two membranes without either fusion or an aqueous space. SCP_2, however, also facilitates the *in vitro* transfer of various phospholipids, sphingomyelin, neutral glycosphingolipids, and gangliosides (Bloj and Zilversmit, 1977; Crain and Zilversmit, 1980), suggesting that it is nonspecific and, accordingly, is also known as nonspecific lipid transfer protein (nsLTP) (see Section 2.2.2).

2.2. Phospholipid Transfer Proteins

Phospholipids are major components of all cellular membranes. The diversity of their chemical structure and their distribution pattern within the membranous structures of the cell are well matched to the existence of various cytosolic phospholipid transfer proteins (PLTPs). Wirtz and Zilversmit (1968) showed that some cytosolic proteins of rat liver stimulated phospholipid movement between membranes. These proteins, which have been previously designated as "phospholipid exchange proteins" or, because of their broad specificity, simply as "lipid transfer proteins" have various specificities and physical characteristics. They have now been purified from animal and plant tissues, yeasts, and bacteria.

2.2.1. Functions and Specificities

Little is known about the precise physiological role of PLTPs. It has been proposed that they are involved in the transport of newly synthesized membrane lipids from their site of synthesis to their final destination in the cell. This proposal is supported by the fact that the rapid intracellular movement of lipids cannot be explained by either spontaneous transfer or vesicle delivery (Sleight, 1987). For example, substantial increases in PC transfer activity occur at the onset of lung surfactant synthesis and these increases correlate with increased PC synthetic activity (Teerlink *et al.*, 1982). Likewise, at the onset of myelogenesis, increases in PI transfer activity are also observed, but it is apparent that transfer proteins are not required for the integration of all membrane lipids (Ruenwongsa *et al.*, 1979). Rothman and Kennedy (1977) reported that PE synthesized on the cytoplasmic side of the plasma membrane of *Bacillus megaterium* is rapidly translocated to the external leaflet, indicating that transmembrane flip-flop must be a catalyzed process in membranes of living cells. The existence of flippases in higher eukaryotic cells has been demonstrated (Backer and Dawidowicz, 1987). Furthermore, no correlation is observed between rates of PC and PE movement in cultured cells and the cellular content or the respective enzymatic activities (Yaffe and Kennedy, 1983). According to Voelker (1985), intracellular movement of phosphatidylserine (PS) in BHK cells is energy dependent. Thus, many controversies exist regarding the exact role and involvement of PLTPs in cellular functions.

2.2.1a. PLTPs from Mammals and Plants. The molecular specificity of PLTPs has been determined under conditions where transfer rates of phospholipids differing in polar headgroup and/or fatty acyl residues have been compared (Wetterau and Zilversmit, 1984). They can be categorized into one of three groups: (1) PC transfer proteins (around 16 kDa), which are specific for PC and have been purified from bovine and rat liver; (2) PI transfer proteins, which have a marked preference for PI but also catalyze the transfer of PC and have been purified as 36-kDa proteins from rat brain and testis, bovine heart and brain, and human platelets (Helmkamp, 1985); and (3) nsLTPs, which are relatively nonspecific for headgroup composition and have been purified from rat, bovine, and human liver; rat lungs; castor bean seeds; and spinach leaf.

The presence of both acyl groups is required for transfer; i.e., transfer of lyso-PC is not accelerated by PC transfer protein (Kamp *et al.*, 1977). Although this protein has a broad specificity for acyl chain composition, unsaturated PC is transferred preferentially between vesicles and between vesicles and erythrocyte membranes (Helmkamp, 1985, 1986).

2.2.1b. PLTPs from Microorganisms. The presence of a PC-transferring protein in the soluble fraction from the filamentous fungus *Mucor mucedo* was shown by Chavant and Kader (1982). Bozzate and Tinker (1987) isolated a 33.7-kDa protein from *Saccharomyces cerevisiae* with a rather broad specificity for the transfer of phospholipids, whereas Daum and Paltauf (1984) were able to identify a PC/PI transfer protein (35 kDa) in the same organism. Phospholipid transfer activity is highest during the exponential phase of growth, indicating that in *S. cerevisiae,* as in higher eukaryotes, PI is essential for growth.

By analogy with eukaryotic cells (Helmkamp, 1986), the cytosol of prokaryotes contains proteins that can catalyze phospholipid transfer between membranes. Tai and Kaplan (1984) purified a PLTP from extracts of *Rhodopseudomonas sphaeroides*. The transfer proteins from two fractions, i.e., cytoplasmic and periplasmic, are different with respect to their molecular mass and their substrate specificity. The cytoplasmic protein that has been purified to homogeneity has a molecular mass of 27 kDa and a preference for phosphatidylglycerol (PG), whereas the periplasmic protein has a molecular mass of 56 kDa and no preference for specific phospholipids. Proteins contained in the soluble fraction of *Bacillus subtilis* homogenates accelerated the transfer of phospholipids from isolated mesosomes to protoplasts.

2.2.2. Nonspecific Lipid Transfer Protein

nsLTP was first isolated from the pH 5.1 supernatant of rat liver cytosol by its ability to stimulate PE transfer from liposomes to mitochondria (Bloj and Zilversmit, 1977). nsLTP is unique in stimulating the intermembrane transfer of a wide range of lipid molecules including diacylglycerols, gangliosides, glycosphingolipids, and cholesterol. Meanwhile, nsLTP has been shown to be identical to SCP_2 (Section 2.1.2). The acyl specificity of nsLTP from rat lung and bovine liver proteins has been examined, and in neither case was a preference for saturated or unsaturated PC observed, although increasing the concentrations of saturated PC in a vesicle of mixed saturated and unsaturated PC inhibited the transfer of both molecular species (Read and Funkhouser, 1984).

In higher plants, in which cellular uptake of exogenous lipids is not observed, the discovery of a new category of proteins able to transfer phospholipids (Kader, 1985) has introduced new concepts for intracellular lipid trafficking. *In vitro,* these proteins have been demonstrated to facilitate a nonspecific intermembrane transfer of phospholipids from sites of active synthesis, such as the endoplasmic reticulum, to organelles unable to synthesize these lipids, such as mitochondria or chloroplasts. These plant proteins are remarkably similar in their biochemical properties; they have molecular masses around 9 kDa, pIs between 8.8 and 10.5, and high stabilities and they cross-react immunologically. Owing to a broad ligand specificity, they are now referred to as lipid transfer proteins (LTPs). They have been purified and characterized from cells of maize and spinach (Kader, 1985) and from castor bean seedlings (Watanabe and Yamada, 1986). Similar properties of plant LTPs and of cytosolic plant proteins characterized by their ability to bind fatty acids and long-chain acyl-CoAs, have led to the suggestion that the ability to transfer phospholipids and to bind fatty acids can be ascribed to the same protein (Rickers *et al.*, 1985).

2.2.3. Application of Transfer Proteins

The availability of purified PLTPs, nsLTPs, and LTPs has made them highly useful for the study of membrane structure and function, in particular lipid transbilayer movement and their role in membrane biogenesis and renewal.

Initial studies examined the kinetics of lipid efflux from uniformly labeled unilamellar vesicles or biological membranes. The efflux of phospholipid present in the inner leaflet is limited by the rate of flip-flop. PC flip-flop has been investigated in cells and vesicles modified by introduction of labeled PC into the outer membrane leaflet by using PC-specific transfer protein and was found to be very slow (Op den Kamp *et al.*, 1985). Rat erythrocytes (Crain and Zilversmit, 1980) and virus membranes (Van Meer *et al.*, 1981) have faster transbilayer movement for PE and PS, probably catalyzed by a flippase. nsLTP has been used by Middelkoop *et al.* (1988) to introduce PS into erythrocyte membranes. Both an interaction of PS with cytoskeleton proteins and an ATP-dependent translocation of PS are involved in maintaining its asymmetric distribution in the erythrocyte membrane (Middelkoop *et al.*, 1988). The results of all these studies established the asymmetric distribution of lipids in outer and inner leaflets of the erythrocyte membrane in particular and of plasma membranes in general. In contrast, nearly all the phospholipids in rat liver microsomes are available for rapid transfer, possibly caused by a specific membrane protein transporter (Kawashima and Bell, 1987).

Plant LTPs have been used for the manipulation of the lipid composition of spinach chloroplasts (Miquel *et al.*, 1987). When chloroplasts of spinach leaf, containing PC rich in linolenic acid, were incubated in the presence of spinach LTP with liposomes made from PC rich in oleic acid, the linolenic acid content in chloroplast PC decreased from 45 to 29% within 30 min, accompanied by an increase in oleic acid content from 10 to 21%; i.e., plastidial PC was exchanged for liposomal PC. The potential for phospholipid exchange was confirmed by using maize mitochondria and human erythrocytes, indicating that plant LTPs can be used to modify the lipid composition of plant as well as animal membranes (J.-C. Kader, unpublished results).

2.3. Glycolipid Transfer Proteins

Glycolipid transfer protein (GLTP), which facilitates the transfer of various glycosphingolipids and glycoglycerolipids between membranes *in vitro*, has been purified to homogeneity from pig brain cytosol. GLTPs from other sources (e.g., bovine spleen and brain) are also localized in the cytosolic fraction. The 22-kDa protein from pig brain has an isoelectric point of 8.3, and its amino acid composition is different from those of PC transfer protein and nsLTP (Abe and Sasaki, 1985). Possible roles of GLTP in the intracellular traffic of glycosphingolipids have been studied by several workers. Thus, a special activator protein is functionally active in the lysosomal degradation of ganglioside (Conzelmann *et al.*, 1982), whereas other data indicate that GLTP may participate in the intracellular traffic of glucosylceramide, since glucosylceramide is the only glycosphingolipid exposed to the cytoplasmic face of the cellular membrane and may thus serve as a ligand for cytosolic GLTP. It has been suggested that plant LTPs may be involved in galactolipid synthesis. Nishida and Yamada (1986) partially purified a GLTP from spinach leaf.

Dower *et al.* (1982) studied the kinetics of ganglioside transport in cultured cells. A variety of drugs, including inhibitors of protein synthesis and energy metabolism, modulators of the cytoskeleton, and monensin, had no effect on the transport of ganglioside to the plasma membrane. Hence, the vesicular transport mechanism (Farquhar, 1985) is the most likely one for the transport of this class of lipids from the site of synthesis to the plasma membrane.

2.4. Binding Proteins for Fatty Acids and Their CoA Esters

The cytosol of certain mammalian cells contains 14–15-kDa proteins that bind fatty acids with high affinity. These fatty acid-binding proteins (FABPs) have low affinities, if at all, for acyl-CoAs, and they are distinct from the recently discovered 10-kDa acyl-CoA-binding proteins, which, in turn, bind acyl-CoAs but not fatty acids (see Section 2.4.2). Sequence analyses up to now reveal three types of FABPs with homologies to other intracellular proteins that also bind hydrophobic ligands. These are the cellular retinoid-binding proteins (see Section 2.5), myelin P2 protein, adipocyte P2 protein, mammary-derived growth inhibitor, and gastrotropin (Spener *et al.*, 1989; Walz *et al.*, 1988).

2.4.1. Fatty Acid-Binding Proteins

FABPs constitute up to 7% of cytosolic proteins in cells specialized for transport and metabolization of exogenously supplied fatty acids. Examples are liver, heart, and intestinal cells, from which the hepatic (hFABP), cardiac (cFABP), and intestinal (iFABP) types, respectively, were first isolated (Sweetser *et al.*, 1987a). Homologies within each type of FABP are of the order of 80–90% based on conserved amino acids, and homologies between types are around 30–40% (Bass, 1988; Spener *et al.*, 1989); also known are isoforms of hepatic and cardiac FABPs (Spener *et al.*, 1990). The cardiac-type FABP is about 60% homologous to myelin P2 protein and the cellular retinoid-binding proteins and as high as 96% homologous to a bovine mammary-derived growth inhibitor. The latter protein appears to arrest cell growth of the mammary epithelium in the course of differentiation (Spener *et al.*, 1990).

2.4.1a. Subcellular Distribution. Immunoelectron microscopy and biochemical evidence reveal the occurrence of FABPs outside the cytosol as well. Interestingly, hFABP and cFABP are present in the nuclei of liver and heart cells, respectively. hFABP is also attached to certain intracellular membranes of hepatocytes, whereas cFABP is found in the mitochondrial matrix of myocytes (Bordewick *et al.*, 1989; Börchers *et al.*, 1989). This difference in the subcellular distribution of hFABP and cFABP must have a functional reason, although this reason is unknown at present.

2.4.1b. Regulation and Functions. The synthesis of hFABP and iFABP, which are expressed together with gastrotropin in enterocytes, is regulated by hormones and can be influenced by hypolipidemic drugs. hFABP is more responsive, whereas iFABP exhibits more constitutive properties (Bass, 1988). Although hFABP is expressed in cells committed equally to anabolic and catabolic lipid paths, iFABP is found solely in cells devoted to fat absorption. cFABP is widely expressed in cells in which β-oxidation of fatty acids prevails, yet it is also found in the brain. The long-term regulation of all FABP types appears to indicate a housekeeping role for these proteins (Sweetser *et al.*, 1987a; Bass, 1988).

The protection of enzymes and cellular structures from the detergent effects of fatty acids, the enhancement of cellular uptake of fatty acids, and the intracellular utilization of fatty acids are consequences of changes in FABP concentrations but cannot be ascribed to a single FABP type (Spener *et al.*, 1989).

Differences in modulating enzyme activities and targeting fatty acids to specific metabolic paths can be inferred from *in vitro* experiments with different FABP types. Microsomal biosynthesis of phosphatidic acid was considerably stimulated by hFABP, and this result may be explained by direct interaction of this protein with a putative receptor in microsomal membranes (Bordewick *et al.*, 1989). The activity of acyl-CoA synthetase in the outer membrane of heart mitochondria, in contrast, was lowered upon addition of cFABP, as this protein competitively bound substrate fatty acid in the soluble phase (J. Hassink and F. Spener, unpublished data). FABPs stimulate activities, however, by competitively removing inhibitory fatty acids from glucose-6-phosphate dehydrogenase (Das *et al.*, 1988) and other enzymes (Bass, 1988).

2.4.1c. Binding and Structure. There is general agreement that all three types of FABP bind C_{16}–C_{20} fatty acids, saturated and unsaturated, with dissociation constants in the range of 0.1–1 μM. With regard to the stoichiometry of fatty acid binding, however, conflicting evidence exists. Ratios of 2 : 1 and 1 : 1 for the binding of fatty acids to hepatic- and cardiac-type FABP are reported in the literature. In the laboratory of F. Spener, ratios of 2 : 1 for one isoform and 1 : 1 for the other isoform of bovine hFABP were determined by independent methods (Schulenberg-Schell *et al.*, 1988). A more detailed insight for iFABP became possible when the three-dimensional structure at 2.5Å resolution was published by Sacchettini *et al.* (1988). The β-strands form a barrel domain with a clam shell-like appearance, which contains one fatty acid in the central cavity.

Little is known with regard to specific van der Waals interactions of the different FABP types with the hydrocarbon chain of the fatty acid; however, ionic binding of its carboxylate anion to one of the arginines near the C terminus of bovine hFABP was shown (Schulenberg-Schell *et al.*, 1988), a feature that is also observed for the other types (Sacchettini *et al.*, 1988; Cistola *et al.*, 1989).

Taking all the arguments together, differences exist for FABP types with respect to (1) subcellular distribution, (2) regulation of expression, (3) modulation of enzyme activities, and (4) ligand binding, although the different types command common tertiary structures. Functional consequences relating to these differences are elusive at present.

2.4.2. Acyl-CoA-Binding Proteins

Although FABP has been suggested by a number of groups to participate in the intracellular transfer of acyl-CoA esters (Bass, 1988), the rather surprising observation was made that hepatic FABP could not induce goat mammary gland fatty acid synthase to produce medium-chain acyl-CoAs by acting as their supposed acceptor. Partially purified hFABP, however, contained a 10-kDa protein which could indeed induce medium-chain fatty acid synthesis. This protein was subsequently purified and named acyl-CoA-binding protein (ACBP) (Mogensen *et al.*, 1987). Earlier reports on high affinities of acyl-CoAs for hFABP may be due to ACBP contamination.

Recently, ACBP has been found to have an identical amino acid sequence to that of a protein called diazepam-binding inhibitor (DBI) (Knudsen, 1990). This protein has been addressed as a natural neurotransmitter that acts through the benzodiazepine-binding site of the γ-aminobutyric acid (GABA) receptor complex (Guidotti *et al.*, 1983). There is no clear-cut evidence, however, for such an extracellular function (Knudsen, 1990). It is safe to say that ACBP/DBI is an intracellular carrier protein with a much higher affinity for acyl-CoAs than for the corresponding fatty acids (Rasmussen *et al.*, 1989).

2.5. Intracellular Retinoid-Binding Proteins

Retinol, retinal, and retinoic acid are members of the vitamin A group. Either directly or in precursor–product relationships, they play essential roles in epithelial growth, differentiation, and vision. Because of the hydrophobic character of these retinoids, a protein-mediated transport mechanism has been envisaged. In fact, the routes of the vitamin from the diet to the nucleus of a target cell, which is the main locus of action, are well characterized and the structure of several extra- and intracellular retinoid-binding proteins are known.

The retinol-binding protein (RBP) of human plasma is a 21-kDa protein; its 182 amino acids are highly conserved. Only minor homologies to the intracellular retinoid-binding protein exist, yet the clam shell-like β-barrel—the common motif of the tertiary structure of intracellular and extracellular hydrophobic molecule transporters (Godovac-Zimmermann, 1988)—is also the special feature of this protein.

The intracellular retinoid-binding proteins are acidic 15- to 16.5-kDa proteins. Cellular RBP type I (CRBP I) binds retinol only, in contrast to the type II protein (CRBP II), which binds both retinol and retinal (MacDonald and Ong, 1987). Cellular retinoic acid-binding protein (CRABP) binds retinoic and straight-chain fatty acids. The complete amino acid sequence of rat liver CRBP I has been determined (Sundelin *et al.*, 1985a). The polypeptide chain is 134 amino acids long and shows striking homologies to intestinal CRBP II (Li *et al.*, 1986) and to bovine CRABP, which is 136 amino acids long (Sundelin *et al.*, 1985b).

The metabolism of retinal and retinol in enterocytes involves CRBP II, which makes up 1% of the total cytosolic protein of the jejunal mucosa and is as abundant as FABPs. Considerable evidence indicates that the protein is involved in absorption as well as esterification of retinol (Ong *et al.*, 1987).

The generated retinyl esters are incorporated into chylomicrons together with triacylglycerols, cholesterol esters, and phospholipids and are transported via the vascular system. Subsequent hydrolysis by lipoprotein lipase yields chylomicron remnants that are taken up by parenchymal liver cells via receptor-mediated endocytosis. By using radioactive retinyl esters, it has been shown that 80% of the total radioactivity is recovered from stellate cells, whereas only minute amounts are found in endothelial cells and Kupffer cells. In well-nourished animals, at least 80% of total hepatic stores of vitamin A are localized in the stellate cells. The remaining store is in the hepatocytes; one-third of this occurs as retinol bound to CRBP I, and the rest is esterified as in stellate cells. The routing of retinol in the hepatocyte is dependent on the vitamin A status. Under conditions of vitamin A deficiency, a major portion of the absorbed retinol is secreted from hepatocytes as complexes with RBP (Blomhoff *et al.*, 1982), and very little is directed to stellate cells.

During embryogenesis, free diffusion of retinoic acid is important to establish retinoic acid gradients which might be involved in determining anterior-posterior specification during limb morphogenesis (Thaller and Eichele, 1987). It is assumed that vitamin A-requiring cells express a specific cell surface receptor for RBP. RBP receptors have been identified from many cells, e.g., bovine retinal pigment epithelial cells, monkey small intestinal epithelial cells, and interstitial cells of rat testis and placenta. Further study on the structure and regulation of the RBP receptor will be helpful in understanding the uptake of retinol and in linking it to the subsequent transport of retinol, retinal, and retinoic acid.

2.6. Dolichols: Lipid Carriers for Carbohydrates

In many tissues dolichols are found as free alcohols, whose transfer through the cell may be mediated by nsLTP (Van Dessel *et al.*, 1989). Dolichols belong to a family of polymeric lipids with the prenyl unit as the repeating building block, leading to C_{85}–C_{110} molecules in higher eukaryotes. Although the existence of dolichols was reported in the early 1960s, it took more than 10 years before this class of polyprenyl alcohols was fully recognized and considered to be lipids. This was possible mainly owing to the discovery by Leloir's group in 1970 that phosphorylated dolichols act as carriers, or rather as coenzymes, in the N-glycosylation of proteins (Behrens and Leloir, 1970). Indeed, the only known function of dolichols in animal tissues to date is the role of essential coenzyme and glycosyl carrier in the synthesis of N-linked glycoproteins.

Dolichol and its derivatives do not behave as the classical membrane lipids, i.e., phospholipids and cholesterol. Minor amounts of dolichols occur esterified with long-chain fatty acids or conjugated with carbohydrate components. Recently, a waxlike substance, dolichyl dolichoate, has been recovered from bovine thyroid (Steen *et al.*, 1984) and many other tissues. Marked differences exist in the total dolichol concentration found among vertebrate species. Human tissues are particularly rich in dolichol; e.g., the human pituitary gland contains more dolichol than phospholipid (Tollbom and Dallner, 1986). It is also interesting that cellular dolichol levels increase upon aging (Andersson *et al.*, 1987). Dolichol concentrations and/or homolog profiles also change considerably under pathological conditions (Rip *et al.*, 1986).

3. CONCLUDING REMARKS—GENETIC APPROACHES

Major advances have undoubtedly been made in our understanding of the functional role of nonenzymatic proteins in intracellular lipid transport. However, direct evidence for the several hypothetical transport mechanisms is difficult to demonstrate unequivocally. Progress in cloning the structural genes coding for some of these carrier or transfer proteins provides new tools for solving the problem.

Isolation and characterization of cDNA clones enabled Tchang *et al.* (1988) to demonstrate that a signal peptide is involved in the synthesis of LTPs from maize and castor bean. The presence of a signal peptide, which is usually characteristic for proteins destined for organelle import, indicates that LTP may be partly bound to membranes or pass across membranes. Preliminary studies (V. Arondel and J.-C. Kader, unpublished data) show that maize LTP is synthesized in membrane-bound polysomes. Isolation of cDNA clones for maize LTP provides probes for studying the levels of LTP mRNA in maize. New information on the role of LTP in plants may be available if maize LTP-cDNA is inserted into other plants.

For SCP_2 in rat adrenal cells, a higher-molecular-mass precursor than the mature protein appears to be synthesized first. Recently a cDNA of SCP_2 has been isolated by Moncecchi *et al.* (1987). Insertion of a DNA construct coding for the antisense mRNA of SCP_2 would produce an SCP_2-negative cell which should give evidence of the ability of a cell to transport cholesterol to membranes in the absence of SCP_2.

cDNA sequences and FABP genes have been characterized for most of the mammalian FABPs. The coding sequences of cDNA for hepatic, intestinal, and cardiac FABPs have been expressed in *Escherichia coli* through appropriate vectors (Lowe *et al.*, 1984, 1987; Oudenampsen *et al.*, 1990). The nucleotide sequences of rat liver and human intestinal FABP genes have been established (Sweetser *et al.*, 1986, 1987b). Genes for murine adipocyte P2 protein and CRBP II were also characterized (Demmer *et al.*, 1987). The genes encoding these proteins have a similar pattern of 4 exons and 3 introns.

The localization and regulation of the genes encoding FABPs and other hydrophobic ligands may provide new insight into their pathophysiological significance. For example, in diabetes a decline of hFABP is observed and cFABP-expression in the aortic wall is lowered by experimentally induced blood hypertension (Sarzani *et al.*, 1988). Developmental changes of FABP expression pre- and postpartum in several rat tissues have been studied. The variations observed in different organs, e.g., an increase in cFABP mRNA levels postpartum in the heart, brain, and testes, but a decrease in the kidneys, are functionally not understood at present (Heuckeroth *et al.*, 1987).

FABPs and related intracellular hydrophobic molecule transporters probably originate from a common ancestor gene (Chan *et al.*, 1985). The genes encoding these proteins are dispersed throughout the genome in both mice and humans. Only CRBP I and II genes are closely linked to the same chromosome. cDNA probes and gene transfer techniques will be helpful in expression studies in different cell types and in site-directed mutagenesis. This offers new information concerning the functions of different FABP types and their relation to other members of this multigene family.

With the discovery of nuclear receptors for retinoic acid (Petkovich *et al.*, 1987), it will be of interest to identify the genes that are regulated by this compound. This may indicate more widespread functions of retinoids in cellular physiology than currently understood, a problem that applies to all lipid carrier proteins of the cell.

ACKNOWLEDGMENTS. The work carried out at the University of Münster and described here was supported by a grant from the Deutsche Forschungsgemeinschaft (SFB 310) and the Fonds der Chemischen Industrie. M.M. is the recipient of a guest professorship at the University of Münster.

4. REFERENCES

Abe, A., and Sasaki, T., 1985, Purification and some properties of the glycolipid transfer protein from pig brain, *J. Biol. Chem.* **260**:11231–11239.

Andersson, M., Appelkvist, E., Kristensson, K., and Dallner, G., 1987, Distribution of dolichol and dolichyl phosphate in human brain, *J. Neurochem.* **49**:685–691.

Backer, J. A., and Dawidowicz, E. A., 1987, Reconstitution of a phospholipid flippase from rat liver microsomes, *Nature* **327**:341–343.

Bass, N. M., 1988, The cellular fatty acid binding proteins: Aspects of structure, regulation and function, *Int. Rev. Cytol.* **111**:143–184.

Behrens, N., and Leloir, L., 1970, Dolichol monophosphate glucose: An intermediate in glucose transfer, *Proc. Natl. Acad. Sci. U.S.A.* **66**:153–159.

Bloj, B., and Zilversmit, D. B., 1977, Rat liver proteins capable of transferring phosphatidylethanolamine. Purification and transfer activity for other phospholipids and cholesterol, *J. Biol. Chem.* **252**:1613–1619.

Blomhoff, R., Helgerud, P., Rasmussen, M., Berg, T., and Norman, K. R., 1982, *In vitro* uptake of chylomicron [^{3}H]retinyl esters by rat liver: Evidence for retinol transfer from parenchymal to nonparenchymal cells, *Proc. Natl. Acad. Sci. U.S.A.* **79**:7326–7330.

Börchers, T., Unterberg, C., Rüdel, H., Robenek, H., and Spener, F., 1989, Subcellular distribution of cardiac fatty acid binding protein in bovine heart muscle and quantitation with an enzyme linked immunosorbent assay, *Biochim. Biophys. Acta* **1002**:54–61.

Bordewick, U., Heese, M., Börchers, T., Robenek, H., and Spener, F., 1989, Compartmentation of hepatic fatty acid binding protein in liver cells and its effect on microsomal phosphatidic acid biosynthesis, *Biol. Chem. Hoppe-Seyler* **370**:229–238.

Bozzato, R. P., and Tinker, D. O., 1987, Purification and properties of two phospholipid transfer proteins from yeast, *Biochem. Cell Biol.* **65**:195–202.

Chan, L., Wei, C.-F., Li, W.-H., Yang, C.-Y., Ratner, P., Pownall, H., Gotto, A. M., Jr., and Smith, L. C., 1985, Human liver fatty acid binding protein cDNA and amino acid sequence, *J. Biol. Chem.* **260**:2629–2632.

Chanderbhan, R., Noland, B. J., Scallen, T. J., and Vahouny, G. V., 1982, Sterol carrier protein 2, delivery of cholesterol from adrenal lipid droplets to mitrochondria for pregnenolone synthesis, *J. Biol. Chem.* **257**:8928–8934.

Chavant, L., and Kader, J.-C., 1982, The presence of phospholipid transfer proteins in filamentous fungi, in *Biochemistry and Metabolism of Plant Lipids* (J. F. G. M. Wintermans and P. J. C. Kuiper, eds.), pp. 125–128, Elsevier Biomedical Press, Amsterdam.

Cistola, D. P., Sacchettini, J. C., Banaszak, L. J., Walsh, M. T., and Gordon, J. I., 1989, Fatty acid interactions with rat intestinal and liver fatty acid-binding proteins expressed in *Escherichia coli*, *J. Biol. Chem.* **264**:2700–2710.

Conzelmann, E., Burg, J., Stephan, G., and Sandhoff, K., 1982, Complexing of glycolipids and their transfer between membranes by the activator protein for degradation of lysosomal ganglioside GM_2, *Eur. J. Biochem.* **123**:455–464.

Crain, R. C., and Zilversmit, D. B., 1980, Net transfer of phospolipid by the nonspecific phospholipid exchange proteins from beef liver, *Biochim. Biophys. Acta* **620**:37–48.

Das, T., Sa, G., and Mukherjea, M., 1989, Human fetal liver fatty acid binding proteins. Role on glucose-6-phosphate dehydrogenase activity, *Biochim. Biophys. Acta* **1002**:164–172.

Daum, G., and Paltauf, F., 1984, Phospholipid transfer in yeast. Isolation and partial characterization of a phospholipid transfer protein from yeast cytosol, *Biochim. Biophys. Acta* **794**:385–391.

Demmer, L. A., Birkenmeier, E. H., Sweetser, D. A., Levin, M. S., Zollman, S., Sparkes, R. S., Mohandas, T., Lusis, A. J., and Gordon, J. I., 1987, The cellular retinol binding protein II gene, *J. Biol. Chem.* **262**:2458–2467.

Dower, S., Miller-Podraza, H., and Fischman, P. H., 1982, Translocation of newly synthesized gangliosides to the cell surface, *Biochemistry* **21**:3265–3270.

Farquhar, M. G., 1985, Progress in unraveling pathways of Golgi traffic, *Annu. Rev. Cell Biol.* **1**:447–488.

Gavey, K. L., Noland, B. J., and Scallen, T. J., 1981, The participation of sterol carrier protein 2 in the conversion of cholesterol to cholesterol ester by rat liver microsomes, *J. Biol. Chem.* **256**:2993–2999.

Godovac-Zimmermann, J., 1988, The structural motif of β-lactoglobulin and retinol-binding protein: A basic framework for binding and transport of small hydrophobic molecules? *Trends Biochem. Sci.* **13**:64–66.

Guidotti, A., Forchetti, C. M., Corda, M. C., Konkel, D., Bennet, C. D., and Costa, E., 1983, Isolation, characterization and purification to homogeneity of an endogenous polypeptide with agonistic action on benzodiazepine receptors, *Proc. Natl. Acad. Sci. U.S.A.* **80**:3531–3535.

Helmkamp, G. M., Jr., 1985, Phosphatidylinositol transfer proteins: Structure, catalytic activity, and physiological functions, *Chem. Phys. Lipids* **38**:3–16.

Helmkamp, G. M., Jr., 1986, Phospholipid transfer proteins: Mechanisms of action, *J. Bioenerget. Biomembr.* **18**:71–91.

Heuckeroth, R. O., Birkenmeier, E. H., Levin, M. S., and Gordon, J. I., 1987, Analysis of the tissue-specific expression, developmental regulation, and linkage relationships of a rodent gene encoding heart fatty acid binding protein, *J. Biol. Chem.* **262**:9709–9717.

Kader, J.-C., 1985, Lipid-binding proteins in plants, *Chem. Phys. Lipids* **38**:51–62.

Kamp, H. H., Wirtz, K. W. A., Baer, P. R., Slotboom, A. J., Rosenthal, A. F., Paltauf, F., and Van Deenen, L. L. M., 1977, Specificity of the phosphatidylcholine exchange protein from bovine liver, *Biochemistry* **16**:1310–1316.

Kawashima, Y., and Bell, R. M., 1987, Assembly of the endoplasmic reticulum phospholipid bilayer: Transporters for phosphatidylcholine and metabolites, *J. Biol. Chem.* **262**:16495–16502.

Keller, G. A., Scallen, T. J., Clarke, D., Maher, P. A., Krisans, S. K., and Singer, S. J., 1989, Subcellular localization of sterol carrier protein$_2$ in rat hepatocyte: its primary localization to peroxisomes, *Cell Biol.* **105**:1353–1361.

Knudsen, J., 1990, Acyl-CoA-binding protein (ACBP) and its relation to fatty acid-binding protein (FABP): An overview, *Mol. Cell. Biochem.* in press.

Li, E., Demmer, L. A., Sweetser, D. A., Ong, D. E., and Gordon, J. I., 1986, Rat cellular retinol-binding protein II: Use of a cloned cDNA to define its primary structure, tissue specific expression and developmental regulation, *Proc. Natl. Acad. Sci. U.S.A.* **83**:5779–5783.

Lowe, J. B., Strauss, A. W., and Gordon, J. I., 1984, Expression of a mammalian fatty acid-binding protein in *Escherichia coli*, *J. Biol. Chem.* **259**:12696–12704.

Lowe, J. B., Sacchettini, J. C., Laposata, M., McQuillan, J. J., and Gordon, J. I., 1987, Expression of rat intestinal fatty acid-binding protein in *Escherichia coli*, *J. Biol. Chem.* **262**:5931–5937.

MacDonald, P. N., and Ong, D. E., 1987, Binding specificities of cellular retinol-binding protein and cellular retinol-binding protein, type II, *J. Biol. Chem.* **262**:10550–10556.

Middelkoop, E., Lubin, B. H., Beevers, E. M., Op den Kamp, J. A. F., Comfurius, P., Chin, D. T.-Y., Zwaal, R. F. A., Van Deenen, L. L. M., and Roelofsen, B., 1988, Studies on sickled erythrocytes provide evidence that the asymmetric distribution of phosphatidylserine in the red cell membrane is maintained by both ATP-dependent translocation and interaction with membrane skeletal proteins, *Biochim. Biophys. Acta* **937**:281–288.

Miquel, M., Block, M. A., Joyard, J., Dorne, A. J., Dubacq, J. P., Kader, J.-C., and Douce, R., 1987, Protein mediated transfer of phosphatidylcholine from liposomes to spinach chloroplast envelope membranes, *Biochim. Biophys. Acta* **937**:219–228.

Mogensen, I. B., Schulenberg-Schell, H., Hansen, H. O., Spener, F., and Knudsen, J., 1987, A novel acyl-CoA binding protein from bovine liver. Effect on fatty acid synthesis, *Biochem. J.* **241**:189–192.

Moncecchi, D., Keightley, J. A., Simmons, P. C., and Scallen, T. J., 1987, Isolation and nucleotide sequence of mouse liver sterol carrier protein$_2$ cDNA, *Fed. Proc.* **46**:2188.

Morris, H. R., Larsen, B. S., and Billheimer, J. T., 1988, A mass spectrometric study of the structure of sterol carrier protein SCP_2 from rat liver, *Biochem. Biophys. Res. Commun.* **154:**476–482.

Nishida, I., and Yamada, M., 1986, Semisynthesis of spin-labeled monogalactosyldiacylglycerol and its application in the assay for galactolipid transfer activity in spinach leaves, *Biochim. Biophys. Acta* **813:**298–306.

Noland, B. J., Arebalo, R. E., Hansbury, E., and Scallen, T. J., 1980, Purification and properties of sterol carrier $protein_2$, *J. Biol. Chem.* **255:**4282–4289.

Ong, D. E., Kakkad, B., and MacDonald, P. N., 1987, Acyl-CoA-independent esterification of retinol bound to cellular retinol-binding protein (type II) by microsomes from rat small intestine, *J. Biol. Chem.* **262:**2729–2736.

Ono, T., and Bloch, K., 1975, Solubilization and partial characterization of rat liver squalene epoxidase, *J. Biol. Chem.* **250:**1571–1579.

Op den Kamp, J. A. F., Roelofsen, B., and Van Deenen, L. L. M., 1985, Structural and dynamic aspects of phosphatidylcholine in the human erythrocyte membrane, *Trends Biochem. Sci.* **10:**320–323.

Oudenampsen, E., Kupsch, E.-M., Wissel, T., Spener, F., and Lezius, A., 1990, Expression of fatty acid binding protein from bovine heart in *Escherichia coli, Mol. Cell. Biochem.*, in press.

Petkovich, M., Brand, N. J., Krust, A., and Chambon, P., 1987, A human retinoic acid receptor which belongs to the family of nuclear receptors, *Nature* **330:** 444–450.

Rasmussen, J. T., Börchers, T., and Knudsen, J., 1989, Comparison of binding affinities of acyl-CoA-binding protein (ACBP) and fatty acid binding protein (FABP) for long-chain acyl-CoA esters, *Biochem. J.*, in press.

Read, R. J., and Funkhouser, J. D., 1984, Acyl-chain specificity and membrane fluidity. Factors which influence the activity of a purified phospholipid-transfer protein from lung, *Biochim. Biophys. Acta* **794:**9–17.

Rickers, J., Spener, F., and Kader, J.-C., 1985, A phospholipid transfer protein that binds long-chain fatty acids, *FEBS Lett.* **180:**29–32.

Rip, J., Rupar, C., Ravi, K., and Carroll, K., 1985, Distribution, metabolism and function of dolichol and polyprenols, *Prog. Lipid Res.* **24:**269–309.

Rothman, J. E., and Kennedy, E. P., 1977, Rapid transmembrane movement of newly synthesized phospholipids during membrane assembly, *Proc. Natl. Acad. Sci. U.S.A.* **74:**1821–1825.

Ruenwongsa, P., Singh, H., and Jungalwala, F. B., 1979, Protein-catalyzed exchange of phosphatidylinositol between rat brain microsomes and myelin, *J. Biol. Chem.* **254:**9358–9363.

Sacchettini, J. C., Gordon, J. I., and Banaszak, L. J., 1988, The structure of crystalline *Escherichia coli*-derived rat intestinal fatty acid-binding protein at 2.5-Å resolution, *J. Biol. Chem.* **263:**5815–5819.

Sarzani, R., Claffey, K. P., Chobanian, A. V., and Brecher, P., 1988, Hypertension induces tissue-specific gene suppression of a fatty acid binding protein in rat aorta, *Proc. Natl. Acad. Sci. U.S.A.* **85:**7777–7781.

Schulenberg-Schell, H., Schäfer, P., Keuper, H. J. K., Stanislawski, B., Hoffmann, E., Rüterjans, H., and Spener, F., 1988, Interactions of oleic acid with neutral fatty-acid-binding protein from bovine liver, *Eur. J. Biochem.* **170:**565–574.

Scow, O. R., and Blanchette-Mackie, E. J., 1985, Why fatty acids flow in cell membranes, *Prog. Lipid Res.* **24:**197–241.

Sleight, R. G., 1987, Intracellular lipid transport in eukaryotes, *Annu. Rev. Physiol.* **49:**193–208.

Spener, F., Börchers, T., and Mukherjea, M., 1989, On the role of fatty acid binding proteins in fatty acid transport and metabolism, *FEBS Lett.* **244:**1–5.

Spener, F., Börchers, T., Unterberg, C., and Grosse, R., 1990, Characteristics of fatty acid binding proteins and their relation to mammary derived growth inhibitor, *Mol. Cell. Biochem.*, in press.

Steen, L., Van Dessel, G., De Wolf, M., Lagrou, A., Hilderson, H. J., De Keukeleire, D., Pinkse, F., Fokkens, R., and Dierick, W., 1984, Identification and characterization of dolichol-dolichoate, a novel isoprenoic derivative in bovine thyroid, *Biochim. Biophys. Acta* **796**:294–303.

Sundelin, J., Anundi, H., Trägårdh, L., Eriksson, U., Lind, P., Ronne, H., Peterson, P. A., and Rask, L., 1985a, The primary structure of rat liver cellular retinol-binding protein, *J. Biol. Chem.* **260**:6488–6493.

Sundelin, J., Das, S., Eriksson, U., Rask, L., and Peterson, P. A., 1985b, The primary structure of bovine cellular retinoic acid-binding protein, *J. Biol. Chem.* **260**:6494–6499.

Sweetser, D. A., Lowe, J. B., and Gordon, J. I., 1986, The nucleotide sequence of the rat liver fatty acid-binding protein gene. *J. Biol. Chem.* **261**:5553–5561.

Sweetser, D. A., Heuckeroth, R. O., and Gordon, J. I., 1987a, The metabolic significance of mammalian fatty-acid-binding proteins: Abundant proteins in search of a function. *Annu. Rev. Nutr.* **7**:337–359.

Sweetser, D. A., Birkenmeier, E. H., Klisak, I. J., Zollman, S., Sparkes, R. S., Mohandas, T., Lusis, A. J., and Gordon, J. I., 1987b, The human and rodent intestinal fatty acid binding protein genes, *J. Biol. Chem.* **262**:16060–16071.

Tai, S.-P., and Kaplan, S., 1984, Purification and properties of a phospholipid transfer protein from *Rhodopseudomonas sphaeroides, J. Biol. Chem.* **259**:12178–12183.

Taylor, F. R., and Kandutsch, A. A., 1985, Oxysterol binding protein, *Chem. Phys. Lipids* **38**:187–194.

Tchang, F., This, P., Stiefel, V., Arondel, V., Morch, M.-D., Pages, M., Puigdomenech, P., Grellet, F., Delseny, M., Bouillon, P., Huet, J.-C., Guerbette, F., Beauvais-Cante, F., Duranton, H., Pernollet, J.-C., and Kader, J.-C., 1988, Phospholipid transfer protein: Full-length cDNA and amino acid sequence in maize, *J. Biol. Chem.* **263**:16849–16855.

Teerlink, T., Van der Krift, T. P., Post, M., and Wirtz, K. W. A., 1982, Tissue distribution and subcellular localization of phosphatidylcholine transfer protein in rats as determined by radioimmunoassay, *Biochim. Biophys. Acta* **713**:61–67.

Thaller, C., and Eichele, G., 1987, Identification and spatial distribution of retinoids in the developing chick limb bud, *Nature* **327**:625–626.

Tollbom, O., and Dallner, G., 1986, Dolichol and dolichyl phosphate in human tissues, *Br. J. Exp. Pathol.* **67**:757–764.

Trzeciak, W. H., Simpson, E. R., Scallen, T. J., Vahouny, G. V., and Waterman, M. R., 1987, Studies on the synthesis of sterol carrier protein$_2$ in rat adrenocortical cells in monolayer culture, *J. Biol. Chem.* **262**:3713–3717.

Tsuneoka, M., Yamamoto, A., Fujiki, Y., and Tashiro, Y., 1988, Nonspecific lipid transfer protein (sterol carrier protein$_2$) is located in rat liver peroxisomes, *J. Biochem.* **104**:560–564.

Van Amerongen, A., Van Noort, M., Van Beckhoven, J. R. C. M., Rommerts, F. F. G., Orly, T., and Wirtz, K. W. A., 1989, The subcellular distribution of the nonspecific lipid transfer protein (sterol carrier protein$_2$) in rat liver and adrenal gland, *Biochim. Biophys. Acta* **1001**:243–248.

Van Dessel, G., De Wolf, M., Lagrou, A., Hilderson, H. J., and Dierick, W., 1989, Intra- and extracellular transport of dolichol. Abstract P11, *Academy Workshop on Intracellular and Intravascular Lipid Transport,* Amsterdam.

Van Meer, G., Simons, K., Op den Kamp, J. A. F., and Van Deenen, L. L. M., 1981, Phospholipid asymmetry in Semliki forest virus grown on baby hamster kidney (BHK-21) cells, *Biochemistry* **20**:1974–1981.

Van Noort, M., Rommerts, F. F. G., Van Amerongen, A., and Wirtz, K. W. A., 1988, Intracellular redistribution of SCP_2 in Leydig cells after hormonal stimulation may contribute to increased pregnenolone production, *Biochem. Biophys. Res. Commun.* **154**:60–65.

Voelker, D. R., 1985, Disruption of phosphatidylserine translocation to the mitochondria in baby hamster kidney cells, *J. Biol. Chem.* **260**:14671–14676.

Walz, D. A., Wider, M. D., Snow, J. W., Dass, C., and Desiderio, D. M., 1988, The complete amino acid sequence of porcine gastrotropin, an ideal protein which stimulates gastric acid and pepsinogen secretion, *J. Biol. Chem.* **263:**14189–14195.

Watanabe, S., and Yamada, M., 1986, Purification and characterization of a non-specific lipid transfer protein from geminated castor bean endosperms which transfer phospholipids and galactolipids, *Biochim. Biophys. Acta* **876:**116–123.

Wetterau, J. R., and Zilversmit, D. B., 1984, Quantitation of lipid transfer activity, *Methods Biochem. Anal.* **30:**119–226.

Wetterau, J. R., and Zilversmit, D. B., 1986, Localization of intracellular triacylglycerol and cholesteryl ester transfer activity in rat tissues, *Biochim. Biophys. Acta* **875:**610–617.

Wirtz, K. W. A., and Zilversmit, D. B., 1968, Exchange of phospholipids between liver mitochondria and microsomes *in vitro, J. Biol. Chem.* **243:**3596–3602.

Yaffe, M. P., and Kennedy, E. P., 1983, Intracellular phospholipid movement and the role of phospholipid transfer proteins in animal cells, *Biochemistry* **22:**1497–1507.

Chapter 2

Application of Fluorescent Phospholipid Analogues to Studies on Phospholipid Transfer Proteins

P. J. Somerharju, P. A. van Paridon, and
K. W. A. Wirtz

1. INTRODUCTION

Fluorescence is one of the most widely used spectroscopic techniques in the research of various biological phenomena; owing to its sensitivity and specificity, it has solved problems that otherwise would have been extremely difficult to approach. Especially in membrane research, fluorescence spectroscopy has be-

Abbreviations used in this chapter: PC-TP, phosphatidylcholine transfer protein; PI-TP, phosphatidylinositol transfer protein; nsL-TP, nonspecific lipid transfer protein; PC, phosphatidylcholine; PA, phosphatidic acid; PI, phosphatidylinositol; PG, phosphatidylglycerol; PE, phosphatidylethanolamine; 2-PnA-PC, 2-*cis*-parinaroylphosphatidylcholine; PyrPC, pyrenylacyl-containing phosphatidylcholine; C(16)Pyr(*x*)PC, 1-palmitoyl-2-pyrenylacyl-*sn*-glycero-3-phosphocholine; Pyr(*x*)C(16)PC, 1-pyrenylacyl-2-palmitoyl-*sn*-glycero-3-phosphocholine (see Fig. 5 for nomenclature of the PyrPC species); 2-PnA-PI, 2-*cis*-parinaroyl-phosphatidylinositol; PyrPI, pyrenylacyl-containing phosphatidylinositol; C(18)Pyr(*x*)PI, 1-stearoyl-2-pyrenylacyl-*sn*-glycero-3-phosphoinositol; C(16)Pyr(*x*)PI, 1-palmitoyl-2-pyrenylacyl-*sn*-3-phosphoinositol; C(16 : 1)Pyr(*x*)PI, 1-palmitoleoyl-2-pyrenylacyl-*sn*-3-phosphoinositol; NBD, *N*-4-nitrobenzo-2-oxa-1,3-diazole; ESR, electron spin resonance; RET, resonance energy transfer; *N*-Rh-PE, *N*-(lissamine rhodamine B sulfonyl)phosphatidylethanolamine.

P. J. Somerharju Department of Medical Chemistry, University of Helsinki, Helsinki 17, Finland. **P. A. van Paridon and K. W. A. Wirtz** Center for Biomembranes and Lipid Enzymology, University of Utrecht, 3508 TB Utrecht, The Netherlands.

come very popular. A variety of probes have been used in this field (Lacowicz, 1983), and the fluorescent phospholipid analogues are among the most promising ones. They have been used to investigate phenomena such as lateral diffusion and dynamics of lipids (Galla and Hartman, 1980; Pugh *et al.*, 1982; Jones and Lentz, 1986), lateral distribution of lipids (Welti and Silbert, 1982; Welti, 1982; Somerharju *et al.*, 1985; Wiener *et al.*, 1985; Hresko *et al.*, 1987), transbilayer movement (Homan and Pownall, 1988), phase transitions (Sklar *et al.*, 1977), and lipid–protein interactions (Jones and Lenz, 1986; Mustonen *et al.*, 1987). Moreover, these fluorescent probes are very useful in studies of the spontaneous (Roseman and Thompson, 1980; Massey *et al.*, 1982; Frank *et al.*, 1983; Arvinte and Hildenbrand, 1984; Nichols, 1987, 1988) and protein-mediated (Somerharju *et al.*, 1981, 1983, 1987; Somerharju and Wirtz, 1982; Nichols and Pagano, 1983; Massey *et al.*, 1985; van Paridon *et al.*, 1987a, 1988a) transfer of phospholipids.

Fluorescent phospholipids are well suited for studies of the structure and function of phospholipid transfer proteins. An obvious reason for their use in these studies is the high detection sensitivity; typically 1 nmol or less of a labeled lipid is sufficient to determine rates of transfer or binding (Somerharju *et al.*, 1987; van Paridon *et al.*, 1988a). Another important property is that the fluorescence properties (i.e., lifetime, intensity, anisotropy, peak position, and shape of the spectrum) of many of the reporter groups are sensitive to the environment. This allows one to obtain information on the structural properties of the lipid-binding site or the size and shape of the transfer protein molecule itself (Berkhout *et al.*, 1984; van Paridon *et al.*, 1987b). The fluorescent phospholipids can be detected by a light microscope, which makes these probe molecules promising tools for studies of lipid transfer in living cells (Pagano and Sleight, 1985a,b; Simons and van Meer, 1988).

This review concentrates on recent studies in which fluorescent phospholipid analogues have been used to obtain novel information both on various aspects of protein-mediated phospholipid transfer and on the structural properties of three mammalian phospholipid transfer proteins. These proteins and the various methods used to elucidate their structure and function have been reviewed in detail (Wirtz *et al.*, 1985, 1986; Helmkamp, 1986, Scallen *et al.*, 1985). In addition, a comprehensive review on the theory and techniques of fluorescence spectroscopy has been published (Lacowicz, 1983; note from editor: see also volume 13 of this series). The reader is referred to these reviews for further detailed information.

2. PHOSPHOLIPID TRANSFER PROTEINS

So far, fluorescent phospholipids have been used to study the structure and function of three mammalian phospholipid transfer proteins. These proteins are

(i) the phosphatidylcholine transfer protein (PC-TP), which is specific for PC; (ii) the phosphatidylinositol transfer protein (PI-TP), which shows marked preference for PI but is also able to transfer PC and, to a lesser extent, phosphatidylglycerol (PG); and (iii) the nonspecific lipid transfer protein (nsL-TP), which transfers most phospholipids, as well as glycolipids and cholesterol, and is now known to be identical to sterol carrier protein 2 (Trzaskos and Gaylor, 1983). Fluorescent lipid analogues have also been used to investigate other transfer proteins such as glycolipid transfer proteins (see Chapter 5) as well as spontaneous lipid transfer (see Chapter 11).

PC-TP and PI-TP act as phospholipid carriers (Demel *et al.*, 1973, 1977). They contain an endogenous phospholipid molecule, which, upon collision with a membrane interface, can be released and replaced by another lipid molecule. Upon dissociation from the membrane, the complex may then collide with another membrane and release its bound lipid. Continuation of this transfer process will lead to a complete equilibration of exchangeable phospholipid molecules between separate interfaces. The mode of action of nsL-TP has not been firmly established yet. Recent studies with fluorescent phospholipids suggest that this protein may also form a phospholipid–protein complex (Nichols, 1987, 1988). However, studies with lipid monolayers have failed to provide any evidence for the binding of cholesterol to nsL-TP (R. A. Demel, unpublished observation).

3. FLUORESCENT PHOSPHOLIPID ANALOGUES

As the name implies, a fluorescent phospholipid molecule contains a moiety (a fluorophore) capable of emitting light when excited by light of suitable energy. The fluorophores that have been used in studies of the transfer proteins are the following: parinaroyl (*cis* or *trans*), pyrenylacyl, and *N*-4-nitrobenzo-2-oxa-1,3-diazole (NBD)-aminoacyl moieties coupled to the *sn*-1 or *sn*-2 hydroxyl of the glycerol moiety of a phospholipid molecule. Each of these fluorophores has distinct properties, which are briefly discussed here.

cis- and *trans*-parinaric acids are naturally occurring, conjugated polyenoic fatty acids, which sterically resemble mammalian unsaturated and saturated fatty acids, respectively (Sklar *et al.*, 1975, 1977), and are thus expected to be minimally perturbing. Another advantage of these reporter molecules is that in a membrane bilayer, they may be self-quenching as a result of probe–probe interactions. In addition, their fluorescence lifetime is on a scale suitable for the determination of rotational correlation times. Consequently, the correlation times of parinaroyl-labeled phospholipids incorporated in the lipid-binding site of the transfer proteins may give direct information on the structural parameters of these proteins (Berkhout *et al.*, 1984; van Paridon *et al.*, 1987b). The major disadvantages of the parinaroyl fluorophores are that they have a relatively low quantum yield in many different environments (Sklar *et al.*, 1977) and that they are more sensitive to photochemical destruction than the two other fluorophores.

The pyrenylacyl derivatives of phospholipids have proved to be very useful tools in studies of phospholipid transfer proteins (Massey *et al.*, 1985; Somerharju *et al.*, 1987; van Paridon *et al.*, 1987a, 1988a). Aside from the high quantum yield, these probes have the important advantage that the length of the acyl chain can be varied and thus the structure of the chain can be modified extensively. Despite the bulkiness of the pyrenyl moiety, the pyrenylacyl chain is well accommodated in the transfer proteins. In fact, one can make use of this bulkiness to probe the acyl chain binding sites (Somerharju *et al.*, 1987; van Paridon *et al.*, 1988a).

The NBD moiety also has a high quantum yield and, similar to the pyrenylacyl derivatives, the length of the NBD-aminoacyl chain can be varied. The excitation and emission maxima of the NBD moiety are in the visible wavelength region. This makes the NBD lipids easily detectable by normal fluorescence microscopy and is one of the reasons why such lipids have been used to study lipid transfer *in vivo* (Pagano and Sleigt, 1985a,b). It is important to note, however, that the NBD-amine moiety is quite polar and ionizable, which may result in an anomalous behavior (Chattopadhyay and London, 1987, 1988).

Preparation of phospholipid analogues of the fluorophores described above is straightforward. Typically, the synthesis involves (1) removal of the *sn*-2 acyl chain from a PC molecule by phospholipase A_2, (2) reacylation with the anhydride of the fluorescent acid, and (3) purification of the fluorescent phospholipid by thin-layer or column chromatography (Gupta *et al.*, 1977; Somerharju *et al.*, 1981, 1985). Fluorescent derivatives of other glycerophospholipids, except PI, can then be prepared from the PC by phospholipase D-catalyzed transphosphatidylation (Comfurius and Zwaal, 1977). Synthesis of the corresponding PI derivatives is somewhat more complicated owing to the reactivity of the hydroxyls on the inositol moiety (Somerharju and Wirtz, 1982; Somerharju *et al.*, 1985).

Although the choice of the fluorophore may be limited by the particular kind of application, in most cases the pyrenylacyl derivatives are suitable for studying the phospholipid transfer proteins. This is due to their stability, as well as to their efficient transfer and binding by these proteins (see Sections 4 and 5). Of course, there may exist other fluorophores, not yet tested, which are even more suitable than the pyrenylacyl chain.

4. TRANSFER RATE ASSAYS

The assays used to determine the rate of protein-mediated or passive lipid transfer can be divided in two main types. One type makes use of radiolabeled lipids and separable donor and acceptor membranes. These assays are time-consuming because they involve several steps. Furthermore, it is often difficult to obtain true initial rates. The other type comprises the spectroscopic assays. They do not require separation of the donor and acceptor vesicle membranes, they give true initial rates, and they are often flexible in terms of membrane composition. In the first spectroscopic transfer assay, electron spin resonance (ESR) spectroscopy and spin-labeled phospholipids were used (Rousselet *et al.*, 1976; Machida and Ohnishi, 1978). Although this method has all the benefits of a continuous assay, it suffers from the fact that ESR spectrometers are uncommon in most biochemical laboratories. Moreover, the availability of spin-labeled phospholipids is limited at present. The other, more common spectroscopic transfer assays make use of fluorescent phospholipid analogues. All assays are based on the principle that the labeled lipid molecules have a different fluorescence intensity in the donor than in the acceptor membranes. In most cases, probe fluorescence in the donor is quenched by either probe–probe interactions (self-quenching) or resonance energy transfer to an acceptor molecule present in the donor.

One of the first fluorescence assays to measure rates of transfer involved the use of *cis*-parinaroyl-PC (PnA-PC) (Somerharju *et al.*, 1981). Vesicles prepared from this lipid display very low fluorescence intensity as a result of self-quenching. When an excess of acceptor vesicles consisting of unlabeled PC and a catalytic amount of PC-TP are added, a fast increase of fluorescence is observed. This increase is due to the transfer of PnA-PC molecules from the donor to the acceptor vesicles, where no self-quenching takes place, owing to low PnA-PC surface concentration. The increase of fluorescence is proportional to the amount of PnA-PC transferred and hence is a direct measure of the rate of transfer.

The fluorescence of the NBD chromophore is also strongly quenched at high local concentrations of the labeled lipid (Nichols and Pagano, 1981; Tanaka and Schroit, 1983). Thus, a transfer rate assay similar to the one described above can be devised. Alternatively, NBD fluorescence in the donor vesicles can be quenched by including a resonance energy transfer (RET) acceptor [e.g., *N*-(lissamine rhodamine B sulfonyl)phosphatidylethanolamine] (*N*-Rh-PE)]. The advantage of this modification is that only small amounts of the labeled lipids are required and that the donor lipid composition can therefore be varied in a wide range (Nichols and Pagano, 1983).

The fluorescence intensity of pyrene derivatives in a membrane bilayer is also concentration dependent, but instead of simple quenching, an increase in the concentration results in the appearance of excimer fluorescence, which is emitted by a complex formed between an excited and a ground-state pyrene moiety (Galla and Hartman, 1980). This excimer emission is observed as a broad band around 470 nm, whereas the fine-structured monomer emission appears close to 380 nm. Because the monomer and excimer intensities and the ratio of these intensities are concentration dependent, any of them can be used to estimate the transfer rates (Doody *et al.*, 1980; Roseman and Thompson, 1980; Massey *et al.*,

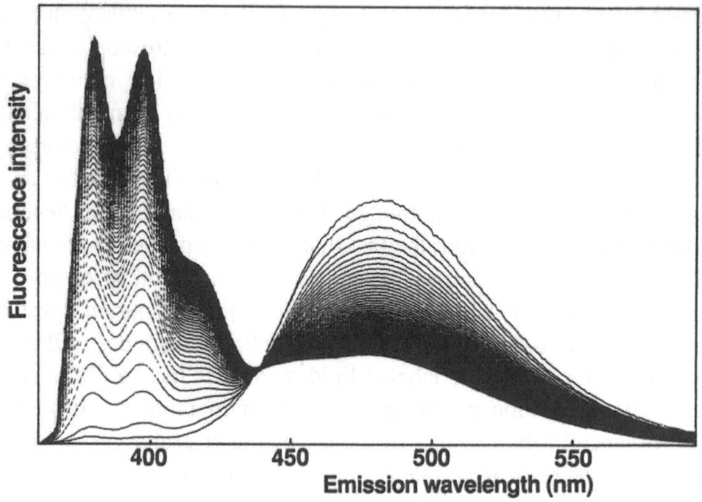

FIGURE 1. Time-dependent change of the pyrene fluorescence emission spectrum as a result of the PI-TP-mediated transfer of PyrPC from donor vesicles [C(16)Pyr(10)PC, 1 nmol; see Figure 5] to acceptor vesicles (PC : PA, 85 : 15 mole%, 100 nmol) in 2 ml of 20 mM Tris hydrochloride–5 mM EDTA–100 mM NaCl (pH 7.4) buffer. Transfer was initiated by the addition of PI-TP (2 μg); the incubation temperature was 37°C. The progress of transfer was monitored by recording emission spectra at 2-min intervals (van Paridon *et al.*, 1988a).

1985; van Paridon *et al.*, 1988a). An example of such application is shown in Figure 1 in which donor vesicles prepared from pyrenyl-labeled PC (PyrPC) display the typical excimer spectrum. Addition of an excess of acceptor vesicles and a catalytic amount of PI-TP results in a time-dependent increase of the monomer spectrum and decrease of the excimer spectrum owing to the transfer of PyrPC from the donor to the acceptor vesicles. Alternatively, a RET acceptor such as *N*-(trinitrophenyl)PE can be included in the donor membrane to quench pyrene monomer fluorescence efficiently. Similar to the corresponding NBD-lipid assay, transfer of the pyrenyl lipids to the acceptor vesicles produces a large increase in monomer fluorescence as the nontransferable quencher remains in the donor vesicles (Somerharju *et al.*, 1987). The major advantage of the use of RET quenching is that the dynamic range of the assay is independent of the concentration of the fluorescent lipid in the donor.

With a fluorescence assay it is also possible to obtain an approximate estimation of the transfer rates of unlabeled lipids. This is based on the observation that when used in catalytic amounts, bovine PC-TP and PI-TP mediate a one-to-one exchange of phospholipids between donor and acceptor membranes (Helmkamp, 1980; Kasper and Helmkamp, 1981). Thus, by assuming that the formation of the lipid–protein complex is the rate-limiting step in the exchange process, the rate of transfer of PnA-PC to acceptor vesicles consisting of different PC analogues reflects the transfer rates of these unlabeled PC species in the opposite direction (Somerharju *et al.*, 1981). In some cases, information on the transfer of unlabeled phospholipid species can be derived from the shape of the progress curves, i.e., the increase of PnA-PC fluorescence intensity with time (Somerharju *et al.*, 1983).

5. BINDING ASSAYS

Binding of phospholipids by the transfer proteins can be readily studied by using the fluorescent analogues (Somerharju *et al.*, 1983, 1987; Berkhout *et al.*, 1984; van Paridon *et al.*, 1987a,b, 1988a; Nichols, 1987). To do this, quenched (by probe–probe interactions or by RET) vesicles containing the labeled lipid are titrated with a transfer protein. If the labeled lipid is incorporated into the transfer

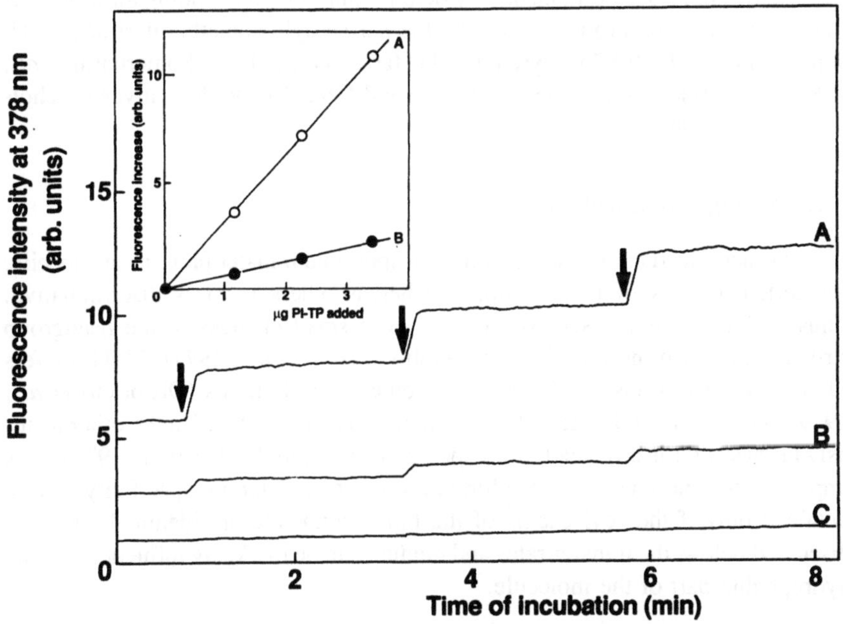

FIGURE 2. Binding of PyrPI and PyrPC to PI-TP. Vesicles consisting of C(16 : 1)Pyr(8)PI, *N*-(trinitrophenyl)PE (quencher), yeast PI, and egg PC (10 : 10 : 20 : 60 mol%, 2 nmol of total lipid) (curve A) and of C(16)Pyr(8)PC, *N*-(trinitrophenyl)PE, PA, and egg PC (10 : 10 : 30 : 50 mol%, 2 nmol of total lipid) (curve B) were titrated with PI-TP (aliquots of 1 µg, see arrows). Curve C represents a titration in the absence of vesicles. The increments of fluorescence yield as a function of the amount of PI-TP added are presented in the inset (van Paridon *et al.*, 1988a).

protein, an increase in fluorescence will be observed as a result of removal of the labeled lipid molecule from the quenched donor membranes in exchange for the unlabeled lipid molecule released by the protein (Figure 2). Under these conditions of exchange, the slope of the fluorescence intensity increase versus the amount of protein added is a measure of the affinity of the transfer protein for the probe lipid. This method allows a determination of the relative binding constants for the label lipids. Importantly, the relative binding constants for unlabeled phospholipid molecules may also be found, by determining their ability to compete with the labeled lipid for the lipid-binding site of the protein (van Paridon *et al.*, 1987a).

6. SPECIFICITY OF PHOSPHOLIPID TRANSFER PROTEINS

The specificity of phospholipid transfer proteins can be conveniently studied by using fluorescent phospholipid analogues. The headgroup and acyl chain specificities have been established by determining both the binding affinities and transfer rates of the probes, including the parinaroyl (Somerharju *et al.*, 1983; van Paridon *et al.*, 1987b), pyrenylacyl (Massey *et al.*, 1985; Somerharju *et al.*, 1987; van Paridon *et al.*, 1987a, 1988a), and NBD-*N*-acyl derivatives (Nichols and Pagano, 1983).

6.1. Headgroup Specificity

As mentioned above, some of the phospholipid transfer proteins are specific for certain classes of phospholipids, whereas others are less discriminative. Thus, PC-TP is highly specific for PC; even small changes in the headgroup structure diminish the transfer considerably (Kamp *et al.*, 1977). PI-TP prefers PI, but can also transfer PC and some other phospholipids (DiCorleto *et al.*, 1979; Zborowski and Demel, 1982; Somerharju *et al.*, 1983). On the other hand, nsL-TP has little if any headgroup specificity (Bloj and Zilversmit, 1977). It is important to note that reliable information on the headgroup specificity can be obtained only if the acyl chains of the lipids compared are identical, since as discussed below, the transfer rates and binding can be markedly influenced by the hydrophobic part of the molecule.

6.1.1. Phosphatidylcholine Transfer Protein

Nichols and Pagano (1983) studied the transfer of NBD-acyl-PC, -PE, -phosphatidic acid (-PA), and -diglyceride by PC-TP and found that only the PC derivative was transferred. This is in agreement with the earlier data obtained with radiolabeled phospholipids (Kamp *et al.*, 1973, 1977; Demel *et al.*, 1977) and indicates that the NBD-acyl moiety does not influence the specificity of this protein.

6.1.2. Phosphatidylinositol Transfer Protein

An interesting aspect of PI-TP is its ability to transfer both PI and PC. The relative affinity of PI-TP for these lipids has been studied recently with the corresponding fluorescent derivatives (PyrPI and PyrPC) that carry a pyrenyldecanoyl chain at the *sn*-2 position (van Paridon *et al.*, 1987a). This was done by determining how efficiently either unlabeled yeast PI or egg PC can compete with PyrPI or PyrPC for the lipid-binding site. Dilution of PyrPI with an equimolar amount of PI resulted in an approximately twofold reduction in PyrPI binding, indicating a similar binding affinity for both yeast and probe PI (Figure 3). A closely similar binding affinity was also observed for egg and probe PC. On the other hand, PyrPC is replaced very efficiently by unlabeled PI, whereas a large excess of unlabeled PC is required to replace PyrPI from the binding site. Analysis of the data showed that the affinity of PI-TP for PI was about 16 times higher than its affinity for PC. Intriguingly, calculations for a model cell indicated that this relative affinity constant of 16 would be optimal for PI-TP to restore the plasma membrane PI levels following stimulus-induced PI breakdown (van Paridon *et al.*, 1987a).

The polar headgroup specificity of PI-TP has also been studied by using an indirect-fluorescence assay in which the donor vesicles consisted of PnA-PC and relatively high levels of a negatively charged phospholipid (25 mol%). When the donor contains a transferable phospholipid such as PI, the progress curve of PnA-PC transfer is highly sigmoidal because PI-TP preferentially transfers PI to the acceptor vesicles (Figure 4, curve A). In this process, donor PI is replaced by acceptor PC, which results in a decrease of the donor vesicle charge and consequently, in an increase of Pna-PC transfer. Normal progress curves are obtained if the charged lipid (e.g., cardiolipin) is not transferred by PI-TP (Figure 4, curve C). It was shown by such an assay that the phospholipid requires an intact inositol ring to be efficiently transferred by PI-TP (Somerharju *et al.*, 1983).

6.1.3. Nonspecific Lipid Transfer Protein

NBD–acyl phospholipids have been used to confirm that nsL-TP has little if any specificity toward the polar headgroup of a phospholipid (Nichols and Pagano, 1983). The fact that this protein does not transfer cardiolipin (Crain and Zilversmit, 1980) may not be due to the structure of the headgroup, but could result from the relatively high hydrophobicity of this phospholipid, which contains four acyl chains. As discussed below, an increase in the length of the acyl

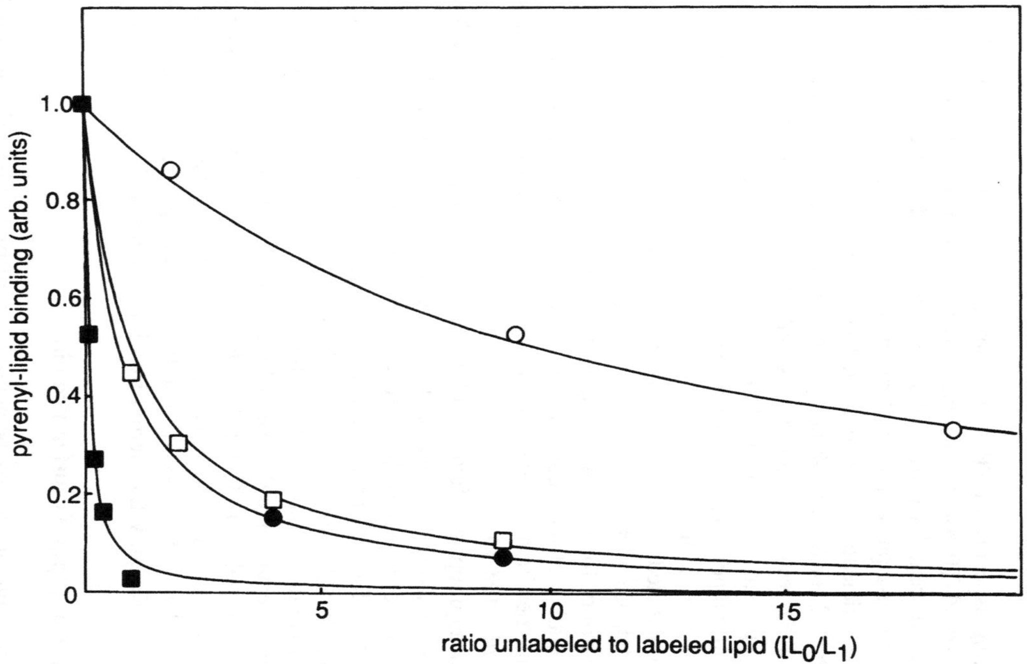

FIGURE 3. Competition of unlabeled PC and PI with pyrene-labeled PC and PI for the binding site on PI-TP. Vesicles consisting of Pyr(10)-PI (●, ○) or Pyr(10)-PC (■, □), various amounts of unlabeled egg PC (○, □) or yeast PI (●, ■), and *N*-(trinitrophenyl)PE (10 mol%) were titrated with PI-TP. Binding of the pyrenyl lipids is represented normalized to the value obtained with vesicles consisting of pure pyrenyl lipid, as a function of the ratio of unlabeled (L_0) to labeled (L_1) lipid (van Paridon *et al.*, 1987b).

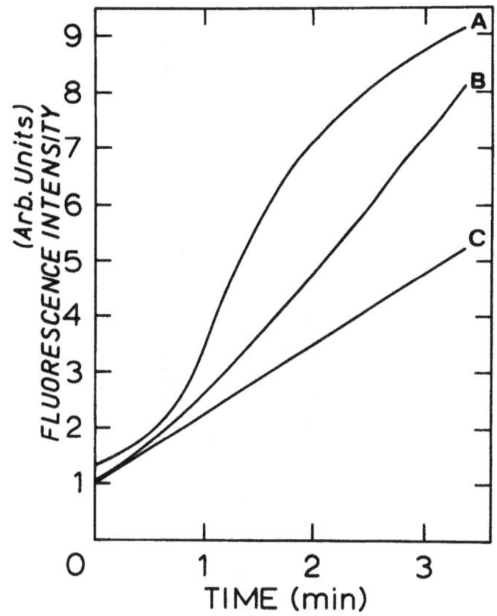

FIGURE 4. Effect of acidic phospholipids in the donor vesicles on the transfer of PnA-PC. Donor vesicles consisted of PnA-PC (3.2 nmol) and 25 mol% of either unlabeled PI (curve A), PG (curve B), or cardiolipin (curve C). Transfer was initiated by subsequent additions of acceptor vesicles (PC : PA, 98 : 2 mol%, 250 nmol of lipid) and of PI-TP (3 μg) in 2 ml of 20 mM Tris–5 mM EDTA (pH 7.4) buffer (Somerharju *et al.*, 1983).

chain (i.e., an increase of hydrophobicity) leads to a decrease in the rate of transfer by nsL-TP. Alternatively, the cardiolipin molecule could be too big to form a complex with nsL-TP; such a (transient) complex could possibly form during the transfer cycle (Nichols, 1987).

6.2. Acyl Chain Specificity

The pyrenyl phospholipids are well suited to studies of acyl chain specificity, since, in addition to the unlabeled chain, the length of the pyrenylacyl hain can be varied. Presumably, variation of the length of this chain also places the bulky pyrene moiety in different positions in the corresponding acyl-binding site of a transfer protein and thus allows additional information to be obtained on the accommodative properties of this site. The structures of the PC analogues used to probe the lipid-binding site of PC-TP and PI-TP are displayed in Figure 5. Similar sets of PI analogues, but with the pyrenylacyl moiety only in the *sn*-2 position, have been used to investigate the acyl-binding sites of PI-TP.

6.2.1. Phosphatidylcholine Transfer Protein

Binding studies showed that the affinity of PC-TP for C(16)Pyr(*x*)PC and Pyr(*x*)C(16)PC species was strongly dependent on both the length and the *sn* position of the pyrenylacyl moiety (Figure 6A). In general, the species with the pyrenylacyl moiety in the *sn*-2 position are better accommodated, with an optimal affinity observed for the C(16)Pyr(10)PC species. The large difference in

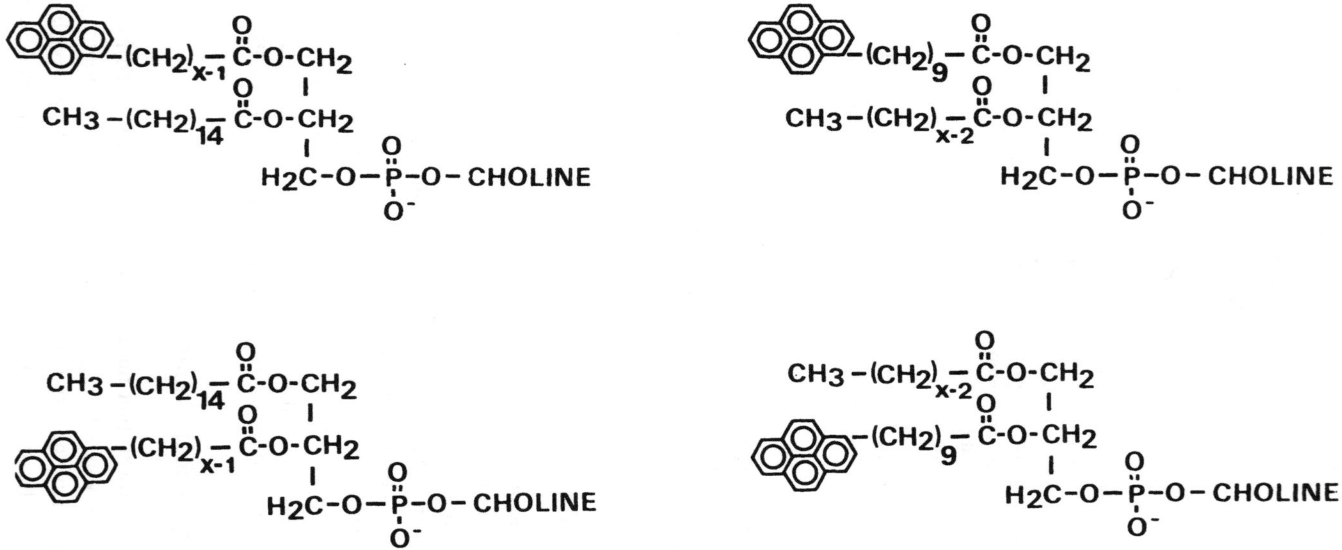

FIGURE 5. Structures and nomenclature of the pyrenyl-PC species. x indicates the total number of carbon units (including the carbonyl one) in the aliphatic chain.

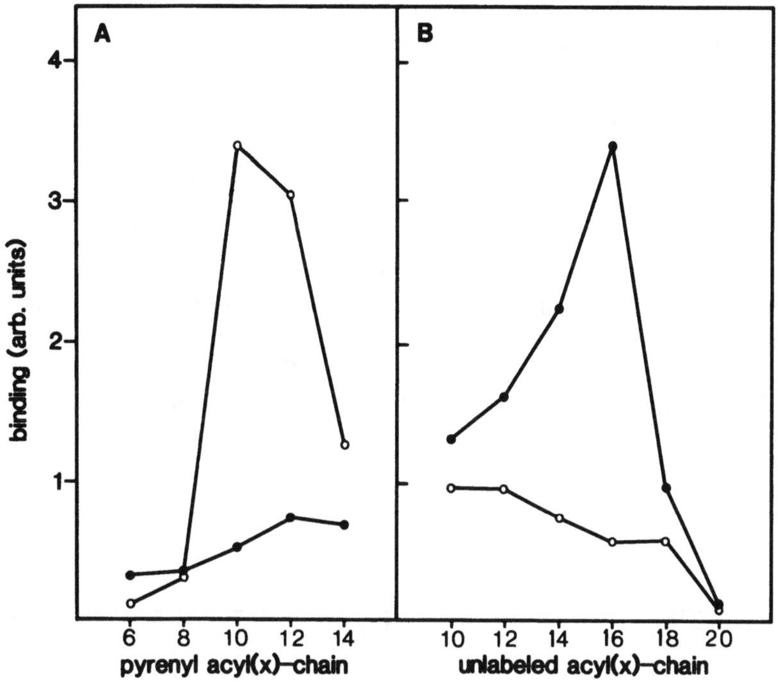

FIGURE 6. Effect of acyl chain length on the binding of PyrPC species by PC-TP. (A) Binding of C(16)Pyr(*x*)PC (○) and Pyr(*x*)C(16)PC (●). (B) Binding of C(*x*)Pyr(10)PC (●) and Pyr(10)C(*x*)PC (○). The relative affinity of PC-TP for PyrPC species was determined by titration of quenched PyrPC–egg PC–*N*-(trinitrophenyl)PE vesicles (18:72:10 mol%, 5.5 nmol of lipid) by PC-TP (Somerharju *et al.*, 1987).

affinity between this species and C(16)Pyr(8)PC suggests that the limited binding of the latter species may be due to a poor fit of the pyrenyl residue in the acyl-binding site of PC-TP. Variation of the unlabeled chain in either the *sn*-1 or *sn*-2 position also influenced the binding of the pyrenyl PC molecule considerably, with the highest affinity observed for the C(16)Pyr(10)PC species (Figure 6B). Also in this instance, different affinities were found for the positional isomers of the C(*x*)Pyr(10)PC and Pyr(10)C(*x*)PC sets (Somerharju *et al.*, 1987).

The marked discrimination observed for most positional isomer pairs strongly indicates that PC-TP has specific binding sites for the *sn*-1 and *sn*-2 acyl chains of PC. This is in accordance with results obtained from time-resolved fluorescence measurements (Berkhout *et al.*, 1984). Furthermore, it appears that the 1-acyl-binding site prefers saturated acyl chains, whereas the 2-acyl-binding site is more flexible in being able to accommodate the pyrenylacyl chains. Studies with positional isomers of radioactive PC derivatives have shown that in some instances PC-TP prefers PC with oleic acid at the *sn*-2 position over PC with oleic acid at the *sn*-1 position (van Loon *et al.*, 1986).

6.2.2. Phosphatidylinositol Transfer Protein

Studies of the binding and transfer of the various PyrPC species by PI-TP have clearly demonstrated that this protein can also discriminate between positional isomers, which implies the presence of specific binding sites for the *sn*-1 and *sn*-2 acyl chains (van Paridon *et al.*, 1988a). However, in contrast to PC-TP, this protein displayed higher affinity toward species containing the pyrenyl chain in the *sn*-1 position. Similarly, C(16)-, C(16 : 1)-, and C(18)Pyr(*x*)PI species were used to investigate the binding and transfer behavior of PI-TP (Figure 7). For each set of species the variation of the length of the pyrenyl chain in the *sn*-2 position

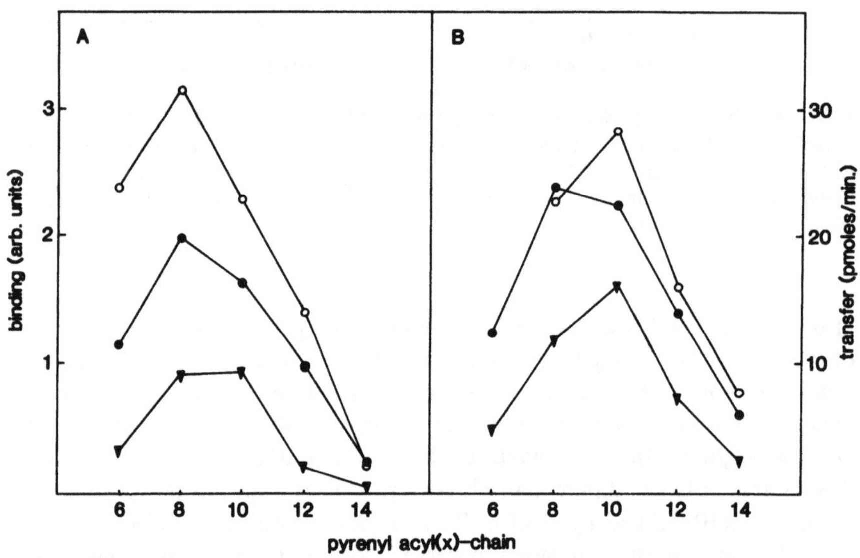

FIGURE 7. Effect of acyl chain composition on binding and transfer of PyrPI species of PI-TP. (A) Binding of C(16)Pyr(*x*)PI (●), C(16 : 1)Pyr(*x*)PI (○), and C(18)Pyr(*x*)PI (▼); *x* varied from 6 to 14 carbon atoms. Binding of PyrPI species was measured by titration of quenched PyrPI–*N*-(trinitrophenyl)PE–yeast PI–egg PC vesicles (10 : 10 : 20 : 60 mol%, 2 nmol of lipid) by PI-TP. (B) Transfer of PyrPI species by PI-TP. Transfer was measured from the quenched donor vesicles (2 nmol of lipid) used in (A) to an excess of egg PC–PA acceptor vesicles (85 : 15 mol%, 100 nmol of lipid) (van Paridon *et al.*, 1988a).

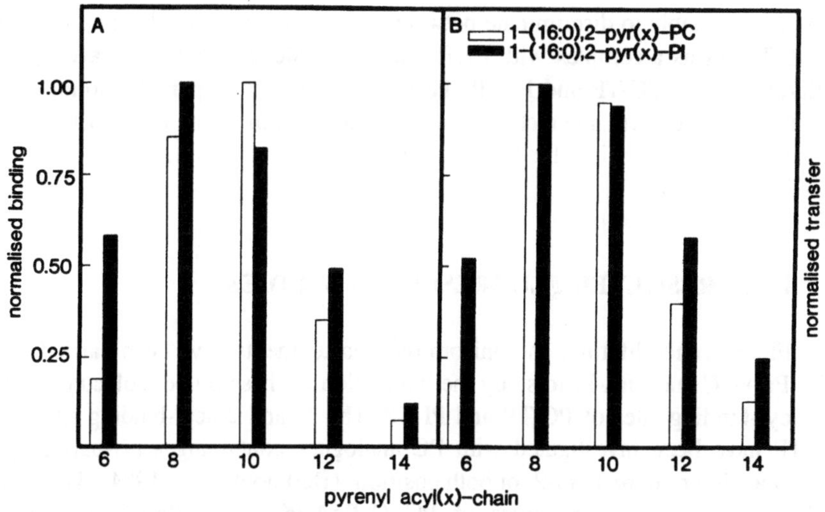

FIGURE 8. Binding and transfer of C(16)Pyr(*x*)PC as compared with C(16)Pyr(*x*)PI by PI-TP. For further details, see van Paridon *et al.* (1988a).

influenced both these parameters to the same degree, with an optimum for the species carrying the Pyr(8) and Pyr(10) chains. Among these sets, the affinities and rates increased in the order C(18)Pyr(*x*)PI < C(16)Pyr(*x*)PI < C(16 : 1) Pyr(*x*)PI, indicating that the length as well as the presence of a double bond contributes to the interactions of the 1-acyl chain with its binding site.

An important question related to the function of PI-TP is the relationship between the sites that bind PI and PC. This was investigated by comparing both the binding and transfer rate versus chain length profiles obtained for C(16)Pyr(*x*)PI and C(16)Pyr(*x*)PC species (Figure 8). The similarity of the profiles strongly suggests that at least the *sn*-2 acyl chains of PI and PC bind to the same site in the protein. This is in agreement with the findings that there is one lipid-binding site in PI-TP and that binding of PI and PC is mutually exclusive (Zborowski and Demel, 1982; van Paridon *et al.*, 1987b).

6.2.3. Nonspecific Lipid Transfer Protein

There is little information available on the acyl chain specificity of this protein. Preliminary experiments with C(16)Pyr(*x*)PC species have shown that the nsL-TP-mediated rate of transfer decreased monotonously as a function of the acyl chain length (A. van Amerongen, personal communication). A similar correlation was reported for the spontaneous rate of transfer (Massey *et al.*, 1984). Preliminary studies with the PyrPC species have shown that nsL-TP is completely unable to discriminate between positional isomers. This implies that if nsL-TP forms a complex with PyrPC in the transfer process, it behaves quite differently from PC-TP and PI-TP. These observations support the notion that specific lipid–protein interactions rather than differences in physicochemical properties are responsible for the discrimination between positional isomers by PC-TP and PI-TP.

7. TIME-RESOLVED FLUORESCENCE STUDIES

Fluorescence lifetime and anisotropy measurements have been carried out with PnA-PC and PnA-PI to study the interaction of these probe molecules with the acyl-binding sites of PC-TP and PI-TP. The 1- and 2-acyl-binding sites on PC-TP have been investigated with PC analogues containing a *cis*-parinaroyl moiety at either the *sn*-1, *sn*-2, or both positions (Berkhout *et al.*, 1984). The *sn*-1 and *sn*-2 parinaroyl derivatives of PC (1-PnA-PC and 2-PnA-PC) yielded high initial anisotropies as well as long rotational correlation times (26 and 11 nsec, respectively), implying that both acyl chains are largely immobilized in the protein. For comparison, the correlation time for 2-PnA-PC in an egg PC vesicle is 1.9 nsec. The difference in correlation times for the 1-PnA-PC/PC-TP and 2-PnA-PC/PC-TP complexes leads to the following conclusions: (1) independent of the shape of the protein, the *sn*-1 and *sn*-2 chains of the bound PC molecule are not parallel; (2) the correlation time of 11 nsec for the 2-parinaroyl chain is characteristic for that of a spherical PC-TP (mol. wt., 25,000); (3) the correlation time of 26 nsec for the 1-parinaroyl chain indicates that PC-TP cannot be spherical but is elongated. From these data it could be concluded that the protein is an ellipsoid (with an axial ratio of 2.50) in which the 1-acyl chain is parallel to the long symmetry axis and the 2-acyl chain is more or less perpendicular (an estimated angle of 60–90°) to this axis (see Figure 9 for the model). In support of this arrangement, a correlation time of 15 nsec was recorded for the PC deriva-

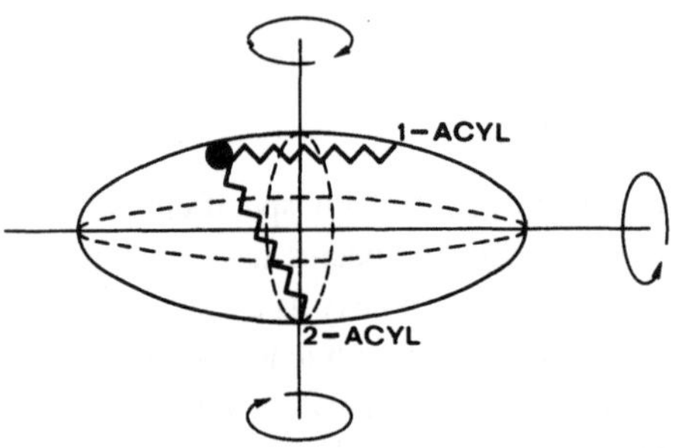

FIGURE 9. Tentative orientation of the 1-acyl and 2-acyl chains of PC bound to PC-TP (Berkhout *et al.*, 1984).

tive carrying a parinaroyl chain on both hydroxyls. This time agrees very well with the harmonic mean of the correlation times of the 1-PnA-PC and 2-PnA-PC bound to PC-TP. This strongly suggests that the two chromophores are independently photoselected and behave as isolated probes. On the other hand, the polarity of the 1- and 2-acyl-binding sites appears to be rather similar, as the average lifetimes of the parinaroyl fluorescence for the 1-PnA-PC- and 2-PnA-PC/PC-TP complexes are very similar (2.1 and 2.5 nsec, respectively).

To obtain further information on the 2-acyl-binding site of PI-TP, time-resolved fluorescence studies have been carried out with the *sn*-2 parinaroyl derivatives of PI and PC bound to this protein (van Paridon *et al.*, 1987b). From measuring the fluorescence anistropy decay, it was estimated that 2-PnA-PI and 2-PnA-PC had long correlation times (16.3 and 17.4 nsec, respectively) and high initial anisotropies; this supports the conclusion that the 2-parinaroyl chains of both PI and PC are strongly immobilized in the protein. However, the average fluorescence lifetimes of bound 2-PnA-PI and 2-PnA-PC were significantly different (4.6 and 6.3 nsec, respectively). Most probably, the specific interactions of the polar headgroup of PI and PC with PI-TP affect the orientation of the 2-PnA chain in the acyl-binding site, giving rise to an environmental microheterogeneity (i.e., variations in polarity). The possibility that the *sn*-2 acyl chains of PI and PC bind to different sites in the protein is rendered unlikely by the similarity of the binding and transfer versus pyrenylacyl chain length profiles for both phospholipids (Figure 8).

8. FLUORESCENT LIPID ANALOGUES AND INTRACELLULAR LIPID TRAFFIC

Fluorescent phospholipids appear to be very useful tools for studying intracellular transport of phospholipids, as demonstrated by the pioneering studies of Pagano and collaborators (for reviews, see Pagano and Sleight, 1985a,b; Simons and van Meer, 1988). The major advantage of such lipids is that microscopic, spectroscopic, and biochemical studies can be combined to obtain data on the localization, rate of transport, environment, and metabolism of the labeled lipid.

The phospholipid analogues used so far in the *in vivo* studies are the NBD-acyl derivatives. Because of the short acyl chain (six carbons) and the polarity of the NBD moiety (Chattopadhay and London, 1987, 1988), these lipid probes are quite soluble in the aqueous phase and are thus readily translocated from one membrane to another (Pagano *et al.*, 1981). Owing to this property, NBD-labeled lipids are efficiently incorporated into the outer surface of the plasma membrane when vesicles containing these lipids are incubated with cells (Struck and Pagano, 1980; Pagano *et al.*, 1983; Sleight and Pagano, 1985; Tanaka and

Schroit, 1983; van Meer *et al.*, 1987). Another advantage of the NBD derivatives is that they can be readily observed in a standard fluorescence microscope (Uster and Pagano, 1986).

The studies carried out with NBD-acyl lipids *in vivo* have shown that the polar headgroup has a remarkable influence on the intracellular distribution as well as metabolism of these lipid analogues. PC and sphingomyelin analogues are largely confined to a pool consisting of the outer surface of the plasma membrane and the luminal side of the Golgi apparatus (Pagano and Sleight 1985b; van Meer *et al.*, 1987). However, PE and PA analogues distribute rapidly to various intracellular membranes, as a result of which PA is extensively metabolized to the corresponding triglyceride and PC derivatives. These results are explained by the ability of PE and PA derivatives to "flip" (by different mechanisms) from the outer to the inner leaflet of the plasma membrane and then, owing to their solubility, translocate to the intracellular membranes. On the other hand, sphingomyelin and PC are not translocated over the plasma membrane and are thus retained in the Golgi–plasma membrane pool (Sleight and Pagano, 1985; Simons and van Meer, 1988).

Although the NBD-lipids used to date have provided plenty of novel data on the compartmentalization of phospholipids in living cells, they do not give any information about the role of lipid transfer proteins in the intracellular lipid traffic. This is simply because the spontaneous rate of translocation of the NBD lipids used in these studies is several orders of magnitude higher than that of natural phospholipids (Nichols and Pagano, 1981; see also Chapter 11). Hence, these lipids will bypass the possible route(s) using transfer proteins. To obtain information on these routes, one obviously has to use analogues with spontaneous transfer rates similar to those of the natural lipids. In addition, the analogues should be transferable by the transfer proteins. Both of these requirements are fulfilled by the parinaroyl and pyrenylacyl derivatives described above. These lipids could be introduced into the cell by fusion of vesicles (van Meer and Simons, 1986), by transfer proteins (van Meer *et al.*, 1980), or by detergent-assisted incorporation (Bülow *et al.*, 1988). The recently developed quantitative microscopic methods, as reviewed by Arndt-Jovin *et al.* (1985), are expected to be very useful in studies of intracellular lipid traffic (van Meer *et al.*, 1987).

9. FLUORESCENT PHOSPHOLIPID ANALOGUES AS MEMBRANE PROBES

Fluorescent phospholipid analogues appear to be very useful in studying a variety of membrane-related phenomena (see Section 1). Until now, such probe lipids have been applied mostly to reconstituted systems. However, these lipids are equally useful for investigations of similar phenomena in natural membranes. With the phospholipid transfer proteins available, these lipids can now be introduced into natural membranes without disturbing the membrane. So far, this strategy has found very limited use. Christiansson *et al.* (1984) used PC-TP to introduce 2-PnA-PC into both microsomal and erythrocyte membranes to see how the motional properties of this lipid are influenced by different cholesterol and acyl chain compositions as measured by steady-state fluorescence spectroscopy. In another study 2-PnA-PC and 2-PnA-PI were introduced by transfer proteins into the rat skeletal muscle sarcolemmal membranes and the acetylcholine receptor-rich membranes from *Torpedo marmorata* to analyze the molecular order and dynamics of these probe lipids by time-resolved fluorescence techniques (van Paridon *et al.*, 1988b). Recent studies have shown that PyrPC species can be readily introduced into the human low- and high-density lipoproteins by PC-TP Vauhkonen and Somerharju, (1989, and unpublished data). Undoubtedly, the combined use of fluorescent phospholipids and transfer proteins will find further applications in the future.

10. REFERENCES

Arndt-Jovin, D. J., Robert-Nicoud, M., Kaufman, S. J., and Jovin, T. M., 1985, Fluorescence digital imaging microscopy in cell biology, *Science* **230**:247–256.

Arvinte, T., and Hildenbrand, K., 1984, N-NBD-L-α-dilauroylphosphatidylethanolamine, a new fluorescent probe to study spontaneous lipid transfer. *Biochim. Biophys. Acta* **775**:86–94.

Berkhout, T. A., Visser, A. J. W. G., and Wirtz, K. W. A., 1984, Static and time-resolved fluorescence studies of fluorescent phosphatidylcholine bound to the phosphatidylcholine transfer protein of bovine liver, *Biochemistry* **23**:1505–1513.

Bloj, B., and Zilversmit, D. B., 1977, Rat liver proteins capable of transferring phosphatidylethanolamine. Purification and transfer activity for other phospholipids and cholesterol, *J. Biol. Chem.* **252**:1613–1619.

Bülow, R., Overath, P., and Davoust, J., 1988, Rapid lateral diffusion of the variant glycoprotein in the coat of *Trypanosoma brucei*, *Biochemistry* **27**:2384–2388.

Chattopadhyay, A., and London, E., 1987, Parallax method for direct measurement of membrane penetration depth utilizing fluorescence quenching by spin-labeled phospholipids, *Biochemistry* **26**:39–45.

Chattopadhyay, A., and London, E., 1988, Spectroscopic and ionization properties of N-(7-nitrobenz-2-oxa-1,3-diazol-4-yl)-labeled lipids in model membranes. *Biochim. Biophys. Acta* **938**:24–34.

Christiansson, A., Kuypers, F. A., Roelofsen, B., Wirtz, K. W. A., and Op den Kamp, J. A. F., 1984, A comparative fluorescence polarization study of *cis*-parinaroyl-phosphatidylcholine and diphenylhexatriene in membranes containing different amounts of cholesterol, *Chem. Phys. Lipids* **35**:247–258.

Comfurius, P., and Zwaal, R. F. A., 1977, The enzymatic synthesis of phosphatidylserine and purification by CM-cellulose column chromatography, *Biochim. Biophys. Acta* **488**:36–42.

Crain, R. C., and Zilversmit, D. B., 1980, Two non-specific phospholipid exchange proteins from beef liver. 1. Purification and characterization, *Biochemistry* **19**:1433–1439.

Damel, R. A., Wirtz, K. W. A., Kamp, H. H., Geurts van Kessel, W. S. M., and van Deenen, L. L.

M., 1973, Phosphatidylcholine exchange protein from beef liver, *Nature New Biol.* **246:**102–105.

Demel, R. A., Kalsbeek, R., Wirtz, K. W. A., and van Deenen, L. L. M., 1977, The protein-mediated net transfer of phosphatidylinositol in model systems, *Biochim. Biophys. Acta* **466:** 10–22.

DiCorleto, P. E., Warach, J. B., and Zilversmit, D. B., 1979, Purification and characterization of two phospholipid exchange proteins from bovine heart, *J. Biol. Chem.* **254:**7795–7802.

Doody, M. C., Pownall, H. J., Kao, Y. J., and Smith, L. C., 1980, Mechanism and kinetics of transfer of a fluorescent fatty acid between single-walled phosphatidylcholine vesicles, *Biochemistry* **19:**108–116.

Frank, A., Barenholtz, Y., Lichtenberg, D., and Thompson, T. E., 1983, Spontaneous transfer of sphingomyelin between phospholipid vesicles, *Biochemistry* **22:**5647–5651.

Galla, H.-J., and Hartman, E., 1980, Excimer-forming lipids in membrane research, *Chem. Phys. Lipids* **27:**199–219.

Gupta, C. M., Radhakrishnan, R., and Khorana, H. G., 1977, Glycerophospholipid synthesis: improved general method and new analogs containing photoactivable groups, *Proc. Natl. Acad. Sci. U.S.A.* **74:**4315–4319.

Helmkamp, G. M., 1980, Concerning the mechanism of action of bovine liver phospholipid exchange protein: exchange or net transfer, *Biochem. Biophys. Res. Commun.* **97:**1091–1096.

Helmkamp, G. M., 1986, Phospholipid transfer proteins: Mechanism of action, *J. Bioenerg. Biomembr.* **18:**71–91.

Homan, R., and Pownall, H. J., 1988, Transbilayer diffusion of phospholipids: Dependence on headgroup structure and acyl chain length, *Biochim. Biophys. Acta* **938:**155–166.

Hresko, R. C., Sugar, I. P., Barenholtz, Y., and Thompson, T. E., 1987, The lateral distribution of pyrene-labeled sphingomyelin and glucosylceramide in phosphatidylcholine bilayers, *Biophys. J.* **51:**725–733.

Jones, M. E., and Lentz, B. R., 1986, Phospholipid lateral organization in synthetic membranes as monitored by pyrene-labeled phospholipids: Effects of temperature and prothrombin fragment 1 binding, *Biochemistry* **25:**567–574.

Kamp, H. H., Wirtz, K. W. A., and van Deenen, L. L. M., 1973, Some properties of phosphatidylcholine exchange protein purified from beef liver, *Biochim. Biophys. Acta* **318:**313–325.

Kamp, H. H., Wirtz, K. W. A., Baer, P. R., Slotboom, A. J., Rosentahl, A. F., Paltauf, F., and van Deenen, L. J. M., 1977, Specificity of the phosphatidylcholine exchange protein from bovine liver, *Biochemistry* **16:**1310–1316.

Kasper, A. M., and Helmkamp, G. M., 1981, Intermembrane phospholipid fluxed catalyzed by bovine brain phospholipid exchange protein, *Biochim. Biophys. Acta* **646:**22–32.

Lacowicz, J. R., 1983, *Principles of Fluorescence Spectroscopy*, Plenum Press, New York.

Machida, K., and Ohnishi, S., 1978, A spin-labeled study of phosphatidylcholine exchange protein. Regulation of the activity by phosphatidylserine and calcium ion, *Biochim. Biophys. Acta* **507:**156–164.

Massey, J. B., Gotto, A. M., and Pownall, H. J., 1982, Kinetics and mechanism of the spontaneous transfer of fluorescent phospholipids between a polipoprotein-phospholipid recombinants. Effect of the polar headgroup, *J. Biol. Chem.* **257:**5444–5448.

Massey, J. B., Hickson-Bick, D., Hoyan, S. S., Sparrow, J. T., Via, D. P., Gotto, A. M., and Pownall, H. J., 1984, Measurement and prediction of the rates of spontaneous transfer of phospholipids between plasma lipoproteins, *Biochim. Biophys. Acta* **794:**274–280.

Massey, J. B., Hickson-Bick, D., Via, D. P., Gotto, A. M., and Pownall, H. J., 1985, Fluorescence assay of the specificity of human plasma and bovine liver phospholipid transfer proteins, *Biochim. Biophys. Acta* **835:**124–131.

Mustonen, P., Virtanen, J. A., Somerharju, P. J., and Kinnunen, P. K. J., 1987, Binding of cytochrome C to liposomes as revealed by the quenching of fluorescence from pyrene-labeled phospholipids, *Biochemistry* **26:**2991–2997.

Nichols, J. W., 1987, Binding of fluorescent-labeled phosphatidylcholine to rat liver nonspecific lipid transfer protein, *J. Biol. Chem.* **262:**14172–14177.

Nichols, J. W., 1988, Kinetics of fluorescent-labeled phosphatidylcholine transfer between nonspecific lipid transfer protein and phospholipid vesicles, *Biochemistry* **27:**1889–1896.

Nichols, J. W., and Pagano, R. E., 1981, Kinetics of soluble lipid monomer diffusion between vesicles, *Biochemistry* **20:**2783–2789.

Nichols, J. W., and Pagano, R. E., 1983, Resonance energy transfer assay of protein-mediated lipid transfer between vesicles, *J. Biol. Chem.* **258:**5368–5371.

Pagano, R. E., and Sleight, R. G., 1985a, Defining lipid transport pathways in animal cells, *Science* **229:**1051–1057.

Pagano, R. E., and Sleight, R. G., 1985b, Emerging problems in the cell biology of lipids, *Trends Biochem. Sci.* **10:**421–425.

Pagano, R. E., Martin, O. C., Schroit, A. J., and Struck, D. K., 1981, Formation of asymmetric phospholipids membranes via spontaneous transfer of fluorescent lipid analogues between vesicle populations, *Biochemistry* **20:**4920–4927.

Pagano, R. E., Longmuir, K. J., and Martin, O. C., 1983, Intracellular translocation and methabolism of a fluorescent phosphatidic acid analogue in cultured fibroblasts, *J. Biol. Chem.* **258:**2034–2040.

Pugh, E. L., Kates, M., and Szabo, A. G., 1982, Studies on fluorescence polarization of 1-acyl-2-*cis*- or *trans*-parinaroyl *sn*-3-glycerophospholrylcholines in model systems and microsomal membranes, *Chem. Phys. Lipids* **30:**55–69.

Roseman, M. A., and Thompson, T. E., 1980, Mechanism of the spontaneous transfer of phospholipids between bilayers, *Biochemistry* **19:**439–444.

Rousselet, A., Colbeau, A., Vignais, P. M., and Devaux, P. F., 1976, Study of the transverse diffusion of spin-labeled phosphatidylcholine in biological membranes. I. Human red blood cell, *Biochim. Biophys. Acta* **426:**372–384.

Scallen, T. J., Pastuszyn, A., Noland, B. J., Chanderbhan, R., Kharroubi, A., and Vahouny, G. V., 1985, Sterol carrier and lipid transfer proteins, *Chem. Phys. Lipids* **38:**239–261.

Simons, K., and van Meer, G., 1988, Lipid sorting in epithelial cells, *Biochemistry* **27:**6197–6202.

Sklar, L. A., Hudson, B. S., and Simoni, R. D., 1975, Conjugated polyene fatty acids as membrane probes: Preliminary characterization, *Proc. Natl. Acad. Sci. U.S.A.* **72:**1649–1654.

Sklar, L. A., Hudson, B. S., and Simoni, R. D., 1977, Conjugated polyene fatty acids as fluorescent probes: Spectroscopic characterization, *Biochemistry* **16:**819–828.

Sleight, R. G., and Pagano, R. E., 1985, Transbilayer movement of a fluorescent phosphatidylethanolamine analogue across the plasma membranes of cultured mammalian cells, *J. Biol. Chem.* **260:**1146–1154.

Somerharju, P. J., and Wirtz, K. W. A., 1982, Semisynthesis of a fluorescent phosphatidylinositol containing a parinaroyl chain, *Chem. Phys. Lipids* **30:**81–91.

Somerharju, P. J., Brockerhoff, H., and Wirtz, K. W. A., 1981, A new fluorometric method to measure protein-catalyzed phospholipid transfer using 1-acyl-2-parinaroylphosphatidylcholine, *Biochim. Biophys. Acta* **649:**521–528.

Somerharju, P. J., van Paridon, P. A., and Wirtz, K. W. A., 1983, Phosphatidylinositol transfer protein from bovine brain: Substrate specificity and membrane binding properties, *Biochim. Biophys. Acta* **731:**186–195.

Somerharju, P. J., Virtanen, J. A., Eklund, K. K., Vainio, P., and Kinnunen, P. K., 1985, 1-Palmitoyl-2-pyrenedecanoyl glycerophospholipids as membrane probes: Evidence for regular distribution in liquid-crystalline phosphatidylcholine bilayers, *Biochemistry* **24:**2773–2781.

Somerharju, P. J., van Loon, D., and Wirtz, K. W. A., 1987, Determination of the acyl chain specificity of the bovine liver phosphatidylcholine transfer protein. Application of pyrene-labeled phosphatidylcholine species, *Biochemistry* **26**:7193–7199.

Struck, D. K., and Pagano, R. E., 1980, Insertion of fluorescent phospholipids into the plasma membrane of a mammalian cell, *J. Biol. Chem.* **255**:5404–5410.

Tanaka, Y., and Schroit, A. J., 1983, Insertion of fluorescent phosphatidylserine into the plasma membrane of red blood cells, *J. Biol. Chem.* **258**:11335–11343.

Trzaskos, J. M., and Gaylor, J. L., 1983, Cytosolic modulators of activities of microsomal enzymes of cholesterol biosynthesis. Purification and characterization of a non-specific lipid-transfer protein, *Biochim. Biophys. Acta* **751**:52–65.

Uster, P. S., and Pagano, R. E., 1986, Resonance energy transfer microscopy: Observations of membrane-bound fluorescent probes in model membranes and in living cells, *J. Cell Biol.* **103**:1221–1234.

van Loon, D., Demel, R. A., and Wirtz, K. W. A., 1986, The phosphatidylcholine transfer protein from bovine liver discriminates between phosphatidylcholine isomers. A monolayer study, *Biochim. Biophys. Acta* **856**:482–487.

van Meer, G., and Simons, K., 1986, The function of tight junctions in maintaining differences in lipid composition between the apical and the basolateral cell surface domains of MDCK cells, *EMBO J.* **5**:1455–1464.

van Meer, G., Poorthuis, B. J. H. M., Wirtz, K. W. A., Op den Kamp, J. A. F., and van Deenen, L. L. M., 1980, Transbilayer distribution and mobility of phosphatidylcholine in intact erythrocyte membranes. A study with phosphatidylcholine exchange protein, *Eur. J. Biochem.* **103**:283–288.

van Meer, G., Stelzer, E. H. L., Wijnaendts-van-Resandt, R. W., and Simons, K., 1987, Sorting of sphingolipids in epithelial (Madin-Darby canine kidney) cells, *J. Cell Biol.* **105**:1623–1635.

van Paridon, P. A., Gadella, T. W. J., Somerharju, P. S., and Wirtz, K. W. A., 1987a, On the relationship between the dual specificity of the bovine brain phosphatidylinositol transfer protein and membrane phosphatidylinositol levels, *Biochim. Biophys. Acta* **903**:68–77.

van Paridon, P. A., Visser, A. J. W. G., and Wirtz, K. W. A., 1987b, Binding of phospholipids to the phosphatidylinositol transfer protein from bovine brain as studied by steady-state and time-resolved fluorescence spectroscopy, *Biochim. Biophys. Acta* **898**:172–180.

van Paridon, P. A., Gadella, T. W. J., Somerharju, P. J., and Wirtz, K. W. A., 1988a, Determination of the acyl chain specificity of the bovine brain phosphatidylinositol transfer protein. Application of pyrene-labeled phospholipids, *Biochemistry* **27**:6208–6214.

van Paridon, P. A., Shute, J. K., Wirtz, K. W. A., and Visser, A. J. W. G., 1988b, A fluorescence decay study of parinaroylphosphatidylinositol incorporated into artificial and natural membranes, *Eur. J. Biophys.* **16**:53–63.

Vauhkonen, M. V., and Somerharja, P. J., 1989, Parinaroyl and pyrenl phospholipids as probes for the lipid surface layer of human low density lipoproteins, *Biochem. Biophys. Acta* **984**:81–87.

Welti, R., 1982, Partition of parinaroyl phospholipids in mixed head group systems, *Biochemistry* **21**:5690–5695.

Welti, R., and Silbert, D. F., 1982, Partition of parinaroyl phospholipid probes between solid and fluid phosphatidylcholine phases, *Biochemistry* **21**:5685–5689.

Wiener, J. R., Pal, R., Barenholtz, Y., and Wagner, R. R., 1985, Effect of the vesicular stomatitis virus matrix protein on the lateral organization of lipid bilayers containing phosphatidylglycerol: use of fluorescent phospholipid analogues, *Biochemistry* **24**:7651–7658.

Wirtz, K. W. A., Teerlink, T., and Akeroyd, R., 1985, Properties and function of phosphatidylcholine transfer proteins, in *The Enzymes of Biological Membranes* (A. N. Martonosi, ed.), Vol. 2, pp. 111–138, Plenum Press, New York.

Wirtz, K. W. A., Op den Kamp, J. A. F. and Roelofsen, B., 1986, Phosphatidylcholine transfer protein: properties and applications in membrane research, in *Progress in Protein–Lipid Interactions* (A. Watts and J. J. H. H. M, de Pont, eds.), Vol. 2, pp. 221–265, Elsevier Science Publishing, Amsterdam.

Zborowski, J., and Demel, R. A., 1982, Transfer properties of the bovine brain phospholipid transfer protein. Effect of charged phospholipids and of phosphatidylcholine fatty acid composition, *Biochim. Biophys. Acta* **688:**381–387.

Chapter 3

Phosphatidyl Transfer Proteins as Probes of Membrane Structure and Function

Richard C. Crain

1. INTRODUCTION

The spontaneous redistribution of phospholipid between classes of unilamellar vesicles (Kornberg and McConnell, 1971; Ehnholm and Zilversmit, 1972) or between microsomal vesicles and mitochondria (Wirtz and Zilversmit, 1968) is very slow for most phospholipids (reviewed by Sleight, 1987; Dawidowicz, 1987a). Wirtz and Zilversmit (1968) and McMurray and Dawson (1969) demonstrated that phospholipid movement was stimulated by rat liver cytosol and later showed that proteins were responsible for this activity (reviewed by Wirtz, 1982). Lipid transfer proteins with various specificities and physical characteristics have now been purified from a number of biological sources, including animal tissues and plasma, plants, yeasts, and bacteria (see Section 2.1).

Little is known about the precise physiological role of these proteins (Sleight, 1987). It was initially proposed that they are involved in the transport of newly synthesized membrane lipids from their sites of synthesis to their final

Abbreviations used in this chapter: PC, phosphatidylcholine; PE, phosphatidyl ethanolamine; PI, phosphatidylinositol; PS, phosphatidylserine; PG, phosphatidylglycerol; PA, phosphatidic acid.

Richard C. Crain Department of Molecular and Cell Biology, The University of Connecticut, Storrs, Connecticut 06268.

destinations in the cell (Wirtz, 1982). Support for this hypothesis has come from the finding of rapid intracellular movement of lipid, a movement that could not readily be explained by either spontaneous transfer or vesicle delivery (Sleight, 1987). Further indirect evidence supporting this conclusion was obtained from studies examining phosphatidylcholine (PC) and phosphatidylinositol (PI) transfer activities in lung and brain during development. These studies found that substantial increases of PC transfer activity occur at the onset of lung surfactant synthesis and that these increases correlate with increased PC synthetic activity (Engle *et al.*, 1978; Teerlink *et al.*, 1982). Increases of PI transfer activity also were observed at the onset of myelogenesis (Brophy and Aitken, 1979; Ruenwongsa *et al.*, 1979). However, because transfer activities for other myelin phospholipids and glycolipids remained absent during the onset of myelogenesis (Ruenwongsa *et al.*, 1979), it is apparent that transfer proteins are not required for the integration of all membrane lipids. Furthermore, Yaffe and Kennedy (1983), comparing hepatocytes and baby hamster ovary cells, did not observe a correlation between rates of phosphatidylethanolamine (PE), *N,N'*-dimethylPE, and PC movement in cells in culture and the cellular content of activities capable of accelerating transfer of these lipids. Voelker (1985) has shown that metabolic energy is required for phosphatidylserine (PS) movement from its site of synthesis, in the endoplasmic reticulum, to its site of decarboxylation, in the mitochondria of BHK cells. Although this observation does not rule out lipid transfer proteins as playing a role, it does suggest that the lipid transport process examined *in vitro* does not necessarily reflect the cellular situation (Voelker, 1985). More recently, Rusinol *et al.* (1987) showed that transfer activity decreases rather than increases during early embryo development, which suggests that it is at least not limiting at this stage of development.

Although the role of lipid transfer proteins in the biosynthesis, intermembrane transport, and metabolism of cellular lipids has not yet been established, the availability of lipid transfer proteins of well-defined specificity, which can be readily purified in large amounts, has made them highly useful for the study of membrane structure and function. Their use is particularly appealing because it provides a means for changing composition without introducing alterations in the native membrane structure. This chapter summarizes the use of lipid transfer proteins in studies examining (1) lipid asymmetry and transbilayer movement (2) factors that influence lipid asymmetry and transbilayer movement, (3) lipid-metabolizing enzymes that require water insoluble substrates, (4) the effect of phospholipid composition on membrane morphology, and (5) the effect of membrane lipid composition on activities of membrane-bound enzymes.

Readers are also referred to previous reviews on the use of lipid transfer proteins in the study of artificial and natural membranes (Zilversmit, 1978, 1983, 1984; Van Deenen, 1981; Bloj and Zilversmit, 1981a; Crain, 1982). Other recent reviews have appeared that cover the mechanism of action of phospholipid transfer proteins (Helmkamp, 1986), intracellular lipid transport (Sleight, 1987), dynamics of membrane lipid metabolism and turnover (Dawidowicz, 1987a), spontaneous and protein-mediated movement of lipids (Dawidowicz, 1987b) and a survey of nonenzymic proteins involved in lipophilic compound metabolism (Bernier and Jolles, 1987).

2. PROPERTIES OF LIPID TRANSFER PROTEINS

2.1. Substrate Specificity

Since lipid transfer proteins of diverse origins, specificity, and physical characteristics have been purified, the properties of these proteins should be considered when selecting the appropriate probe for membrane studies. A compilation of some of the major purified proteins is presented in Table I.

The molecular specificity of these proteins has been determined for phospholipids differing in polar headgroup or fatty acid content (Zilversmit and Hughes, 1977; Wetterau and Zilversmit, 1984). Although these proteins differ considerably in polar headgroup specificity, many of them can be categorized into one of three groups: (1) PC transfer proteins, which are specific for PC and have been purified from both bovine and rat liver; (2) PI transfer proteins, which have a marked preference for PI but also catalyze transfer of PC and phosphatidylglycerol (PG) and have been purified from bovine heart and brain, human platelets, and yeasts; and, finally (3) nonspecific lipid transfer proteins, which are relatively nonspecific for headgroup composition and have been purified from rat, bovine, and human liver, rat lung, castor bean seeds, and spinach leaf. Most membrane studies have been conducted with proteins that can be purified easily in large amounts, such as the PC transfer protein from bovine liver, the PI transfer protein from bovine brain or heart, or the nonspecific lipid transfer protein from rat or bovine liver.

The acyl specificity of the PC transfer protein has been examined for transfer between vesicles (Kamp *et al.*, 1977; Kasper and Helmkamp, 1981; Welti and Helmkamp, 1984) and between erythrocyte membranes and lipid vesicles (Child *et al.*, 1985a; Kuypers *et al.*, 1986). The presence of both acyl groups is required for transfer; i.e., transfer of lyso-PC is not accelerated by this protein (Kamp *et al.*, 1977). Although the protein has a broad specificity for acyl chain composition, unsaturated PC is transferred preferentially. Kamp *et al.* (1977) found that 1-palmitoyl-2-oleoyl was transferred at three times the rate of dipalmitoyl-PC. Similarly, Welti and Helmkamp (1984) found that dioleoyl-PC was transferred at three times the rate of distearoyl-PC and at six times the rate of dimyristoyl-PC, whereas Child *et al.* (1985a) found a 10-fold greater rate of transfer for soybean PC than for dipalmitoyl-PC. Comparison of the efflux of individual molecular

Table I
Specificity of Purified Phospholipid Transfer Proteins

Source	Type[a]	Substrate specificity[a]	Molecular weight	pI	Reference
Bovine liver	PC-TP	PC	24,681	5.8	Kamp *et al.* (1973)
	nsL-TP[b]	PC, PI, PA, PG, PS, PE, Cho, Sph, Gm_1, $GbOse_4Cer$	13,900	9.6	Crain and Zilversmit (1980a)
					Bloj and Zilversmit (1981b)
Rat liver	PC-TP	PC	28,000	8.4	Poorthuis *et al.* (1980)
	nsL-TP[b]	PC, PI, PS, PE, Sph, Cho	12,400	8.8	Bloj and Zilversmit (1977)
Human liver	nsL-TP	Not determined	14,000	—	van Amerongen *et al.* (1987)
Rat hepatoma	nsL-TP	PC, PI, PE, Sph	11,200	5.2	Dyatlovitskaya *et al.* (1978)
Bovine heart	PI-TP[b]	PI > PC	33,400	5.4	DiCorleto *et al.* (1979)
Bovine brain	PI-TP[b]	PI > PC	32,000	5.4	Helmkamp *et al.* (1974)
Rat lung	nsL-TP	PI, PC, PE, PS	33,000	5.6	Read and Funkhouser (1983)
Human platelets	PI-TP[b]	PI > PC > PG	28,000	5.6	George and Helmkamp (1985)
Human plasma	CE-TP	CE > TG > PC	64,000	5.0	Albers *et al.* (1984)
Castor bean seeds	nsL-TP	PI, PG, PE, PA, MGDG, DGDG	9,000	10.5	Watanabe and Yamada (1986)
Spinach leaf	nsL-TP	PC, PI, PG, PE	9,000	9.0	Kader *et al.* (1984)
Maize seedlings	nsL-TP	PC, PE, PI	9,000	8.8	Kader (1985)
Saccharomyces cerevisiae	PI-TP	PI, PC	35,000	4.6	Daum and Paltauf (1983)
	nsL-TP	PC > PE > PI > PS	33,400	6.2	Bozzato and Tinker (1987)
Rhodopseudomonas sphaeroides	nsL-TP	PG > PC > PE	27,000	4.9	Tai and Kaplan (1984)

[a]Abbreviations: PC-TP, PC-specific transfer protein; PI-TP, PI-specific transfer protein; nsL-TP, nonspecific lipid transfer protein; CE-TP, cholesterol ester transfer protein; Cho, cholesterol; Sph, sphingomyelin; Gm_1, II^3-alpha-*N*-acetylneuraminosyl-gangliotetraglycosyceramide; $GbOse_4Cer$, globotetraglycosylceramide; MGDG, monogalactosyldiacylglyceride; DGDG, digalactosyldiacylglyceride; CE, cholesterol ester; TG, triacylglycerol.
[b]Phospholipid transfer proteins that exist in two forms differing slightly in molecular weight and pI.

species of PC showed that the rate of transfer increased with increasing unsaturation: 1-palmitoyl-2-lineoyl-PC > 1-palmitoyl-2-oleoyl-PC > 1,2-dipalmitoyl-PC (Child *et al.*, 1985a; Kuypers *et al.*, 1986).

The acyl specificity of nonspecific lipid transfer protein has been examined by Read and Funkhouser (1984) for the rat lung protein and by Crain (1982) for the bovine liver protein. In both cases, little specificity for saturated versus unsaturated PC was observed, although both investigators found that increasing concentrations of saturated PC in a vesicle of mixed saturated and unsaturated PC inhibited the transfer of both molecular species. Furthermore, transfer was greatly reduced from vesicles whose lipid was in the gel state (Read and Funkhouser, 1984).

2.2. Net Transfer and Exchange of Phospholipid

The use of phospholipid transfer proteins in the study of membrane structure and function is dependent on their ability to facilitate the replacement of an acceptor membrane phospholipid by that from a donor. We can describe two general classes of transfer: exchange and net transfer (Crain and Zilversmit, 1980c; Helmkamp, 1980, 1986). The generic term "exchange" describes replacement of a lipid in an acceptor membrane by one from a donor membrane (Helmkamp, 1986). Either specific or nonspecific lipid transfer proteins may accelerate the bidirectional movement of an individual class of phospholipids between membrane fractions; this movement is termed "homo-exchange" (Crain and Zilversmit, 1980c). Under different conditions, a relatively nonspecific lipid transfer protein may promote the transfer of one lipid from donor to acceptor and a different lipid in the opposite direction (Helmkamp, 1986); this process is termed "hetero-exchange." Although homo-exchange and hetero-exchange may involve the net movement of molecular species of a phospholipid class or net transfer of a class of phospholipid from donor to acceptor membrane, they do not cause a change in the total phospholipid content of either the donor or acceptor membrane. For example, bovine PtdIns transfer protein has been shown by a number of investigators to stimulate the *net movement* of PI and/or PC between membranes (Demel *et al.*, 1977; Kasper and Helmkamp, 1981), but *net transfer* of phospholipid mass is not observed (Helmkamp, 1986).

In contrast to the exchange of lipids, net transfer involves the unidirectional movement of lipid. Both donor and acceptor membranes are altered in both content and composition of phospholipid.

The capacity of the bovine PC-specific transfer protein to effect net transfer of PC from vesicles composed primarily of PC either to vesicles composed of PE and PA (Wirtz *et al.*, 1980; Berkhout *et al.*, 1984;) or to vesicles composed solely of PA (Xu *et al.*, 1983) has been demonstrated. Under a number of other conditions, this protein was found to stimulate the exchange of PC, which results

Table II
Net Transfer and Hetero-Exchange of Phospholipids Catalyzed by Phospholipid Transfer Proteins

Transfer protein[a]	Donor membrane[a]	Acceptor membrane[a]	Phospholipid transfer[a]	Reference
PC-TP	PC/PA (75 : 25, mole %)	PE/PA (81 : 19, mole %)	Net transfer of PC to acceptor vesicle	Wirtz *et al.* (1980)
PC-TP	DMPC	DMPA	Net transfer of DMPC to acceptor vesicle	Xu *et al.* (1983)
PC-TP	PC	PA	Net transfer of PC to acceptor vesicle	Berkhout *et al.* (1984)
PC-TP	Egg PC	DMPC	Exchange of PC	Helmkamp (1980)
PC-TP	DPPC	Rat microsomes	Exchange of microsomal PC by DPPC	Crain (1982)
PC-TP	PC	HDL	Exchange	Crain and Zilversmit (1980c)
PC-TP	PC	Mitochondria	Exchange	Crain and Zilversmit (1981)
PI-TP	PC/PI/LacCer (82 : 10 : 8, mole %)	PC/PA (95 : 5, mole %)	Hetero-exchange	Kasper and Helmkamp (1981)
PI-TP	PC/CL (9 : 1, mole %)	HDL	Exchange of PC	Crain and Zilversmit (1980c)
nsL-TP	PC/PE (1 : 1, mole %)	Synaptic vesicles	No net change in phospholipid content or composition	Baba *et al.* (1986)
nsL-TP	PC/PI (9 : 1, mole %)	Apo-HDL	Net transfer of both PC and PI to delipidated HDL	Crain and Zilversmit (1980c)
nsL-TP	Egg PC	Synaptic plasma membranes	Net transfer of lipid at high nsL-TP	North and Fleischer (1983)
nsL-TP	PC	Mitoplasts	Net transfer of PC to mitoplasts	Crain and Zilversmit (1980c)
nsL-TP	DPPC/RL-PC (3 : 1, mole %)	Microsomes	Hetero-exchange: DPPC to microsomes, PE and PI to vesicles	Crain (1982)
nsL-TP	Phospholipid	*Salmonella typhimurium*	Net transfer of phospholipid to cells	Bishop (1983)

[a]Abbreviations: DMPC, dimyristoylphosphatidylcholine; DMPA, dimyristoylphosphatidic acid; DPPC, dipalmitoylphosphatidylcholine; LacCer, lactosylceramide; HDL, high-density lipoprotein; apo-HDL, delipidated high-density lipoprotein; RL-PC, rat liver phosphatidylcholine; CL, cardiolipid. For other abbreviations, see Table I, Footnote *a*.

in a change in the fatty acid composition of membrane phospholipid without any alteration of the phospholipid content (Table II).

The nonspecific lipid transfer protein from liver also catalyzes net transfer of phospholipid under a variety of conditions (Table II), whereas no net change in phospholipid or cholesterol content is observed under other conditions (Table II). This property of the nonspecific lipid transfer protein, i.e., its capacity for either exchange or net transfer, is particularly important when it is used as a probe of membrane structure and function (see Section 4.1).

3. MEMBRANE STRUCTURE: MEASUREMENT OF MEMBRANE ASYMMETRY

Early studies examining protein-mediated phospholipid transfer from lipid vesicles and biological membranes established that phospholipid is transferred almost exclusively between the outer leaflets of the donor (Johnson *et al.*, 1975; Rothman and Dawidowicz, 1975; Sandra and Pagano, 1979) and acceptor (Sandra and Pagano, 1979) membranes. This vectorial action of phospholipid transfer proteins has therefore made them appropriate for use as probes of lipid asymmetry in vesicles and biological membranes (reviewed by Op den Kamp, 1979; Bloj and Zilversmit, 1981a; Krebs, 1982).

Asymmetry has generally been monitored by measuring the kinetics of transfer of labeled lipid from a donor membrane or vesicle to an acceptor membrane or vesicle. Frequent changes of acceptor vesicle allow the transfer to be examined kinetically (Bloj and Zilversmit, 1976; Crain and Zilversmit, 1980b). The initial rapid phase of phospholipid transfer is dependent upon the concentration of transfer protein used (Zilversmit and Hughes, 1977; Crain and Zilversmit, 1980b) and represents the transfer of phospholipid from the outer membrane leaflet to the acceptor membrane (Sandra and Pagano, 1979). The second pool of phospholipid, which is transferred at a lower rate, is localized on the inner leaflet and becomes available for exchange only after translocation to the outer leaflet (see Section 4.1).

The use of phospholipid transfer proteins for measurements of membrane asymmetry is limited by the specificity of the protein and its ability to facilitate phospholipid movement from the membrane being studied. Furthermore, accurate results are dependent upon the following. (1) Membrane integrity must be maintained during incubation, which could be a problem in studies in which a net transfer of phospholipid occurs (Crain and Zilversmit, 1980c). (2) Uniform labeling of the phospholipids in the membrane under study must exist for the biphasic exchange kinetics to be interpreted in terms of asymmetry. (3) The transbilayer movement of phospholipid must be slow compared with the rate of transfer (Zilversmit and Hughes, 1977). (4) Nonspecific lipid adsorption must be mini-

Table III
Use of Phospholipid Transfer Proteins in Studies of Membrane Asymmetry

Transfer protein[a]	Membrane studied	Phospholipids examined[a]	Reference
PC-TP, PI-TP	Influenza virus membranes	PC, PI	Rothman *et al.* (1976)
nsL-TP	Erythrocyte ghosts	PC, SM, PE, PI + PS	Bloj and Zilversmit (1976)
PI-TP	Rat liver microsomes	PI	Brophy *et al.* (1978)
PC-TP	Vesicular stomatitis virus	PC	Shaw *et al.* (1979)
PC-TP	Erythrocyte membranes on polylysine-coated beads	PC	Kramer and Branton (1979)
nsL-TP	Erythrocytes	PC, SM, PE	Crain and Zilversmit (1980b)
PC-TP	Erythrocytes	PC	Van Meer *et al.* (1980)
PC-TP	Semliki Forest virus membranes	PC	Van Meer *et al.* (1981)
PC-TP	Intestinal brush border membranes	PC	Barsukov *et al.* (1986)
PC-TP	*Plasmodium knowlesi*-infected erythrocytes	PC	Van der Schaft *et al.* (1987)

[a]For abbreviations, see Table I, Footnote *a*.

mal (Barsukov *et al.*, 1986). The phospholipid asymmetry of a number of biological membranes has been examined (Table III).

4. DYNAMIC ASPECTS OF PHOSPHOLIPID MOVEMENT IN BIOLOGICAL MEMBRANES

4.1. Phospholipid Transbilayer Movement

The application of transfer proteins to the measurement of the rate of transbilayer movement of phospholipids (often referred to as flip-flop) was established early (reviewed by Bloj and Zilversmit, 1981a). Few other methods have been developed that can be used to measure flip-flop of phospholipid species normally found in biological membranes, while not introducing perturbations of membrane structure that might be expected to alter the rate of phospholipid transbilayer movement. Initial studies examined the kinetics of lipid efflux from uniformly labeled unilamellar vesicles or biological membranes to acceptor membranes (Johnson *et al.*, 1975; Bloj and Zilversmit, 1976; DiCorleto and Zilversmit, 1977). As discussed in Section 3.1, transfer of phospholipid in the inner leaflet is limited by the rate of flip-flop, and provided that this rate is low compared with the rate of phospholipid efflux, it can be calculated (Bloj and

Zilversmit, 1976; Crain and Zilversmit, 1980b). The rate of transbilayer movement of phospholipid in lipid vesicles was found to be low; the half times of transbilayer equilibration are on the order of days (Van den Besselaar *et al.*, 1978; DiCorleto and Zilversmit, 1979; Low and Zilversmit, 1980). Rat erythrocyte (Crain and Zilversmit, 1980b; Van Meer *et al.*, 1980) and virus (Shaw *et al.*, 1979; Van Meer *et al.*, 1981) membranes had higher rates of transbilayer movement, with half times for equilibration on the order of 5–10 hr. In contrast, nearly all of the phospholipid in a rat liver microsomal fraction (Zilversmit and Hughes, 1977; Van den Besselaar *et al.*, 1978; Hutson *et al.*, 1985) and intestinal brush border membranes (Barsukov *et al.*, 1986) is available for rapid transfer, which indicates rapid transbilayer equilbration. The rapid translocation of phospholipid across the microsomal bilayer has recently been shown to be caused by a specific membrane protein transporter (Bishop and Bell, 1985; Kawashima and Bell, 1987).

More recently, PC flip-flop in cells and vesicles modified by introduction of exogenous, labeled PC into the outer membrane leaflet has been examined by using the PC-specific transfer protein. The time-dependent decrease in accessibility of the labeled PC to hydrolysis by phospholipase A_2 in the presence of sphingomyelinase (reviewed by Op den Kamp *et al.*, 1985) was used to determine the rate of PC transbilayer equilibration. In human erythrocytes, the rate of transbilayer movement of this exogenously replaced PC depends upon fatty acid composition, with half times for equilibration of 26, 14, 2.9, and 9.7 hr for dipalmitoyl-PC, dioleoyl-PC, 1-palmitoyl-2-linoleoyl-PC, and 1-palmitoyl-2-Δ^4arachidonoyl-PC, respectively (Middelkoop *et al.*, 1986).

This approach has also been applied to studies of the relationship between phospholipid organization, as measured by rates of flip-flop, and conditions that alter cell morphology and cytoskeleton–membrane interactions. An increase in PC transbilayer movement, reflecting disorganization of the lipid bilayer, has been observed to occur in (1) erythrocytes of patients with hereditary pyropoikilocytosis, a congenital hemolytic anemia that is caused by enhanced thermal sensitivity of the membrane cytoskeleton (Franck *et al.*, 1985b); (2) reversibly sickled erythrocytes under deoxygenated conditions (Franck *et al.*, 1983), which is a consequence of uncoupling of the cytoskeleton from the bilayer (Franck *et al.*, 1985a); and (3) erythrocytes infected by *Plasmodium knowlesi* (Van der Schaft *et al.*, 1987), which indicates that membrane destabilization is caused by the parasite infection.

Nonspecific lipid transfer protein has also been used to introduce PtdSer into erythrocyte membranes (Middelkoop *et al.*, 1988). In oxygenated reversibly sickled cells, 90% of the labeled PS was inaccessible to phospholipase A_2 after 1 hr. This fraction was dramatically decreased by subsequent deoxygenation of the cells. On the basis of the effects of sickling and ATP depletion on accessibility of the PS introduced by phospholipid transfer, Middelkoop *et al.* (1988) conclude that both an interaction of PS with cytoskeletal proteins and an ATP-dependent translocation of PS are involved in maintaining its asymmetric distribution in the erythrocyte membrane.

Other investigators have used transfer protein-stimulated efflux of phospholipid as a measure of the lipid disorganization introduced by chemical cross-linking of protein and lipid in erythrocyte membranes. Diamide-induced cross-linking of the erythrocyte cytoskeletal proteins resulted in a twofold increase in the rate of phospholipid efflux (Bittman *et al.*, 1985) and an increase in the extent of exchangeable PC (Franck *et al.*, 1982). This indication of enhanced transbilayer migration is consistent with proposals by Haest *et al.* (1978) and Marinetti and Crain (1978) that the asymmetric distribution of phospholipids in the erythrocyte membrane is maintained in part by interactions between the membrane skeleton and the negatively charged phospholipids.

A third method for assessing phospholipid transbilayer movement has recently been explored by Van der Meer *et al.* (1987). Asymmetric large unilamellar vesicles containing pyrene-labeled PC exclusively on the inner leaflet were prepared by incubating pyrene-labeled PC large unilamellar vesicles with an excess of PC small unilamellar vesicles in the presence of PC-specific lipid transfer protein to exchange the pyrene-labeled PC in the outer leaflet for unlabeled PC. These asymmetrically labeled vesicles, after binding of complement (C5b67), were treated with complement C8. The increased transbilayer movement caused by complement C8 was determined from the dose-dependent decrease in the eximer/monomer fluorescence intensity of the pyrene-labeled PC, which resulted from its dilution as it migrated to the outer leaflet.

It should be pointed out that artifactual results can be induced by membrane perturbations. Under numerous experimental conditions, PC-specific transfer protein has been shown to catalyze a one-for-one exchange of PC (Table II), allowing the introduction of traces of labeled PC (Middelkoop *et al.*, 1986) or a complete exchange of PC in the outer membrane leaflet (Van der Meer *et al.*, 1987) without altering the structure of the membrane, provided that appropriate molecular species of PC are substituted (Kuypers *et al.*, 1984a,b). Nonspecific lipid transfer protein, on the other hand, although ideal for the introduction of a small amount of labeled lipid into a membrane (Tilley *et al.*, 1986; Middelkoop *et al.*, 1988), may cause membrane perturbations (B. W. Van der Meer, personal communication) and vesicle aggregation (Crain and Zilversmit, 1980c) owing to the net transfer of phospholipid to or from the membrane (see Section 2.2).

4.2. Metabolism and Intracellular Movement of Exogenously Added Phospholipid

As discussed in the previous section, transfer proteins have been used to introduce exogenous phospholipids into isolated membranes and intact cells. D'Souza *et al.* (1983), who introduced PC and sphingomyelin into murine neuroblastoma cells in the presence of rat liver nonspecific lipid transfer protein, found that sphingomyelin was preferentially (compared with PC) incorporated into the plasma membrane and microsomal membrane fractions, whereas PC was incorporated into the mitochondrial membrane fraction to a 128-fold greater extent than sphingomyelin was. Similar studies have been pursued recently by Pagano and co-workers, who have been able to incorporate relatively water-soluble lipid analogues into intact cells in the absence of lipid transfer protein and have also examined their subsequent cellular distribution (Pagano and Sleight, 1985).

Mechanisms of phospholipid translocation in human erythrocytes into which PE, PC, and PS are introduced in the presence of nonspecific lipid transfer protein have also been examined. Rapid transbilayer migration of PE and PS is observed, whereas PC is translocated to the inner leaflet only slowly (Tilley *et al.*, 1986). ATP depletion, either by glucose starvation or by metabolic inhibitors, decreases the migration of PE and PS to the inner leaflet. This finding supports the notion that an ATP-dependent, phospholipid headgroup-specific translocase (Tilley *et al.*, 1986; Middelkoop *et al.*, 1988) is involved in the maintenance of amino phospholipid asymmetry in the erythrocyte membrane.

4.3. Lipid–Lipid and Lipid–Protein Interactions

Several investigators have used nonspecific lipid transfer protein to introduce fluorescently labeled phospholipid derivatives into membrane fragments to study properties of the lipid phase and effects of protein on lipid order. Muczynski and Stahl (1983) introduced dansylated PE and PS and dehydroergosterol into electroplax plasma membrane enriched in Na^+,K^+-ATPase and used these lipids as probes of membrane organization of the lipid associated with the ATPase. More recently, Harris (1985) introduced 8-(dimethylamino)-naphthalene-1-sulfonyl-PS into purified lamb kidney Na^+,K^+-ATPase, also by using nonspecific lipid transfer protein. The lipid probe was used as a reporter of protein-induced order in ATPase-associated lipid and in an examination of the relationship between ATPase activity and membrane fluidity.

Nonspecific lipid transfer protein has also been used recently to modify vesicles and membranes for fluorescence detection of fusion. Walter *et al.* (1986) made asymmetric vesicles containing *N*-4-nitrobenzo-2-oxa-1,3-diazole-PE and rhodamine sulfonyl-PE only on the inner membrane leaflet. This was done by using the transfer protein to exchange fluorescently labeled PE in the outer leaflet of small unilamellar vesicles with unlabeled phospholipid from a 10-fold excess of large unilamellar vesicles. Polylysine-induced fusion of the asymmetric small unilamellar vesicles with an excess of vesicles not containing the fluorescent probes was monitored by fluorescence energy transfer or, more specifically, by the increase in fluorescence resulting from reduced energy transfer between the probes that occurred owing to fusion-dependent dilution. The presence of the probes on the inner leaflet ruled out their spontaneous transfer as the mechanism of increased fluorescence. Similarly, Bamford *et al.* (1987) introduced 1-palmitoyl-2-pyrenedecanoyl-PC and 1-palmitoyl-2-pyrenedecanoyl-PG into bacteriophage Φ6 by using nonspecific lipid transfer protein. After the phage PE had been labeled with trinitrobenzenesulfonate, which quenches the pyrene emission, phage fusion with *Pseudomonas syringae* could be assayed by monitoring the dilution-induced reduction in quenching of pyrene-labeled phospholipid.

Cholesterol–phospholipid interactions have been estimated on the basis of the rates of spontaneous cholesterol efflux from erythrocyte membranes, modified with PC of defined fatty acid composition by using PC transfer protein. Cholesterol efflux was influenced by the fatty acid composition of the membrane PC; disaturated PC enhanced efflux, whereas polyunsaturated PC reduced efflux (Child *et al.*, 1985b).

5. LIPID BIOSYNTHESIS AND MEMBRANE ASSEMBLY

Transfer proteins have recently been added to assays in which water-insoluble lipids act as substrates. Voelker and Kennedy (1983), studying sphingolipid biosynthesis, added partially purified PC transfer protein to incubations containing exogenous [^{3}H]PC and plasma membrane fractions to introduce labeled PC substrate into the membrane domain of the phosphocholine transferase. Similarly, partially purified nonspecific lipid transfer protein has been used to introduce radiolabeled PC into synaptosomal membranes to characterize the oleic acid stimulation of phospholipase D (Hattori *et al.*, 1987), to introduce [^{14}C]PtdEtn into synaptic vesicles for studies of the stimulation of phospholipase A_2 by depolarization (Baba *et al.*, 1986), and to introduce lactosylceramide into microsomal and Golgi apparatus vesicles for studies on the properties of CMP-*N*-acetylneuraminate : lactosylceramide-α2,3-sialyltransferase (Kadowaki *et al.*, 1988).

The effect of nonspecific lipid transfer protein on *in vitro* biosynthesis and metabolism of cholesterol has been investigated by a number of laboratories (Table IV). The fact that nonspecific lipid transfer protein is distributed in organs involved in cholesterol metabolism, e.g., the intestine (Kharroubi *et al.*, 1988; Van Amerongen *et al.*, 1985), liver (Van Amerongen *et al.*, 1985), and adrenal gland (Kharroubi *et al.*, 1986), has led to the suggestion that it may play a physiological role in the regulation of the last steps of cholesterol biosynthesis, in sterol utilization for steroid hormone synthesis, and in cholesterol esterification (Scallen *et al.*, 1985a,b). This possibility has been explored in hepatomas, which have markedly reduced levels of nonspecific lipid transfer protein (Crain *et al.*, 1983; Teerlink *et al.*, 1984). Hepatomas have elevated cholesterol levels (Crain

Table IV
Effect of Nonspecific Lipid Transfer Protein (Sterol Carrier Protein$_2$) on Membrane Enzyme Activities

Enzymatic assay	Effect	Reference
Microsomal formation of cholesterol from 7-dehydrocholesterol	3-fold stimulation	Trzaskos and Gaylor (1983)
	>50-fold stimulation	Scallen *et al.* (1985a,b)
Microsomal formation of 7-dehydrocholesterol from Δ^7-cholesten-3β-ol	3-fold stimulation	Trzaskos and Gaylor (1983)
Microsomal formation of cholesterol ester from exogenous cholesterol by acyl CoA:cholesterol acyltransferase	>30-fold stimulation	Gavey *et al.* (1981)
	7-fold stimulation	Poorthuis and Wirtz (1982)
	6.8-fold stimulation	Trzaskos and Gaylor (1983)
Pregnenolone formation from exogenous cholesterol (adrenal lipid droplets) by adrenal mitochondria	>100-fold stimulation	Scallen *et al.* (1985a,b)
	4-fold stimulation	Kharroubi *et al.* (1988)
Release of cholesterol from adrenal lipid droplets	Enhancement	Scallen *et al.* (1985a)

et al., 1983), effectively incorporate [^{3}H]mevalonate into cholesterol formed *in vivo,* and exhibit similar kinetics of cholesterol ester formation (Van Amerongen *et al.*, 1985). These results have led Van Amerongen *et al.* (1985) to suggest that the intracellular biosynthesis and esterification of cholesterol is not critically dependent upon nonspecific lipid transfer protein.

6. EFFECT OF PHOSPHOLIPID ACYL CHAIN COMPOSITION ON ERYTHROCYTE MORPHOLOGY

The discoid shape of the erythrocyte is maintained primarily by membrane cytoskeleton (Lange *et al.*, 1982) but may be perturbed by treatments affecting the lipid bilayer (Lange and Slayton, 1982). Morphological changes have been induced in erythrocytes by phospholipases (Allan *et al.*, 1978; Fuji and Tamura, 1979) and are postulated to be caused by changes in the bilayer distribution (Allan *et al.*, 1978) or structure (Fuji and Tamura, 1979) of the lipid products of phospholipase action.

Recently, PC-specific transfer protein from bovine liver has been used to introduce various PC species of differing geometries into human erythrocytes (Kuypers *et al.*, 1984a,b; Christiansson *et al.*, 1985; Op den Kamp *et al.*, 1985). Since only PC in the erythrocyte outer leaflet is exchanged (Van Meer *et al.*, 1980), changes in molecular species of membrane PC are accompanied by the introduction of an asymmetry of PC species. When the concentration of disaturated PC in the outer leaflet was increased, the cells became echinocytic (crenated), whereas dilinoleoyl-PC caused them to become stomatocytic (cup formed) (Kuypers *et al.*, 1984a). Replacement by either dipalmitoyl-PC or the highly unsaturated PC increased the osmotic fragility and hemolysis of the cells (Kuypers *et al.*, 1984b). In contrast, replacement of virtually all of the PC in the outer leaflet of the erythrocyte by 1-palmitoyl-2-oleoyl-PC or 1-palmitoyl-2-linoleoyl-PC did not alter the morphology or osmotic fragility. The relationship between lipid geometry and morphological changes in erythrocytes has been investigated further by successively replacing the membrane PC with species of various geometries followed by hydrolysis by phospholipases (Christiansson *et al.*, 1985). A combination of the two approaches for changing the molecular geometry of the membrane lipid has strengthened the proposal that lipid molecular shape affects erythrocyte morphology.

7. DEPENDENCE OF MEMBRANE-ASSOCIATED ENZYME ACTIVITIES ON COMPOSITION AND PHYSICAL PROPERTIES OF THE LIPID BILAYER

The interaction between lipids and proteins is essential to the activity of many membrane activities. A considerable number of techniques exist for modification of the lipid patterns of both intact cells and isolated membranes, including (1) phospholipase cleavage of membrane phospholipid, (2) replacement of membrane phospholipid by exogenous phospholipid by using transfer proteins, (3) fusion of membranes with lipid vesicles, (4) alterations of fatty acyl chains by changes in diet or cell culture medium composition, and (5) reconstitution of membrane activities into lipid vesicles of defined composition (reviewed by Deuticke and Haest, 1987).

The *in vitro* modification of the lipid composition of biological membranes by transfer protein-stimulated replacement with exogenously supplied lipid has recently been developed as a method for correlating membrane lipid composition with function (see Section 4.3). Early studies showed a relationship between microsomal PE and glucose-6-phosphatase (Dyatlovitskaya *et al.*, 1979; Crain and Zilversmit, 1981; Crain, 1982). Minimal effects were seen upon 50% enrichment with PC or replacement of up to 45% of the PC with dipalmitoyl-PC (Crain and Zilversmit, 1981). More recently, the transfer of exogenous phospholipids has been shown to restore activities of membrane-associated enzymes following inhibition by phospholipase or lysoPC treatment (Dyatlovitskaya *et al.*, 1982a,b). A number of studies have been performed that have used the nonspecific lipid transfer protein from bovine liver to introduce lipid into a variety of isolated membranes (Table V). The use of this transfer protein in studies of phospholipid

Table V
Application of Phospholipid Transfer Proteins to Investigations of the Relationship between Lipid Composition and Membrane-Associated Ezyme Activity

Transfer protein[a]	Membrane	Lipid modification[a]	Effect on membrane-associated enzyme activity	Reference
nsL-TP	Rat liver microsomes	40% of membrane PC replaced by DPPC	15% decrease in glucose-6-phosphatase activity	Crain and Zilversmit (1981); Crain (1982)
		30% decrease in PE	40% inhibition of activity	Crain and Zilversmit (1981); Crain (1982)
nsL-TP	Synaptic membranes	65% decrease in cholesterol	80% decrease in GABA uptake	North and Fleischer (1983)
		240% increase in cholesterol	No change in GABA uptake	
nsL-TP	Turkey erythrocytes	Increased PI content	40% inhibition of isoproterenol-stimulated adenylate cyclase activity	McOsker *et al.* (1983)
PC-TP	Synaptosomal membranes	Up to 50% replacement of PC by DMPC, DOPC, or DEPC	Little change in membrane potential, choline uptake, or GABA uptake	North and Fleischer (1984)
nsL-TP	Fibroblast microsomes	Replacement of microsomal PC by exogenous DPPC	5-fold decrease in HMG-CoA reductase activity; reversible by replacement of DPPC by egg PC	Davis and Poznansky (1987)
nsL-TP	Thyroid plasma membranes	Increased PI	70% inhibition of TSH-stimulated adenylate cyclase; no effect on forskolin-stimulated activity	Depauw *et al.* (1988)
		Increased PC	No change in TSH-stimulated activity	Depauw *et al.* (1988)
		Increased PS, PA	25% inhibition in TSH-stimulated activity	Depauw *et al.* (1988)

[a]Abbreviations: GABA, sodium gradient-dependent γ-aminobutyric acid; DOPC, dioleoylphosphatidylcholine; DEPC, dielaidoylphosphatidylcholine. For other abbreviations, see Tables I and II, Footnote *a*.

localization and transbilayer movement has been limited because straightforward interpretation of the data is complicated by the possibility of net transfer of lipid (see Sections 2.2 and 4.1). However, the broad specificity of this protein allows it to be used to effect large changes in phospholipid composition, fatty acyl chain composition, and even cholesterol content (Crain and Zilversmit, 1981; North and Fleischer, 1983; Franck *et al.*, 1984).

ACKNOWLEDGMENTS. I would like to express my gratitude to Dr. M. J. Morse and Dr. Alvah Phillips for their helpful suggestions concerning this chapter.

8. REFERENCES

Albers, J. J., Tollefson, J. H., Chen, C.-H., and Steinmetz, A., 1984, Isolation and characterization of human plasma lipid transfer proteins, *Arteriosclerosis* **4**:49–58.

Allan, D., Thomas, P., and Michell, R. H., 1978, Rapid transbilayer diffusion of 1,2-diacylglycerol and its relevance to control of membrane curvature, *Nature* **276**:289–290.

Baba, A., Onoe, H., Ohta, A., and Iwata, H., 1986, Assay of phospholipase A_2 activity of synapatic membranes using a phospholipid transfer protein: Stimulation by depolarization, *Biochim. Biophys. Acta* **878**:25–31.

Bamford, D. H., Romantschuk, M., and Somerharju, P., 1987, Membrane fusion in prokaryotes: bacteriophage Φ6 membrane fuses with the *Pseudomonas syringae* outer membrane, *EMBO J.* **6**:1467–1473.

Barsukov, L. I., Bergelson, L. D., Spiess, M., Hauser, H., and Semenza, G., 1986, Phospholipid topology and flip-flop in intestinal brush-border membrane, *Biochim. Biophys. Acta* **862**:87–99.

Berkhout, T. A., Van den Bergh, C., Mos, H., De Kruijff, B., and Wirtz, K. W. A., 1984, Regulation of the activity of phosphatidylcholine transfer protein by vesicle phosphatidic acid and membrane curvature: A fluorescence study using 2-parinaroyl phosphatidylcholine, *Biochemistry* **23**:6894–6900.

Bernier, I., and Jolles, P., 1987, A survey on cytosolic non-enzymatic proteins involved in the metabolism of lipophilic compounds: From organic anion binders to new protein families, *Biochimie* **69**:1127–1152.

Bishop, W. R., 1983, Phospholipid organization and dynamics in *Salmonella typhimurium*. Ph.D. Thesis, The University of Connecticut.

Bishop, W. R., and Bell, R. M., 1985, Assembly of the endoplasmic reticulum phospholipid bilayer: The phosphatidylcholine transporter, *Cell* **42**:51–60.

Bittman, R., Clejan, S., Robinson, B. P., and Witzke, N. M., 1985, Kinetics of cholesterol and phospholipid exchange from membranes containing cross-linked proteins or cross-linked phosphatidylethanolamines, *Biochemistry* **24**:1403–1409.

Bloj, B., and Zilversmit, D. B., 1976, Asymmetry and transposition rates of phosphatidylcholine in rat erythrocyte ghosts, *Biochemistry* **15**:1277–1283.

Bloj, B., and Zilversmit, D. B., 1977, Rat liver proteins capable of transferring phosphatidylethanolamine. Purification and transfer activity for other phospholipids and cholesterol, *J. Biol. Chem.* **252**:1613–1619.

Bloj, B., and Zilversmit, D. B., 1981a, Lipid transfer proteins in the study of artificial and natural membranes, *Mol. Cell. Biochem.* **40**:163–172.

Bloj, B., and Zilversmit, D. B., 1981b, Accelerated transfer of neutral glycosphingolipids and ganglioside Gm1 by a purified lipid transfer protein, *J. Biol. Chem.* **256:**5988–5991.

Bozzato, R. P., and Tinker, D. O., 1987, Purification and properties of two phospholipid transfer proteins from yeast, *Biochem. Cell Biol.* **65:**195–202.

Brophy, P. J., and Aitken, J. W., 1979, Phosphatidylinositol transfer activity in rat cerebral hemispheres during development, *J. Neurochem.* **33:**355–356.

Brophy, P. J., Burbach, P., Nelemans, S. A., Westerman, J., Wirtz, K. W. A., and Van Deenen, L. L. M., 1978, The distribution of phosphatidylinositol in microsomal membranes from rat liver after biosynthesis *de novo:* Evidence for the existence of different pools of microsomal phosphatidylinositol by the use of phosphatidylinositol-exchange protein, *Biochem. J.* **174:**413–420.

Child, P., Myher, J. J., Kuypers, F. A., Op den Kamp, J. A. F., Kuksis, A., and Van Deenen, L. L. M., 1985a, Acyl specificity in the transfer of molecular species of phosphatidylcholines from human erythrocytes, *Biochim. Biophys. Acta* **812:**321–332.

Child, P., Op den Kamp, J. A. F., Roelofsen, B., and Van Deenen, L. L. M., 1985b, Molecular species composition of membrane phosphatidylcholine influences the rate of cholesterol efflux from human erythrocytes and vesicles of erythrocyte lipid, *Biochim. Biophys. Acta* **814:**237–246.

Christiansson, A., Kuypers, F. A., Roelofsen, B., Op Den Kamp, J. A. F., and Van Deenen, L. L. M., 1985, Lipid molecular shape affects erythrocyte morphology: a study involving replacement of native phosphatidylcholine with different species followed by treatment of cells with sphingomyelinase C or phospholipase A_2, *J. Cell Biol.* **101:**1455–1462.

Crain, R. C., 1982, Nonspecific lipid transfer proteins as probes of membrane structure and function, *Lipids* **17:**935–943.

Crain, R. C., and Zilversmit, D. B., 1980a, Two nonspecific phospholipid exchange proteins from beef liver. I. Purification and characterization, *Biochemistry* **19:**1433–1439.

Crain, R. C., and Zilversmit, D. B., 1980b, Two nonspecific phospholipid exchange proteins from beef liver. II. Use in studying the asymmetry and transbilayer movement of phosphatidylcholine, phosphatidylethanolamine, and sphingomyelin in intact rat erythrocytes, *Biochemistry* **19:**1440–1449.

Crain, R. C., and Zilversmit, D. B., 1980c, Net transfer of phospholipid by the nonspecific phospholipid exchange proteins from beef liver, *Biochim. Biophys. Acta* **620:**37–48.

Crain, R. C., and Zilversmit, D. B., 1981, The lipid dependence of glucose-6-phosphate phosphohydrolase: a study using purified phospholipid transfer proteins and phosphatidylinositol-specific phospholipase C, *Biochemistry* **20:**5320–5326.

Crain, R. C., Clark, R. W., and Harvey, B. E., 1983, Role of lipid transfer proteins in the abnormal lipid content of Morris hepatoma mitochondria and microsomes, *Cancer Res.* **43:**3197–3202.

Daum, G., and Paltauf, F., 1983, Phospholipid transfer in yeast. Isolation and partial characterization of a phospholipid transfer protein from yeast cytosol, *Biochim. Biophys. Acta* **794:**385–391.

Davis, P. J., and Poznansky, M. J., 1987, Modulation of 3-hydroxy-3-methylglutaryl-CoA reductase by changes in microsomal cholesterol content or phospholipid composition, *Proc. Natl. Acad. Sci. U.S.A.* **84:**118–121.

Dawidowicz, E. A., 1987a, Dynamics of membrane lipid metabolism and turnover, *Annu. Rev. Biochem.* **56:**43–61.

Dawidowicz, E. A., 1987b, Lipid exchange: transmembrane movement, spontaneous movement, and protein-mediated transfer of lipids and cholesterol, *Curr. Top. Membr. Transp.* **29:**175–202.

Demel, R. A., Kalsbeek, R., Wirtz, K. W. A., and Van Deenen, L. L. M., 1977, The protein-mediated net transfer of phosphatidylinositol in model systems, *Biochim. Biophys. Acta* **466:**10–22.

Depauw, H., de Wolf, M., van Dessel, G., Hilderson, H. J., Lagrou, A., and Dierick, W., 1988, Modification of TSH-stimulated adenylate cyclase activity of bovine thyroid by manipulation of membrane phospholipid composition with a nonspecific lipid transfer protein, *Biochim. Biophys. Acta* **937**:359–368.

Deuticke, B., and Haest, C. W. M., 1987, Lipid modulation of transport proteins in vertebrate cell membranes, *Annu. Rev. Physiol.* **49**:221–236.

DiCorleto, P. E., and Zilversmit, D. B., 1977, Protein-catalyzed exchange of phosphatidylcholine between sonicated liposomes and multilamellar vesicles, *Biochemistry* **16**:2145–2150.

DiCorleto, P. E., and Zilversmit, D. B., 1979, Exchangeability and rate of flip-flop of phosphatidylcholine in large unilamellar vesicles, cholate dialysis vesicles, and cytochrome oxidase vesicles, *Biochim. Biophys. Acta* **552**:114–119.

DiCorleto, P. E., Warach, J. B., and Zilversmit, D. B., 1979, Purification and characterization of two phospholipid exchange proteins from bovine heart, *J. Biol. Chem.* **254**:7795–7802.

D'Souza, C., Clarke, J. T. R., Cook, H. W., and Spence, M. W., 1983, Phospholipid transfer protein-mediated incorporation and subcellular distribution of exogenous phosphatidylcholine and sphingomyelin in cultured neuroblastoma cells, *Biochim. Biophys. Acta* **729**:1–8.

Dyatlovitskaya, E. V., Timofeeva, N. G., and Bergelson, L. D., 1978, A universal lipid exchange protein from rat hepatoma, *Eur. J. Biochem.* **82**:463–471.

Dyatlovitskaya, E. V., Lemenovskaya, A. F., and Bergelson, L. D., 1979, Use of protein-mediated lipid exchange in the study of membrane-bound enzymes, *Eur. J. Biochem.* **99**:605–612.

Dyatlovitskaya, E. V., Petkova, D. K., and Bergel'son, L. D., 1982a, Study of lipid dependence of cytochrome P-450 activity of rat liver microsomes using phosphatidylcholine-transporting protein from bovine liver, *Biokhimiya* **47**:1366–1369.

Dyatlovitskaya, E. V., Yaronskaya, E. B., and Bergel'son, L. D., 1982b, Investigations of the lipid dependence of the pyrophosphatase activity of liver microsomes and rat hepatoma using phospholipase C, *Biokhimiya* **47**:1222–1229.

Ehnholm, C., and Zilversmit, D. B., 1972, Use of Forssman antigen in the study of phosphatidylcholine exchange between liposomes, *Biochim. Biophys. Acta* **274**:652–657.

Engle, M. J., Van Golde, L. M. G., and Wirtz, K. W. A., 1978, Transfer of phospholipids between subcellular fractions of the lung, *FEBS Lett.* **86**:277–281.

Franck, P. F. H., Roelofsen, B., and Op den Kamp, J. A. F., 1982, Complete exchange of phosphatidylcholine from intact erythrocytes after protein crosslinking, *Biochim. Biophys. Acta* **687**:105–108.

Franck, P. F. H., Chiu, D. T.-Y., Op den Kamp, J. A. F., Lubin, B., Van Deenen, L. L. M., and Roelofsen, B., 1983, Accelerated transbilayer movement of phosphatidylcholine in sickled erythrocytes: A reversible process, *J. Biol. Chem.* **258**:8435–8442.

Franck, P. F. H., De Ree, J. M., Roelofsen, B., and Op den Kamp, J. A. F., 1984, Modification of the erythrocyte membrane by a non-specific lipid transfer protein, *Biochim. Biophys. Acta* **778**:405–411.

Franck, P. F. H., Bevers, E. M., Lubin, B. H., Comfurius, P., Chiu, D. T.-Y., Op den Kamp, J. A. F., Zwaal, R. F. A., Van Deenen, L. L. M., and Roelofsen, B., 1985a, Uncoupling of the membrane skeleton from the lipid bilayer: The cause of accelerated phospholipid flip-flop leading to an enhanced procoagulant activity of sickled cells, *J. Clin. Invest.* **75**:183–190.

Franck, P. F. H., Op den Kamp, J. A. F., Lubin, B., Berendsen, W., Joosten, P., Briet, E., Van Deenen, L. L. M., and Roelofsen, B., 1985b, Abnormal transbilayer mobility of phosphatidylcholine in hereditary pyropoikilocytosis reflects the increased heat sensitivity of the membrane skeleton, *Biochim. Biophys. Acta* **815**:259–267.

Fuji, T., and Tamura, A., 1979, Asymmetric manipulation of the membrane lipid bilayer of intact erythrocytes with phospholipase A, C, or D induces a change in cell shape, *J. Biochem.* **86**:1345–1352.

Gavey, K. L., Noland, B. J., and Scallen, T. J., 1981, The participation of sterol carrier protein$_2$ in the conversion of cholesterol to cholesterol ester by rat liver microsomes, *J. Biol. Chem.* **256**:2993–2999.

George, P. Y., and Helmkamp, G. M., Jr., 1985, Purification and characterization of a phosphatidylinositol transfer protein from human platelets, *Biochim. Biophys. Acta* **836**:176–184.

Haest, C. W. M., Plasa, G., Kamp, D., and Deuticke, B., 1978, Spectrin as a stabilizer of the phospholipid asymmetry in the human erythrocyte membrane, *Biochim. Biophys. Acta* **509**:21–32.

Harris, W. E., 1985, Modulation of Na^+,K^+,-ATPase activity by the lipid bilayer examined with dansylated phosphatidylserine, *Biochemistry* **24**:2873–2883.

Hattori, H., Kanfer, J. N., and Massarelli, R., 1987, Stimulation of phospholipase D activity and indication of acetylcholine synthesis by oleate in rat brain synaptosomal preparations, *Neurochem. Res.* **12**:687–692.

Helmkamp, G. M., Jr., 1980, Concerning the mechanism of action of bovine liver phospholipid exchange protein: Exchange or net transfer, *Biochem. Biophys. Res. Commun.* **97**:1091–1096.

Helmkamp, G. M., Jr., 1986, Phospholipid transfer proteins: Mechanism of action, *J. Bioenerg. Biomembr.* **18**:71–82.

Helmkamp, G. M., Jr., Harvey, M. S., Wirtz, K. W. A., and Van Deenen, L. L. M., 1974, Phospholipid exchange between membranes. Purification of bovine brain proteins that preferentially catalyze the transfer of phosphatidylinositol, *J. Biol. Chem.* **249**:6382–6389.

Hutson, J. L., Higgins, J. A., and Wirtz, K. W. A., 1985, Microsomal membranes contain phosphatidylcholine that equilibrates across the bilayer, and phosphatidylcholine that does not, *FEBS Lett.* **183**:145–150.

Johnson, L. W., Hughes, M. E., and Zilversmit, D. B., 1975, Use of phospholipid exchange protein to measure inside-outside transposition in phosphatidylcholine liposomes, *Biochim. Biophys. Acta* **375**:176–185.

Kader, J.-C., 1985, Lipid-binding properties in plants, *Chem. Phys. Lipids* **38**:51–62.

Kader, J.-C., Julienne, M., and Vergnolle, C., 1984, Purification and characterization of a spinach-leaf protein capable of transferring phospholipids from liposomes to mitochondria or chloroplasts, *Eur. J. Biochem.* **139**:411–416.

Kadowaki, H., Symanski, L. A., and Koff, R. S., 1988, Nonspecific lipid transfer protein in the assay of a membrane-bound enzyme CMP-N-acetyl-neuraminate : lactosylceramide sialyltransferase, *J. Lipid Res.* **29**:52–62.

Kamp, H. H., Wirtz, K. W. A., and Van Deenen, L. L. M., 1973, Some properties of phosphatidylcholine exchange protein purified from beef liver, *Biochim. Biophys. Acta* **318**:313–325.

Kamp, H. H., Wirtz, K. W. A., Baer, P. R., Slotboom, A. J., Rosenthal, A. F., Paltauf, P., and Van Deenen, L. L. M., 1977, Specificity of the phosphatidylcholine exchange protein from bovine liver, *Biochemistry* **16**:1310–1316.

Kasper, A. M., and Helmkamp, G. M., Jr., 1981, Intermembrane phospholipid fluxes catalyzed by bovine brain phospholipid exchange protein, *Biochim. Biophys. Acta* **664**:22–32.

Kawashima, Y., and Bell, R. M., 1987, Assembly of the endoplasmic reticulum phospholipid bilayer: Transporters for phosphatidylcholine and metabolites, *J. Biol. Chem.* **262**:16495–16502.

Kharroubi, A., Chanderbhan, R., Fiskum, G., Noland, B. J., Scallen, T. J., and Vahouny, G. V., 1986, Distribution of sterol carrier protein$_2$, SCP$_2$, in rat tissues and evidence for slow turnover in liver and adrenal cortex, *Fed. Proc.* **45**:1025.

Kharroubi, A., Wadsworth, J. A., Chanderbhan, R., Wiesenfeld, P., Noland, B., Scallen, T., Vahouny, G. V., and Gallo, L. L., 1988, Sterol carrier protein$_2$-like activity in rat intestine, *J. Lipid Res.* **29**:287–292.

Kornberg, R. D., and McConnell, H. M., 1971, Inside-outside transitions of phospholipids in vesicle membranes, *Biochemistry* **10:**1111–1120.

Kramer, R. M., and Branton, D., 1979, Retention of lipid asymmetry in membranes on polylysine-coated polyacrylamide beads, *Biochim. Biophys. Acta* **556:**219–232.

Krebs, J. J. R., 1982, The topology of phospholipids in artificial and biological membranes, *J. Bioenerg. Biomembr.* **14:**141–157.

Kuypers, F. A., Berendsen, W., Roelofsen, B., Op den Kamp, J. A. F., and Van Deenen, L. L. M., 1984a, Shape changes in human erythrocytes induced by replacement of the native phosphatidylcholine with species containing various fatty acids, *J. Cell Biol.* **99:**2260–2267.

Kuypers, F. A., Roelofsen, B., Op den Kamp, J. A. F., and Van Deenen, L. L. M., 1984b, The membrane of intact human erythrocytes tolerates only limited changes in the fatty acid composition of phosphatidylcholine, *Biochim. Biophys. Acta* **769:**337–347.

Kuypers, F. A., Andriesse, X., Child, P., Roelofsen, B., Op den Kamp, J. A. F., and Van Deenen, L. L. M., 1986, The rate of uptake and efflux of phosphatidylcholine from human erythrocytes depends on the fatty acyl composition of the exchanging species, *Biochim. Biophys. Acta* **857:**75–84.

Lange, Y., and Slayton, J. M., 1982, Interaction of cholesterol and lysophosphatidylcholine in determining red cell shape, *J. Lipid Res.* **23:**1121–1127.

Lange, Y., Hadesman, R. A., and Steck, T. L., 1982, Role of the reticulum in the stability and shape of the isolated human erythrocyte membrane, *J. Cell Biol.* **92:**714–721.

Low, M. G., and Zilversmit, D. B., 1980, Phosphatidylinositol distribution and translocation in sonicated vesicles: A study with exchange protein and phospholipase C, *Biochim. Biophys. Acta* **596:**223–234.

Marinetti, G. V., and Crain, R. C., 1978, Topology of amino-phospholipids in the red cell membrane, *J. Supramol. Struct.* **8:**191–213.

McMurray, W. C., and Dawson, R. M. C., 1969, Phospholipid exchange reactions within the liver, *Biochem. J.* **112:**91–108.

McOsker, C. C., Weiland, G. A., and Zilversmit, D. B., 1983, Inhibition of hormone-stimulated adenylate cyclase activity after altering turkey erythrocyte phospholipid composition with a nonspecific lipid transfer protein. Phosphatidylinositol uncouples catecholamine binding from adenylate cyclase activation, *J. Biol. Chem.* **258:**13017–13026.

Middelkoop, E., Lubin, B. H., Op den Kamp, J. A. F., and Roelofsen, B., 1986, Flip-flop rates of individual molecular species of phosphatidylcholine in the human red cell membrane, *Biochim. Biophys. Acta* **855:**421–424.

Middelkoop, E., Lubin, B. H., Bevers, E. M., Op den Kamp, J. A. F., Comfurius, P., Chiu, D. T.-Y., Zwaal, R. F. A., Van Deenen, L. L. M., and Roelofsen, B., 1988, Studies on sickled erythrocytes provide evidence that the asymmetric distribution of phosphatidylserine in the red cell membrane is maintained by both ATP-dependent translocation and interaction with membrane skeletal proteins, *Biochim. Biophys. Acta* **937:**281–288.

Muczynski, K. A., and Stahl, W. L., 1983, Incorporation of dansylated phospholipids and dehydroergosterol into membranes using a phospholipid exchange protein, *Biochemistry* **22:**6037–6048.

North, P., and Fleischer, S., 1983, Use of a nonspecific lipid transfer protein to modify the cholesterol content of synaptic membranes, *Methods Enzymol.* **98:**599–613.

North, P., and Fleischer, S., 1984, Protein mediated exchange of synthetic phosphatidylcholines into synaptosomal membranes, *Biochim. Biophys. Acta* **772:**65–76.

Op den Kamp, J. A. F., 1979, Lipid asymmetry in membranes, *Annu. Rev. Biochem.* **48:**47–71.

Op den Kamp, J. A. F., Roelofsen, B., and Van Deenen, L. L. M., 1985, Structural and dynamic aspects of phosphatidylcholine in the human erythrocyte membrane, *Trends Biochem. Sci.* **10:**320–323.

Pagano, R. E., and Sleight, R. G., 1985, Defining lipid transport pathways in animal cells, *Science* **229:**1051–1057.

Poorthuis, B. J. H. M., and Wirtz, K. W. A., 1982, Increased cholesterol esterification in rat liver microsomes by purified non-specific phospholipid transfer protein, *Biochim. Biophys. Acta* **710:**99–105.

Poorthuis, B. J. H. M., Van der Krift, T. P., Teerlink, T., Akeroyd, R., Hostetler, K. Y., and Wirtz, K. W. A., 1980, Phospholipid transfer activities in Morris hepatomas and the specific contribution of the phosphatidylcholine exchange protein, *Biochim. Biophys. Acta* **600:**376–386.

Read, R. J., and Funkhouser, J. D., 1983, Properties of a non-specific phospholipid-transfer protein purified from rat lung, *Biochim. Biophys. Acta* **752:**118–126.

Read, R. J., and Funkhouser, J. D., 1984, Acyl-chain specificity and membrane fluidity. Factors which influence the activity of a purified phospholipid-transfer protein from lung, *Biochim. Biophys. Acta* **794:**9–17.

Rothman, J. E., and Dawidowicz, E. A., 1975, Asymmetric exchange of vesicle phospholipids catalyzed by the phosphatidylcholine exchange protein: Measurement of inside-outside transitions, *Biochemistry* **14:**2809–2816.

Rothman, J. E., Tsai, D. K., Dawidowicz, E. A., and Lenard, J., 1976, Transbilayer phospholipid asymmetry and its maintenance in the membrane of influenza virus, *Biochemistry* **15:**2361–2370.

Ruenwongsa, P., Singh, H., and Jungalwala, F. B., 1979, Protein-catalyzed exchange of phosphatidylinositol between rat brain microsomes and myelin, *J. Biol. Chem.* **254:**9385–9393.

Rusinol, A., Salomon, R. A., and Bloj, B., 1987, Phospholipid transfer activities in toad oocytes and developing embryos, *J. Lipid Res.* **28:**100–107.

Sandra, A., and Pagano, R. E., 1979, Liposome-cell interactions: studies of lipid transfer using isotopically asymmetric vesicles, *J. Biol. Chem.* **254:**2244–2249.

Scallen, T. J., Noland, B. J., Gavey, K. L., Bass, N. M., Ockner, R. K., Chanderbhan, R., and Vahouny, G. V., 1985a, Sterol carrier protein$_2$ and fatty acid-binding protein: Separate and distinct physiological functions, *J. Biol. Chem.* **260:**4733–4739.

Scallen, T. J., Pastuszyn, A., Noland, B. J., Chanderbhan, R., Kharroubi, A., and Vahouny, G. V., 1985b, Sterol carrier and lipid transfer proteins, *Chem. Phys. Lipids* **38:**239–361.

Shaw, J. M., Moore, N. F., Patzer, E. J., Correa-Freire, M. C., Wagner, R. R., and Thompson, T. E., 1979, Compositional asymmetry and transmembrane movement of phosphatidylcholine in vesicular stomatitis virus membranes, *Biochemistry* **18:**538–543.

Sleight, R. G., 1987, Intracellular lipid transport in eukaryotes, *Annu. Rev. Physiol.* **49:**193–208.

Tai, S. P., and Kaplan, S., 1984, Purification and properties of a phospholipid transfer protein from *Rhodopseudomonas sphaeroides*, *J. Biol. Chem.* **259:**12178–12183.

Teerlink, T., Van der Krift, T. P., Post, M., and Wirtz, K. W. A., 1982, Tissue distribution and subcellular localization of phosphatidylcholine transfer protein in rats as determined by radioimmunoassay, *Biochim. Biophys. Acta* **713:**61–67.

Teerlink, T., Van der Krift, T. P., Van Heusden, G. P., and Wirtz, K. W. A., 1984, Determination of nonspecific lipid transfer protein in rat tissue and Morris hepatoma by enzyme immunoassay, *Biochim. Biophys. Acta* **793:**251–259.

Tilley, L., Cribier, S., Roelofsen, B., Op den Kamp, J. A. F., and Van Deenen, L. L. M., 1986, ATP-dependent translocation of amino phospholipids across the human erythrocyte membrane, *FEBS Lett.* **194:**21–27.

Trzaskos, J. M., and Gaylor, J. L., 1983, Cytosolic modulators of activities of microsomal enzymes of cholesterol biosynthesis: Purification and characterization of a non-specific lipid-transfer protein, *Biochim. Biophys. Acta* **751:**52–65.

Van Amerongen, A., Teerlink, T., Van Heusden, G. P. H., and Wirtz, K. W. A., 1985, The nonspecific lipid transfer protein, sterol carrier protein$_2$, from rat and bovine liver, *Chem. Phys. Lipids* **38:**195–204.

Van Amerongen, A., Helms, J. B., Van der Krift, T. P., Schutgens, R. B. H., and Wirtz, K. W. A., 1987, Purification of nonspecific lipid transfer protein, sterol carrier protein$_2$, from human liver and its deficiency in livers from patients with cerebro-hepato-rena, Zellweger, syndrome, *Biochim. Biophys. Acta* **919:**149–155.

Van Deenen, L. L. M., 1981, Topology and dynamics of phospholipids in membranes, *FEBS Lett.* **123:**3–15.

Van den Besselaar, A. M. H. P., De Kruijff, B., Van den Bosch, H., and Van Deenen, L. L. M., 1978, Phosphatidylcholine mobility in liver microsomal membranes, *Biochim. Biophys. Acta* **510:**242–255.

Van der Meer, B. W., Fugate, R. D., Tilford, K. P., and Sims, P. J., 1987, Complement proteins C5b-9 induce transbilayer migration of membrane phospholipid, *Bull. Am. Phys. Soc.* **32:**1423.

Van der Schaft, P. H., Beaumelle, B., Vial, H., Roelofsen, B., Op den Kamp, J. A. F., and Van Deenen, L. L. M., 1987, Phospholipid organization in monkey erythrocytes upon *Plasmodium knowlesi* infection, *Biochim. Biophys. Acta* **901:**1–14.

Van Meer, G., Poorthuis, B. J. H. M., Wirtz, K. W. A., Op den Kamp, J. A. F., and Van Deenen, L. L. M., 1980, Transbilayer distribution and mobility of phosphatidylcholine in intact erythrocyte membranes: A study with phosphatidylcholine exchange protein, *Eur. J. Biochem.* **103:**283–288.

Van Meer, G., Simons, K., Op den Kamp, J. A. F., and Van Deenen, L. L. M., 1981, Phospholipid asymmetry in Semliki Forest virus grown on baby hamster kidney (BHK-21) cells, *Biochemistry* **20:**1974–1981.

Voelker, D. R., 1985, Disruption of phosphatidylserine translocation to the mitochondria in baby hamster kidney cells, *J. Biol. Chem.* **260:**14671–14676.

Voelker, D. R., and Kennedy, E. P., 1983, Phospholipid exchange protein-dependent synthesis of sphingomyelin, *Methods Enzymol.* **98:**596–598.

Walter, A., Steer, C. J., and Blumenthal, R., 1986, Polylysine induces pH-dependent fusion of acidic phospholipid vesicles: a model for polycation-induced fusion, *Biochim. Biophys. Acta* **861:**319–330.

Watanabe, S., and Yamada, M., 1986, Purification and characterization of a nonspecific lipid transfer protein from germinated castor bean endosperms which transfers phospholipids and galactolipids, *Biochim. Biophys. Acta* **876:**116–123.

Welti, R., and Helmkamp, G. M., Jr., 1984, Acyl specificity of phosphatidylcholine transfer protein from bovine liver, *J. Biol. Chem.* **259:**6937–6941.

Wetterau, J. R., and Zilversmit, D. B., 1984, Quantitation of lipid transfer activity, *Methods Biochem. Anal.* **30:**199–226.

Wirtz, K. W. A., 1982, Phospholipid transfer proteins, in: *Lipid Protein Interactions,* (P. C. Jost and O. H. Griffith, eds.), Vol. 1, pp. 151–222, John Wiley & Sons, Inc., New York.

Wirtz, K. W. A., and Zilversmit, D. B., 1968, Exchange of phospholipids between liver mitochondria and microsomes *in vitro, J. Biol. Chem.* **243:**3596–3602.

Wirtz, K. W. A., Devaux, P. F., and Bienvenue, A., 1980, Phosphatidylcholine exchange protein catalyzes the net transfer of phosphatidylcholine to model membranes, *Biochemistry* **19:**3395–3399.

Xu, Y.-H., Gietzen, K., Galla, H.-J., and Sackmann, E., 1983, Protein-mediated lipid transfer: the effects of lipid-phase transition and of charged lipids, *Biochem. J.* **213:**21–24.

Yaffe, M. P., and Kennedy, E. P., 1983, Intracellular phospholipid movement and the role of phospholipid transfer proteins in animal cells, *Biochemistry* **22:**1497–1507.

Zilversmit, D. B., 1978, Phospholipid-exchange proteins as membrane probes, *Ann. N.Y. Acad. Sci.* **308:**149–163.

Zilversmit, D. B., 1983, Lipid transfer proteins: overview and applications, *Methods Enzymol.* **98:**565–573.

Zilversmit, D. B., 1984, Lipid transfer proteins, *J. Lipid Res.* **25:**1563–1569.

Zilversmit, D. B., and Hughes, M. E., 1977, Extensive exchange of rat liver microsomal phospholipids, *Biochim. Biophys. Acta* **469:**99–110.

Chapter 4

Intracellular Transfer of Phospholipids, Galactolipids, and Fatty Acids in Plant Cells

Jean-Claude Kader

1. INTRODUCTION

The biosynthesis of lipids in plant cells has been extensively studied (Mazliak *et al.*, 1982; Moore, 1984; Heemskerk and Wintermans, 1987; Harwood, 1988). This formation of lipids is essential for membrane biogenesis, which allows plant growth and development. These studies on lipid metabolism revealed several needs for intracellular fluxes of lipids. It has been established that some phospholipids, such as phosphatidylcholine (PC), are found in all membranes of higher plants, including chloroplasts; however, the biosynthesis of PC is con-

Abbreviations used in this chapter: LTP, lipid transfer protein; PLTP, phospholipid transfer protein; PLEP, phospholipid exchange protein; FABP, fatty-acid-binding protein; GLTP, galactolipid transfer protein; nsPL-TP, nonspecific phospholipid transfer protein; PC-TP, phosphatidylcholine transfer protein; PI-TP, phosphatidylinositol transfer protein; PC, phosphatidylcholine; PI, phosphatidylinositol; PG, phosphatidylglycerol; PE, phosphatidylethanolamine; PS, phosphatidylserine; PA, phosphatidic acid; MGDG, monogalactosyldiacylglycerol; DGDG, digalactosyldiacylglycerol; pI, isoelectric point; FPLC, fast protein liquid chromatography; ESR, electron spin resonance; IgG, immunoglobulin G; CoA, coenzyme A; SDS, sodium dodecyl sulfate; ELISA, enzyme-linked immunosorbent assay.

Jean-Claude Kader Laboratoire de Physiologie Cellulaire, Unité de Recherches Associée au CNRS 1180, Université Pierre et Marie Curie, 75005 Paris, France.

fined to the membranes of the endoplasmic reticulum. A mechanism for intracellular transfer of this phospholipid from endoplasmic reticulum to organelles unable to perform this biosynthesis, such as mitochondria or chloroplasts, is therefore needed.

Another example of the need for an intracellular movement of lipids in plant cells is the presence of a cooperative pathway between chloroplasts and endoplasmic reticulum for the biosynthesis of galactolipids containing linolenic acid (the major fatty acid of photosynthetic membranes). This cooperative pathway suggests that oleoyl-coenzyme A (CoA) is exported from chloroplasts toward endoplasmic reticulum and that PC, desaturated in the endoplasmic reticulum, is imported toward chloroplasts, where galactolipids are synthesized. Also, lipids must move within the chloroplast stroma, since galactolipids are formed in the envelope and accumulate in thylakoids.

All these requirements for lipid transfer explain the considerable interest in the transfer process. Three different mechanisms can be envisaged: (1) spontaneous movements of lipids, (2) intracellular transfer of membrane vesicles, and (3) participation of proteins able to transfer lipids (for reviews, see Wirtz, 1982; Bernier and Jollès, 1987; Dawidowicz, 1987). The discovery of this category of proteins several years ago (for reviews, see Kader, 1977; Mazliak and Kader, 1989; Kader *et al.*, 1982; Kader, 1985) introduced new concepts for the intracellular transfer of lipids in plants. This review will present recent data about these proteins, which have been designated phospholipid exchange proteins (PLEP) or phospholipid transfer proteins (PLTP). In connection with recent observations about the broad specificity of these proteins for phospholipids, they will be designated as lipid transfer proteins (LTPs). The properties of plant LTPs will be compared with those of other cytosolic proteins from animal cells interacting with lipids.

The properties of plant LTPs will also be compared with those of cytosolic proteins characterized by their ability to bind fatty acids and designated as fatty-acid-binding proteins (FABPs) (Rickers *et al.*, 1984). Some common features recently established between plant LTPs and FABPs have led to the suggestion that the abilities to transfer and to bind lipids can be associated with the same protein.

2. LIPID TRANSFER AND BINDING ASSAYS

2.1. Lipid Transfer Assay

The determination of the activity of lipid transfer proteins is based on the use of two different types of membranes, one of which contains a labeled lipid. This lipid could be radiolabeled, spin labeled, or fluorescent. The lipid transfer is determined when membranes are incubated with LTP. When the two membrane categories are separated after incubation, the assay is known as "discontinuous" (Wirtz, 1982); a "continuous" assay does not involve a separation of the membranes.

2.1.1. Discontinuous Assay Methods

The assays involve either membrane fractions isolated from various plant tissues or artificial membranes prepared from pure lipids (Table I). The first assay demonstrating that LTPs are present in potato tuber cytosols involved the incubation of microsomes containing ^{32}P-labeled phospholipids with unlabeled mitochondria (Kader, 1975). The microsome–mitochondrion assay has been used with other LTPs isolated from castor bean (Yamada *et al.*, 1978). However, this type of assay is quite laborious and requires separation and analysis of the transferred lipids.

These assays were improved by replacement of one of the two membranes with liposomes obtained by sonicating a mixture of labeled lipids; the liposome–membrane assay is now widely used for determination of the activity of plant LTPs. Mitochondria are often used as receptor membranes. The liposomes usually contain, in addition to the lipid to be studied (labeled with ^{3}H or ^{14}C), a nontransferrable marker (for instance, [^{14}C]cholesteryl oleate). This assay was used to monitor the purification of LTPs from plant tissues. The main advantage

Table I
Transfer Assays for the Detection of LTPs in Higher Plants

Source	Assay	Reference
Potato tubers	Microsome–mitochondrion	Kader (1975)
Castor bean seeds	Microsome–mitochondrion	Yamada *et al.* (1978)
Castor bean seeds	Liposome–mitochondrion	Douady *et al.* (1978)
Castor bean seeds	Liposome–microsome	Tanaka and Yamada (1979)
Castor bean seeds	Liposome–mitochondrion	Boussange *et al.* (1980)
Castor bean seeds	Liposome–liposome	Yamada *et al.* (1980)
Maize seeds	Liposome–liposome	Guerbette *et al.* (1981)
Maize seeds	Liposome–mitochondrion	Douady *et al.* (1982)
Maize seeds	Liposome–mitochondrion	Douady *et al.* (1985)
Spinach leaves	Liposome–mitochondrion	Kader *et al.* (1984)
Spinach leaves	Liposome–liposome	Nishida and Yamada (1986)
Oat leaves	Liposome–liposome	Yamada *et al.* (1980)
Jerusalem artichoke	Liposome–mitochondrion	Douady *et al.* (1978)
Cauliflower florets	Liposome–mitochondrion	Douady *et al.* (1978)
Barley seeds	Liposome–mitochondrion	Kader (unpublished)
Wheat seeds	Liposome–mitochondrion	Kader (unpublished)

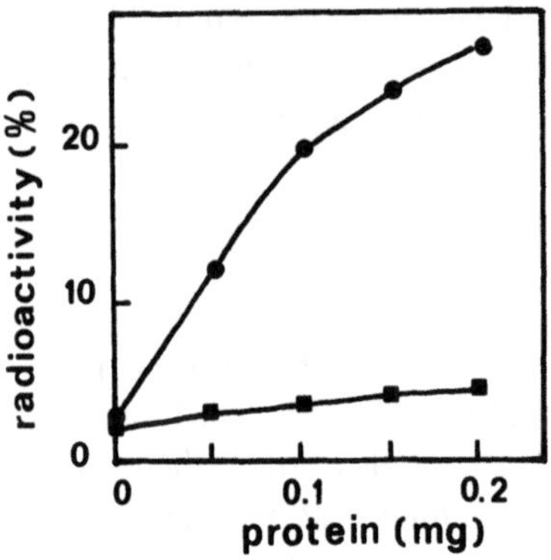

FIGURE 1. Determination of the lipid transfer activity by the liposome-mitochondrion assay. Liposomes made from [^{3}H]PC and cholesteryl-[1^{14}C]oleate were incubated at 30°C for 30 min with mitochondria (2 mg of protein) and increasing amounts of purified spinach leaf LTP. The increase in ^{3}H label (●) (expressed as a percentage of the radioactivity of the initial liposomes) indicates the transfer of PC from liposomes to mitochondria. The slight increase in ^{14}C label (■) indicates a low cross-contamination of mitochondria by liposomes. From Kader *et al.* (1987).

of this type of assay is that the extent of cross-contamination of the acceptor membrane by the artificial donor vesicles can be easily estimated. A typical experiment is presented in Figure 1. It is of interest that the assays used to determine the activity of LTPs isolated from other living cells (animal cells, for example) are very similar (Zilversmit and Hughes, 1976; Wirtz, 1982).

Two other types of discontinuous assays have been developed. In these assays, both membranes are lipid vesicles, differing either in their size or by the presence of a glycolipid. In one type of assay, also used with nonspecific phospholipid transfer protein (nsL-TP) from beef liver (DiCorleto *et al.*, 1979), multilamellar hand-shaken vesicles were made from PC. These vesicles are easily separated from unilamellar liposomes, obtained after sonication, by centrifugation at 40,000 × *g* for 15 min. The transfer of [^{3}H]PC introduced into unilamellar liposomes can thus be monitored between liposomes and multilamellar vesicles (F. Guerbette and J.-C. Kader, unpublished data). In the other assay, one category of liposomes, prepared by sonication, was made reactive to concanavalin A. This assay has been adapted to LTP isolated from maize (Guerbette *et al.*, 1981).

2.1.2. Continuous Assay Methods

In these assays, one of the two populations of liposomes (donor membrane) contains a spin-labeled lipid. The movement of this lipid toward the other membrane (acceptor membrane) is monitored by electron spin resonance (ESR) spectroscopy. Tanaka *et al.* (1980) used this method to establish the occurrence of LTPs in spinach leaf extracts (Figure 2). They prepared liposomes containing spin-labeled PC (with a 12-nitroxide stearic acid at the glycerol *sn*-2 position) (Machida and Ohnishi, 1978). When these donor liposomes were incubated with acceptor membranes in the presence of LTP, the single broad line observed initially by ESR spectroscopy changed to a three-line spectrum, owing to the

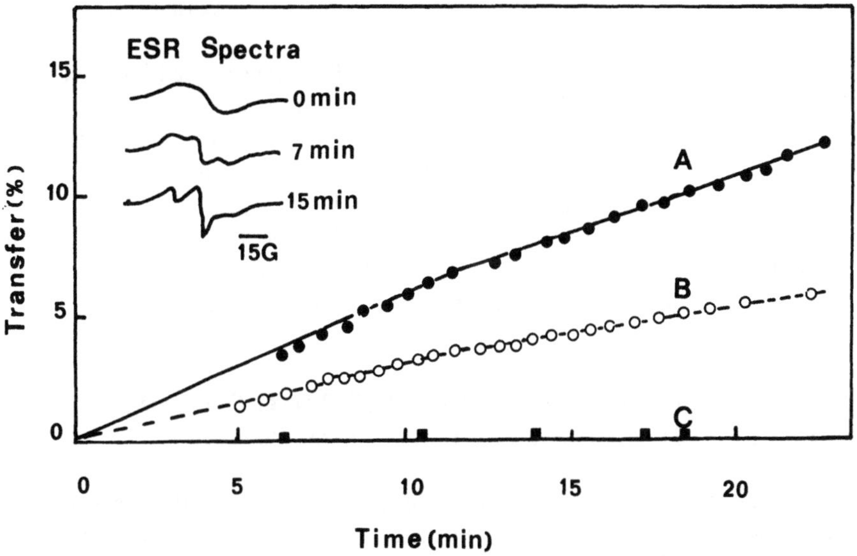

FIGURE 2. Time course of ESR spectrum changes in the transfer of spin-labeled PC from donor liposomes to acceptor liposomes in the presence of spinach leaf extract. Donor liposomes with spin-labeled PC (30 μg) were incubated with acceptor liposomes with egg yolk PC (600 μg) and spinach leaf protein (500 μg) in 20 mM Tris hydrochloride buffer (pH 7.5) (in a final volume of 80 μl) at room temperature. Lines: A, the complete system; B, the complete system without spinach protein; C, the complete system without acceptor liposomes. From Tanaka *et al.* (1980).

incorporation of spin-labeled PC into acceptor membranes. This method, developed for animal LTPs (Machida and Ohnishi, 1978), has also been used to monitor the transfer of PC toward human erythrocytes that is mediated by a pure LTP from maize seeds (P. Devaux and J.-C. Kader, unpublished data).

In addition to phospholipid transfer, galactolipid movement was studied by the spin-label method. Nishida and Yamada (1986) synthesized a spin-labeled monogalactosyldiacylglycerol (MGDG) and studied the transfer of this lipid between liposomes by ESR spectroscopy. This method helped to demonstrate that spinach leaf extract contains galactolipid transfer proteins (GLTPs).

2.2. Fatty-Acid-Binding Assays

Various methods are available to detect FABPs. All are based on the incubation of proteins with a labeled fatty acid; the complex formed by the protein with the fatty acid is separated by different techniques. Rickers *et al.* (1984) established for the first time the occurrence of FABPs in plant cells by incubating cytosolic proteins extracted from etiolated seedlings with [1-^{14}C]oleic acid. The labeled fatty acid–protein complex was separated by successive steps of gel filtration and cation exchange chromatography.

The binding properties of the isolated FABPs were studied by isoelectric focusing, which separates lipid–protein complexes from unbound fatty acids; the complexes were detected by autoradiography in parallel with protein staining. This method has also been used to establish that spinach leaf protein possesses binding ability for fatty acids and oleoyl-CoA (Rickers *et al.*, 1985). The binding characteristics of these plant FABPs were studied by the dextran-coated charcoal method, consisting of the separation of bound fatty acids from unbound ones by precipitation of the latter with the adsorbent (Rickers *et al.*, 1984).

3. PURIFICATION OF LIPID TRANSFER PROTEINS AND FATTY-ACID-BINDING PROTEINS FROM PLANTS

Although the first demonstrations of *in vivo* phospholipid transfer and *in vitro* intermembrane lipid transfer mediated by cytosolic proteins from plants were given as early as 1970 (Abdelkader and Mazliak, 1970), the purification of several LTPs and FABPs has been achieved only in the last 6 years. This is in contrast with the PC transfer protein (PC-TP) of bovine liver, which was purified as early as 1973 (Kamp *et al.*, 1973). One of the reasons is related to the fact that the aqueous extracts from plant tissues contain active lipases and proteases, as well as phenolic compounds. It is important to rapidly separate the contaminating enzymes by gel filtration (Kader, 1975) or to inactivate them by heat treatment (Rickers *et al.*, 1984). After a partial purification of a protein from potato tuber, the purification to homogeneity of LTPs was achieved from three plant tissues and FABP from one plant. In addition, GLTPs have been purified from spinach leaves.

3.1. Partial Purification of LTP from Potato Tuber

The demonstration that the postmitochondrial supernatant from potato tuber (*Solanum tuberosum* L.) is able to enhance the *in vitro* transfer of phospholipids has led to the use of this supernatant for the purification of eventual LTPs. For this purification, the postmitochondrial supernatant obtained from potato tuber cytosol was treated by adjusting the pH to 5.1. Centrifugation at 15,000 × *g* for 10 min eliminated residual membranes (Wirtz and Zilversmit, 1969), whereas the supernatant, after adjustment to neutral pH, still contained the transfer activity. The proteins of this supernatant, called "pH 5.1. supernatant," were then precipitated with ammonium sulfate and dialyzed. Gel filtration separated several peaks of absorbance, one containing fractions able to enhance the transfer of ^{32}P-labeled phospholipids between microsomes and mitochondria (Kader, 1975).

This result clearly established, for the first time, the presence of an LTP in plant tissues. However, only a partial purification has been obtained from potato tuber. The molecular mass, calculated from gel filtration mobilities, was about 22 kDa.

3.2. LTPs from Maize Seeds

The protein extracts were prepared from 3-day-old etiolated maize (*Zea mays* L.) seedlings as already indicated for extraction from potato tuber. These proteins, which are able to actively mediate *in vitro* lipid transfer between liposomes and mitochondria (Douady *et al.*, 1982), were first separated on a Sephadex G75 column. The low-molecular-mass fractions, containing the phospholipid transfer activity, were chromatographed on DEAE-Trisacryl. Interestingly, the major part of the activity, associated with the basic proteins, was not retained on the column equilibrated at pH 7.0. The unbound proteins were then loaded onto a CM-Sepharose column and eluted by a gradient of phosphate at pH 5. One major peak of activity was detected, corresponding to a protein with an apparent molecular mass of 20 kDa, as indicated by sodium dodecyl sulfate (SDS)-electrophoresis. This protein was purified by chromatofocusing.

The purification procedure for maize LTP was shortened and improved by eliminating the DEAE-Trisacryl and chromatofocusing steps, by introducing a heat treatment, and by raising the pH of the cationic column from 5 to 7.9 (Douady *et al.*, 1985). With this procedure, several LTPs were detected in maize cytosol, one of these proteins (the more active one) being highly purified (designated CM2) (Figure 3). From this CM2 fraction, an additional gel filtration step gave a pure maize LTP, with an apparent molecular mass of 9 kDa. It should be noted that relatively large amounts of proteins are isolated from maize seeds: 60 mg of pure LTP was obtained from 2 kg of dry seeds. The same is true for spinach leaves.

3.3. LTPs From Castor Bean Endosperm

As previously indicated, the cytosolic extracts from castor bean (*Ricinus communis* L.) endosperms excised from 4-day-old seedlings are able to facilitate *in vitro* transfer of phospholipids between mitochondria and microsomes (Yamada *et al.*, 1978) and between liposomes and mitochondria (Douady *et al.*, 1980; Boussange *et al.*, 1980). Starting from the pH 5.1 supernatants, Yamada *et al.* (1980) and Tanaka and Yamada (1982) used DEAE-Sepharose chromatography to separate two fractions, I (actively transferring phosphatidylinositol [PI]) and II (more specific for PC). Fraction II, when separated by Sephacryl S-200, gave five peaks, corresponding to molecular masses between 11.1 and 69.2 kDa. All these LTPs are acidic, since they are retained by the anion exchange column.

In addition to these acidic LTPs, the castor bean endosperm contains basic

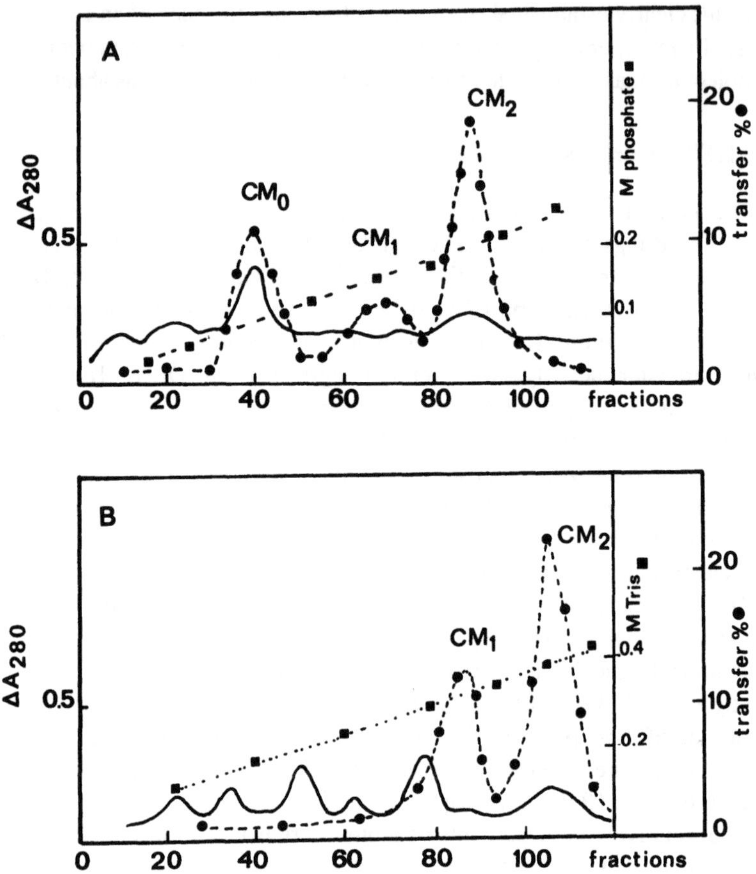

FIGURE 3. Elution profile of spinach leaf (A) and maize seed (B) LTPs, from CM-Sepharose chromatography. The active cytosolic proteins, partially purified by gel filtration on Sephadex G-75, were applied to a CM-Sepharose column and eluted (■) by a gradient of phosphate at pH 7.2 (A) or Tris at pH 7.9 (B). The absorbance (———) was measured at 280 nm. The transfer activity (●) is expressed as a percentage of liposomal PC transferred toward mitochondria in the presence of 2 ml of each fraction. From Kader *et al.* (1987) and Douady *et al.* (1985).

ones, which have been recently isolated by Watanabe and Yamada (1986). The unbound proteins from a DEAE-Sepharose column were separated by two successive CM-Sepharose columns with a phosphate gradient (Figure 4). The active fraction was purified by successive chromatography on Sephadex G-50 and hydroxyapatite. The final protein preparation exhibited a single band corresponding to a molecular mass of 9 kDa after SDS-electrophoresis. The purification factor (Table II) was 172, and the recovery was 2.6%. This protein preparation will be designated in the following sections as "castor bean LTP."

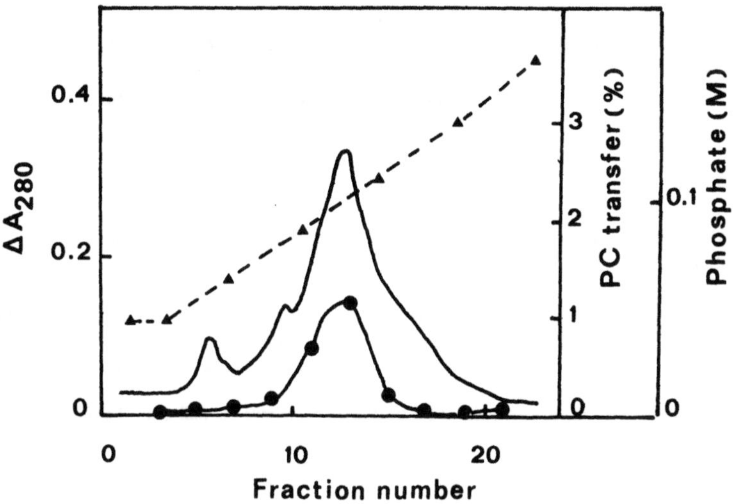

FIGURE 4. Purification of castor bean LTP on CM-Sepharose. The partially purified castor bean proteins were applied to a column of CM-Sepharose at pH 7.2 and eluted by a gradient of phosphate (▲). The absorbance (———) was measured at 280 nm. The transfer activity (●) is expressed as the percentage of PC transferred from liposomes to mitochondria in the presence of 5 μl of each fraction. From Watanabe and Yamada (1986).

In addition to this protein isolated from the endosperm, other LTPs having different molecular masses (7 and 8 kDa) have recently been isolated in Yamada's laboratory (Yamada *et al.*, 1988).

3.4. LTPs from Spinach Leaves

The spinach (*Spinacia oleracea* L.) leaf has been widely used for the isolation of intact chloroplasts (Douce *et al.*, 1984). This explains its interest as a source of LTPs which could interact with chloroplasts. However, this material is not easy to use for lipid transfer experiments, since it contains several compounds that interfer with the lipid transfer assays. This explains why Murphy and Kuhn (1981) did not succeed in detecting a transfer activity mediated by spinach leaf cytosol, although it contains a high lipid transfer activity (Julienne *et al.*, 1981).

By using successive steps of gel filtration, DEAE-Trisacryl chromatography, and CM-Sepharose chromatography, Kader *et al.* (1984) obtained a highly purified protein which is basic and has a molecular mass of 9 kDa. In a typical preparation, 11 mg of pure LTP was obtained from 5 kg of spinach leaves, with a purification factor of 197. In addition to this protein, which eluted from the CM-Sepharose column as CM2, other active fractions (CM1 and CM0) were detected

Table II
Purification of LTP from Castor Bean Seedlings[a]

Purification steps	Protein (mg)	Recovery of activity (%)	Specific activity[a]	Purification factor
1. pH 5.1 supernatant	6,280	100	0.38	–
2. $(NH_4)_2SO_4$ precipitate	5,440	95	0.42	1.1
3. DEAE-Sepharose unbound	1,250	59	1.12	3.0
4. CM-Sepharose (1st)	15.5	4.6	7.05	18.5
5. CM-Sepharose (2nd)	6.7	4.9	17.6	46.2
6. Sephadex G-50	1.9	3.2	40.5	107
7. Hydroxyapatite	0.9	2.6	65.6	172

[a]From Watanabe and Yamada (1986).
[b]Specific activity is expressed as nanomoles of PC transferred per minute per milligram of protein from liposomes to mitochondria.

(Kader *et al.*, 1987). This gave the first indication of the presence of isoforms of LTPs in leaf tissues.

The occurrence of basic spinach LTP was confirmed by Nishida and Yamada (1986), who isolated a 11-kDa protein from spinach leaf extracts that was able to mediate not only the transfer of PC between liposomes and mitochondria but also the transfer of spin-labeled MGDG between liposomes. This coelution of a phospholipid and galactolipid transfer activity is addressed in Section 4.3.

3.5. Other LTPs

In addition to these pure LTPs, other partially purified proteins with a lipid-transferring activity have been extracted from various plant tissues: barley seeds (P. Coutos-Thevenot, unpublished data), sunflower seedlings (V. Arondel, unpublished data), and wheat seeds (F. Guerbette, unpublished data). It is of interest that on SDS-electrophoresis these partially purified LTPs exhibit a major band corresponding to a molecular mass of 9 kDa.

3.6. FABPs from Oat Seedlings

Rickers *et al.* (1984) were the first to establish the presence of a FABP in plant tissues. A cytosolic extract prepared from 4-day-old oat (*Avena sativa* L.) seedlings was centrifuged and heat treated to eliminate membranes and nonbinding proteins. After $(NH_4)_2SO_4$ treatment (35–65% saturation) and dialysis, the protein extract was delipidated with butanol and lyophilized. At this step, the proteins were incubated with the labeled fatty acid [^{14}C]oleic acid and chromatographed on Sephacryl S-200. The ^{14}C-labeled proteins were loaded onto a CM-Sephadex column and eluted with a linear gradient of KCl. The radioactive peaks were designated as CMI, CMII, and CMIII (Figure 5). On SDS-electrophoresis, the CMI fraction exhibited a single band corresponding to an apparent molecular mass of 8.7 kDa.

3.7. GLTPs from Spinach Leaves

Partially purified GLTPs, detected by their ability to transfer spin-labeled MGDG between liposomes, were isolated from spinach leaf, by extracting either the total leaf or the stroma of chloroplasts (Nishida and Yamada, 1986). When the pH 5.1 supernatant prepared from the whole spinach leaf was salted out by 75% saturation of $(NH_4)_2SO_4$, one broad peak of GLTP activity, corresponding to a molecular mass of about 28 kDa was obtained after Sephadex G-100 chromatography. The fractions from this broad peak were then separated on DEAE-Sepharose and CM-Sepharose; these separations gave, respectively, one and two peaks of activity, corresponding to one acidic and two basic GLTPs.

However, when the salting-out procedure was performed with a 35–75% saturation of $(NH_4)_2SO_4$, the Sephadex G-100 chromatography gave no peak at 28 kDa but two peaks at 11 and 5.1 kDa (Figure 6). Interestingly, these peaks of GLTP activity are also able to transfer PC. The 11-kDa protein could correspond to one of the two basic proteins separated by CM-Sepharose. This 11-kDa protein is probably identical to the 9-kDa LTP purified from the same leaf (Kader *et al.*, 1984). The differences in molecular mass could be linked to difficulties in determining the sizes of these basic proteins by gel filtration.

This 11-kDa protein seems not to be present in the stroma of chloroplasts. When spinach leaf chloroplast stroma was separated by Sephadex G-100 chromatography, only a peak of GLTP around 30 kDa (probably corresponding to the 28-kDa protein detected in the whole leaf) was found. This is the first demonstration that an LTP is present in the chloroplast stroma.

3.8. Conclusions

LTPs have been detected in all plant tissues studied so far; this indicates a universal presence of LTPs in the plant kingdom. However, no LTP able to transfer PC seems to be present within the chloroplast stroma (Schwitzguebel *et al.*, 1984). Despite the problems encountered in the isolation of LTPs and lipid-binding proteins, several highly purified proteins have been obtained. These proteins were isolated by considering their properties of transferring phosphospholipids or binding fatty acids. In addition to these pure proteins, partially purified LTPs, FABPs, or GLTPs have been detected. These proteins, which in some cases are isoforms, must then be purified. As an example, isoforms of the

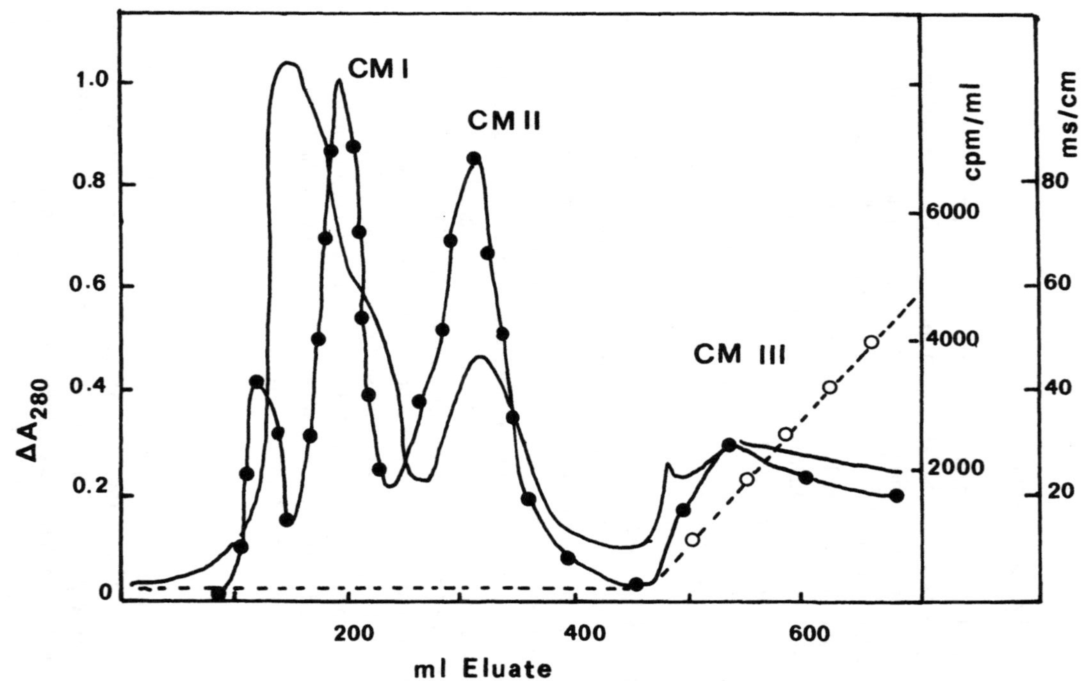

FIGURE 5. Purification of FABPs from oat seedlings by CM-Sephadex. The labeled proteins, after incubation with [1-^{14}C]oleic acid and separation over Sephacryl S-200, were applied to a CM-Sephadex column at pH 6.0 and eluted by a gradient of KCl indicated by conductivity measurements (in milliseconds per centimeter) (○). FABPs were detected by radioactivity bound (in cpm per milliliter) (●); the absorbance (———) was determined at 280 nm. From Rickers *et al.* (1984).

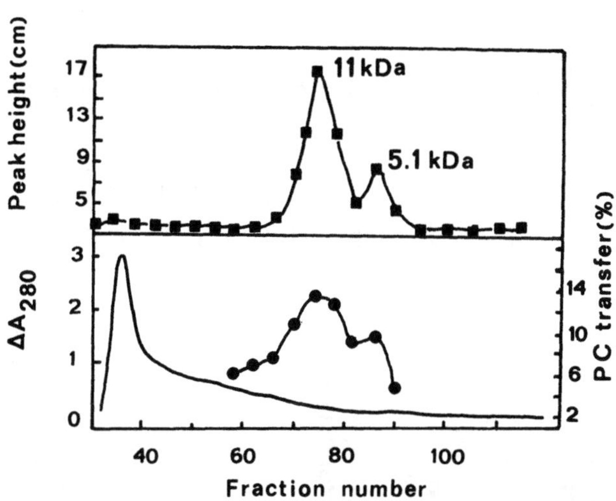

FIGURE 6. Detection of GLTPs in spinach leaf. Spinach leaf proteins were applied to a Sephadex G-100 column. The galactolipid transfer activity was detected by the spin-label method and expressed as peak height after 15 min of incubation (■), whereas the transfer of PC (●) was determined after 20 min of incubation. The molecular masses were estimated relative to standard proteins. ———, Absorbance at 280 nm. From Nishida and Yamada (1986).

castor bean LTP have recently been purified (Yamada *et al.*, 1988). Various plant sources have been used for these purifications: either leaves or etiolated seedlings of monocotyledonous (maize) or dicotyledonous (castor bean, spinach) plants. It is thus of a considerable interest to compare the physical and chemical properties of these proteins to search for common features or conserved domains in their structure.

4. PROPERTIES OF LIPID TRANSFER PROTEINS AND FATTY-ACID-BINDING PROTEINS

The availability of highly purified LTPs and FABPs from higher plants has allowed studies of their properties, which will be compared with those of proteins isolated from other organisms (animals, yeasts). Some other information concerning partially purified proteins will also be discussed.

4.1. Molecular Mass

The molecular masses, first established by SDS-electrophoresis of the basic LTPs, are remarkably similar; the three proteins isolated from maize, spinach, and castor bean exhibit a major band corresponding to 9 kDa (Table III). These

Table III
Properties of LTPs, FABPs, and GLTPs from Higher Plants

Source	Protein	Molecular mass (kDa)	pI	Specificity	Reference
Maize seeds	LTP	20[a]	8.8	PC, PE, PI	Douady *et al.* (1982)
Maize seeds	CM2	9[a]	8.8	PC, PE, PI	Douady *et al.* (1985)
Maize seeds	CM2	9.054[b]	8.8	PC, PE, PI	Tchang and Kader, unpublished
Spinach leaves	LTP	9[a]	9	PC, PE, PI, PG	Kader *et al.* (1984)
Spinach leaves	LTP	8.832[b]	9	PC, PE, PI, PG	Bouillon *et al.* (1987)
Spinach leaves	LTP	11[c], 5.1[c]	Basic	PC[d], MGDG	Nishida and Yamada (1986)
Spinach leaves	GLTP	28[c], 11[c], 5.1[c]	Acidic and basic	MGDG[d]	Nishida and Yamada (1986)
Spinach chloroplast stromata	GLTP	30[c]	–	MGDG[d]	Nishida and Yamada (1986)
Castor beans	LTP	9[a]	10.5	PC, PI, PE, PG	Watanabe and Yamada (1986)
Castor beans	LTP	9.313[b]		PA, MGDG, DGDG	
Castor beans	nsL-TP9B	9.847[b]			Yamada *et al.* (1988)
Castor beans	nsL-TP9C	9.593[b]			Yamada *et al.* (1988)
Castor beans	LTP-Ia	–	Acidic	PC	Tanada and Yamada (1982)
Castor beans	LTP-Ib	90	Acidic	PI	Tanaka and Yamada (1982)
Castor beans	LTP-II a	69.2	6.6	PC	Tanaka and Yamada (1982)
Castor beans	LTP-II b	30.3	5.8	PC	Tanaka and Yamada (1982)
Castor beans	LTP-II c	22.1	5.8	PC	Tanaka and Yamada (1982)
Castor beans	LTP-II d	15.8	5.5–5.8	PC	Tanaka and Yamada (1982)
Castor beans	LTP-II e	11.1	5.4–5.6	PC	Tanaka and Yamada (1982)
Sunflowers	LTP	9	8	PC, PE, PI	Arondel and Kader (unpublished)
Oats	FABP-CMI	8.7[a]	8.4 (4.8[e])	Oleic acid, all C_{16}–C_{20} fatty acids, palmitoyl-CoA, oleoyl-CoA	Rickers *et al.* (1984)

[a]Determined by SDS-electrophoresis.
[b]Determined by amino acid composition.
[c]Determined by gel filtration.
[d]Assays performed only with this lipid.
[e]Change of pI upon lipidation.

values were confirmed by the amino acid composition and determination of the complete amino acid sequence (Bouillon *et al.*, 1987; Takishima *et al.*, 1986; F. Tchang and J.-C.-Kader, unpublished data), which indicated that the molecular weights vary from 8832 to 9313 Da. However, other values have been found for basic LTPs. For example, the pure maize LTP exhibited a band with a molecular mass of 22 kDa. This might correspond to the formation of dimers. Dimers (18 kDa) and trimers (27 kDa) of spinach leaf LTP have also been observed under nonreducing conditions. In addition, lower molecular masses (5.1 kDa) have been determined for a basic LTP and GLTP in whole spinach leaf extract (Nishida and Yamada, 1986).

In the case of acidic LTPs isolated from castor bean, the formation of high-molecular-mass polymers has been suggested by Tanaka and Yamada (1982); values as high as 90 kDa have been determined. Tanaka and Yamada (1982) proposed that the combination of subunits of 5.5 and 6.5 kDa could lead to high molecular masses, up to 69.2 kDa.

The molecular masses established for plant LTPs, and especially the 9-kDa values, are close to those of some LTPs isolated from animal cells: rat hepatoma nsPL-TP (11.2 kDa; Dyatlovitskaya *et al.*, 1978); rat liver nsPL-TP, which is identical to sterol carrier protein 2 (12.5 kDa; Bloj and Zilversmit, 1977); bovine liver nsPL-TP (14.5 kDa; Crain and Zilversmit, 1980). However, higher values have been found for LTPs specific for PC or PI: 24.68 kDa for bovine liver PC-TP (Akeroyd *et al.*, 1981), 28 kDa for rat liver PC-TP (Lumb *et al.*, 1976), 32.3 for bovine brain PI transfer protein (PI-TP) (Helmkamp, 1985), and 33.5 kDa for bovine heart PI-TP (DiCorleto *et al.*, 1979). The molecular mass of LTPs isolated from yeasts is higher than that of LTPs isolated from higher plants. Daum and Paltauf (1984) and Bozzato and Tinker (1987) found, respectively, molecular masses 35 and 33.4 kDa.

The two other categories of plant proteins involved in the transfer or binding of lipids, i.e., GLTPs and FABPs, have various sizes. The spinach leaf GLTPs have molecular masses varying from 5.1 to 28 kDa. The latter protein seems to be present within the chloroplast stroma. By comparison, GLTPs purified by Sasaki's group from pig or rat brain have a molecular mass of about 20 kDa (Sasaki, 1985; see Chapter 5).

The only FABP isolated from higher plants is the 8.7-kDa protein from oat seedlings (Rickers *et al.*, 1984). This value is lower than those determined for FABPs purified from animal cells, e.g., 14 kDa for rat liver FABP (see Chapter 7). However, the value determined for oat FABP is very close to those of plant LTPs, raising the question of a possible identity between FABPs and LTPs in higher plants.

4.2. Isoelectric Point

All LTPs purified to homogeneity from higher plants are basic (Table III). The determination of isoelectric point (pI) by chromatofocusing indicated pI values of 8.8 for maize LTP (Douady *et al.*, 1982) and 9 for spinach LTP (Kader *et al.*, 1984). Values higher than 10.5 kDa have been found for castor bean LTP, as indicated by two-dimensional electrophoresis (Watanabe and Yamada, 1986). The pI of the isoforms of the basic LTPs detected in these three plants remain to be established.

These high isoelectric points for the major plant LTPs are similar to those found for rat liver nsPL-TP (8.6–9, Bloj and Zilversmit, 1977) and for the isoforms of bovine liver nsPL-TP (9.55 and 9.75; Crain and Zilversmit, 1980). In addition to these major plant LTPs, acidic ones are present, with pI values (varying from 5.4 to 6.6) close to those determined for animal LTPs: 5.8 for bovine liver PC-TP (Kamp *et al.*, 1973), and 5.3 and 5.6 for isoforms of bovine brain PI-TP (Helmkamp, 1985) and bovine heart PI-TP (DiCorleto *et al.*, 1979).

Since basic and acidic LTPs are present in plant cells, it could be suggested that the intracellular localizations of these proteins are different. One possibility, for the acidic LTPs, is a localization within the stroma of chloroplasts. Indeed, a GLTP, retained by the DEAE-Sepharose column and hence acidic, has been found in the stroma of spinach chloroplasts (Nishida and Yamada, 1986).

The determination of the pI of the FABP isolated from oat seedlings revealed surprising properties of this protein. When oat FABP was delipidated, its pI was 8.4 as determined by isoelectric focusing. Lipidation of FABP with fatty acid resulted in a shift of the pI to 4.8. This shift could be due to the addition of anionic charges introduced by the lipidation, but conformational changes of the protein, leading to the formation of negative charges at the surface of the FABP, might also be involved (Rickers *et al.*, 1984).

4.3. Specificity toward Lipids

The major LTPs isolated from plants have a broad specificity for phospholipids. All three LTPs are able to transfer PC, PI, and phosphatidylethanolamine (PE); in addition, spinach LTP transfers phosphatidylglycerol (PG), whereas castor bean LTP transfers phosphatidic acid (PA), PG, and also MGDG and digalactosyldiacylglycerol (DGDG). This wide specificity for lipids explains the use of the term "nonspecific lipid transfer protein (nsL-TP)" for the major LTPs from plants. However, it is interesting that not all lipids are transferred with the same efficiency; PE, for example, is less rapidly transferred than the other lipids (Figure 7) (Kader *et al.*, 1984). Also, diacyl- and triacylglycerols are not transferred by castor bean LTP (Watanabe and Yamada, 1986). This broad specificity of plant LTPs is similar to that observed for bovine nsPL-TP, which is able to transfer a wide range of lipids, including phospholipids, diacylglycerols, gangliosides, and cholesterol (Bloj and Zilversmit, 1977).

An intriguing point concerning plant LTPs is their unexpected property either to bind fatty acids (spinach leaf LTP; Rickers *et al.*, 1985) or to transfer

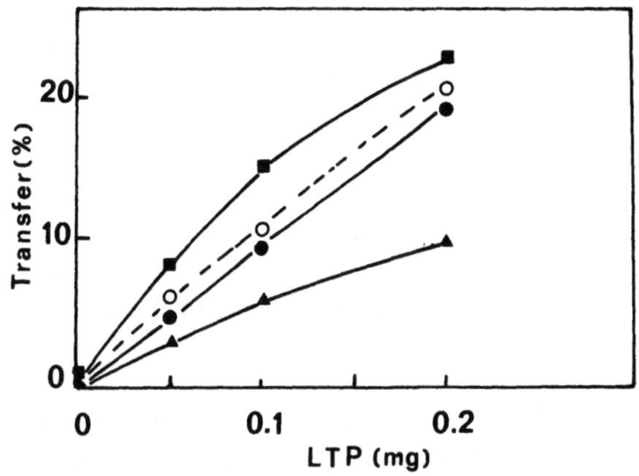

FIGURE 7. Phospholipid transfer mediated by spinach LTP. Liposomes made from [^{3}H]PC (■), [^{3}H]PG (●), [^{3}H]PI (○), and [^{3}H]PE (▲) were incubated for 30 min at 30°C with mitochondria (2 mg of protein) and increasing amounts of purified spinach leaf LTP. From Kader *et al.* (1984).

galactolipids (spinach leaf LTP; Nishida and Yamada, 1986). This raises the question of whether the same protein is involved in these various processes. This point must be carefully examined in future investigations. Although FABPs and LTPs are clearly different proteins in animal cells, it has been found that the properties of transferring phospholipids and sterolic compounds are shared by the same protein, isolated from rat liver and called nsPL-TP (Bloj and Zilversmit, 1977) or sterol carrier protein 2 (Noland *et al.*, 1980). The amino acid determinations confirmed that the same protein is involved in these different properties.

4.4. Stability

Both LTPs and FABPs are remarkably stable. When stored at 4°C in the presence of sodium azide, maize LTP keeps its activity for several months (Douady *et al.*, 1982). Under the same conditions, spinach LTP loses 25% of its activity per month (Kader *et al.*, 1984), whereas the activity of castor bean LTP is stable for at least 1 month (Watanabe and Yamada, 1986). It seems important to store LTPs at a concentration not higher than 0.5 mg ml^{-1} to avoid aggregation (Kader *et al.*, 1984). The same is true for animal PC-TP (Wirtz, 1982). Lower temperatures (−20°C) can be used for plant LTPs if glycerol (final concentration, of 50%) is added to the protein solution (Kader *et al.*, 1984). LTPs from seeds are also remarkably resistant to elevated temperatures. The activity of castor bean LTP remains unchanged after the purified protein has been heated for 30 min at 90°C (Watanabe and Nishida, 1986). However, spinach leaf LTP is very sensitive to heat treatment: the activity is lost after 10 min at 90°C or 30 min at 70°C (Kader *et al.*, 1984). The fact that maize LTP and oat FABP keep their activity after heating has led to the introduction of heat treatment as a purification step (Rickers *et al.*, 1984).

4.5. Amino Acid Composition

The availability of sufficient amounts of highly purified LTPs from the three plants maize, spinach, and castor bean has allowed the determination of their amino acid composition and, more recently, of their complete amino acid sequence (Takishima *et al.*, 1986; Bouillon *et al.*, 1987; Tchang and Kader, unpublished).

Comparison of the amino acid composition of the three plant LTPs revealed several common points (Table IV). The total number of amino acid residues,

Table IV
Amino Acid Composition of LTPs and FABPs from Higher Plants and Comparison with nsPL-TPs from Animal Tissues[a]

	Number of residues					
Amino acid	Spinach LTP	Castor bean LTP	Maize LTP	Oat FABP	Bovine nsPL-TP	Rat nsPL-TP
Asx	7	9	8	9	16	12
Thr	5	6	5	5	4	5
Ser	7	10	14	5	5	7
Glx	0	5	2	6	13	13
Pro	6	6	4	4	4	4
Gly	15	5	10	9	15	14
Ala	14	15	17	8	9	11
Cys	6	8	8	4	2	4
Val	3	5	5	5	5	8
Met	2	0	0	1	5	5
Ile	7	7	6	3	5	5
Leu	6	5	3	4	11	10
Tyr	3	0	2	3	0	0
Phe	0	2	0	3	7	6
Lys	7	6	2	4	17	19
His	2	0	0	3	0	0
Arg	1	3	6	4	0	0
Trp	0	0	0	1	1	0
Total	91	92	93	81	125	117

[a]References: spinach LTP, Bouillon *et al.* (1987); castor bean LTP, Takishima *et al.* (1986); maize LTP, Tchang and Kader (unpublished); oat FABP, Rickers and Spener (unpublished); bovine liver nsPL-TP, Crain and Zilversmit (1980); rat liver nsPL-TP, Bloj and Zilversmit (1977).

which is similar, varies from 91 to 93. All proteins have a larger content of basic than acidic residues. The ratio of asparagine to aspartic acid is 7 : 1 for maize, 5 : 4 for castor bean, and 5 : 2 for spinach. The ratio of glutamine to glutamic acid was 2 : 0 for maize LTP and 4 : 1 for castor bean LTP (Glx is lacking in spinach LTP). This extensive amidination of acidic residues explains the high isoelectric point of these purified LTPs. The isoelectric points calculated from amino acid composition are 8.97 for maize, 9.36 for spinach, and 8.56 for castor bean.

A common feature of the three proteins, which have alanine as the major amino acid, is the presence of several cysteine residues (six to eight). The number of disulfide bridges is known only for castor bean protein, which contains two bridges (M. Yamada, unpublished data). All proteins lack various amino acids: tryptophane is absent in the three proteins, methionine is absent in castor bean and maize proteins, phenylalanine is absent in spinach and maize proteins, histidine is absent in castor bean and maize proteins, and tyrosine is absent in castor bean protein. The very low level of aromatic amino acids is also observed with animal nsPL-TPs from bovine or rat liver (Akeroyd *et al.*, 1981; Westerman and Wirtz, 1985), which contain respectively, 125 and 117 residues, whereas rat liver sterol carrier protein 2 contains 122 amino acids (Pastuszyn *et al.*, 1987).

4.6. Primary Structure

Comparison of the complete amino acid sequences reveals that the LTPs are very homologous to each other (Figure 8). A deletion of two residues is necessary at positions 25 and 26 for aligning the spinach protein to the maize protein, and a single residue shift toward the C-terminal end is needed to align the castor bean protein with the others. The percent match between the aligned sequences (51.6%, 47.8%, 41.3%, and 32.3% for the comparisons of maize versus spinach, maize versus castor bean, spinach versus castor bean, and of all sequences, respectively) and the homology scores (−260, −253, and −212 for the comparisons of maize versus spinach, maize versus castor bean, and spinach versus castor bean, respectively) indicate the very high homology between these proteins. Most of the alterations correspond to conservative replacements, which occur all along the polypeptide chain, with the exception of two regions of lower homology (residues 21–31 and residues 58–67). The positions of cysteinyl residues and of charged amino acids are remarkably conserved.

Analysis of hydropathy and flexibility profiles of the proteins revealed similar features (Figure 9): (1) the presence of a hydrophilic region in the central core of the sequence; (2) the presence of two hydrophobic domains surrounding the central polar zone; and (3) the highly hydrophobic character of the N-terminal end. The possible interaction of these domains with lipids is discussed in Section 5.1.1.

					5					10					15					20
M	Ala	Ile	Ser	Cys	Gly	Gln	Val	Ala	Ser	Ala	Ile	Ala	Pro	Cys	Ile	Ser	Tyr	Ala	Arg	Gly
S	Gly	Ile	Thr	Cys	Gly	Met	Val	Ser	Ser	Lys	Leu	Ala	Pro	Cys	Ile	Gly	Tyr	Leu	Lys	Gly
R		Val	Asp	Cys	Gly	Gln	Val	Asn	Ser	Ser	Leu	Ala	Ser	Cys	Ile	Pro	Phe	Leu	Thr	Gly

					25					30					35					40
M	Gln	Gly	Ser	Gly	Pro	Ser	Ala	Gly	Cys	Cys	Ser	Gly	Val	Arg	Ser	Leu	Asn	Asn	Ala	Ala
S	Gly	Pro	Leu	Gly			Gly	Gly	Ser	Ser	Gly	Gly	Ile	Lys	Ala	Leu	Asn	Ala	Ala	Ala
R	Gly	Val	Ala	Ser	Pro	Ser	Ala	Ser	Cys	Cys	Ala	Gly	Val	Gln	Asn	Leu	Lys	Thr	Leu	Ala

					45					50					55					60
M	Arg	Thr	Thr	Ala	Asp	Arg	Arg	Ala	Ala	Cys	Asn	Cys	Leu	Lys	Asn	Ala	Ala	Ala	Gly	Val
S	Ala	Thr	Thr	Pro	Asp	Arg	Lys	Thr	Ala	Cys	Asn	Cys	Leu	Lys	Ser	Ala	Ala	Asn	Ala	Ile
R	Pro	Thr	Ser	Ala	Asp	Arg	Arg	Ala	Ala	Cys	Glu	Cys	Ile	Lys	Ala	Ala	Ala	Ala	Arg	Phe

					65					70					75					80
M	Ser	Gly	Leu	Asn	Ala	Gly	Asn	Ala	Ala	Ser	Ile	Pro	Ser	Lys	Cys	Gly	Val	Ser	Ile	Pro
S	Lys	Gly	Ile	Asn	Tyr	Gly	Lys	Ala	Ala	Gly	Leu	Pro	Gly	Met	Cys	Gly	Val	His	Ile	Pro
R	Pro	Thr	Ile	Lys	Gln	Asp	Ala	Ala	Ser	Ser	Leu	Pro	Lys	Lys	Cys	Gly	Val	Asp	Ile	Asn

					85					90			
M	Tyr	Thr	Ile	Ser	Thr	Ser	Thr	Asp	Cys	Ser	Arg	Val	Asn
S	Tyr	Ala	Ile	Ser	Pro	Ser	Thr	Asn	Cys	Asn	Ala	Val	His
R	Ile	Pro	Ile	Ser	Lys	Thr	Thr	Asn	Cys	Gln	Ala	Ile	Asn

FIGURE 8. Comparison of the amino acid sequences of plant LTPs. The three sequences presented are those of maize seed LTP (M) (Tchang and Kader, unpublished), spinach leaf LTP (S) (Bouillon *et al.*, 1987), and castor bean LTP (R) (Takishima *et al.*, 1986).

From the amino acid sequence, it can be predicted that plant LTPs are all β proteins, since they have a high percentage of β structure (extended structure and turns). It is of interest that the hydrophobic binding domain of bovine liver PC-TP is predicted to have a β-sheet conformation with β turns (Wirtz, 1982).

Comparison of the three amino acid sequences of plant LTPs with those established for bovine liver PC-TP (Akeroyd *et al.*, 1981) and bovine liver nsPL-

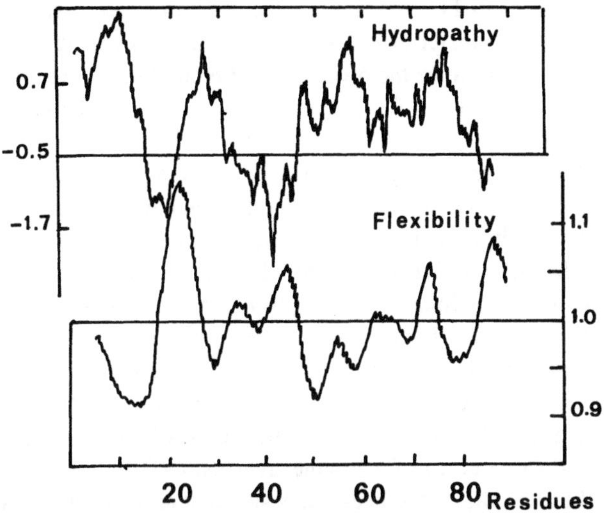

FIGURE 9. Structural parameters (hydropathy and flexibility) of maize LTP. The vertical scale for hydrophobicity indicates the average free energy in kilocalories per mole per amino acid for transfer from a hydrophobic to a hydrophilic environment. The vertical scale for flexibility indicates the temperature factors of the C_α atoms of proteins (B_{norm} coefficients). The horizontal scale indicates the number of amino acid residues.

TP (Westerman and Wirtz, 1985) has not revealed a high homology, despite their common property to transfer lipids.

5. MODE OF ACTION

All the information obtained so far about the activity and molecular structure of LTPs should be able to explain these proteins are able to transfer and bind lipids. However, the mechanism of action of LTP is still unknown. The ability of LTPs to transfer lipids necessarily involves an interaction with lipids. Is this interaction similar to that implied in the binding of fatty acids on FABPs? The study of the mode of action of both categories of plant proteins has revealed interesting features.

5.1. Lipid Transfer Process Facilitated by LTPs

5.1.1. LTPs as Lipid Carriers

According to the schemes elaborated for bovine liver PC-TP (Wirtz, 1982), it is supposed that a reversible complex between PC-TP and the PC molecule is formed. PC-TP, to which PC is bound, forms a collision complex with membranes and exchanges its PC with a PC molecule from a membrane. This leads to an overall process of transfer of PC between membranes (Wirtz, 1982; Van Den Besselaar *et al.*, 1975). Since the process is bidirectional, an exchange (although not strictly one-for-one) is, in fact, observed. According to the kinetic model of Van Den Besselaar *et al.* (1975), the rate constant of association and dissociation of PC-TP with the membrane governs the intensity of the transfer.

By using PC-TP as a model, it can be postulated that plant LTPs are also able to bind phospholipid molecules. Indeed, experiments carried out with maize LTP showed that radioactive phospholipid–LTP complexes were isolated by chromatofocusing when liposomes containing radioactive phospholipids were incubated with maize LTP (Douady *et al.*, 1982); this observation was confirmed by using spinach leaf LTP (Kader *et al.*, 1984). However, the percentage of the initial radioactivity bound to LTP was very weak relative to the results obtained with bovine liver PC-TP. Moreover, no endogenous lipid has been detected in plant LTPs. The same is true for bovine liver nsPL-TP (Crain and Zilversmit, 1980). From these results, it has been suggested that animal nsPL-TPs, as well as plant LTPs, do not act by carrying lipids between membranes but, rather, by modifying the lipid–protein interactions within membranes. However, recent information (Nichols, 1988) has indicated that rat liver nsPL-TP can indeed bind phospholipids; in these experiments, Nichols used a fluorescent lipid, 1-palmitoyl-2-(12-((7-nitro-2,1,3-benzoxadiazol-4-yl)amino)dodecanoyl)-PC and monitored the binding of this phospholipid by spectrofluorimetry. This successful demonstration of the binding of a phospholipid containing a short-chain fatty acid can be explained by the competition for lipid binding between the protein and the vesicles; this competition is in favor of the vesicles as the acyl chain length increases. Preliminary experiments, performed with maize LTP, confirmed that a similar binding of this fluorescent phospholipid occurs (Guerbette and Kader, unpublished).

Another demonstration of a binding of phospholipids (PC and PS) on plant LTPs has been provided by Yamada *et al.* (1980) in a study of acidic castor bean LTPs.

5.1.2. The Different Steps of the Transfer Process

The different steps of the lipid transfer mediated by plant LTPs can be described in a scheme similar to those elaborated for animal proteins (Wirtz, 1982; Kader *et al.*, 1982) (Figure 10). The following steps might be postulated. (1) The phospholipid molecule bound to LTP is embedded in a cavity; this was suggested by the prediction of the secondary structure of maize LTP. The polar headgroup of the phospholipid could interact with the central, positively charged part of the molecule, while the acyl chains of the phospholipid interact with the hydrophobic domains of the LTP surrounding the central part. However, it is

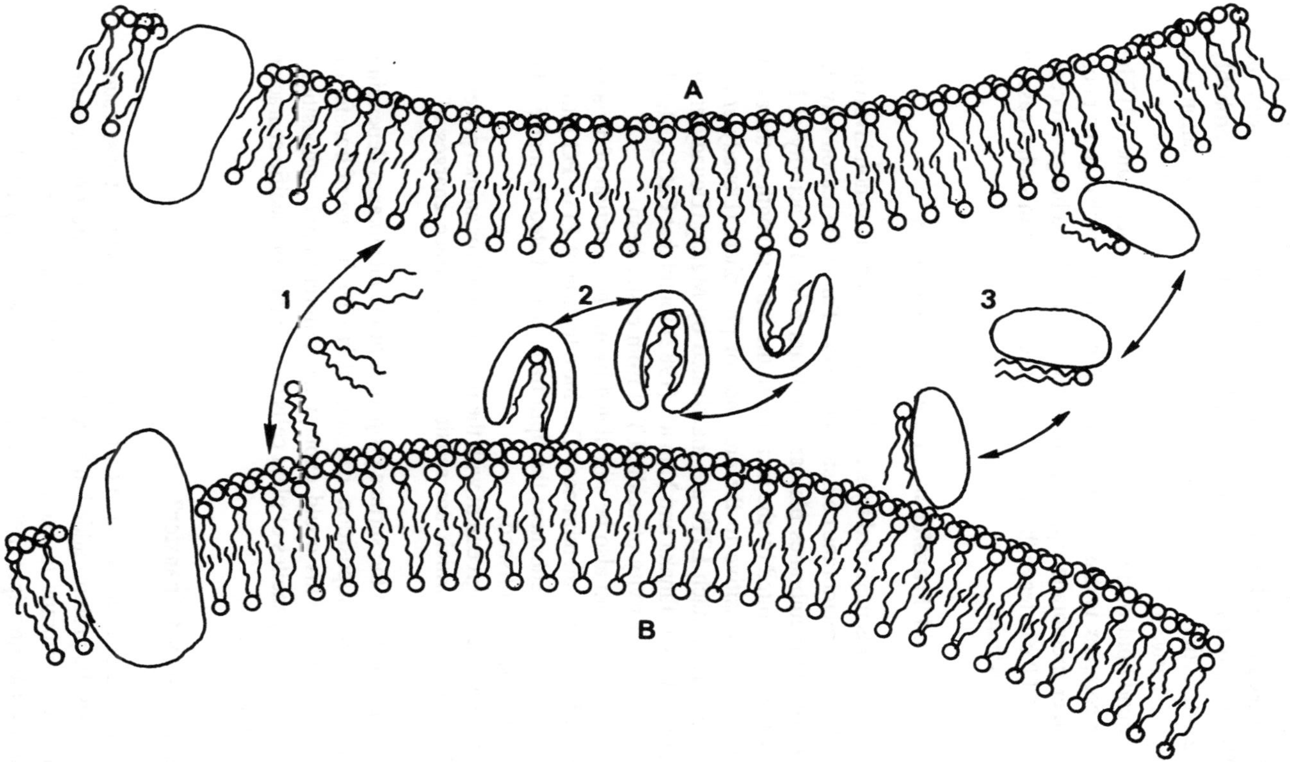

FIGURE 10. Postulated scheme for the mode of action of plant LTPs. In addition to spontaneous movements of phospholipids (1), a transfer between membranes A and B can be mediated by LTP. The phospholipid molecule, interacting with LTP, is either embedded within a cavity (2) or bound to the external surface of the protein (3). Electrostatic charges are involved in the interaction of the polar head of the phospholipid molecule with LTP, whereas the hydrophobic domains of the LTP react with the acyl chains. Depending on the surface charges of the membranes A and B, as well as on the amounts of lipids present in the two membranes, an exchange (replacement of one phospholipid molecule from A by one molecule of B) or a net transfer (leading to an increase in the amount of one of the two membranes, if the exchange is not a one-for-one process) occurs.

possible that the phospholipid molecule bound to the LTP is located on the outer surface of the protein. (2) The phospholipid–LTP complex interacts with the membrane. The N-terminal end of the LTP, which is hydrophobic, is probably involved in the interaction of the protein with the membrane. The formation of a collision complex between LTP, phospholipid, and membrane allows an exchange of the bound lipid with lipid molecules present in the membrane. (3) The LTP–membrane complex is dissociated. The LTP can again exchange its phospholipid with other membrane lipids, leading to an overall exchange process. Alternatively, the LTP can leave the membrane devoid of lipid; in such a case, a net transfer process occurs.

5.1.3. Control of Lipid Transfer by Membrane Properties

The transfer of lipids mediated by a LTP is a surface phenomenon involving the formation of a LTP–membrane collision complex. This is the reason why the surface charges of the membranes play an important role. The activity of LTP from various plants is inhibited by the increase in the ionic strength of the assay mixture: maize LTP is sensitive to the presence of phosphate (Douady *et al.*, 1982); and maize and spinach LTPs are inactivated by $MgCl_2$ (Douady *et al.*, 1982; Kader *et al.*, 1984). Tanaka and Yamada (1982) found that $CaCl_2$, $MgCl_2$, and $LaCl_3$ inhibited the activity of acidic LTPs from castor bean.

The activity of LTPs can also be stimulated by the introduction of negative charges at the surface of the membranes. The introduction of PI or PG into PC liposomes in different ratios enhances the lipid transfer mediated by maize LTP (Guerbette and Kader, unpublished).

All observations are in agreement with the results found for animal proteins, either PC-TP or nsPL-TP, which are sensitive to the surface charges of the membranes (Wirtz, 1982).

It may be suggested that the surface properties of the membranes depending on their lipid and protein composition, can control *in vivo* the rate of transfer of lipids mediated by LTPs. This could explain the different lipid compositions observed in the various organelles of plant cells despite the nonspecific character of the major plant LTPs.

5.1.4. Exchange or Net Transfer?

In the first experiments on intermembrane lipid movement, it was established that the cytosolic proteins added in assays facilitate a bidirectional transfer of lipids from one membrane to another (Kader, 1975; Yamada *et al.*, 1978). These proteins, exchanging lipids between membranes, were designated as phospholipid exchange proteins (PLEP). This is true for all proteins transferring lipids isolated from plant tissues. The precise study of the pure basic LTPs isolated from spinach or maize indicated that these proteins act by equilibrating the lipid pools of the two categories incubated together. As an example, this has been clearly shown in experiments carried out with liposomes and chloroplast envelope vesicles (Miquel *et al.*, 1987).

However, this exchange is not strictly a one-for-one process. Depending on the experimental conditions, the amount of lipids present in one of the two membrane increases, indicating a net transfer process. This accumulation of lipid was noted when chloroplast envelopes were incubated with PC liposomes for longer than 10 min. The possibility of a net transfer, also found with other types of membranes, is particularly important in establishing the role of LTPs in membrane biogenesis.

The fact that an exchange or net transfer process can occur is dependent on the membrane properties as well as the relatuve amounts of lipids present in the membranes. It is not known what really happens *in vivo* in the plant cell. It can be assumed that the relative amounts of membrane lipids and proteins, as well as the intracellular level of LTP, play crucial roles.

5.2. Binding Properties of FABPs

The information provided by the studies on oat FABP indicated its binding ability for all straight-chain fatty acids (from C_{16} to C_{20}), saturated or unsaturated, as well as oleoyl-CoA and palmitoyl-CoA. The binding characteristics of this protein have been carefully analyzed, and the value of the overall dissociation constant for the binding of oleic acid (3.5×10^{-6} M) (rate constant of association versus dissociation) on oat FABP indicated a tight binding of the fatty acid to the protein.

Another interesting property of oat FABP is the pI shift upon lipidation, which can be provoked by the accumulation of ionizable acidic groups at the surface of the protein.

An unexpected property of spinach leaf LTP is its ability to tightly bind oleic acid and oleoyl-CoA as observed by isoelectric focusing (Rickers *et al.*, 1985). In this case, however, no pI shift was observed in the titration curves.

An additional property shared by oat FABP and spinach FABP is that they are unable to bind PC as indicated by separations performed by isoelectric focusing.

In the light of all these observations, it could be supposed that LTPs and FABPs correspond to the same proteins in plant cells. However, additional information concerning the properties of both categories of proteins must be provided before it can be decided whether these proteins belong to the same category. In future experiments the number of sites of binding of fatty acids and the location of these sites must be determined. Knowledge of the complete primary structure of plant LTPs could help this study.

6. LIPID TRANSFER PROTEINS AS PROBES FOR MEMBRANE STUDIES

Since plant LTPs have a high activity that remains stable for months and are broadly specific for phospholipids, they can be used as probes for membrane studies and for manipulation of the membrane lipid composition. Similar experiments are described for animal nsPL-TP (see Chapter 3).

The possibility of using plant LTPs to manipulate the lipid composition of membranes has been realized for the first time with spinach chloroplasts (Miquel *et al.*, 1987). When chloroplast envelopes from spinach leaf (rich in linolenic acid) were incubated with liposomes made from PC (rich in oleic acid) in the presence of spinach LTP, the fatty acid composition of PC molecules extracted from chloroplast envelope was modified. Within 30 min, there was a decrease in the linolenic acid level from 45 to 29%, accompanied by an increase in the oleic acid level from 10 to 21%. These results can be explained by the exchange of liposomal PC with the plastidial PC.

This experiment was confirmed by the recent observation (Guerbette and Kader, unpublished) that when dioleoyl-PC liposomes were incubated with maize mitochondria (rich in linoleic acid), an increase in oleic acid content was observed. These investigations indicated that plant LTPs can be used to modify the lipid composition of plant membranes. It has also been predicted that the lipid composition of animal membranes can be modified by plant LTPs, since these proteins are able to transfer phospholipids to human erythrocytes (Devaux and Kader, unpublished). The effects of lipid changes upon membrane functions must be further examined in future experiments.

Plant LTPs can also be used as probes for studying the organization of lipids within membranes. These proteins can provide informations about the asymmetry of the lipid pools, as well as about the rates of flip-flop of lipids within membranes. These approaches must be actively developed in the future with plant cell membranes.

7. PHYSIOLOGICAL FUNCTIONS OF LIPID TRANSFER PROTEINS AND FATTY-ACID-BINDING PROTEINS

LTPs and FABPs, interacting *in vitro* with various lipids and membranes, are assumed to play important roles *in vivo* in the intracellular fluxes of lipids and in membrane biogenesis.

7.1. Role in Membrane Biogenesis and Renewal

Since LTPs can convey phospholipids between any type of membrane, it is postulated that in the plant cell, LTPs carry phospholipids from their sites of biosynthesis toward membranes that do not contain the enzymes needed for the synthesis of lipids. This is the case for mitochondria (Moore, 1984) and chloroplasts (Douce *et al.*, 1984), which are unable to synthesize PC. According to Mazliak *et al.* (1982) and Moore (1984), PC is synthesized in the endoplasmic reticulum in plant cells. Few studies, however, have been devoted to the study of transmembrane movements of PC and PC metabolites, which have been analyzed in animal cells (Kawashima and Bell, 1987).

7.1.1. Transfer to Mitochondria

The transfer *in vitro* of PC toward mitochondria, as already described (see Section 2.1.1), is accelerated by LTPs from plants. Only a weak movement of phospholipids from liposomes to mitochondria or from endoplasmic reticulum to mitochondria is observed in the absence of LTP. This could be explained either by a spontaneous movement of lipids or by the action of LTP bound to membranes. This latter hypothesis is reinforced by the experiments of Megli *et al.* (1986), who reduced almost totally the spontaneous transfer between liposomes and mitochondria by eliminating with a saline wash the LTP bound to these membranes.

However, the question of a transfer between outer and inner mitochondrial membranes remains unsolved. It is not known whether LTPs are present in the intermembrane phase. Wojtczak *et al.* (1971) suggested that a spontaneous transfer of PA occurs between these two membranes in rat liver.

7.1.2. Transfer towards Chloroplasts

The demonstration that spinach leaf LTP is able to transfer PC toward intact chloroplasts (Kader *et al.*, 1984; Miquel *et al.*, 1987) or isolated chloroplast envelopes (Miquel *et al.*, 1987) reinforces the hypothesis of a participation of LTP in the biogenesis of chloroplast membranes (Figure 11). The transfer of PC observed *in vitro* is rather high, since in a typical experiment, within 30 min of incubation 75 nmol of liposomal PC was added to envelope membranes, corresponding to 1 mg of protein; this value represents about 35% of the PC present in the envelope. This important transfer of PC is consistent with an *in vivo* role of LTP. In addition, it has been observed that a net transfer of PC from liposomes to envelope membranes, either normal or depleted in phospholipids by phospholipase C, was possible.

It has been found that PC molecules transported toward intact chloroplasts are located mainly in the outer surface of the outer membranes (Miquel *et al.*, 1987). This is in agreement with the observation that PC molecules are associated mainly with the external leaflet of the outer envelope membrane (Douce *et al.*, 1984). An active transfer of PC mediated by LTP and an absence of flip-flop could explain this asymmetry.

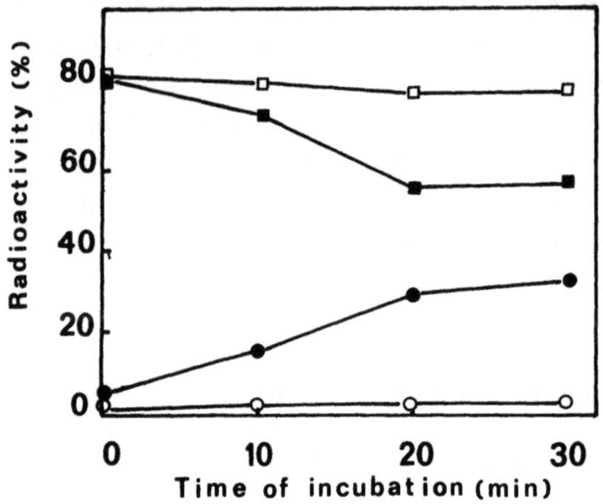

FIGURE 11. Time course of transfer of PC from liposomes toward chloroplast envelope vesicles. ^{3}H- labeled and ^{14}C-liposomes, prepared as indicated in the legend to Figure 1, were incubated with envelope vesicles from spinach leaf chloroplasts (corresponding to 370 μg of proteins) in the presence of 50 μg of spinach leaf LTP. The ^{3}H radioactivity in the envelope pellet (●) and in the supernatant (■), and the ^{14}C radioactivity in the envelope pellet (○) and in the supernatant (■) are expressed as percentages of the initial radioactivity of liposomes. From Miquel *et al.* (1987).

However, some questions are still subjects of debate. What is the explanation of the absence of PE from the chloroplast envelope, even though LTP is able to carry this phospholipid? How are lipids transported inside chloroplasts? Intrachloroplastid transfer of PC from the outer membrane envelope toward thylakoids seems to be absent (Miquel *et al.*, 1987); this result was confirmed by the absence of protein able to transfer PC in the stroma of spinach chloroplasts (Schwitzguebel *et al.*, 1984). However, a need for transfer of galactolipid molecules is clear, since galactolipids are synthesized in the chloroplast envelope and accumulate within the thylakoid membranes (Douce *et al.*, 1984). As already indicated, the presence of proteins able to transfer galactolipids (Nishida and Yamada, 1986) provides an argument in favor of a protein-mediated process. Intrachloroplastidial transfers of membrane fragments are also possible (Carde *et al.*, 1982).

7.1.3. Transfer to Other Membranes

Few membranes, other than mitochondria and chloroplasts, have been involved in *in vitro* studies of lipid transfer. Some preliminary experiments concerned plasmalemma isolated from maize coleoptiles by the two-phase partition method (Guerbette and Kader, unpublished).

7.1.4. Conclusions

The concept that LTPs play important roles is reinforced by *in vitro* experiments, but has not directly demonstrated *in vivo*. The two other processes postulated to explain the lipid movements linked to membrane biogenesis, i.e., spontaneous lipid transfer and movement of membrane fragments which fuse with other membranes, must also be considered. The latter process has been analyzed by Moreau *et al.* (1984), who monitored the intracellular transfer of very-long-chain fatty acids through various membrane compartments (Golgi, plasma membrane, endoplasmic reticulum) of *Allium porrum* cells.

It could be suggested that the three processes—spontaneous movement, protein-mediated transfer, and membrane transfer—are all present in plant cells in various degrees of importance. However, although spontaneous movements are too slow to be of physiological importance, the membrane flow theory (Morré, 1975) is unable to explain the different lipid compositions observed in plant cells and especially in chloroplast membranes.

A direct demonstration *in vivo,* which is still lacking in animal cells (Wirtz, 1982), would be either to use the molecular biology approach (see Section 9) or to study mutants of plants such as *Arabidopsis thaliana.*

7.2. Role in Lipid Metabolism

Various roles can be attributed to LTPs and FABPs in plant lipid metabolism. One major point concerns the cooperative pathway (called "eukaryotic") for the biosynthesis of galactolipids containing linolenic acid (Heemskerk and Wintermans, 1987; Harwood, 1988). This complex pathway comprises various steps (Figure 12): (1) in plant cells, the unique site for the synthesis of fatty acids is the plastid, which contains the enzymes of the fatty acid synthetase complex; (2) plastid is able to synthesize oleoyl-CoA, which is exported toward the cytosol; (3) oleoyl-CoA is used in the endoplasmic reticulum for the acylation of lyso-PC, which forms oleoyl-PC; (4) Oleoyl-PC is desaturated to linoleoyl-PC by oleate desaturase associated with the endoplasmic reticulum; (5) linoleoyl-PC is then transported to the chloroplast envelope, in which an additional desaturation to linolenic acid takes place. Linoleoyl-MGDG is formed from linoleoyl-PC and used as a precursor for the synthesis of "eukaryotic" MGDG, which contains linolenic acid in the *sn*-1 and *sn*-2 positions. In addition to this eukaryotic pathway, another pathway, called "prokaryotic," having all its steps within the chloroplast, is able to synthesize "prokaryotic" MGDG containing hexadecatrienoic acid at the *sn*-2 position and linolenic acid at the *sn*-1 position (Roughan, 1987). The relative importance of these two pathways depends on the nature of the plants: for example, the prokaryotic pathway is more highly developed in spinach, whereas the eukaryotic pathway is the major one in pea.

It is clear that FABPs and LTPs are good candidates to mediate the flux of lipid molecules between the different compartments of plant cells.

FABPs, defined by their properties to bind oleoyl-CoA, could play a major role in regulating the intracellular level of oleoyl-CoA exported from plastids. This role could also be played by LTPs, such as those of spinach leaf, which have the ability to bind oleoyl-CoA (Rickers *et al.*, 1985). In some experiments, made

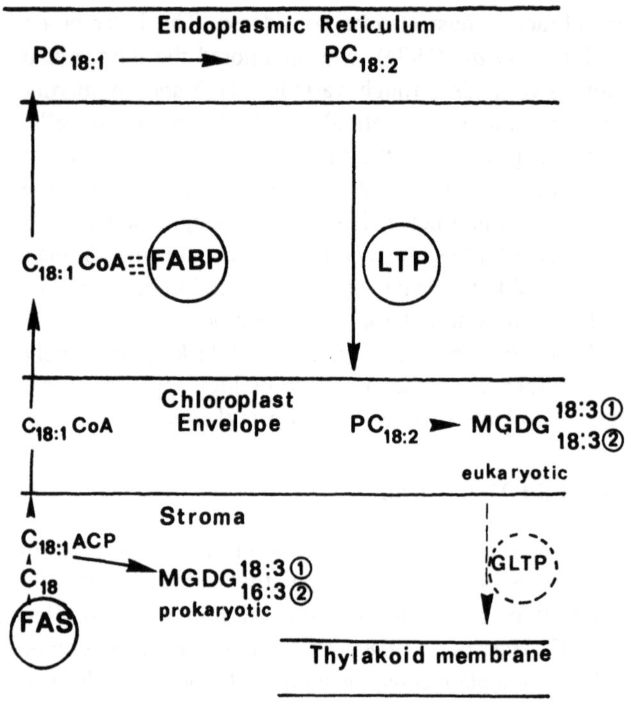

FIGURE 12. Postulated prokaryotic and eukaryotic pathways for the synthesis of galactolipids. Oleoyl acyl carrier protein ($C_{18:1}$ ACP) formed within the chloroplastidial stroma by the fatty acid—synthetase complex (FAS) is used to synthesize "prokaryotic" MGDG esterified in the *sn*-1 position by linolenic acid ($C_{18:3}$) and in the *sn*-2 position by hexadecatrienoic acid ($C_{16:3}$). In the eukaryotic pathway, $C_{18:1}$ ACP is converted to oleoyl-CoA ($C_{18:1}$ CoA), which is exported from the chloroplast and integrated into PC molecules which are desaturated in the endoplasmic reticulum. Linoleoyl-PC is transported toward the chloroplast, where it serves as an acyl chain donor for the synthesis of "eukaryotic" MGDG containing $C_{18:3}$ in the two positions. This galactolipid is transported to thylakoids, where it is accumulated. The three proteins studied in this review (FABP, LTP, and GLTP) might be involved in the different lipid movements postulated in this scheme either by conveying phospholipids and galactolipids (LTP and GLTP) or by interacting with oleoyl-CoA (FABP).

by incubating spinach chloroplasts, microsomes, and spinach LTP in the presence of [1-^{14}C]acetate, an increase in fatty acid labeling has been observed (Dubacq *et al.*, 1984). Also, in preliminary experiments, the addition of maize LTP can stimulate the desaturation of linoleoyl-PC, determined by incubating potato tuber microsomes in the presence of oleoyl-CoA (C. Demandre and J.-C. Kader, unpublished data). These effects might be related to the binding properties of LTPs for oleoyl-CoA.

More generally, in plant cells FABPs and LTPs could play several roles related to the modulation of the intracellular levels of free fatty acids and oleoyl-CoA. Additional experiments *in vitro* are needed to provide arguments in favor of the participation of these proteins. It may also be supposed that these proteins are implicated in the transport of hydrophobic compounds, which are used as substrates for membranous enzymes.

The participation of LTPs in the cooperative pathway for galactolipid biosynthesis is suggested by the observation that active transfer of PC toward chloroplasts can be mediated by LTP (Miquel *et al.*, 1987). Moreover, reconstitution experiments have been performed by Ohnishi and Yamada (1982), who observed a synthesis of dilinolenoyl-MGDG when dilinoleoyl-PC liposomes were incubated with oat plastids and castor bean LTP. Oursel *et al.* (1987) incubated spinach chloroplasts with PC-liposomes in the presence of spinach LTP and then treated PC molecules transported to the chloroplasts with phospholipase C to form diacylglycerol; synthesis of MGDG by galactosylation of diacylglycerol was observed. All these experiments provided some arguments in favor of the participation of LTPs in the cooperative "eukaryotic" scheme. Further arguments might be provided by the study of mutants of *Arabidopsis thaliana;* a mutant of this plant, devoid of the prokaryotic pathway and supposed to contain an more active eukaryotic pathway, has been recently discovered (Kunst *et al.*, 1988).

As previously indicated, plant LTPs are able to transfer PI. In the light of recent hypotheses involving the cycle of polyphosphoinositides in plant cells, it may be supposed that plant LTPs participate in the PI cycle (Sommarin and Sandelius, 1988). It is assumed that in response to external stimuli such as hormonal signals, PI from the plasma membrane is phosphorylated in PI derivatives which are hydrolyzed by phospholipase C, and inositol trisphosphate is formed; this compound, acting as a second messenger, affects the intracellular calcium level, which can control various physiological responses. For an efficient functioning of the PI cycle, it is necessary to replace the hydrolyzed PI by new PI. Since PI is synthesized exclusively in the endoplasmic reticulum (Moore, 1984), a transfer of PI toward the plasma membrane, probably mediated by LTP, is necessary. The same hypothesis has been suggested by Van Paridon *et al.* (1987) for animal cells.

7.3. Conclusions

The presumptive roles of FABPs and LTPs are numerous and important for the physiology of plant cells. However, direct arguments in favor of the participation of these proteins in membrane biogenesis and renewal, as well as in lipid metabolism, are needed.

8. IMMUNOCHEMICAL CHARACTERIZATION OF LIPID TRANSFER PROTEINS

A monospecific antibody was raised in rabbits against maize LTP (CM2 fraction; Douady *et al.*, 1986; Grosbois *et al.*, 1987). This antibody is able to progressively inhibit the activity of LTP when maize LTP is incubated with increasing amounts of immunoglobulin G (IgG) (Figure 13). A complete inhibition of the transfer activity was observed at an IgG-to-LTP ratio of 10 : 50 (w/w). Results from controls with preimmune IgG confirmed that the IgG against maize LTP was truly reactive with LTP.

8.1. Qualitative Studies

The antibody was also used to detect LTP by immunodiffusion (Grosbois *et al.*, 1987). The experiments have been done with pure LTP, with crude extracts obtained from the whole maize seedling, and with extracts prepared from various parts (root, shoot, embryo) of etiolated maize seedlings. These results indicate that the isoform CM2 is present in all organs of maize seedlings.

The antibody allowed immunoblotting approaches. After SDS-electrophoresis of either pure maize LTP or crude extract from maize seedlings and transfer to nitrocellulose sheets, the immunoreactive compounds were detected with anti-LTP-IgG followed by goat anti-rabbit IgG conjugated to peroxidase. In addition to the LTP, another compound corresponding to a slightly lower molecular mass was detected. The significance of this compound (which may be an isoform of LTP or perhaps partially degraded LTP) is unknown.

Since plant LTPs present structural homologies, it was of interest to examine whether maize LTP IgG cross-reacts with castor bean or spinach LTP. No clear-

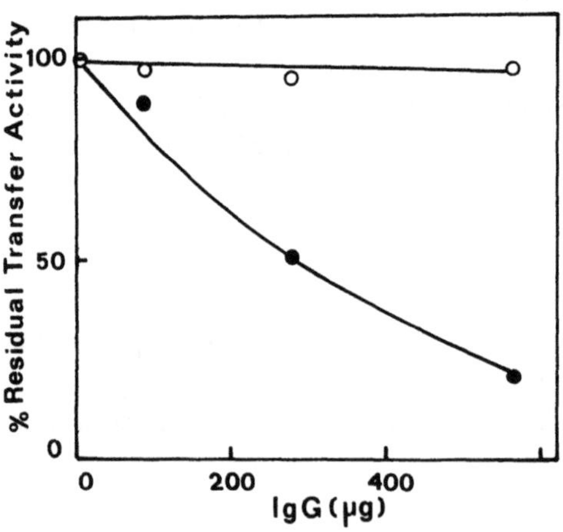

FIGURE 13. Effect of anti-maize LTP IgG on phospholipid transfer activity. Maize LTP (140 μg) was incubated with increasing amounts of IgG and assayed for the activity of transferring PC (see the legend to Figure 1 for the experimental conditions) (●). The same experiments was carried out with preimmune IgG (○). Results are expressed as percent residual activity. From Grosbois *et al.* (1987).

cut results have been obtained with these two LTPs. However, maize antibody reacts positively with oat or pea crude extracts (J.-C. Kader and F. Spener, unpublished data). This suggests interesting possibilities for future studies of biogenesis and molecular biology of LTPs in these tissues. Also, an eventual identity between oat FABP and maize LTP will be examined by using antibodies against oat FABP in immunoblotting studies with maize LTP.

8.2. Quantitative Studies

The monospecific character of IgG against maize LTP has allowed the development of an enzyme-linked immunosorbent assay (ELISA) able to determine the LTP levels in the range from 1 to 20 ng (Grosbois *et al.*, 1987) (Figure 14). The ELISA was used in combination with fast protein liquid chromatography (FPLC) to detect maize LTP in crude extracts and to measure its concentration. Relatively large amounts of LTP have been found in the postmicrosomal proteins (2%) and in the proteins precipitated by $(NH_4)_2SO_4$ (4.1%). This observation provided the first evidence that maize cytosol is rich in LTP. This is in agreement with the following observations: (1) large amounts of pure LTP are extracted from maize seeds (60 mg from 2 kg of dry seeds); (2) LTP appears as a clearly visible band in SDS-electrophoresis performed on crude extracts; and (3) a purification factor of about 100 is sufficient to obtain a pure protein.

In contrast to plant LTPs, nsL-TP from rat liver is a minor protein (0.08% of the total 105,000 × *g* supernatant proteins). However, this abundance of plant LTPs is similar to that found for FABPs in animal cells: 5% of rat liver proteins found in cytosols are FABPs (Glatz and Veerkamp, 1985).

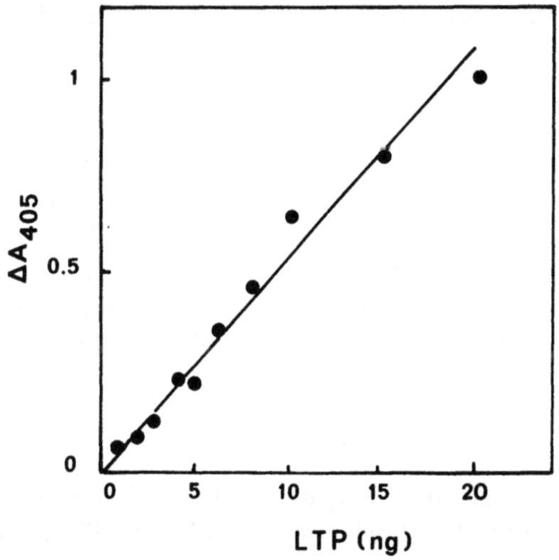

FIGURE 14. ELISA standard curve for maize LTP. Maize LTP was immobilized on plastic plates and incubated with the anti-LTP antibody. A second antibody (anti-rabbit IgG conjugated with alkaline phosphatase) was used to indirectly detect the protein. The absorbance was measured at 405 nm. From Grosbois *et al.* (1987).

8.3. Study of the Intracellular Localization of LTPs

All the purification procedures for the isolation of LTPs and FABPs from plant tissues are based on the assumption that these proteins are located in the cytosolic phase of the cell. However, these proteins seem to be partly associated with intracellular membranes. This has been shown by immunoblotting and immunodiffusion combined with FPLC (Douady *et al.*, 1986). In these approaches, purified mitochondria or crude microsomal fractions extracted from maize seedlings were solubilized with sodium deoxycholate. The membrane proteins were then separated either by SDS-electrophoresis and immunoblotting or by FPLC and immunodiffusion. With the first method, a band corresponding to the maize LTP was detected among microsomal and mitochondrial proteins. This was confirmed by the second approach, in which the immunodiffusion step showed the presence of an immunoreactive fraction.

These experiments established an association between LTPs and intracellular membranes. However, the physiological significance of this observation remains to be elucidated. Even though this binding of LTPs on membranes might be artifactual and could occur during the homogenization procedure, it may play a role in the functioning of LTPs. Immunocytochemical techniques could provide additional information about the intracellular localization of plant LTPs.

Animal LTPs are also partly membrane bound. Teerlink *et al.* (1984), in an ELISA study of the intracellular distribution of nsPL-TP in rat liver, found that 30% of the total transfer proteins are associated with membranes. Immunocytochemical techniques indicated that the proteins are located mainly in the peroxisomes.

9. MOLECULAR BIOLOGY OF LIPID TRANSFER PROTEINS

Plant LTPs are assumed to play an important role in membrane biogenesis by conveying phospholipids from their site of biosynthesis to membranes in formation. This hypothesis is based on *in vitro* experiments, and it is necessary to provide more direct *in vivo* demonstrations. This is why recent investigations have studied the biogenesis of plant LTPs to (1) establish a link between membrane formation and biogenesis of LTP, (2) study the expression of genes involved in the biogenesis of LTP and search for tissue specificity, and (3) isolate cDNA and genomic clones which will allow site-directed mutagenesis.

9.1. *In Vitro* Synthesis

The *in vitro* synthesis of an LTP in maize has been studied by Tchang *et al.* (1985, 1987), who isolated poly(A)$^+$ RNAs from 3-day–old etiolated seedlings by affinity chromatography on oligo(dT)-cellulose. Poly(A)$^+$ RNAs were translated in a rabbit reticulocyte lysate system in the presence of [^{35}S]methionine. The newly synthesized proteins were separated by SDS-electrophoresis. To detect the LTP among the newly synthesized proteins, the protein mixture was injected into a reverse-phase column of an FPLC. The different fractions collected were separated by SDS-electrophoresis. Fluorography detected a radioactive band with a molecular mass of 9 kDa comigrating with a chemically radiolabeled maize LTP. This result was confirmed by monitoring the radioactivity of the effluent from the reverse-phase column. These observations indicate that LTP is synthesized as a 9-kDa compound. From these experiments, it can be concluded that maize coleoptile contains active mRNAs coding for the LTP and that a 9-kDa LTP is detectable among the synthesized proteins. However, these approaches do not indicate whether LTP is synthesized with a precursor of higher molecular mass. Preliminary experiments have recently been performed by immunoprecipitating the proteins newly synthesized from [^{35}S]methionine or [^{35}S]cysteine by the anti-maize LTP antibody. In addition to the mature 9-kDa LTP, a higher-molecular-mass compound is immunoprecipitated, corresponding to a 12- to 13-kDa protein (C. Vergnolle and J.-C. Kader, unpublished data). This gave a first indication that maize LTP is synthesized in a precursor form. Recent indications from Yamada's group suggest that castor bean LTPs are also synthesized as precursors with a molecular mass 3.5 to 4 kDa higher than that of the mature LTP (Yamada *et al.*, 1988). Interestingly, a high-molecular-mass polypeptide (58 kDa) is detected in rat liver, in addition to the mature nsPL-TP (14.8 kDa) by immunoblotting with the antibody against nsPL-TP (Van Der Krift *et al.*, 1985). Also, the synthesis of rat adrenal nsPL-TP revealed a band corresponding to 14.4 kDa; this molecular mass is higher than that of the mature protein (11.3 kDa), suggesting the presence of a signal peptide (Trzeciak *et al.*, 1987).

9.2. Isolation and Characterization of cDNA Clones

The antibody against maize LTP was used to screen cDNA libraries issued from maize coleoptiles for poly(A)$^+$ RNAs. A first cDNA library was constructed by inserting total cDNA into the vector pBR322 at the *Pst* site by deoxyguanine/deoxycytosine (DG/DC) tailing and using *Escherichia coli* as a host. This library was screened by using anti-maize IgG purified by affinity chromatography. From 600 recombinant clones, 1 positive clone (6B6) was identified. The nucleotide sequence of the cDNA from 6B6 (677 bases) was determined. The deduced amino acid sequence from this cDNA corresponded to the major part of the amino acid sequence determined on the pure maize LTP.

To search a full-length cDNA, a second library was prepared with size-selected cDNAs (0.5 to 2.5 kb). From 1000 recombinant colonies screened, 4

```
AGCTGCCATTGCCATCTACTGACCTGTGACCTCCGCATTCCCAAACAGCC          50

CGAGCGAGCGAGCTAGCAGAGCGGCGGCAGGCACCTCCCTCCTCAAGGAAC        101

ATG GCC CGC ACG CAG CAG TTG GCA GTA GTG GCG ACC GCC        140
Met Ala Arg Thr Gln Gln Leu Ala Val Val Ala Thr Ala         13

GTG GTG GCC TTG GTG CTG CTG GCG GCG GCG ACC TCG GAG        179
Val Val Ala Leu Val Leu Leu Ala Ala Ala Thr Ser Glu         26

GCG GCC ATC AGC TGC GGG CAG GTG GCG TCG GCC ATC GCG        218
Ala Ala Ile Ser Cys Gly Gln Val Ala Ser Ala Ile Ala         39

CCC TGC ATC TCC TAC GCG CGC GGC CAG GGC TCG GGG CCC        257
Pro Cys Ile Ser Tyr Ala Arg Gly Gln Gly Ser Gly Pro         52

TCC GCC GGC TGC TGC AGC GGC GTC AGG AGC CTC AAC AAC        296
Ser Ala Gly Cys Cys Ser Gly Val Arg Ser Leu Asn Asn         65

GCC GCC CGC ACC ACC GCC GAC CGC CGC GCC GCC TGC AAC        335
Ala Ala Arg Thr Thr Ala Asp Arg Arg Ala Ala Cys Asn         78

TGC CTC AAG AAC GCC GCC GCC GGC GTC AGC GGC CTC AAC        374
Cys Leu Lys Asn Ala Ala Ala Gly Val Ser Gly Leu Asn         91

GCC GGT AAC GCC GCC AGC ATC CCC TCC AAG TGC GGC GTC        413
Ala Gly Asn Ala Ala Ser Ile Pro Ser Lys Cys Gly Val        104

AGC ATC CCC TAC ACC ATC AGC ACC TCC ACC GAC TGC TCC        452
Ser Ile Pro Tyr Thr Ile Ser Thr Ser Thr Asp Cys Ser        117

AGG GTG AAC TGA ACCCTAAACGACGACGACGGCGCCGGCGGCCGCCT        499
Arg Val Asn  -                                             120

GAAGACGATCCATATCTATCCCGCCGTCGTCGGCGCCCGCGGATCGGACCG        550

TCATGACCATATCTACCTACGCATGATCACTACTTGAATGTCTACTACTGC        601

TACGTACTACTACTGAATAAAAAACCTCTCCGCGCACATCGTCCTTGTGTT        652

TGTGTGTGTGAGAGAGAGAGGAGACGGAGAGCCGGCCATGTCCTTTTGGCT        703

GTTTGTTTGGTTTATTTTTGAGTTCATGCGGTCTTTATTTATTTCCATGTA        754

CGCTGCTTCCTCGTACCCACTTTAAATTTGTACCGATGCATGCATGTTCTT        805

GATGATACAAAAAAAAA                                          822
```

FIGURE 15. Nucleotide sequence of 9C2 cDNA corresponding to maize LTP. The deduced amino acid sequence corresponds that determined for maize LTP (see Section 4.6). The sequence of the signal peptide, at the N-terminal end of the protein, is underlined (Tchang and Kader, unpublished).

positive clones were obtained, and the longest one (9C2) was further investigated. The 9C2 clone, with 822 bases, has been completely sequenced (Figure 15). It corresponds to the full-length cDNA of the maize LTP, since the deduced amino acid sequence of 9C2 is identical to the primary structure of the purified protein. In addition, an extra peptide of 27 amino acids is present at the N terminus. This fragment possesses the characteristics of a signal peptide.

This observation provides, for the first time, a clear demonstration that a signal peptide is involved in the synthesis of a LTP. The presence of this extra peptide explains the previous findings indicating that LTPs from maize or castor bean and sterol carrier protein 2 from rat adrenal cells are synthesized *in vitro* as precursors with higher molecular mass than the mature protein.

However, the physiological significance of the presence of the signal peptide remains to be established. This signal peptide, which is usually characteristic of exported proteins, may indicate that LTP is bound to membranes or passes across membranes, although the major part of LTP is recovered in the cytosol. Indeed, it has been shown that maize LTP is partly associated with membranes (see Section 8.3). It would be interesting to study the localization of the synthesis of LTPs; preliminary experiments indicate that maize LTP is preferentially synthesized by membrane-bound polysomes (V. Arondel and J.-C. Kader, unpublished). The post translational process leading to the mature LTP must be further examined.

The availability of cDNA coding for maize LTP—the first to be obtained for a lipid transfer protein—opens new perspectives for the study of the synthesis of this protein as well as for the isolation of LTP genes. This cDNA could be used as a probe for studying the levels of LTP mRNA in maize tissues or for *in situ* hybridization, indicating the cells where the LTP genes are expressed. Future investigations will involve site-directed mutagenesis and transformation of plants to establish the physiological roles of LTPs in plant cells.

10. CONCLUSIONS AND PROSPECTS

Our knowledge about LTPs and FABPs has increased in the last 5 years. The main reason is the purification to homogeneity of LTPs from various plants (which allowed the precise study of their primary structure), the preparation of monospecific antibodies, and the isolation and characterization of cDNA clones corresponding to a LTP.

The major part of the lipid transfer activity found in plant cytosols is associated with a category of proteins which have in common the following properties: (1) a molecular of about 9 kDa (8.8 to 9.8 kDa), (2) a high pI (8.8 to 10.5), (3) a broad specificity for phospholipids, and (4) a remarkable stability. It remains to generalize the observations, made with some LTPs, that these proteins are also able to transfer galactolipids or to bind fatty acids. A precise study is needed to establish the possible multiples roles of a plant LTP.

Some of the properties described for plant LTPs have also been found for animal nsPL-TPs, which are also basic and nonspecific for phospholipids. However, no homology has been found between the primary structures of animal nsPL-TP and plant LTPs. In contrast, comparison of the structures of LTPs of plant origin has revealed conserved domains which could be important for the functioning of these proteins. However, precise determination of the structures of plant LTP by crystallographic approaches, as well as the sites of interaction of phospholipids and fatty acids, must be performed.

The remarkable stability and the high activity of plant LTPs have allowed them to be used as tools for modifying the lipid composition of membranes of either plant or animal origin. It could be anticipated that their use will develop in the future in order to study the role of the lipid environment on the function of membrane-bound enzymes.

As in animal cells, plant LTPs are assumed to be involved in the distribution of phospholipids not only to various types of membranes also found in animal cells, but also organelles characteristic of plants, such as chloroplasts, which contain galactolipid-rich thylakoids. A precise analysis of intrachloroplastidial movement of lipids, mediated by specific proteins, must be developed.

Plant LTPs are also assumed to be involved in a cooperative pathway for the synthesis of lipids. If the participation of LTPs in mediating the fluxes of lipids (mainly oleoyl-CoA from chloroplasts to endoplasmic reticulum and PC from endoplasmic reticulum to chloroplasts) is clearly established *in vivo*, this will indicate that the dynamics of lipids in plant cells is dependent on LTP activity. The involvement of LTPs in the intracellular lipid fluxes in plants could be demonstrated *in vivo* by using molecular biology. The isolation of a cDNA clone corresponding to maize LTP will allow, in the near future, site-directed mutagenesis based on the knowledge of the different domains of a LTP molecule needed for its activity. It will be possible to insert maize cDNA (or a LTP gene, in case of its isolation) into other plants (tobacco, for example). The transformation of tobacco plants by maize cDNA, either untreated or modified, could give novel information about the role of LTP within plant cells. This might lead to genetic engeenering of plants to improve their productivity and their agronomical properties linked to lipid metabolism.

ACKNOWLEDGMENTS. I sincerely thank Drs. M. Yamada, C. Somerville, and F. Spener for communicating preprints of their work or unpublished informations, as well as for stimulating discussions. I am grateful to my colleagues V. Arondel, M. Grosbois, F. Guerbette, F. Tchang, A. Jolliot, and C. Vergnolle for helping my progress in the study of plant LTPs. I am indebted to J. C. Pernollet, M.

Delseny, H. Duranton, A. L. Haenni, and P. Puigdomenech for their cooperation in the studies of the structure and molecular biology of LTPs.

REFERENCES

Abdelkader, A. B., and Mazliak, P., 1970, Echange de lipides entre mitochondries, microsomes et surnageant cytoplasmique de pomme de terre ou de chou-fleur, *Eur. J. Biochem.* **15:**250–262.

Akeroyd, R., Moonen, P., Westerman, J., Puijk, W. C., and Wirtz, K. W. A., 1981, The complete primary structure of the phosphatidylcholine transfer protein from bovine liver, *Eur. J. Biochem.* **114:**385–391.

Bernier, I., and Jollès, P., 1987, A survey on cytosolic non-enzymic proteins involved in the metabolism of lipophilic compounds: From organic anion binders to new protein families, *Biochimie* **69:**1127–1152.

Bloj, B., and Zilversmit, D. B., 1977, Rat liver proteins capable of transferring phosphatidylethanolamine. Purification and transfer activity for other phospholipids and cholesterol, *J. Biol. Chem.* **252:**1613–1619.

Bouillon, P., Drischel, C., Vergnolle, V., Duranton, H., and Kader, J. C., 1987, Complete primary structure of spinach leaf phospholipid transfer protein, *Eur. J. Biochem.* **166:**387–391.

Boussange, J., Douady, D., and Kader, J. C., 1980, Stimulation of phosphatidylethanolamine exchange, *Plant Physiol.* **65:**355–358.

Bozzato, R. P., and Tinker, D. O., 1987, Purification and properties of two phospholipid transfer proteins from yeast, *Biochem. Cell Biol.* **65:**195–202.

Carde, J. P., Joyard, J., and Douce, R., 1982, Biogenesis of plastid membranes, *Biol. Cell* **44:**315–324.

Crain, R. C., and Zilversmit, D. B., 1980, Two nonspecific phospholipid exchange proteins from beef liver. 1. Purification and characterization, *Biochemistry* **19:**1433–1439.

Daum, G., and Paltauf, F., 1984, Phospholipid transfer in yeast. Isolation and partial characterization of a phospholipid transfer protein from yeast cytosol, *Biochim. Biophys. Acta* **784:**385–391.

Dawidowicz, E. A., 1987, Dynamics of membrane lipid metabolism and turnover, *Annu. Rev. Biochem.* **56:**43–61.

DiCorleto, P. E., Warach, J. B., and Zilversmit, D. B., 1979, Purification and characterization of two phospholipid exchange proteins from bovine heart, *J. Biol. Chem.* **254:**7795–7802.

Douady, D., Kader, J. C., and Mazliak, P., 1978, Movements of phospholipids involving plant mitochondria. Regulation by phospholipid-exchange proteins, in *Plant Mitochondria* (G. Ducet and C. Lance, eds.), pp. 357–364, Elsevier Biomedical Press, Amsterdam.

Douady, D., Kader, J. C., and Mazliak, P., 1980, Kinetics of the exchange of phosphatidylcholine mediated by plant cytosol proteins, *Plant Sci. Lett.* **17:**295–301.

Douady, D., Grosbois, M., Guerbette, F., and Kader, J. C., 1982, Purification of basic phospholipid transfer protein from maize seedlings, *Biochim. Biophys. Acta* **710:**143–153.

Douady, D., Guerbette, F., and Kader, J. C. 1985, Purification of phospholipid transfer protein from maize seeds using a two-step chromatographic procedure, *Physiol. Veg.* **23:**373–380.

Douady, D., Grosbois, M., Guerbette, F., and Kader, J. C., 1986, Phospholipid transfer protein from maize seedlings is partly membrane-bound, *Plant Sci.* **45:**151–156.

Douce, R., Block, M. A., Dorne, A. J., and Joyard, J., 1984, The plastid envelope membranes: their structure, composition and role in chloroplast biogenesis, in *Subcellular Biochemistry,* (D. B. Roodyn, ed.), Vol. 10, pp. 1–84, Plenum Press, New York.

Dubacq, J. P., Drapier, A., Trémolières, A., and Kader, J. C. 1984, Role of phospholipid transfer protein in the exchange of phospholipids between microsomes and chloroplasts. *Plant Cell Physiol.* **25:**1197–1200.

Dyatlovitskaya, E. V., Timofeeva, N. G., and Bergelson, L. D., 1978, A universal lipid exchange protein from rat hepatoma, *Eur. J. Biochem.* **82:**463–471.

Glatz, J. F. C., and Veerkamp, J. H., 1985, Intracellular fatty acid-binding proteins, *Int. J. Biochem.* **17:**13–22.

Grosbois, M., Guerbette, F., Douady, D., and Kader, J. C., 1987, Enzyme immunoassay of a plant phospholipid transfer protein, *Biochim. Biophys. Acta* **917:**162–168.

Guerbette, F., Douady, D., Grosbois, M., and Kader, J. C., 1981, Transfer of phosphatidylcholine between liposomes mediated by phospholipid transfer proteins. Use of concanavalin A for the separation of liposomes, *Physiol. Veg.* **19:**467–472.

Harwood, J. L., 1988, Fatty acid metabolism, *Annu. Rev. Plant Physiol. Plant Mol. Biol.* **39:**101–138.

Heemskerk, J. W. M., and Wintermans, J. F. G. M., 1987, Role of the chloroplast in leaf acyl-lipid synthesis, *Physiol. Plant.* **70:**558–568.

Helmkamp, G. M., 1985, Phosphatidylinositol transfer proteins: Structure, catalytic activity and physiological function, *Chem. Phys. Lipids* **38:**3–16.

Julienne, M., Vergnolle, C., and Kader, J. C., 1981, Activity of phosphatidylcholine transfer protein from spinach leaves with mitochondria and chloroplasts, *Biochem. J.* **197:**763–766.

Kader, J. C., 1975, Proteins and the intracellular exchange of lipids. 1. Stimulation of phospholipid exchange between mitochondria and microsomal fractions by protein isolated from potato tuber, *Biochim. Biophys. Acta* **380:**31–44.

Kader, J. C., 1977, Exchange of phospholipids between membranes, *in Dynamic Aspects of Cell Surface Organization* (G. Poste and G. L. Nicolson, eds.), pp. 127–204, Elsevier Biomedical Press, Amsterdam.

Kader, J. C., 1985, Lipid-binding proteins in plants, *Chem. Phys. Lipids* **38:**51–62.

Kader, J. C., Douady, D., and Mazliak, P., 1982, Phospholipid transfer proteins, in *Phospholipids, a Comprehensive Treatise* (J. N. Hawthorne and G. B. Ansell, eds.), pp. 279–311, Elsevier Biomedical Press, Amsterdam.

Kader, J. C., Julienne, M., and Vergnolle, C., 1984, Purification and characterization of a spinach leaf protein capable of transferring phospholipids from liposomes to mitochondria or chloroplasts, *Eur. J. Biochem.* **139:**411–416.

Kader, J. C., Vergnolle, C., and Mazliak, P., 1987, Movements of phospholipids between membranes. Purification of a phospholipid transfer protein from spinach leaf, *Methods Enzymol.* **148:**661–666.

Kamp, H. H., Wirtz, K. W. A., and Van Deenen, L. L. M., 1973, Some properties of phosphatidylcholine exchange protein purified from bovine liver, *Biochim. Biophys. Acta* **318:**313–325.

Kawashima, Y., and Bell, R. M., 1987, Assembly of the endoplasmic reticulum phospholipid bilayer. Transporters for phosphatidylcholine and metabolites, *J. Biol. Chem.* **262:**16495–16502.

Kunst, L., Browse, J., and Somerville, C., 1988, Altered regulation of lipid biosynthesis in a mutant of *Arabidopsis* deficient in chloroplast glycerol phosphate acyltransferase activity, *Proc. Natl. Acad. Sci. U.S.A.*, *85:*4143–4147.

Lumb, R. H., Kloosterman, A. D., Wirtz, K. W. A., and Van Deenen, L. L. M., 1976, Some properties of phospholipid exchange proteins from rat liver, *Eur. J. Biochem.* **698:**15–22.

Machida, K., and Ohnishi, S., 1978, A spin-label study of phosphatidylcholine exchange protein. Regulation of the activity by phosphatidylserine and calcium ion, *Biochim. Biophys. Acta* **507:**156–164.

Mazliak, P., and Kader, J. C., 1980, Phospholipid exchange systems, in *The Biochemistry of Plants* (P. K. Stumpf and E. E. Conn, eds.), pp. 283–300, Academic Press, New York.

Mazliak, P., Jolliot, A., and Bonnerot, C., 1982, Biosynthesis and metabolism of phospholipids, in *Biochemistry and Metabolism of Plant Lipids* (J. F. G. M. Wintermans and P. J. C. Kuiper, eds.), pp. 89–98, Elsevier Biomedical Press, Amsterdam.

Megli, F. M., De Lisi, A., van Amerongen, A., Wirtz, K. W. A., and Quagliariello, E., 1986, Nonspecific lipid transfer protein (sterol carrier protein 2) is bound to rat liver mitochondria: Its role in spontaneous intermembrane phospholipid transfer, *Biochim. Biophys. Acta* **861**:463–470.

Miquel, M., Block, M. A., Joyard, J., Dorne, A. J., Dubacq, J. P., Kader, J. C., and Douce, R., 1987, Protein mediated transfer of phosphatidylcholine from liposomes to spinach chloroplast envelope membranes, *Biochim. Biophys. Acta* **937**:219–228.

Moore, T. S., 1984, Biochemistry and biosynthesis of plant acyl lipids, in *Structure, Function and Metabolism of Plant Lipids* (P. A. Siegenthaler and W. Eichenberger, eds.), pp. 93–96, Elsevier Biomedical Press, Amsterdam.

Moreau, P., Juguelin, H., Lessire, R., and Casagne, C., 1984, Study of the intracellular transfer of lipids to the plasmalemma, in *The Metabolism, Structure, and Function of Plant Lipids* (P. K. Stumpf, J. B. Mudd, and W. D. Nes, eds.), pp. 221–224, Plenum Press, New York.

Morré, D. J., 1975, Membrane biogenesis, *Annu. Rev. Plant Physiol.* **26**:441–481.

Murphy, D. J., and Kuhn, D. N., 1981, Lack of phospholipid transfer protein in spinach leaves, *Biochem. J.* **194**:257–264.

Nichols, J. W., 1988, Kinetics of fluorescent-labeled phosphatidylcholine transfer between non-specific lipid-transfer protein and phospholipid vesicles, *Biochemistry* 27:1889–1896.

Nishida, I., and Yamada, M., 1986, Semisynthesis of a spin-labeled monogalactosyldiacylglycerol and its application in the assay for galactolipid-transfer activity in spinach leaves, *Biochim. Biophys. Acta* **813**:298–306.

Noland, B. J., Arebalo, R. E., Hansbury, E., and Scallen, T. E., 1980, Purification and properties of sterol carrier protein, *J. Biol. Chem.* **255**:4282–4289.

Ohnishi, J. I., and Yamada, M., 1982, Glycerolipid synthesis in *Avena* leaves during greening of etiolated seedlings, *Plant Cell Physiol.* **23**:767–773.

Oursel, A., Escoffier, A., Kader, J. C., Dubacq, J. P., and Trémolières, A., 1987, Last step in the cooperative pathway for galactolipid synthesis in spinach leaves: Formation of monogalactosyldiacylglycerol with C18 polyunsaturated acyl groups at both carbon atoms of the glycerol, *FEBS Lett.* **219**:393–399.

Pastuszyn, A., Noland, B. J., Bazan, J. F., Fletterick, R. J., and Scallen, T. J., 1987, Primary sequence and structural analysis of sterol carrier protein 2 from rat liver: Homology with immunoglobulins, *J. Biol. Chem.* **262**:1319–13227.

Rickers, J., Tober, I., and Spener, F., 1984, Purification and binding characteristics of a basic fatty acid binding protein from Avena sativa seedlings, *Biochim. Biophys. Acta* **794**:313–319.

Rickers, J., Spener, F., and Kader, J. C., 1985, A phospholipid transfer protein that binds long-chain fatty acids, *FEBS Lett.* **180**:29–32.

Roughan, P. G., 1987, Th control of fatty acid compositions of plant glycerolipids, in *The Metabolism, Structure, and Function of Plant Lipids* (P. K. Stumpf, J. B. Mudd, and W. D. Nes, eds.), pp. 247–254, Plenum Press, New York.

Sasaki, T., 1985, Glycolipid binding proteins, *Chem. Phys. Lipids* **38**:63–78.

Schwitzguebel, J. P., Nguyen, T. D., and Siegenthaler, P. A., 1984, Are phospholipid exchange proteins present in the stroma from higher plant chloroplasts? in *Structure, Function and Metabolism of Plant Lipids* (P. A. Siegenthaler and W. Eichenberger, eds.), pp. 299–302, Elsevier Biomedical Press, Amsterdam.

Sommarin, M., and Sandelius, A. S., 1988, Phosphatidylinositol and phosphatidylinositolphosphate kinases in plant plasma membranes, *Biochim. Biophys. Acta* **958**:268–278.

Takishima, K., Watanabe, S., Yamada, M., and Mamiya, G., 1986, The amino acid sequence of the nonspecific lipid transfer protein from germinated castor bean endosperm, *Biochim Biophys. Acta* **870**:248–250.

Tanaka, T., and Yamada, M., 1979, Phospholipid exchange proteins from castor bean seeds, *Plant Cell Physiol.* **20**:533–542.

Tanaka, T., and Yamada, M., 1982, Properties of phospholipid exchange proteins from germinated castor bean endosperms, in *Biochemistry and Metabolism of Plant Lipids* (J. F. G. M. Wintermans and P. J. C. Kuiper, eds.), pp. 99–106, Elsevier Biomedical Press, Amsterdam.

Tanaka, T., Ohnishi, J.-I., and Yamada, M., 1980, The occurrence of phosphatidylcholine exchange protein in leaves, *Biochem. Biophys. Res. Commun.* **96**:394–399.

Tchang, F., Laroche-Raynal, M., Vergnolle, C., Demandre, C., Douady, D., Grosbois, M., Guerbette, F., Delseny, M., and Kader, J. C., 1985, In vitro synthesis of a plant phospholipid transfer protein. A study by HPLC, *Biochem. Biophys. Res. Commun.* **133**:75–81.

Tchang, F., Guerbette, F., Douady, D., Grosbois, M., Vergnolle, C., Jolliot, A., Dubacq, J. P., and Kader, J. C., 1987, Properties and in vitro synthesis of phospholipid transfer proteins, in *The Metabolism, Structure, and Function of Plant Lipids* (P. K. Stumpf, J. B. Mudd, and W. D. Nes, eds.), pp. 353–356, Plenum Press, New York.

Teerlink, T., Van der Krift, T. P., Van Heusden, G. P. H., and Wirtz, K. W. A., 1984, Determination of nonspecific lipid transfer protein in rat tissues and Morris hepatomas by enzyme immunoassay, *Biochim. Biophys. Acta* **793**:251–259.

Trzeciak, W. H., Simpson, E. R., Scallen, T. J., Vahouny, G. V., and Waterman, M. R., 1987, Synthesis of sterol carrier protein 2 in rat adrenal cells, *J. Biol. Chem.* **262**:3713–3717.

Van Den Besselaar, A. M. H. P., Helmkamp, G. M., and Wirtz, K. W. A., 1975, Kinetic model of the protein-mediated phosphatidylcholine exchange between single bilayer liposomes, *Biochemistry* **14**:1852–1858.

Van der Krift, T. P., Leunissen, J., Teerlink, T., Van Heusden, G. P. H., Verkleij, A. J., and Wirtz, K. W. A. 1985, Ultrastructural localization of a peroxisomal protein in rat liver using the specific antibody against the non-specific lipid transfer protein (sterol carrier protein 2), *Biochim. Biophys. Acta* **812**:387–392.

Van Paridon, P. A., Visser, A. J. W. G., and Wirtz, K. W. A., 1987, On the relationship between the dual specificity of the bovine brain phosphatidylinositol transfer protein and membrane phosphatidylinositol levels, *Biochim. Biophys. Acta* **898**:172–180.

Watanabe, S., and Yamada, M., 1986, Purification and characterization of a non-specific lipid transfer protein from germinated castor bean endosperms which transfers phospholipids and galactolipids, *Biochim. Biophys. Acta* **876**:116–123.

Westerman, J., and Wirtz, K. W. A., 1985, The primary structure of bovine liver nonspecific phospholipid transfer protein, *Biochem. Biophys. Res. Commun.* **127**:333–338.

Wirtz, K. W. A., 1982, Phospholipid transfer proteins, in *Lipid-Protein Interactions* (P. C. Jost and O. H. Griffith, eds.), pp. 151–233, Elsevier Biomedical Press, Amsterdam.

Wirtz, K. W. A., and Zilversmit, D. B., 1969, Participation of soluble liver proteins in the exchange of membrane phospholipids, *Biochim. Biophys. Acta* **187**:468–476.

Wojtczak, L., Baranska, J., Zborowski, J., and Drahota, Z., 1971, Exchange of phospholipids between microsomes and outer and inner membranes, *Biochim. Biophys. Acta* **249**:41–52.

Yamada, M., Tanaka, T., Kader, J. C., and Mazliak, P., 1978, Transfer of phospholipids from microsomes to mitochondria in germinating castor bean endosperm, *Plant Cell Physiol.* **19**:173–176.

Yamada, M., Tanaka, T., and Ohnishi, J.-I., 1980, Phospholipid exchange protein from higher

plants, in *Biogenesis and Function of Plant Lipids* (P. Mazliak, P., Benveniste, C. Costes, and R. Douce, eds.), pp. 161–167, Elsevier Biomedical Press, Amsterdam.
Yamada, M., Tsuboi, S., Ozeki, Y., and Takishima, K., 1989, Isoforms of nonspecific lipid transfer protein from germinated castor bean seeds, in *Metabolism and Biological Function of Plant Lipids* (P. Biacs and L. Vigh, eds.), Elsevier Biomedical Press, Amsterdam, in press.
Zilversmit, D. B., and Hughes, M. E., 1976, Phospholipids exchange between membranes, in *Methods in Membrane Biology* (E. D. Korn, ed.), pp. 211–259, Plenum Press, New York.

Chapter 5

Glycolipid Transfer Protein in Animal Cells

Terukatsu Sasaki, Akira Abe, and Frank Roerink

1. PERSPECTIVES AND SUMMARY

The membrane-free cytosol of pig brain contains glycolipid transfer protein (GL-TP), which *in vitro* facilitates the transfer of various glycosphingolipids and glyceroglycolipids between membranes. This protein has been purified to homogeneity from this source. GL-TP from pig brain transfers glycolipids with a β-glucosyl or β-galactosyl residue directly linked to either ceramide or diacylglycerol. Antisera raised against pig GL-TP in rabbits are cross-reactive with rat, mouse, and human GL-TP, GL-TP appears to be present in many animal cells, since it has been found in all cell lines of rat, mouse, pig, and human origins examined so far for the protein. GL-TP has been partially purified from bovine brain and spleen, rat brain and liver, and human spleen. Evidence has been obtained indicating that GL-TP is a cytoplasmic protein.

GL-TP from pig brain functions as a specific carrier of glycolipids, most probably by forming a 1 : 1 molecular complex. However, glycolipid binding

Abbreviations used in this chapter: GL-TP, glycolipid transfer protein; GalCer, galactosylceramide; PyrGalCer (pyrene-labeled galactosylceramide), 1-*O*-(β-galactosyl)-*N*-[10-(1-pyrenyl)-decanoyl]-sphingosine; SDS, sodium dodecyl sulfate; PAGE, polyacrylamide gel electrophoresis; DIDS, 4,4′-diisothiocyano-2,2′-stilbenedisulfonic acid.

Terukatsu Sasaki, Akira Abe, and Frank Roerink Department of Biochemistry, Cancer Research Institute, Sapporo Medical College, Sapporo 060, Japan experiments indicate that purified GL-TP binds at most 0.13 mol of glycolipid per mol of the protein. Under conditions where one of the membranes lacks glycolipids, GL-TP catalyzes a net transfer reaction. It appears to be a distinctive property of GL-TP that it exists both with and without a bound glycolipid. GL-TP without a bound glycolipid represents a state of the protein after net mass transfer of a glycolipid molecule to an acceptor membrane.

Nothing is known about the function of GL-TP. The pH dependence of glycolipid transfer facilitated by GL-TP indicates that the protein is active at neutral pH but not at or below pH 4. The glycolipid transfer facilitated by GM_2-activator, a cofactor of lysosomal *N*-acetyl-β-hexosaminidase A, has a pH optimum at pH 4.5–5, which is consistent with the presumed functional role of the activator protein in the lysosomes.

Current evidence indicates that glycolipid glycosylations occur at the luminal side of the Golgi apparatus, except for the glucosylation of ceramide, which occurs at the cytoplasmic side of the Golgi apparatus. Therefore, it appears most likely that GL-TP participates in the intracellular traffic of glucosylceramide.

2. GLYCOLIPID TRANSFER PROTEIN

2.1. Purification and Properties

Metz and Radin (1980) found and characterized glucosylceramide transfer activity in bovine spleen cytosol which facilitates the transfer of glucosylceramide from liposomes to erythrocytes or to erythrocyte ghosts. These workers partially purified the active protein, which they called cerebroside transfer protein: 3 mg of partially purified protein was obtained from 1.28 kg of bovine spleen pulp at a yield of 2% (Metz and Radin, 1982; Radin and Metz, 1983). The purified preparation obtained by these worker contained four to seven protein bands upon analysis by sodium dodecyl sulfate-polyacrylamide gel electrophoresis (SDS-PAGE). The molecular weight of cerebroside transfer protein, which is probably identical with GL-TP described below, was estimated to be 20,300 by chromatography on Sephadex G-75. The isoelectric point was estimated to be 9.1.

Abe *et al.* (1982) and Abe and Sasaki (1985) purified GL-TP to homogeneity from pig brain: 2.6 mg of GL-TP was obtained from 4 kg of pig brain at a yield of 19%. The transfer activity was assayed by galactosylceramide (GalCer) transfer from donor to acceptor liposomes (Yamada and Sasaki, 1982a; Abe *et al.*, 1984). A seven-step procedure was used in the purification. The final preparation was about 6000-fold purified as compared with the membrane-free supernatant and had a specific activity of 40.6 nmol of GalCer transferred per min per mg of protein under the conditions of $[^3H]$GalCer transfer assay (Abe and Sasaki, 1985) and of 140 nmol of pyrene-labeled GalCer (PyrGalCer) transferred per min per mg of protein under the conditions of the PyrGalCer transfer assay (Abe *et al.*, 1984). The following four lines of evidence indicate the homogeneity of the purified protein: (1) a single band in acidic gel electrophoresis and in SDS-PAGE, (2) a single band in analytical gel isoelectric focusing, (3) exact correspondence between the GL-TP activity and stained protein absorbance in the acidic gel electrophoresis, and (4) coincidence between the transfer activity and protein absorbance at 280 nm in phosphocellulose chromatography and in gel filtration through Ultrogel AcA-54. The protein has an isoelectric point of about 8.3. A molecular weight of 22,000 has been estimated from its behavior on SDS-PAGE. A molecular weight of 15,000 was calculated from AcA-54 (PharmaciaLKB) gel filtration. It appears that the elution of GL-TP from AcA-54 column is delayed owing to a hydrophobic interaction of the protein with the gel matrix. GL-TP is high in aromatic and nonpolar amino acid residues. The amino acid composition of GL-TP (Abe and Sasaki, 1985) is obviously different from those of phosphatidylcholine transfer proteins from bovine liver (Akeroyd *et al.*, 1981) and from rat liver (Poorthuis *et al.*, 1980), nonspecific lipid transfer proteins from bovine liver (Crain and Zilversmit, 1980) and from rat liver (Poorthuis *et al.*, 1981), and GM_2 activator from human liver (Li *et al.*, 1981). No hexosamine was detected in GL-TP by the amino acid analysis. GL-TP separated by SDS-PAGE was not stained by the periodate-Schiff method. The purified GL-TP is a mixture of two forms of a protein with (about 15%) and without (about 80%) an intramolecular disulfide bond. The protein has one additional free sulfhydryl group reactive with 5,5′-dithiobis(2-nitrobenzoic acid) and *N*-ethylmaleimide under nondenaturing conditions. Its activity is inhibited partly (about 60%) by 10 mM *N*-ethylmaleimide and completely by 2.5 mM mercuric chloride. The activity of cerebroside transfer protein from bovine spleen also decreases upon treatment of the protein with *N*-ethylmaleimide (Metz and Radin, 1980).

The GL-TP in pig brain was eluted in the major and minor peaks from the phosphocellulose column used in the final step of the purification procedure (Abe and Sasaki, 1985). No difference has been found so far in the properties of GL-TP in the major and minor peaks. The results described in this chapter have been obtained by the use of GL-TP in the major peak, which contains the two forms of GL-TP with and without an intramolecular disulfide bond. In the purification of cerebroside transfer protein from bovine spleen, Metz and Radin (1982) found that the transfer activity eluted in two or three peaks from the CM-cellulose column.

GL-TP has been partially purified from bovine brain (Wong *et al.*, 1984; Brown *et al.*, 1985; Gammon *et al.*, 1987). A molecular weight of 18,200 was estimated for GL-TP from bovine brain by gel filtration through a Sephadex G-75 column (Wong *et al.*, 1984). GL-TP from bovine brain was separated into two peaks of activity by chromatography on a Mono S cation exchange column: the molecular weight of GL-TP in the major peak was estimated to be about 20,000 by SDS-PAGE (Gammon *et al.*, 1987). GL-TP has also been partially purified from rat brain (Yamada and Sasaki, 1982a); rat liver (Yamada and Sasaki, 1982b); human spleen; and several cell lines of rat, mouse, pig, and human origins. The membrane-free supernatant of rat brain contained about 10-fold-higher specific activity of GL-TP than did the same amount of rat liver supernatant (Yamada and Sasaki, 1982a,b).

2.2. Intracellular Location

The intracellular location of GL-TP in PK(15) cells, a pig kidney cell line, was studied by the immunochemical method (F. Roerink, A. Abe, and T. Sasaki, manuscript in preparation). Cultured PK(15) cells were disrupted by N_2 cavitation, and subcellular fractions were obtained by differential pelleting. Acidic *N*-acetyl-β-hexosaminidase, a lysosomal marker, remained latent after the N_2 cavitation. GL-TP in each subcellular fraction was released by sonication. Almost all the activity of GL-TP was recovered in the cytosol fraction. GL-TP in each fraction was concentrated by phosphocellulose chromatography and separated by SDS-PAGE. Separated GL-TP on the gel was transferred to a nitrocellulose membrane and located by immunochemical staining with an affinity-purified anti-GL-TP antibody. Most of the protein reactive with the anti-GL-TP was found in the cytosol fraction, indicating a cytoplasmic location of GL-TP.

2.3. Characteristics and Specificity of Glycolipid Transfer Facilitated by GL-TP

The pH dependence of glycolipid transfer facilitated by GL-TP from pig brain was compared with that facilitated by GM_2 activator from human liver, a cofactor of lysosomal *N*-acetyl-β-hexosaminidase A, in an assay of PyrGalCer transfer from donor to acceptor vesicles. GL-TP facilitated PyrGalCer transfer at a pH range from 5 to 8 but did not facilitate the transfer at or below pH 4. GM_2 activator facilitated PyrGalCer transfer at pH 4; the transfer activity at neutral pH was lower than the activity at pH 4.5–5. These results are consistent with those reported by Conzelmann *et al.* (1982), who showed that the rate of GM_2 activator-facilitated ganglioside transfer at pH 7.1 was 15% of the rate of that found at pH 4.2. The glycolipid transfer activity of GM_2 activator has been reviewed by Sasaki (1985).

GL-TP from pig brain facilitates the transfer of all glycosphingolipids examined so far (Yamada *et al.*, 1985). These include glucosylceramide, GalCer, lactosylceramide, sulfatide, lactosylceramide-^{3}H-sulfate, globotriaosylceramide, globotetraosylceramide, globopentaosylceramide, sialosyllactosylceramide, and GM_1 ganglioside. The transfer of Manβ1 → 4Glcβ1 → ceramide and Manα1 →

4Manβ1 → 4Glcβ1 → ceramide is also facilitated by GL-TP from pig brain (Yamada *et al.*, 1986). In addition to glycosphingolipids, GL-TP from pig brain facilitates the transfer of 3-[Gal%β1]-*sn*-1,2-diacylglycerol, 3-[Galα1 → 6Galβ1]-*sn*-1,2-diacylglycerol, and 3-[Glcβ1]-*rac*-1,2-dipalmitylglycerol (Yamada *et al.*, 1985, 1986). The protein does not facilitate the transfer of 3-[Manαl → 3Manαl]-*sn*-1,2-diacylglycerol, 3-[Glcα1]-*sn*-1,2-diacylglycerol, phosphatidylcholine, phosphatidylinositol, phosphatidylethanolamine, cholesterol, or cholesteryl oleate (Yamada *et al.*, 1985). GL-TP from pig brain stimulates the transfer of pyrene-labeled sphingomyelin to a very small but significant extent. These results were interpreted as indicating that GL-TP from pig brain transfers glycolipids with a β-glucosyl or β-galactosyl residue directly linked to either ceramide or diacylglycerol. It is obvious that the specificity of GL-TP is not directed to the nonreducing part of glycolipids and that GL-TP has low specificity for the nonpolar portion of glycolipids because the protein facilitates the transfer of both glycosphingolipids and glyceroglycolipids.

Metz and Radin (1982) found that cerebroside transfer protein from bovine spleen facilitates the transfer of GalCer, glucosylceramide, lactosylceramide, and globotriaosylceramide from liposomes to erythrocyte ghosts. They also reported that the protein does not facilitates the transfer of phosphatidylcholine, cholesterol, ceramide, or globotetraosylceramide. However, it appears that the assay method used by these workers was not sensitive enough to measure the small amount of globotetraosylceramide transfer. Brown *et al.* (1985) reported that the transfer of glucosylceramide, lactosylceramide, asialo-GM_1 ganglioside, and GM_1 ganglioside between phospholipid vesicles were facilitated at comparable rates by GL-TP from bovine brain. Under the same conditions, the protein did not facilitate the transfer of phosphatidylcholine, whereas the transfer of sphingomyelin was facilitated by the protein to a very small extent.

It has been shown that GL-TP facilitates the transfer of glycolipids only when both the donor and acceptor membranes are in the liquid-crystalline state (Wong *et al.*, 1984). If either donor or acceptor membranes are in the gel state, GL-TP-mediated transfer is markedly reduced. Furthermore, the rate of glycolipid transfer facilitated by GL-TP is markedly affected by the curvature of the donor and acceptor membranes. A strong preference was found for sonicated unilamellar vesicles over multilamellar liposomes in the GL-TP-mediated transfer (Brown *et al.*, 1985), indicating that the interaction of GL-TP with membranes is severely affected by the packing of lipids in the membranes.

2.4. Binding of GalCer to GL-TP

Binding of a glycolipid to GL-TP has been shown by three different methods. In the first method, a complex formation between [^{3}H]GalCer and GL-TP was shown by PAGE at pH 4.3 (Abe and Sasaki, 1985). GL-TP was incubated with liposomes containing [^{3}H]GalCer, and the mixture was analyzed by PAGE.

^3H radioactivity comigrating with GL-TP was found on the gel after the electrophoresis. About 30 pmol of [^3H]GalCer was located at the position of GL-TP when the incubation mixture containing 230 pmol of the protein was analyzed. The results indicate that about 0.13 mol of [^3H]GalCer was bound per mol of GL-TP under conditions in which the molar ratio of [^3H]GalCer to GL-TP in the incubation mixture was 6.5 : 1. Addition of 4.2 mM *N*-ethylmaleimide to the incubation mixture reduced the binding of [^3H]GalCer to GL-TP to 65% of the control level. When liposomes containing the [^3H]phosphatidylcholine were incubated with GL-TP, no ^3H label was found at the position of GL-TP on analysis of the incubation mixture of PAGE.

In the second method, the formation of GL-TP–glycolipid complex was shown by the binding of PyrGalCer to GL-TP. The binding was measured by an increase in pyrene monomer fluorescence upon addition of GL-TP to phospholipid vesicles containing PyrGalCer. In nonquenched vesicle assay, an immediate drop in the intensity ratio of eximer to excited monomer was observed upon addition of GL-TP to the vesicles. In the quenched vesicle assay, in which the phospholipid vesicles contained *N*-trinitrophenylphosphatidylethanolamine in addition to PyrGalCer to quench the fluorescence due to PyrGalCer present in the vesicles, an increase in excited monomer fluorescence was observed upon addition of GL-TP to the vesicles. These results indicate that GL-TP enhances a release of PyrGalCer monomer from the vesicles, and the results are interpreted as demonstrating the binding of PyrGalCer monomer to GL-TP. Brown *et al.* (1985) also found an increase in monomer fluorescence upon addition of bovine brain GL-TP to phospholipid vesicles containing PyrGalCer. Moreover, these workers showed that such an increase in monomer fluorescence did not occur when pyrene-labeled phosphatidylcholine was used in place of PyrGalCer. In the third method, the GL-TP–[^3H]GalCer complex was isolated by Sephadex G-75 gel filtration, as described in Section 2.5.

The GL-TP–glycolipid complex is formed as a result of removal of a glycolipid molecule from a membrane to a GL-TP molecule. It appears that local perturbation of the structure of the lipid layer is induced by the protein and that this perturbation is an essential step in the glycolipid extraction by the protein. The specificity of GL-TP is consistent with a simple mechanism for glycolipid extraction by GL-TP because the specificity suggests that GL-TP recognizes and binds the sugar residue directly linked to either ceramide or diacylglycerol in addition to the nonpolar portion of glycolipids.

2.5. Isolation of GL-TP–[^3H]GalCer Complex and Transfer of [^3H]GalCer from the Complex to Acceptor Liposomes

The complex between GL-TP and [^3H]GalCer was isolated by Sephadex G-75 gel filtration from the incubation mixture of the protein and liposomes containing [^3H]GalCer (Abe and Sasaki, 1985). The isolated [^3H]GalCer–GL-TP complex contained about 0.12 mol of [^3H]GalCer per mole of GL-TP.

An incubation of the GL-TP–[^{3}H]GalCer complex with liposomes resulted in the transfer of [^{3}H]GalCer from the complex to the acceptor liposomes (Abe and Sasaki, 1985). When the complex was incubated with liposomes containing about 300-fold more GalCer than the [^{3}H]GalCer in the complex, almost all the [^{3}H]GalCer originally present in the complex was transferred to the liposomes during the incubation. The result suggests an extensive mixing of the GL-TP-bound [^{3}H]GalCer with the liposomal GalCer. The result therefore indicates that the GL-TP–glycolipid complex functions as the intermediate in the glycolipid transfer. When the GL-TP–[^{3}H]GalCer complex was incubated with liposomes containing no glycolipid, about half of the [^{3}H]GalCer remained as the GL-TP–[^{3}H]GalCer complex even after 1 h of incubation at 20°C.

Formation of a stable complex between GL-TP and phosphatidylcholine vesicles with and without glucosylceramide was reported by Wong *et al.* (1984) and Metz and Radin (1982) when a GL-TP preparation of low purity was used. Brown *et al.* (1985) were unable to produce such a stable complex when using a GL-TP preparation of higher purity.

2.6. Net Mass Transfer of Glycolipids Facilitated by GL-TP

GL-TP facilitates GalCer transfer from donor liposomes to acceptor liposomes with and without glycolipids. Since liposomes prepared entirely from nontransferable lipids can function as excellent acceptor membranes, it was assumed that the protein most probably facilitates a net mass transfer of glycolipids from the donor to the acceptor membranes. The net mass transfer of [^{14}C]GalCer facilitated by GL-TP was shown by the use of a monomolecular lipid film spread at the air–water interface (Sasaki and Demel, 1985).

It appears that GL-TP binds 1 mol of glycolipid per mol of the protein. However, it has not been possible to saturate the purified GL-TP with glycolipids under several different conditions examined; at most, 0.13 mol of glycolipids bound to 1 mol of GL-TP in the binding experiments. In net mass transfer of a glycolipid molecule from donor to acceptor membranes, GL-TP carrying a glycolipid monomer unloads the bound glycolipid onto an acceptor membrane. Therefore, it appears that the GL-TP without a bound lipid represents a state of the protein after net mass transfer of a glycolipid molecule to an acceptor membrane. It appears to be a distinctive property of GL-TP that the protein exists both with and without a bound glycolipid.

3. POSSIBLE ROLE OF GLYCOLIPID TRANSFER PROTEIN IN GLYCOSPHINGOLIPID METABOLISM

It is important to consider briefly the subcellular location of glycolipid biosynthesis and the topography of glycolipid glycosylation before discussing the role of GL-TP.

3.1. Intracellular Location of Glycosphingolipids

The prevailing view has been that glycosphingolipids are localized predominantly, if not exclusively, in the outer leaflet of the plasma membrane. Miller-Podraza *et al.* (1982) determined the localization of gangliosides in cultured cell lines. Cell-surface-located gangliosides were distinguished from intracellular gangliosides by neuraminidase treatment, by oxidation with periodate followed by reduction with sodium boro[^{3}H]hydride, or by binding of cholera toxin to GM_1 ganglioside. The results obtained by these workers indicate that the major pool of cellular gangliosides is on the cell surface.

3.2. Intracellular Location of Enzymes of Glycosphingolipid Biosynthesis

It is assumed that the endoplasmic reticulum is the site of ceramide biosynthesis, since fatty acid additions generally occur on the endoplasmic reticulum. It is now known that the glycosylation of glycosphingolipids by the sequential addition of monosaccharides from sugar nucleotides to an acceptor is catalyzed by glycosyltransferases located in the Golgi membrane (Keenan *et al.*, 1974; Wilkinson *et al.*, 1976; Richardson *et al.*, 1977; Pacuszka *et al.*, 1978; Siegrist *et al.*, 1979; Eppler *et al.*, 1980). Although the localization of glycolipid glycosyltransferases to specific cisternae in discrete parts of the Golgi apparatus has not been tested, one might expect, by analogy with glycoprotein glycosyltransferases (Roth and Berger, 1982; Dunphy *et al.*, 1985), that such a localization does occur. The sequential glycosylation of one glycosphingolipid to the next higher homolog involves a very small pool of intermediates, which does not mix with the main pool of cellular glycosphingolipids (Miller-Podraza *et al.*, 1982). Current evidence indicates that the small pool of intermediates locates at the Golgi apparatus.

3.3. Topography of Glycolipid Glycosylation in the Golgi Apparatus

Our understanding of the topography of glycolipid glycosylation in the Golgi apparatus is limited. Yusuf *et al.* (1983a,b, 1984) showed that tunicamycin inhibits the synthesis of GM_1 and GM_2 gangliosides in isolated Golgi vesicles; this inhibition was found to be due to a block in carrier-mediated transport of nucleotide sugars across the Golgi vesicles, consistent with the luminal orientation of glycosyltransferases involved in GM_1 and GM_2 synthesis.

Evidence strongly suggesting that the glycosylation of glycolipids occurs in the lumen of the Golgi apparatus has been obtained from analyses of glycosylation mutants of Chinese hamster ovary cells. Mutant cell lines of one complementation group (Stanley, 1983) (Lec2 and Clone 1021) have a 90% reduction in the sialylation of both glycoproteins and glycolipids compared with wild-type cells. Sialosyl-lactosylceramide (GM_3 ganglioside) is the major glycolipid in parent cells, whereas Lec2 and Clone 1021 cells have lactosylceramide as the predominant glycolipid (Briles *et al.*, 1977; Stanley, 1980). The Lec2 cells were found to be deficient in the reaction to translocate CMP-sialic acid across Golgi vesicle membranes from an external compartment (Deutscher *et al.*, 1984). Other biochemical analyses strongly suggest that this translocation deficiency is the primary defect responsible for the mutant phenotype (Briles *et al.*, 1977; Deutscher *et al.*, 1984).

Mutant cells of the second complementation group (Stanley, 1983) (Lec8 and Clone 13) show an 80–90% reduction in both galactosylation and sialylation of their glycoproteins and glycolipids when compared with wild-type cells (Briles *et al.*, 1977; Stanley, 1980, 1985). Clone 13 cells have glucosylceramide as the major glycolipid (Briles *et al.*, 1977). The primary biochemical defect in Lec8 and Clone 13 cells was found to be their inability to translocate UDP-galactose into the lumen of the Golgi apparatus (Deutscher and Hirschberg, 1986). These mutant cells of both complementation groups were found to possess the appropriate nucleotide sugars, glycoprotein and glycolipid acceptors, and glycosyltransferases (Briles *et al.*, 1977).

These results provided strong evidence for the orientation of the catalytic sites of glucosylceramide galactosyltransferase and lactosylceramide sialyltransferase toward the lumen of the Golgi apparatus, since mutants that are unable to translocate UDP-galactose and CMP-sialic acid into the lumen of the Golgi apparatus are defective in the syntheses of lactosylceramide and GM_3 ganglioside.

3.4. Topography of Glucosylceramide Synthesis in the Golgi Apparatus

The topography of ceramide glucosyltransferase and *de novo*-synthesized glucosylceramide has been examined on the Golgi vesicles from porcine submaxillary glands (Coste *et al.*, 1986). UDP-glucose-ceramide glucosyltransferase of this tissue is associated with membranes of the Golgi apparatus (Costa *et al.*, 1985). Two lines of evidence support the hypothesis that ceramide glucosyltransferase is an enzyme whose catalytic site faces the cytoplasmic side of the Golgi membrane.

First, ceramide glucosyltransferase was found to have protease-sensitive sites facing the cytoplasmic side of the Golgi membrane. Vesicles isolated from the Golgi membrane that are sealed that have "right-side-out" orientation (cytoplasmic side out) were treated with either pronase or trypsin. Ceramide glucosyltransferase was inactivated by nearly 100% following treatment of the Golgi vesicles with these proteases. Under the same conditions, ovomucoid-β-D-galactosyltransferase, a protein known to face the lumen of the Golgi apparatus, was found to be protease resistant in intact vesicles and protease sensitive in disrupted vesicles.

Second, the synthesis of glucosylceramide was blocked when intact vesicles were treated with the stilbene derivative DIDS, which seems to interact with the anion-binding sites of glycosyltransferases (Spiro and Spiro, 1985). Under the same conditions, ovomucoid-β-D-galactosyltransferase was resistant to DIDS in intact vesicles and sensitive to DIDS in disrupted vesicles. Since DIDS has been shown to be a membrane-impermeable agent (Cabantchik *et al.*, 1978; Wolosin, 1980; Capasso and Hirschberg, 1984), it is likely that a key part of the catalytic site of ceramide glucosyltransferase faces the cytoplasmic side of the Golgi membrane.

Furthermore, evidence has been obtained (Coste *et al.*, 1986) in support of the cytoplasmic orientation of the newly synthesized glucosylceramide. Using glucosylceramidase from human placenta to probe the orientation of this compound, these workers found that the glucosylceramide synthesized in the isolated Golgi vesicles is equally sensitive in intact and disrupted vesicles to hydrolysis by the enzyme, suggesting that the newly synthesized glucosylceramide is oriented toward the cytoplasmic face of the Golgi membrane.

From these results, it appears that cytoplasmically oriented glucosylceramide is somehow utilized as the acceptor by glucosylceramide galactosyltransferase, which seems to add galactose to the acceptor from a luminal pool of UDP-galactose. How the transmembrane movement of glucosylceramide occurs is not known. The movement could be part of the reaction of glucosylceramide galactosyltransferase or could be due to protein-mediated glucosylceramide flip-flop.

3.5. Site of Glucosylation of Ceramide in the Golgi Apparatus: Studies with Monensin

Through the study of the effect of monensin on glycosphingolipid metabolism in cultured cells, a great deal of evidence has been obtained in support of the hypothesis that the initial glucosylation of ceramide and the subsequent elongation of the oligosaccharide chain of glycosphingolipids occur at different sites within the cells. Saito *et al.* (1984) found that exposure of human fibroblasts to monensin, an ionophore for monovalent cations, caused a marked accumulation of glucosylceramide and, to a much smaller extent, lactosylceramide. The cellular content of more highly glycosylated neutral glycosphingolipids and gangliosides remained the same in the presence or absence of monensin. Monensin greatly increased the incorporation of labeled glucose and galactose into glucosylceramide (and lactosylceramide) and, at the same time, reduced to incorporation into higher glycosylated glycosphingolipids. Since Gaucher's disease fibroblasts, with genetically impared lysosomal glucocerebrosidase activity, accumulated even more glucosylceramide and lactosylceramide than normal cells upon exposure to monensin, Saito *et al.* (1984) concluded that monensin affects biosynthesis rather than merely disrupting lysosomal glycosphingolipid catabolism.

Miller-Prodraza and Fishman (1984) also reported the effect of monensin on the biosynthesis and levels of glycosphingolipids in rat glioma cells and in mouse neuroblastoma cells. In monensin-treated cells, the labeling of glucosylceramide increased severalfold, whereas the labeling of more complex neutral glycosphingolipids and gangliosides decreased; progressively more inhibition of the synthesis was found in more highly glycosylated lipids. An increase in the level of glucosylceramide was observed in the monensin-treated cells, whereas no change was observed in the level of GM_3 ganglioside.

It has been shown that monensin inhibits intra-Golgi transport of glycoproteins, which results in the accumulation of membrane and secretary protein precursors in middle or *trans* Golgi cisternae and in the blocking of terminal glycosylation of preventing exposure of glycoproteins to glycosyltransferases located at *trans* regions of the Golgi apparatus (Farquhar, 1985; Elbein, 1987). Lipsky and Pagano (1985) showed that monensin markedly inhibits the translocation of glucoslyceramide from the Golgi apparatus to the plasma membrane. These workers used a fluorescent analogue of ceramide in a fluorescence microscopy studies of the intracellular synthesis and transport of sphingomyelin and glucosylceramide in Chinese hamster fibroblasts (Lipsky and Pagano, 1983). Treatment of the cells with monensin inhibited the appearance of intracellularly synthesized glucosylceramide analogue at the plasma membrane, resulting in a marked accumulation of the analogue in the Golgi apparatus.

Taken together, these results suggest that glucosylation of ceramide occurs in the proximal face of the Golgi apparatus, whereas higher glycosylation of glycosphingolipids occurs in the distal face of the Golgi apparatus. It seems that monensin stimulates glucosylceramide synthesis, in addition to an accumulation of this lipid caused by the inhibition of the synthesis of more complex glycosphingolipids.

3.6. Possible Role of GL-TP in Intracellular Traffic of Glucosylceramide

Very little is known about the mechanism of glycosphingolipid transport from the site of synthesis to the plasma membrane. Dower *et al.* (1982) examined the kinetics of ganglioside transport from an intracellular site of synthesis to the plasma membrane in cultured cells. These workers distinguished surface and intracellular gangliosides by oxidizing cell surface gangliosides with sodium periodate and by reacting the oxidized gangliosides with dinitrophenylhydrazine. They found that the transfer of gangliosides from the site of synthesis to the cell surface required approximately 20 min. A variety of drugs, including inhibitors of protein synthesis and energy metabolism, cytoskeleton modulators, and monensin, had no effect on the transport of newly synthesized GD_{1a} ganglioside to the plasma membrane (Miller-Prodraza and Fishman, 1984). Only low temperature effectively blocked the translocation. It appears that the vesicular transport mechanism (Morré *et al.*, 1979; Farquhar, 1985) is the most likely one for the transport of glycosphingolipids from the site of synthesis to the plasma membrane (Miller-Prodraza and Fishman, 1984). According to the vesicular transport model, glycosphingolipids are transported to the cell surface via shuttling vesicles, which bud off from one membrane and fuse with another. In vesicular transport, glycosphingolipids always have a luminal orientation and are always inaccessible from the cytosol. Available data indicate that GL-TP is a cytosolic protein. Glucosylceramide is the only glycosphingolipid that seems to be exposed on the cytoplasmic face of the cellular membranes. Therefore, it is most likely that GL-TP participates in the intracellular traffic of glucosylceramide.

4. REFERENCES

Abe, A., and Sasaki, T., 1985, Purification and some properties of the glycolipid transfer protein from pig brain, *J. Biol. Chem.* **260:**11231–11239.

Abe, A., Yamada, K., and Sasaki, T., 1982, A protein purified from pig brain accelerates the intermembranous translocation of mono- and dihexosylceramides, but not the translocation of phospholipids, *Biochem. Biophys. Res. Commun.* **104:**1386–1393.

Abe, A., Yamada, K., Sakagami, T., and Sasaki, T., 1984, A fluorimetric determination of the activity of glycolipid transfer protein and some properties of the protein purified from pig brain, *Biochim. Biophys. Acta* **778:**239–244.

Akeroyd, R., Moonen, P., Westerman, J., Puyk, W. C., and Wirtz, K. W. A., 1981, The complete primary structure of the phosphatidylcholine-transfer protein from bovine liver: Isolation and characterization of the cyanogen bromide peptides, *Eur. J. Biochem.* **114:**385–391.

Briles, E. B., Li, E., and Kornfeld, S., 1977, Isolation of wheat germ agglutinin-resistant clones of Chinese hamster ovary cells deficient in membrane sialic acid and galactose, *J. Biol. Chem.* **252:**1107–1116.

Brown, R. E., Stephenson, F. A., Markello, T., Barenholz, Y., and Thompson, T. E., 1985, Properties of a specific glycolipid transfer protein from bovine brain, *Chem. Phys. Lipids* **38:**79–93.

Cabantchik, Z. I., Knauf, P. A., and Rothstein, A., 1978, The anion transport system of the red blood cell: The role of membrane protein evaluated by the use of 'probes', *Biochim. Biophys. Acta* **515:**239–302.

Capasso, J. M., and Hirschberg, C. B., 1984, Effect of atractylosides, palmitoyl coenzyme A, and anion transport inhibitors on translocation of nucleotide sugars and nucleotide sulfate into Golgi vesicles, *J. Biol. Chem.* **259:**4263–4266.

Conzelmann, E., Burg, J., Stephan, G., and Sandhoff, K., 1982, Complexing of glycolipids and their transfer between membranes by the activator protein for degradation of lysosomal ganglioside GM_2, *Eur. J. Biochem.* **123:**455–464.

Coste, H., Martel, M. B., Azzar, G., and Got, R., 1985, UDPglucose-ceramide glucosyltransferase from porcine submaxillary glands is associated with the Golgi apparatus, *Biochim. Biophys. Acta* **814:**1–7.

Coste, H., Martel, M. B., and Got, R., 1986, Topology of glucosylceramide synthesis in Golgi membranes from porcine submaxillary glands, *Biochim. Biophys. Acta* **858:**6–12.

Crain, R. C., and Zilversmit, D. B., 1980, Two nonspecific phospholipid exchange proteins from beef liver: Purification and characterization, *Biochemistry* **19:**1433–1439.

Deutscher, S. L., and Hirschberg, C. B., 1986, Mechanism of galactosylation in the Golgi apparatus: A Chinese hamster ovary cell mutant deficient in translocation of UDP-galactose across Golgi vesicle membranes, *J. Biol. Chem.* **261:**96–100.

Deutscher, S. L., Nuwayhid, N., Stanley, P., Briles, E. I. B., and Hirschberg, C. B., 1984, Translocation across Golgi vesicle membranes: A CHO glycosylation mutant deficient in CMP-sialic acid transport, *Cell* **39:**295–299.

Dower, S., Miller-Podraza, H., and Fishman, P. H., 1982, Translocation of newly synthesized gangliosides to the cell surface, *Biochemistry* **21:**3265–3270.

Dunphy, W. G., Brands, R., and Rothman, J. E., 1985, Attachment of terminal N-acetylglucosamine to asparagine-linked oligosaccharides occurs in central cisternae of the Golgi stack, *Cell* **40:**463–472.

Elbein, A. D., 1987, Inhibitors of the biosynthesis and processing of N-linked oligosaccharide chains, *Annu. Rev. Biochem.* **56:**497–534.

Eppler, C. M., Morré, D. J., and Keenan, T. W., 1980, Ganglioside biosynthesis in rat liver: Characterization of cytidine-5′-monophospho-*N*-acetylneuraminic acid:hematoside (GM_3) sialyltransferase, *Biochim. Biophys. Acta* **619:**318–331.

Farquhar, M. G., 1985, Progress in unraveling pathways of Golgi traffic, *Annu. Rev. Cell Biol.* **1:**447–488.

Gammon, C. M., Vaswani, K. K., and Ledeen, R. W., 1987, Isolation of two glycolipid transfer proteins from bovine brain: Reactivity toward gangliosides and neutral glycosphingolipids, *Biochemistry* **26:**6239–6243.

Keenan, T. W., Morré, D. J., and Basu, S., 1974, Ganglioside biosynthesis: Concentration of glycosphingolipid glycosyltransferases in Golgi apparatus from rat liver, *J. Biol. Chem.* **249:**310–315.

Li, S.-C., Hirabayashi, Y., and Li, Y.-T., 1981, A protein activator for the enzymic hydrolysis of GM_2 ganglioside, *J. Biol. Chem.* **256:**6234–6240.

Lipsky, N. G., and Pagano, R. E., 1983, Sphingolipid metabolism in cultured fibroblasts: Microscopic and biochemical studies employing a fluorescent ceramide analogue, *Proc. Natl. Acad. Sci. U.S.A.* **80:**2608–2612.

Lipsky, N. G., and Pagano, R. E., 1985, Intracellular translocation of fluorescent sphingolipids in cultured fibroblasts: Endogenously synthesized sphingomyelin and glucocerebroside analogues pass through the Golgi apparatus en route to the plasma membrane, *J. Cell Biol.* **100:**27–34.

Metz, R. J., and Radin, N. S., 1980, Glucosylceramide uptake protein from spleen cytosol, *J. Biol. Chem.* **255:**4463–4467.

Metz, R. J., and Radin, N. S., 1982, Purification and properties of a cerebroside transfer protein, *J. Biol. Chem.* **257:**12901–12907.

Miller-Podraza, H., Bradley, R. M., and Fishman, P. H., 1982, Biosynthesis and localization of gangliosides in cultured cells, *Biochemistry* **21:**3260–3265.

Miller-Prodraza, H., and Fishman, P. H., 1984, Effect of drugs and temperature on biosynthesis and transport of glycosphingolipids in cultured neurotumor cells, *Biochim. Biophys. Acta* **804:**44–51.

Morré, D. J., Kartenbeck, J., and Franke, W. W., 1979, Membrane flow and interconversions among endomembranes, *Biochim. Biophys. Acta* **559:**71–152.

Pacuszka, T., Duffard, R. O., Nishimura, R. N., Brady, R. O., and Fishman, P. H., 1978, Biosynthesis of bovine thyroid gangliosides, *J. Biol. Chem.* **253:**5839–5846.

Poorthuis, B. J. H. M., van der Krift, T. P., Teerlink, T., Akeroyd, R., Hostetler, K. Y., and Wirtz, K. W. A., 1980, Phospholipid transfer activities in Morris hepatomas and the specific contribution of the phosphatidylcholine exchange protein, *Biochim. Biophys. Acta* **600:**376–386.

Poorthuis, B. J. H. M., Glatz, J. F. C., Akeroyd, R., and Wirtz, K. W. A., 1981, A new high-yield procedure for the purification of the nonspecific phospholipid transfer protein from rat liver, *Biochim. Biophys. Acta* **665**:256–261.

Radin, N. S., and Metz, R. J., 1983, Cerebroside transfer protein, *Methods Enzymol.* **98**:613–622.

Richardson, C. L., Keenan, T. W., and Morré, D. J., 1977, Ganglioside biosynthesis: Characterization of CMP-*N*-acetylneuraminic acid:lactosylceramide sialyltransferase in Golgi apparatus from rat liver, *Biochim. Biophys. Acta* **488**:88–96.

Roth, J., and Berger, E. G., 1982, Immunocytochemical localization of galactosyltransferase in HeLa cells: Codistribution with thiamine pyrophosphatase in trans-Golgi cisternae, *J. Cell Biol.* **93**:223–229.

Saito, M., Saito, M., and Rosenberg, A., 1984, Action of monensin, a monovalent cationophore, on cultured human fibroblasts: Evidence that it induces high cellular accumulation of glucosyl- and lactosylceramide (Gluco- and Lactocerebroside), *Biochemistry* **23**:1043–1046.

Sasaki, T., 1985, Glycolipid-binding proteins, *Chem. Phys. Lipids* **38**:63–77.

Sasaki, T., and Demel, R. A., 1985, Net mass transfer of galactosylceramide facilitated by glycolipid transfer protein from pig brain: A monolayer study, *Biochemistry* **24**:1079–1083.

Siegrist, H. P., Burkart, T., Wiesmann, U. N., Herschkowitz, N. N., and Spycher, M. A., 1979, Ceramide-galactosyltransferase and cerebroside-sulphotransferase localisation in Golgi membranes isolated by a continuous sucrose gradient of mouse brain microsomes, *J. Neurochem.* **33**:497–504.

Spiro, M. J., and Spiro, R. G., 1985, Effect of anion-specific inhibitors on the utilization of sugar nucleotides for *N*-linked carbohydrate unit assembly by thyroid endoplasmic reticulum vesicles, *J. Biol. Chem.* **260**:5808–5815.

Stanley, P., 1980, Altered glycolipids of CHO cells resistant to wheat germ agglutinin, *ACS Symp. Ser.* **128**:213–221.

Stanley, P., 1983, Selection of lectin-resistant mutants of animal cells, *Methods Enzymol.* **96**:157–184.

Stanley, P., 1985, Membrane mutants of animal cells: Rapid identification of those with a primary defect in glycosylation, *Mol. Cell. Biol.* **5**:923–929.

Wilkinson, F. E., Morré, D. J., and Keenan, T. W., 1976, Ganglioside bisoynthesis: characterization of uridine diphosphate galactose:GM_2 galactosyltransferase in Golgi apparatus from rat liver, *J. Lipid Res.* **17**:146–153.

Wolosin, J. M., 1980, A procedure for membrane-protein reconstitution and the functional reconstitution of the anion transport system of the human-erythrocyte membrane, *Biochem. J.* **189**:35–44.

Wong, M., Brown, R. E., Barenholz, Y., and Thompson, T. E., 1984, Glycolipid transfer protein from bovine brain, *Biochemistry* **23**:6498–6505.

Yamada, K., and Sasaki, T., 1982a, A rat brain cytosol protein which accelerates the translocation of galactosylceramide, lactosylceramide and glucosylceramide between membranes, *Biochim. Biophys. Acta* **687**:195–203.

Yamada, K., and Sasaki, T., 1982b, Rat liver glycolipid transfer protein: A protein which facilitates the translocation of mono- and dihexosylceramides from donor to acceptor liposomes, *J. Biochem.* **92**:457–464.

Yamada, K., Abe, A., Sasaki, T., 1985, Specificity of the glycolipid transfer protein from pig brain, *J. Biol. Chem.* **260**:4615–4621.

Yamada, K., Abe, A., and Sasaki, T., 1986, Glycolipid transfer protein from pig brain transfers glycolipids with β-linked sugars but not with α-linked sugars at the sugar-lipid linkage, *Biochim. Biophys. Acta* **879**:345–349.

Yusuf, H. K. M., Pohlentz, G., and Sandhoff, K., 1983a, Tunicamycin inhibits ganglioside biosynthesis in rat liver Golgi apparatus by blocking sugar nucleotide transport across the membrane vesicles, *Proc. Natl. Acad. Sci. U.S.A.* **80**:7075–7079.

Yusuf, H. K. M., Pohlentz, G., Schwarzmann, G., and Sandhoff, K., 1983b, Ganglioside biosynthesis in Golgi apparatus of rat liver: Stimulation by phosphatidylglycerol and inhibition by tunicamycin, *Eur. J. Biochem.* **134**:47–54.

Yusuf, H. K. M., Pohlentz, G., and Sandhoff, K., 1984, Ganglioside biosynthesis in Golgi apparatus: New perspectives on its mechanism, *J. Neurosci. Res.* **12**:161–178.

Chapter 6

Transport and Metabolism of Phosphatidylinositol in Eukaryotic Cells

George M. Helmkamp, Jr.

1. INTRODUCTION

Phospholipids are the major lipid component of eukaryotic cell membranes. A diversity of phospholipid structures is reflected in distributions that are unique to specific intracellular membranes, as well as the cytoplasmic and extracytoplasmic surfaces of the membrane bilayer. Matched to the spectrum of phospholipids within the cell are cytosolic phospholipid transfer proteins. As a general class of catalyst, these proteins facilitate the transport of monomeric phospholipid molecules between membrane domains (Kader *et al.*, 1982; Wirtz, 1982; Helmkamp, 1986). Although a number of interesting and often highly suggestive catalytic activities have been demonstrated *in vitro* for phospholipid transfer proteins, it has proved frustratingly difficult to define any *in vivo* function.

Abbreviations used in this chapter: CerPCho, sphingomyelin; LacCer, lactosylceramide; PC, phosphatidylcholine; PE, phosphatidylethanolamine; PG, phosphatidylglycerol; PGP, phosphatidylglycerol phosphate; PI, phosphatidylinositol; PI4P, phosphatidylinositol 4-phosphate; PI(4,5)P_2, phosphatidylinositol 4,5-bisphosphate; PA, phosphatidate; PS, phosphatidylserine; CoA, coenzyme A.

George M. Helmkamp, Jr. Department of Biochemistry and Molecular Biology, School of Medicine, University of Kansas Medical Center, Kansas City, Kansas 66103-8410.

Phosphatidylinositol (PI) transfer protein is a specific class of phospholipid transfer protein. PI transfer proteins are distributed widely in eukaryotic cells and tissues, they share a number of important structural features, and they catalyze selectively the transport of PI and phosphatidylcholine (PC). This chapter is devoted to an examination of the molecular structure and catalytic properties of PI transfer protein, some of which have been discussed previously (Helmkamp, 1985). Considerable emphasis is placed on the various phospholipid transfer activities identified in rats; this serves as a prelude to a description of new information about PI transfer activity and protein in rat tissues. Recent progress with rat brain and testis PI transfer protein is highlighted. Pathways and subcellular localization of PI synthesis, phosphorylation, and degradation in animal cells are reviewed. The chapter concludes with a working hypothesis and supporting evidence for the participation of PI transfer protein in phosphoinositide metabolism in the quiescent and agonist-stimulated eukaryotic cell.

2. STRUCTURAL FEATURES OF PHOSPHATIDYLINOSITOL TRANSFER PROTEINS

2.1. Purification of PI Transfer Proteins from Mammalian Sources

PI transfer proteins have been purified to homogeneity from three mammalian species. Considerable similarity exists among these proteins in cellular concentration, molecular size, electrophoretic mobility, isoform pattern, immunologic reactivity, and catalytic activity. From fresh bovine brain, Helmkamp *et al.* (1974) isolated PI transfer protein with a recovery of activity of 20% and a purification factor of 430–510. Major steps in the purification scheme included anion exchange and molecular sieve chromatography and isoelectric focusing; hydroxylapatite chromatography was used in subsequent purifications (Helmkamp *et al.*, 1976a). Postmitochondrial supernatants of bovine heart homogenates served as the starting material for the isolation of PI transfer protein described by DiCorleto *et al.* (1979). This group used hydrophobic, molecular sieve, and cation exchange chromatography to achieve a 2000- to 2400-fold purification with an overall activity recovery of 5%. George and Helmkamp (1985) applied molecular sieve, cation exchange, and hydroxylapatite chromatography to homogenates of outdated human platelet concentrates and obtained PI transfer protein with a recovery of 9% and a purification factor of 1100. Recently, Venuti and Helmkamp (1988b) isolated PI transfer protein from frozen rat brain at a recovery of 3% and purification factor of 400; they used molecular sieve, anion exchange, and hydroxylapatite chromatography and isoelectric focusing. The principal criterion of purity in all these examples was electrophoretic homogeneity on polyacrylamide gels, in the presence and absence of sodium dodecyl sulfate and 2-mercaptoethanol.

Values of the purification factor for the isolation of PI transfer proteins from mammalian tissues suggest a relatively low abundance in the cell, in the range of 0.05–0.25% of cytosolic protein. PI transfer protein appears to be localized exclusively in the cytosolic compartment, based on the absence of appreciable catalytic activity or immunologic cross-reactivity in any membrane or organelle fraction in rat liver homogenates (S. E. Venuti and G. M. Helmkamp, Jr., unpublished observation). The relatively small differences in tissue purification factor are consistent with the specific activities calculated for unfractionated membrane-free tissue preparations. Thus, for bovine brain and heart, the ratio of PI transfer activity was 1.6 and for rat brain and heart the ratio was 5.4 (Helmkamp *et al.*, 1976a; Venuti and Helmkamp, 1988b). An important assumption in reaching this conclusion is that most, if not all, PI transfer activity in a mammalian tissue can be attributable to PI transfer protein. Indeed, this has been demonstrated for bovine brain, heart, and liver in experiments in which polyclonal rabbit antibodies to PI transfer protein and PC transfer protein were used (Helmkamp *et al.*, 1976a). Nevertheless, without newer means to distinguish among catalytic activities, contributions to PI transfer activity in cell homogenates by other lipid transfer proteins cannot be excluded.

2.2. Molecular Properties

Molecular sieve chromatography and polyacrylamide gel electrophoresis have been used to determine the size of PI transfer protein. The results of these studies have been reviewed (Helmkamp, 1985). The chromatographic technique yielded molecular weights in the range 24,000–28,000 for bovine and human PI transfer protein; the electrophoretic technique gave values in the range 29,000–36,000. A marked discrepancy between the two techniques is obvious and may be attributed an unusual association of PI transfer protein and the cross-linked dextran beads, which retards the elution of protein. Resolution of the discrepancy could be facilitated by using molecular sieve chromatography in the presence of a denaturing concentration of urea or guanidinium chloride or on a support matrix with different chemical characteristics. Fewer artifacts should be anticipated with polyacrylamide gel electrophoresis, and excellent agreement is seen for the molecular mass determination of species and tissue isoforms.

In the three mammalian species studied extensively, two isoforms have been identified and separated by preparative isoelectric focusing. Without exception, the isoelectric points of human, bovine, and rat PI transfer proteins are acidic. Remarkably, the isoelectric points for the two protein forms differed by 0.3–0.4 pH units in each case. Within the limits of experimental error, the isoforms of any species exhibited identical molecular mass. The molecular basis of the isoform pattern has been clarified recently by van Paridon *et al.* (1987b), who showed convincingly that PI was noncovalently associated with the more acidic protein and that PC was associated with the more basic form of bovine PI transfer protein. These two phospholipids are transferred by the protein and have a common substrate-binding site. Such a site most probably accommodates the *sn*-1 and *sn*-2 glycerol fatty acyl chains but somehow distinguishes the structurally diverse phosphorylinositol and phosphorylcholine functions. Moreover, for a given species, the two isoforms generally had similar specific transfer activities toward PI or PC, when determined in a variety of assay systems.

Using a series of *cis*-parinaroyl-PI and *cis*-parinaroyl-PC molecules and dynamic fluorescence anisotropy measurements, van Paridon *et al.* (1987b) have concluded that the *sn*-2 acyl chain of both phospholipid substrates is immobilized to the same extent and therefore experiences the same chemical environment within the protein. The binding of one substrate molecule excludes the simultaneous binding of another phospholipid, although the two isoforms are readily interchangeable (van Paridon *et al.*, 1987b). The distribution of PI transfer protein between the two isoforms is a direct reflection of the relative affinity constants for transferable phospholipid substrates and the phospholipid composition of the tissue from which the proteins are isolted. Van Paridon *et al.* (1987a) calculated that 63% of bovine PI transfer protein should contain endogenous PI, a value close to that found experimentally (van Paridon *et al.*, 1987b). Rat PI transfer protein showed a similar distribution between the acidic and basic isoforms (Venuti and Helmkamp, 1988b).

2.3. Amino Acid Composition

For the bovine brain and heart proteins, amino acid compositions (Helmkamp *et al.*, 1976a; DiCorleto *et al.*, 1979) have been determined. A consensus composition for bovine PI transfer protein, calculated from the published data of the two isoforms from brain and heart and normalized to a mass of 31.9 kDa, can be compared with that determined for rat PI transfer protein (Table I). Reasonably good correlations are noted for the anionic, cationic, aromatic, and sulfur-containing amino and imino acids. The presence of disulfide bonds has not been ascertained. Failure to obtain protein sequence information by chemical methods has led to the tentative conclusion that rat PI transfer protein contains an acylated or otherwise blocked *N*-terminus (G. M. Helmkamp, Jr., unpublished observation). Of interest is the finding of an *N*-acetylmethionyl residue at the *N*-terminus of bovine liver PC transfer protein (Akeroyd *et al.*, 1981).

2.4. Immunologic Cross-Reactivity

PI transfer proteins are similar in size, amino acid composition, and isoelectric point. To correct for differences in experimental design, cytosol fractions of a number of tissues were separated electrophoretically and analyzed by immunoblot with polyclonal rabbit antibody to bovine brain PI transfer protein (Helmkamp *et al.*, 1976a). The results revealed a single, immunologically reac-

Table I
Amino Acid Composition of PI Transfer Proteins

	Amount (residues/mole) in:		
Amino acid	Bovine brain/heart[a]	Rat brain[b]	Rat brain[c]
Alanine	17.1 ± 1.7	23	13
Arginine	11.2 ± 2.7	18	11
Aspartate +	33.9 ± 2.4	26	18
Asparagine			11
Cystine (half)	3.6 ± 0.1	4	4
Glutamate +	44.5 ± 0.5	41	27
Glutamine			13
Glycine	24.0 ± 6.0	29	14
Histidine	6.5 ± 0.9	9	8
Isoleucine	12.6 ± 2.2	5	12
Leucine	26.0 ± 0.7	17	21
Lysine	25.2 ± 1.6	20	29
Methionine	6.3 ± 0.7	4	8
Phenylalanine	7.6 ± 1.2	9	8
Proline	11.9 ± 1.9	11	12
Serine	14.4 ± 2.8	17	7
Threonine	14.4 ± 0.2	14	13
Tryptophan	ND[d]	ND	8
Tyrosine	9.7 ± 1.7	11	13
Valine	17.3 ± 0.4	14	21

[a]The data of Helmkamp *et al.* (1976a) and DiCorleto *et al.* (1979) have been normalized to a protein mass of 31.9 kDa and expressed as mean ± standard deviation.
[b]Calculated from a single amino acid analysis (Helmkamp, unpublished).
[c]Predicted from the sequence of the cDNA encoding rat PI transfer protein (Dickeson *et al.*, 1989).
[d]ND, Not determined.

tive protein of 35–37 kDa (Figure 1). In terms of molecular properties and immunologic activity, PI transfer proteins constitute a highly conserved class of proteins in higher eukaryotes. Interestingly, no cross-reactivity could be demonstrated between this antibody preparation and cytosolic fractions prepared from freshly harvested log-phase *Saccharomyces cerevisiae*, a lower eukaryote that contains a PI transfer protein of similar molecular properties and catalytic activity (Daum and Paltauf, 1984; Bozzato and Tinker, 1987).

2.5. Primary Structure of Rat PI Transfer Protein

The strong immunologic cross-reactivity of a rabbit polyclonal antibody to bovine PI transfer protein permitted the screening of several λgt11 rat brain cDNA libraries. Several clones were isolated, and sequence analysis was carried out on

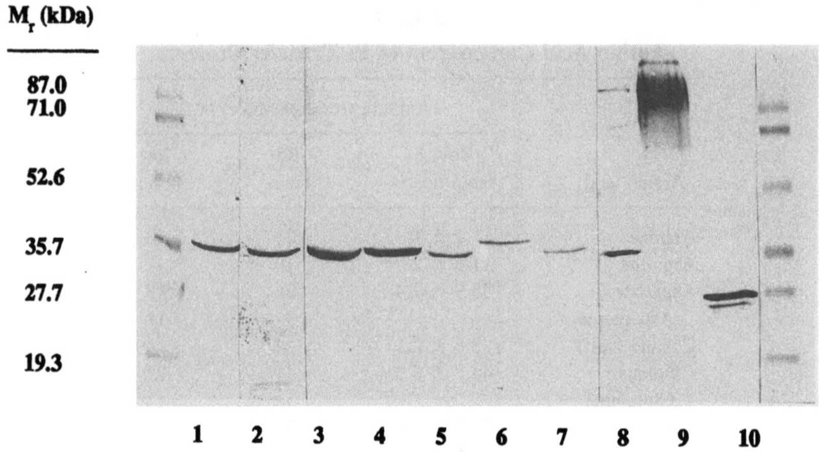

FIGURE 1. Immunoblot of animal brain and other eukaryotic cell cytosolic preparations. Samples (50 μg of protein) were subjected to electrophoresis on gels of 12% polyacrylamide in the presence of sodium dodecyl sulfate and 2-mercaptoethanol. The electrophoretically separated proteins were transferred to nitrocellulose and treated sequentially with reconstituted nonfat dried milk, rabbit anti-bovine PI transfer protein antibody, goat anti-rabbit immunoglobulin G antibody–alkaline phosphatase conjugate, and a substrate mixture of 5-bromo-4-chloro-3-indolyl phosphate and Nitro Blue Tetrazolium (Venuti and Helmkamp, 1988b). The estimated molecular masses of standard proteins are indicated. The animal brain samples are human (1), cow (2), rat (3), mouse (4), chicken (5), frog (6), and lizard (7); the other eukaryotic cells are *Aedes albopictus* (mosquito) C6/36 cultured cell line (8), *S. cerevisiae* (9), and *Neurospora crassa* (10).

both nucleotide strands by standard dideoxynucleotide termination methods (Dickeson *et al.*, 1989). The sequence of 1668 base pairs contained a single open reading frame that encoded a polypeptide of 271 amino acids (Figure 2). The translation initiation codon ATG was present in a classical sequence GACATGG; the polyadenylation signal sequence AATAAA was located 19 nucleotides upstream (5′ direction) of the poly(A) tail.

The deduced primary structure of rat PI transfer protein was compared with *N*-terminal amino acid sequence data obtained from bovine brain PI transfer protein (Dickeson *et al.*, 1989) and a phospholipid transfer protein isolated from rat lung (Funkhouser, 1987). Subjecting the bovine brain protein to gas phase Edman degradation, seventeen residues corresponding to amino acids 4–20 were identified: 12 were identical to amino acids in comparable positions in the rat sequence, and 3 represented conservative substitutions. The first 21 residues at the *N*-terminus of the rat lung protein differed in only two positions from the rat cDNA sequence: the absence of the translation-initiating methionine and histidine rather than serine at amino acid 14. These results provide strong verification that the isolated cDNA clones were indeed PI transfer protein clones. Addi-

GAATTCGGGCGGGAGGTGGCGAG

GCCAGGGGCAGGGTGGCAGAAGCCCGCGCGGCGACCGCACGCAGAGCAGCGGGGACAGAGCAGACCACAACG

AAGGCGCACAGGCAGCGGGGCCGGGCCGGGCCACCGGGAGAGCGGCCACCGGGAAGCGGGCGGGAGGCCGG

GCTGGCAGCTGGCAGCCGCCGAGCCGCGAAGC *<u>GAC</u>* *<u>ATG</u>* *<u>G</u>*TG CTG CTC AAG GAA TAT CGG GTC ATC

Met Val Leu Leu Lys Glu Tyr Arg Val Ile 10

CTG CCT GTG TCT GTA GAT GAG TAT CAA GTG GGG CAG CTG TAC TCT GTG GCT GAA GCC AGT

Leu Pro Val Ser Val Asp Glu Tyr Gln Val Gly Gln Leu Tyr Ser Val Ala Glu Ala Ser 30

AAA AAT GAA ACT GGT GGT GGG GAA GGT GTG GAG GTC CTG GTG AAC GAG CCC TAC GAG AAG

Lys Asn Glu Thr Gly Gly Gly Glu Gly Val Glu Val Leu Val Asn Glu Pro Tyr Glu Lys 50

GAT GAT GGC GAG AAA GGC CAG TAC ACA CAC AAG ATC TAC CAC TTA CAG AGC AAA GTT CCC

Asp Asp Gly Glu Lys Gly Gln Tyr Thr His Lys Ile Tyr His Leu Gln Ser Lys Val Pro 70

ACG TTT GTT CGA ATG CTG GCC CCA GAA GGC GCC CTG AAT ATA CAT GAG AAA GCC TGG AAT

Thr Phe Val Arg Met Leu Ala Pro Glu Gly Ala Leu Asn Ile His Glu Lys Ala Trp Asn 90

GCC TAC CCT TAC TGC AGA ACC GTT ATT ACA AAT GAG TAC ATG AAG GAA GAC TTT CTC ATT

Ala Tyr Pro Tyr Cys Arg Thr Val Ile Thr Asn Glu Tyr Met Lys Glu Asp Phe Leu Ile 110

AAA ATT GAA ACC TGG CAC AAG CCA GAC CTT GGC ACC CAG GAG AAT GTG CAT AAA CTG GAG

Lys Ile Glu Thr Trp His Lys Pro Asp Leu Gly Thr Gln Glu Asn Val His Lys Leu Glu 130

CCT GAG GCA TGG AAA CAT GTG GAA GCT ATA TAT ATA GAC ATC GCT GAT CGA AGC CAA GTA

Pro Glu Ala Trp Lys His Val Glu Ala Ile Tyr Ile Asp Ile Ala Asp Arg Ser Gln Val 150

CTT AGC AAG GAT TAC AAG GCA GAG GAA GAC CCA GCA AAA TTT AAA TCT ATC AAA ACA GGA

Leu Ser Lys Asp Tyr Lys Ala Glu Glu Asp Pro Ala Lys Phe Lys Ser Ile Lys Thr Gly 170

CGA GGA CCA TTG GGC CCG AAT TGG AAG CAA GAA CTT GTC AAT CAG AAG GAC TGC CCA TAT

Arg Gly Pro Leu Gly Pro Asn Trp Lys Gln Glu Leu Val Asn Gln Lys Asp Cys Pro Tyr 190

ATG TGT GCA TAC AAA CTG GTT ACT GTC AAG TTC AAG TGG TGG GGC TTG CAG AAC AAA GTG

Met Cys Ala Tyr Lys Leu Val Thr Val Lys Phe Lys Trp Trp Gly Leu Gln Asn Lys Val 210

GAA AAC TTT ATA CAT AAG CAA GAG AAG CGT CTG TTT ACA AAC TTT CAC AGG CAG CTG TTC

Glu Asn Phe Ile His Lys Gln Glu Lys Arg Leu Phe Thr Asn Phe His Arg Gln Leu Phe 230

TGT TGG CTT GAT AAA TGG GTT GAT CTG ACT ATG GAT GAC ATT CGG AGG ATG GAA GAA GAG

Cys Trp Leu Asp Lys Trp Val Asp Leu Thr Met Asp Asp Ile Arg Arg Met Glu Glu Glu 250

ACG AAG AGA CAG CTG GAT GAG ATG AGA CAA AAG GAC CCC GTG AAA GGA ATG ACA GCA GAT

Thr Lys Arg Gln Leu Asp Glu Met Arg Gln Lys Asp Pro Val Lys Gly Met Thr Ala Asp 270

GAC TAG CGCTACCGCCCTTTCTGCACTTTTTGCAAGACAGTGGTCCACAGGAAGGCCCAGCAGCTCCACCCTCC

Asp *** 271

TGAGTGGCAGGCCATCTTGGACGCAGCCTTTCTGTGCTCATTCTTCAGGCAACTTTCAGTCCCTTACTATAGCTATT

TCTAGATGCCCTTTAACATTGTGAACAAATAGACCGTCTGGTATTATAAAGCCTGTGTGTGCGGGAGCCTGATCCTC

TAAGATATATGGTGTATGCTGCTGTATTTACAGCTCATCCTCTCTCCATTGTGCCCTGACCCATTTCTGTATGCCCTC

CCCCTTTATTTACAAGTGACGTTCTCATTCTTTCAAATGTGCTGCCTTTGTGATGTGATTATTTAAGCTCACTTATTTC

ACGTTTGGGATAAGCTGAGCTATTTTGCTTCCTGGGCAGATCTTTGGCAATGTTGAATATTCACTTGGGGAGAGAGG

AGGAGTTAAAAATACAGAGACACCTACCCTAGTCCCTCACAATAGGAATGTGGCTTCAGAACCCTGACCTCTGCTCC

TAGTAACCTCTTGTTCGCGTGCTCTAAGCACCAAGAAGGCACCCAAGGCCCAGCCAGTCTAAGAAAATTTAGCAAAT

<u>AATAAA</u>TAACCCCTACTCGAAAAAAAAAAAAAAAAAAAAAA

FIGURE 2. Nucleotide sequence and predicted protein sequence of rat PtdIns transfer protein cDNA. The nucleotide sequence begins with the first base of the 8-base pair *Eco*RI linker used in cloning; the numbered amino acid sequence begins with the translation-initiating methionine. The nucleotide signal sequence for translation initiation is in underlined italics and that for polyadenylation is in underlined boldface.

tional verification came from the *in vitro* transcription of one of the cDNA clones, followed by *in vitro* translation in a rabbit reticulocyte lysate, polyacrylamide gel electrophoresis, and immunoblot analysis (Dickeson *et al.*, 1989). A single immunoreactive band at 36 kDa was observed, the intensity of which could be competed by bovine brain PI transfer protein.

2.6. Other Phospholipid Transfer Proteins in Rat Tissues

In addition to PI transfer proteins mammalian tissues contain other lipid transfer proteins. In the rat, for example, three additional catalytic activities have been identified and partially or completely purified (Table II). With the exception of PI transfer protein, some disagreement exists concerning the structural properties of PC and nonspecific lipid transfer proteins. Nonspecific lipid transfer protein is identical to sterol carrier protein 2 (Scallen *et al.*, 1985), a protein that not only binds phospholipids and sterols but also facilitates the enzymatic synthesis and esterification of cholesterol. The recently determined primary struc-

Table II
Phospholipid Transfer Proteins (Partially) Purified from Rat Tissues

Class	Tissue	Substrate specificity	Structural properties	Reference
PC	Liver	PC	M_r 16,000; pI 8.4	Lumb *et al.* (1976)
		PC, not PI, PE	M_r 28,000	Poorthuis *et al.* (1980)
	Small intestine	PC	pI 8–9	Lutton and Zilversmit (1976)
		PC	None reported	Yamada *et al.* (1978)
PI	Brain	PI, PC	pI 5.0, 5.3	Wirtz *et al.* (1976)
		PI, PC	M_r 36,000; pI 4.9, 5.3	Venuti and Helmkamp (1988b)
	Liver	PI, PC	pI 5.1, 5.3	Lumb *et al.* (1976)
	Small intestine	PI, PC	None reported	Yamada *et al.* (1978)
PG	Lung	PG, not PC, PE, CerPCho	M_r 14,000	Van Golde *et al.* (1980)
Nonspecific lipid	Liver	PC, PE, PI, CerPCho, cholesterol	M_r 12,500; pI 8.3–8.7	Bloj and Zilversmit (1977)
		PC, PE, PS	M_r 12,400	Baranska and Grabarek (1979)
		PE, PI	M_r 14,800	Poorthuis *et al.* (1981)
	Hepatoma	PC, PI, PE, CerPCho	M_r 11,200; pI 5.2	Dyatlovitskaya *et al.* (1978)
	Lung	PC, PE, PI, PS	M_r 35,000; pI 5.6–5.7	Read and Funkhouser (1983)

ture of rat liver nonspecific lipid transfer protein (sterol carrier protein 2) exhibits significant sequence homology with bovine liver PC transfer protein and, more remarkably, with the variable domains of immunoglobulin G heavy chain (Pastuszyn *et al.*, 1987).

3. CATALYTIC PROPERTIES OF PHOSPHATIDYLINOSITOL TRANSFER PROTEINS

3.1. Specificity of PI and PC Transport

The most striking characteristic of PI transfer-protein-mediated catalysis is a well-delineated substrate specificity. With few exceptions, the proteins that share the general structural properties discussed in Section 2.2 facilitate the transport of just two phospholipids, PI and PC. The specificity toward these phospholipids has been observed in a variety of experimental transfer systems that use membranes of different physical structure and chemical composition (Helmkamp, 1985). There is a marked preference toward PI, particularly when catalytic activity is calculated as a function of the percentage of phospholipid in the donor membrane transferred to the acceptor membrane. After converting percent transfer per unit time to moles of phospholipid per unit time activity toward PC often exceeds that toward PI. When attempts have been made to maintain equivalent pools of PI and PC, in small unilamellar vesicles or monolayers, a two- to eight-fold-greater specific activity toward PI is measured (Demel *et al.*, 1977; DiCorleto *et al.*, 1979; Helmkamp, 1985; Venuti and Helmkamp, 1988b).

3.2. Transport of Other Phospholipids

In addition to PI and PC, mammalian PI transfer proteins can facilitate the transport of other phospholipids. Detectable rates of phosphatidylglycerol (PG) transfer were found with bovine brain and human platelet PI transfer proteins (Helmkamp, 1985; George and Helmkamp, 1985). The relative rates of transfer, expressed as nanomoles per minute per milligram, for the series PI : PC : PG were 100 : 17 : 5 for the bovine protein and 100 : 38 : 35 for the human protein. Structural similarity between the hydroxylated polar headgroups of PI and PG probably accounts for this activity. Its physiological significance, on the other hand, may be low, since the abundance and membrane distribution of PG are extremely limited (White, 1973). Zborowski and Demel (1982) demonstrated transfer of PG by bovine PI transfer protein; they further showed that neither phosphatidylserine (PS) nor phosphatidate (PA) was a substrate. Some activity of bovine PI transfer protein toward sphingomyelin (CerPCho) has been detected, the relative rates being 100 : 6 for PI : sphingomyelin (CerPCho) (DiCorleto *et al.*, 1979). Phosphatidylethanolamine (PE) is not transported by PI transfer protein (DiCorleto *et al.*, 1979).

Phosphorylated derivatives of PI have been examined in small unilamellar vesicle assay systems as substrates of PI transfer protein. Bovine brain PI transfer protein exhibited no activity toward phosphatidylinositol 4-phosphate (PI4P) (Schermoly and Helmkamp, 1983). Likewise, no protein-catalyzed transfer of PI4P or phosphatidylinositol 4,5-bisphosphate [PI(4,5)P_2] was detected in the presence of human platelet PI transfer protein (Figure 3) (P. Y. George and G. M. Helmkamp, Jr., unpublished data). These observations suggest that once PI is phosphorylated to PI4P and PI(4,5)P_2, these derivatives most probably remain associated with the membrane domains where they were synthesized, with little possibility of intermembrane migration (see Section 6.4). Several PI analogues in which the cyclitol system was partially oxidized and then reduced failed to function as effective substrates of bovine PI transfer protein (Somerharju *et al.*, 1983). Similarly, Demel *et al.* (1982) showed that modifications to the choline moiety of PC converted such phospholipids to inactive substrates for bovine PI

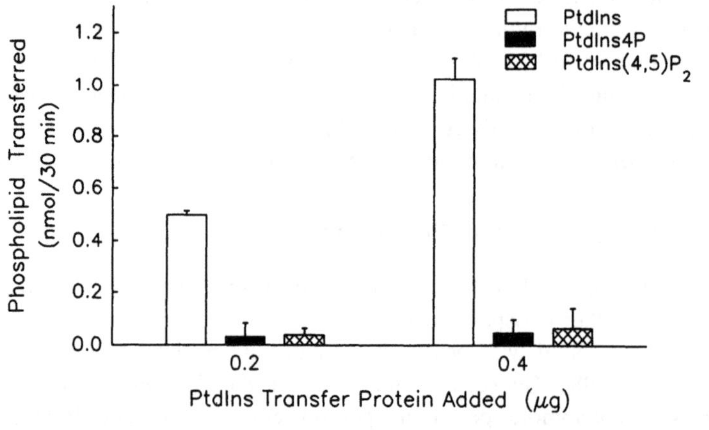

FIGURE 3. Substrate specificity of human platelet PI transfer protein. Donor vesicles contained egg PtdCho (89 mol%), LacCer (6 mol%), and one of the following radiolabeled lipids (5 mol%): rat liver [^{3}H]PI, human erythrocyte [^{32}P]PI4P, or human erythrocyte [^{32}P]PI(4,5)P_2. Acceptor vesicles contained egg PC (98 mol%) and egg PA (2 mol%). Incubations were carried out at 37°C in a total volume of 1.0 ml and included 100 nmol of donor vesicles, 300 nmol of acceptor vesicles, and the indicated amount of human platelet PI transfer protein (George and Helmkamp, 1985). Control incubations contained no transfer protein. Acceptor vesicles were recovered in the supernatant after precipitation of the donor vesicles in the presence of *Ricinus communis* agglutinin (Kasper and Helmkamp, 1981). The results represent the mean ± standard deviation of four incubations.

transfer protein. It would appear, then, that the protein domains that interact with both the PI and PC polar headgroups are highly discriminating.

3.3. Substrate Availability in Membranes

For virtually all biological membranes, PC is the most abundant amphiphilic molecule, whereas PI makes only a minor contribution to the total phospholipid pool (see Section 5.1). This overall distribution often generates molar ratios of PC to PI greater than 15 (White, 1973). Contributions of bilayer asymmetry and limited transbilayer diffusion tend to reduce this ratio in the mammalian plasma membrane owing to a high proportion, if not exclusive localization, of PI on the cytoplasmic surface (Op den Kamp, 1979; Houslay and Stanley, 1982). In contrast, intracellular membranes such as those associated with rat liver endoplasmic reticulum, Golgi vesicles, and lysosomes and the mitochondrial inner membrane exhibit phospholipid asymmetries in which PI is predominantly on extracytoplasmic surfaces (Nilsson and Dallner, 1977). It is also likely that phase separations and domain formation within each monolayer of the membrane bilayer will generate further differences in the distribution of PI and PC (Schroeder, 1985). Thus, the exposure of PI transfer protein to its transferable phospholipid substrates is sensitive to the particular membrane being used. Similar considerations must be kept in mind when artificial bilayer membranes prepared from two or more phospholipids are used in transfer assays. Experimental systems that use homogeneous monomolecular films at air/water interfaces can provide unambiguous information on the magnitude and availability of transferable phospholipids substrates. An alternative approach has been taken by van Paridon *et al.* (1987a), who used pyrene-labeled PI and PC analogs and a continuous fluorometric assay. From the relative rate constants determined for PI and PC transfer, it was concluded that the endogenous substrate levels have a profound effect on the kinetic properties of bovine PI transfer protein.

3.4. Variations in Membrane Composition

Modulation of PI transfer activity by membrane composition has been reviewed in detail (Helmkamp, 1985). Both enhanced and diminished transfer activities can be obtained, with the reference phospholipid mixture being, in general, egg PC plus egg PA (98 : 2, mol/mol). Rates of PI and PC transfer are sensitive to membrane phospholipid polar headgroup and fatty acyl chain composition. Nonspecific ionic forces, rather than any specific protein–lipid interaction, best explain the observed differences in transfer activity and the association of PI transfer protein with membrane surfaces (Zborowski and Demel, 1982; Somerharju *et al.*, 1983; Yoshimura and Helmkamp, 1984). The marked inhibition of bovine PI transfer protein by anionic phospholipids has been investigated further by van Paridon *et al.* (1988), who attribute the effects of PA, PI4P, and PI(4,5)P_2 in vesicle membranes to decreased dissociation rate constants for the protein–membrane complex. The hypothesis that such compositional variation leads to changes in the phase behavior, which may, in turn, exert effects on transfer activity has been examined: marked differences in transfer rate and apparent activation energy occur with small unilamellar vesicles prepared from 1-stearoyl-2-elaidoyl-PC and incubated at temperatures above and below the gel–liquid crystalline transition (Helmkamp, 1980).

When cationic amphiphiles were introduced to microsome–vesicle or vesicle–vesicle assay systems, an inhibition of PI and PC transfer resulted (Badger and Helmkamp, 1982; Mullikin and Helmkamp, 1984). At pharmacological concentrations of local anesthetics and tricyclic antidepressants, inhibition correlated well with water–lipid partition coefficients. In contrast to the generalized effects of surface charge imparted by phospholipid polar headgroup variations described above, the cationic amphiphiles formed electrostatic complexes with the anionic PI or zwitterionic PC and prevented these substrates from interacting with PI transfer protein.

3.5. Catalytic Intermediates of PI Transfer Protein

Bovine PI transfer protein forms a noncovalent complex with a transferable phospholipid. This was first demonstrated with radiolabeled PI and separation of the PI–protein complex from excess vesicles by electrophoresis on polyacrylamide (Helmkamp *et al.*, 1976b). Binding of both PI and PC to bovine PI transfer protein was inferred from increases in subphase radioactivity following the introduction of protein beneath a monolayer prepared from PI and PC in a 1 : 1 molar ratio (Demel *et al.*, 1977). Recently, binary complexes of PI transfer protein and PI or PC were resolved by analytical isoelectric focusing on layers of polyacryamide (van Paridon *et al.*, 1987b). Titration of a limiting quantity of small unilamellar vesicles prepared from 2-*cis*-parinaroyl-PI with increasing amounts of pure bovine PI transfer protein (van Paridon *et al.*, 1987b) led to the conclusion that the protein has one binding site for PI. Altogether, these results indicate that PI transfer protein has the capacity to form a stoichiometric complex with PI or PC and to exist in an aqueous environment with its "solubilized" phospholipid substrate. This complex can be interpreted as a catalytically significant intermediate in protein-mediated transfer; i.e., PI transfer protein functions as a true carrier of phospholipid molecules. A similar conclusion has been reached for bovine PC transfer protein (Wirtz, 1982). The formation of an aqueous complex between rat nonspecific lipid transfer protein and one of its transferable substrates, PC, and the participation of such a complex in the catalytic process has been shown by Nichols (1987, 1988), although other data would suggest that an alternative kinetic mechanism, not involving the binding of lipid to protein, may hold for this transfer protein (Nichols and Pagano, 1983).

3.6. Exchange and Net Transfer of Phospholipids

The demonstration of a membrane-dissociated phospholipid–PI transfer protein complex during the catalytic process strongly favors a one-for-one exchange of phospholipid substrates at membrane surfaces (Helmkamp, 1986). That is, the protein-bound phospholipid molecule is replaced by a membrane-bound phospholipid molecule. Within this framework, PI–PI or PC–PC homologous exchanges or PI–PC heterologous exchanges may occur, at rates that depend upon the intrinsic affinities of PI transfer protein for PI and PC and the proportion of the substrate in the participating membranes (van Paridon *et al.*, 1987a). The two homologous exchanges would lead to no alteration of membrane phospholipid composition based upon polar head group chemistry, although changes in the fatty acyl groups within the PI or PC class could result. On the other hand, the heterologous mode of transfer still represents an exchange of phospholipid molecules, but since the substrates for the opposing routes of transfer (donor to acceptor and acceptor to donor) contain different polar headgroups, a net transfer of phospholipid class will have been catalyzed. This is most commonly seen in experimental systems in which the acceptor membrane is initially devoid of donor membrane PI or PC (Demel *et al.*, 1977; DiCorleto *et al.*, 1979; Venuti and Helmkamp, 1988b).

Whether the net transfers of PI and PC in opposing directions were strictly coupled, such that neither membrane exhibited a net gain or loss in membrane mass in the presence of bovine PI transfer protein, was examined independently in two laboratories. Using a vesicle–vesicle assay in which the donor and acceptor membranes were separated by quantitative agglutination and precipitation, Kasper and Helmkamp (1981) analyzed both membrane populations following incubation and calculated equivalent bidirectional fluxes of PI and PC under homologous conditions of exchange. When net transfer conditions were imposed on the assay system by excluding PI during the preparation of the acceptor vesicles, a net flux of PI occurred, as expected, along the donor-to-acceptor route, but an equivalent and compensatory flux of PC was measured along the acceptor-to-donor route (Figure 4). Similarly, using a monolayer–vesicle assay system, Zborowski and Demel (1982) observed net transfers of equal magnitude for PI movement from monolayer to vesicle and PC movement from vesicle to monolayer. As confirmed by the absence of change in monolayer film pressure, no net transfer of phospholipid mass had taken place. These data suggest that PI transfer protein preferentially catalyzes an exchange of phospholipid substrates.

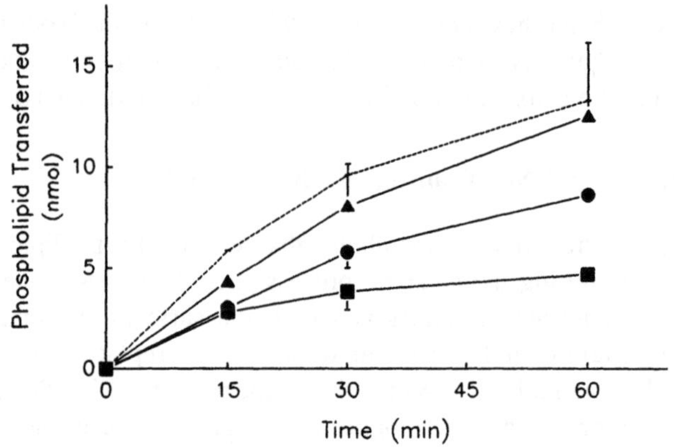

FIGURE 4. Bidirectional flux of phospholipids catalyzed by bovine brain PI transfer protein. Vesicle population I contained egg PC, rat liver PI, and LacCer (82 : 10 : 8, mol%), and vesicle population II contained egg PC and egg PA (95 : 5, mol%). Incubations were performed at 37°C, with 400 nmol of each vesicle population and 126 ng of bovine brain PI transfer protein in a total volume of 1 ml. Vesicles were separated in the presence of *R. communis* agglutinin; vesicle population I was recovered in the pellet, and vesicle population II remained in the supernatant. The datum points represent the mean ± standard deviations of four to six incubations. Individual fluxes were measured in different experiments by using radiolabeled substrates: transfer of PI from vesicle I to vesicle II (■); transfer of PC from vesicle I to vesicle II (●); transfer of PC from vesicle II to vesicle I (▲); and the sum of the observed transfers of PI and PC at the indicated incubation times from vesicle I to vesicle II (----). Reprinted with permission from Kasper and Helmkamp (1981). Copyright (1981) Elsevier Science Publishing.

Net phospholipid class transfer accompanied by net mass transfer occurs less favorably. Comparable constraints apply to the catalytic activity of bovine PC transfer protein (Runquist and Helmkamp, 1988).

3.7. Kinetics of Phospholipid Transfer

With PI transfer protein functioning as a catalyst and the donor and acceptor membrane phospholipids serving as substrate molecules, or, more precisely, as substrate surfaces, a number of kinetic paths are available for intermembrane phospholipid transport. This system of transfer protein and phospholipid is strictly analogous to the enzyme-catalyzed reaction involving two substrates, and therefore it may take advantage of existing kinetic models and experimental approaches. Several special considerations do, however, apply. There is no chemical alteration of the phospholipid substrate during the transfer process, but merely a potential change in the environment of the phospholipid substrate. Accordingly, it is of paramount importance that the donor and acceptor membranes, or, at the very minimum, their respective environments, be readily distinguishable during the course or at the completion of protein-catalyzed phospholipid transport. This necessity has led, in general, to assay systems in which donor and acceptor membranes can be separated by chemical, biochemical, or physical means (Wetterau and Zilversmit, 1984). Table III lists the assay systems used in the investigation of rat PI transfer activity. Methods of separating the two membrane populations include differential pelleting, acidification followed by centrifugation, ion exchange chromatography, and precipitation of glycoconjugate-containing membranes. It is obvious that the donor and acceptor membranes must differ sufficiently to permit separation and that the separation technique itself may be quite time-consuming.

There are several exceptions to the requirement of separable donor and acceptor membranes. Demel and co-workers have perfected the formation and stabilization of monomolecular films for use in protein-mediated phospholipid transport studies (Demel *et al.*, 1977; Zborowski and Demel, 1982). While sharing a common aqueous subphase, the donor and acceptor monolayers are separated by a physical barrier and remain so throughout the experiment. The choice and availability of transferable phospholipid substrates are completely controlled by the investigator. This apparent advantage must, nevertheless, be tempered by the exceedingly slow time course of experiments in which a phospholipid monolayer functions as one or both of the participating membranes. A second exception to the traditional assay procedures involves the use of spectroscopically sensitive phospholipid substrates. The difference between donor and acceptor membrane in the physical or chemical environment should cause measureable changes in spectroscopic properties such as spectral line broadening, chemical shift anisotropy, quenching, or resonance energy transfer. Some particularly elegant experiments by Wirtz and co-workers, using fluorescent pyrene-acyl-labeled and *cis*-parinaroyl-labeled PI and PC, have contributed much new information on the structure and mechanism of PI transfer protein isolated from bovine brain (Somerharju *et al.*, 1983; van Paridon *et al.*, 1987a,b, 1988).

3.8. Mathematical Analysis of Phospholipid Transfer

Evidence for a phospholipid–PI transfer protein complex was taken into consideration in constructing a kinetic model based upon an enzymatic double-displacement mechanism or ping-pong Bi-Bi mechanism (Helmkamp *et al.*, 1976b). In this model the protein catalyst oscillates alternatively between two intermediate states: as a complex with a phospholipid substrate derived from a donor membrane or as a complex with a phospholipid substrate derived from an acceptor membrane. From simple association and dissociation events of the protein with donor and acceptor membranes, an expression for the apparent initial velocity of phospholipid transfer from donor membrane to acceptor membrane was derived:

$$\nu_{\mathrm{app}} = \frac{V[\mathrm{D}][\mathrm{A}]}{K_{\mathrm{m,D}}[\mathrm{A}]\{1 + [\mathrm{A}]/K_{\mathrm{i,A}} + [\mathrm{A}]/K_{\mathrm{s,A}}\} + K_{\mathrm{m,A}}[\mathrm{D}]\{1 + [\mathrm{D}]/K_{\mathrm{i,D}} + [\mathrm{D}]/K_{\mathrm{s,D}}\} + [\mathrm{D}][\mathrm{A}]} \tag{1}$$

where [D] and [A] are the concentrations, usually in terms of lipid phosphorus, of the donor and acceptor membranes, respectively, V is the maximum velocity in the transfer direction from donor to acceptor, $K_{m,D}$ and $K_{m,A}$ are the Michaelis constants for the donor and acceptor membranes, respectively, and K_s and K_i are dissociation constants for protein–membrane complexes in substrate and inhibitor roles, respectively. The general form of this expression permitted several predictions of kinetic behavior of protein-catalyzed phospholipid transfer. Decreased transfer velocity will occur at a sufficiently high concentration of donor or acceptor membrane, leading to an apparent inhibition of transfer activity. In addition, a family of parallel lines will result when the concentration of one membrane is varied while the other remains fixed and the data are analyzed in a double-reciprocal plot. Experimental evidence confirmed both predictions (Helmkamp *et al.*, 1976b).

Other mathematical treatments of protein-catalyzed phospholipid transfer, all based upon a double-displacement mechanism, have been derived and evaluated for bovine PC transfer protein (van den Besselaar *et al.*, 1975; Bozzato and Tinker, 1982). Noteworthy in these two modeling studies is a lack of distinction between K_s and K_i and, consequently, initial velocity expressions with denominators containing fewer terms. These expressions also permitted a graphical estimation of all kinetic and rate constants.

Experimental systems routinely monitor phospholipid transfer in one direction: from the donor membrane to the acceptor membrane. This poses a practical limitation when information about the maximum catalytic activity of a transfer protein is desired. A more comprehensive kinetic model has recently been formulated by Yoshimura *et al.* (1988) that illustrates graphically the partitioning of phospholipid transfer activity among the four possible intermembrane routes: donor to donor, donor to acceptor, acceptor to donor, and acceptor to acceptor. An expression for the total rate of transfer is multiplied by the probability function for transfer along a specific route to yield four separate apparent velocities. For the donor-to-acceptor transfer path, the apparent initial velocity can be written

$$\nu_{\mathrm{app}} = \frac{V_{\mathrm{D}}V_{\mathrm{A}}\{[\mathrm{D}]/K_{\mathrm{D}}\}\{[\mathrm{A}]/\mathrm{K}_{\mathrm{A}}\}}{\{V_{\mathrm{D}}[\mathrm{D}]/K_{\mathrm{D}} + V_{\mathrm{A}}[\mathrm{A}]/K_{\mathrm{A}}\}\{1 + [\mathrm{D}]/K_{\mathrm{D}} + [\mathrm{A}]/K_{\mathrm{A}}\}} \tag{2}$$

where V_D and V_A are the maximum transfer velocities between donor membranes and acceptor membranes, respectively, and K is the dissociation constant for the

Table III
Assay Systems for Measuring PI Transfer Activity in Rat Tissues

Tissue	Protein purity	Donor membrane	Acceptor membrane	Phospholipids transferred	Reference
Brain	High-speed supernatant	Microsome (rat brain)	Mitochondrion (rat brain)	PI, PC, PA	Possmayer (1974)
	Partially purified	Microsome (rat liver)	Small unilamellar vesicle	PI, PC	Wirtz *et al*. (1976)
	pH 5.1 supernatant	Microsome (rat brain)	Myelin (rat brain)	PI, PC	Ruenwongsa *et al*. (1979)
	High-speed supernatant	Microsome (rat brain)	Small unilamellar vesicle	PI	Brophy and Aitken (1979)
	Pure	Small unilamellar vesicle	Small unilamellar vesicle	PI, PC	Venuti and Helmkamp (1988b)
Liver	Partially purified	Microsome (rat liver)	Small unilamellar vesicle	PI, PC	Lumb *et al*. (1976)
	High-speed supernatant	Microsome (rat liver)	Mitochondrion (rat liver)	PI, PC, PE	Carey and Foster (1979)
Intestine	Partially purified	Small unilamellar vesicle	Small unilamellar vesicle	PI, PC	Yamada *et al*. (1978)
Testis	Partially purified	Small unilamellar vesicle	Small unilamellar vesicle	PI, PC	Thomas *et al*. (1989)

indicated protein–membrane complex. As above, the apparent inhibition of transfer observed at relatively high membrane concentrations, donor or acceptor, can be appreciated from the form of the mathematical expression. None of the kinetic models proposed thus far for transfer proteins with multiple substrates (Helmkamp *et al.*, 1976b; Bozzato and Tinker, 1987; van Paridon *et al.*, 1988) has attempted to incorporate selectivity among different phospholipid species when they are present in the participating membranes.

4. DISTRIBUTION OF PHOSPHATIDYLINOSITOL TRANSFER PROTEIN IN RAT TISSUES

4.1. PI and PC Transfer Activities and Protein Levels

Rat tissues have been examined for PI transfer activity by using an assay system composed of small unilamellar vesicles for the donor and acceptor membranes (Venuti and Helmkamp, 1988b). Brain exhibited the highest specific transfer activity and was nearly twice that seen in the next most active tissues: thymus, spleen, and lungs (Table IV). The least active tissues were the two muscle masses investigated, heart and hind limb, and abdominal adipose tissue.

Table IV
PI Transfer Activity in Rat Tissues

Tissue	Transfer activity[a] (nmole/h per mg of protein)	Transfer protein[b] (μg/mg of protein)
Adipose	5.1 ± 1.5	0.3
Brain	86.3 ± 2.5	4.8
Cardiac muscle	14.3 ± 5.3	0.8
Intestinal mucosa	31.0 ± 5.9	1.7
Kidney	37.0 ± 0.4	2.0
Liver	33.0 ± 3.8	1.8
Lung	44.7 ± 8.9	2.5
Ovary	43.0 ± 0.4	2.4
Skeletal muscle	4.4 ± 1.1	0.3
Spleen	51.0 ± 8.0	2.8
Testis	36.9 ± 2.2	2.1
Thymus	52.6 ± 5.4	2.9
Uterus	26.6 ± 10.6	1.5

[a]Activity measurements were performed in triplicate on two different tissue preparations (150,000 × *g* supernatant fractions) by using a vesicle–vesicle assay system.
[b]The tissue protein level was estimated from the quotient of specific PI transfer activity and a specific activity of 18 μmol/h per mg for purified rat PI transfer protein (Venuti and Helmkamp, 1988b).

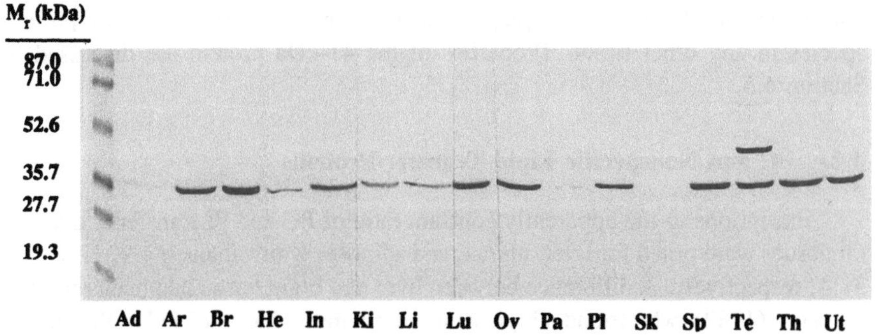

FIGURE 5. Immunoblot of rat tissue cytosolic preparations. The amount of protein analyzed was 40 μg; see Figure 1 for other details. The rat tissues examined were adipose, abdominal (Ad); adrenal gland (Ar); brain, cortex (Br); heart (He); intestinal mucosa, duodenum (In); kidney (Ki); liver (Li); lung (Lu); ovary (Ov); pancreas (Pa); placenta (Pl); skeletal muscle, upper hind limb (Sk); spleen (Sp); testis (Te); thymus (Th); and uterus (Ut).

PC transfer activity was also determined for the high-speed supernatant fractions of these tissues. With the same assay system, in which the molar ratio of PC to PI was 18 in the donor membrane population, a value of 9.3 ± 2.0 was found for the ratio of PC and PI transfer activities in 10 tissues. The ratio of PC and PI transfer activity for the purified rat protein was 10.3 ± 1.3. It is therefore likely that in most rat tissues, a single catalytic protein is responsible for the majority of both PI and PC transfer.

Another means of comparing rat tissues for differences in PI transfer activity is through the use of specific immunoreagents. The straightforward double-diffusion and immunoblot procedures have been used to show that a polyclonal rabbit antibody against bovine PI transfer protein cross-reacts with tissue cytosol preparations from a number of other animal species. An immunoblot analysis of rat tissues indicated a common reactive protein corresponding to a molecular mass of 36 kDa (Figure 5). The size of this protein is indistinguishable from that of PI transfer protein purified to homogeneity from rat brain. An excellent correlation can also be observed between the intensity of the immunologic stain, in this case conjugated alkaline phosphatase activity, and the level of PI transfer activity determined for the same tissue supernatant preparation (Table IV). Brain tissue is the most intensely stained, whereas the weakest-staining tissues are cardiac muscle, skeletal muscle, and adipose. Thus, the PI transfer activity of a tissue can be attributed in most cases to the level of functionally active PI transfer protein. Additional tissues that have been analyzed by immunoblot and found to contain a single reactive protein of 36 kDa include adrenal cortex, epididymus, lactating mammary gland, pancreas, placenta, and seminal vesicle. The comparative tissue immunoblot pointed out a unique situation in rat testis: an additional reactive protein of 41 kDa. Application of larger quantities of protein during polyacrylamide electrophoretic separation failed to reveal this protein species in any other tissue. Properties of the 41-kDa protein are discussed in Section 4.5.

4.2. PC and Nonspecific Lipid Transfer Proteins

Exceptions to the apparently constant ratio of PC and PI transfer activity in rat tissues were noted for liver, uterus, and adipose, with values of 4.9, 17.1, and 31.4, respectively. A difference between liver and brain tissue had been reported by Wirtz (1982), who found a higher ratio, on a molar basis, of PC to PI transfer in liver. However, the tissue preparations were pH 5.1 supernatant fractions, rather than 150,000 × *g* supernatant fractions, and the assay system consisted of rat liver microsome donor membranes and rat liver mitochondrion acceptor membranes. The tissue distribution of other lipid transfer proteins may explain the phospholipid transfer activity in liver (Table II).

PC transfer protein has been purified to homogeneity from rat liver (Poorthuis *et al.*, 1980). Its catalytic activity is exclusively toward PC; neither PI nor PE is transferred to any measurable extent. Despite a sharply defined tissue distribution in liver and intestinal mucosa (Table V), the contribution of PC

Table V
Levels of Phospholipid Transfer Proteins in Cytosolic Fraction of Rat Tissues

Tissue	PC Transfer protein[a] (μg/mg of protein)	Nonspecific lipid transfer protein[b] (μg/mg of protein)
Liver	0.32 ± 0.06	0.78 ± 0.23
Intestinal mucosa	0.16 ± 0.02	0.46 ± 0.03
Kidney	0.09 ± 0.02	0.09 ± 0.01
Spleen	0.09 ± 0.01	0.09 ± 0.01
Lung	0.03	0.10 ± 0.02
Adrenal cortex	0.01	0.09 ± 0.02
Brain	None detected	0.13 ± 0.02
Heart	None detected	0.07 ± 0.01
Ovary	Not determined	0.12 ± 0.02
Testis	Not determined	0.06 ± 0.01

[a]Measurements were performed by radioimmunoassay; the values represent the mean ± standard deviation of 2–14 determinations (Teerlink *et al.*, 1982).

[b]Measurements were performed by enzyme immunoassay; the values represent the mean ± standard deviation of 2–12 determinations (Teerlink *et al.*, 1984).

transfer protein to tissue levels of PC transfer activity may be quite low. Again, because different methods have been used to determine the catalytic properties of phospholipid transfer proteins, a strict comparison cannot be undertaken.

Nonspecific lipid transfer protein in liver was initially described by Bloj and Zilversmit (1977) and then purified to homogeneity by Poorthuis *et al.* (1981); its substrate specificity extends over a wide range of amphiphilic molecules. Bloj and Zilversmit (1977) established the decreasing order PI > PE = PC > CerPCho among phospholipids. The marked preference for PI, coupled with the high abundance of nonspecific lipid transfer protein in liver (Table V), could lead to a disproportionately low ratio of PC to PI transfer activity in unfractionated liver cytosol. This observation must consequently lead to caution in interpreting and comparing the values of PI transfer protein in rat tissues with unusually large amounts of nonspecific transfer protein, i.e., liver and intestinal mucosa. It should be pointed out that the basic character of nonspecific transfer protein, reflected in a pI of >8, virtually ensures its complete separation from PI transfer protein (pI 4.9–5.3), during anion exchange chromatography.

The structural and catalytic properties of PC transfer protein and nonspecific lipid transfer protein distinguish these proteins from PI transfer protein. The dual specificity of PI transfer protein toward PI and PC, and the concomitant lack of activity toward other glycerophospholipids and sphingolipids, is, on the one hand, broader than that of PC transfer protein and, on the other hand, more restrictive than that of nonspecific lipid transfer protein. In the purification schemes of these three proteins, only PI transfer protein remained bound to anion exchange media at neutral pH and low salt concentration. Another significant difference among these proteins is their tissue distribution. PI transfer protein is ubiquitous in the rat: measurable activity is detectable in all tissues investigated, with a range in specific activity (nmol PI transferred/min/mg protein) of approximately 20. A similarly ubiquitous distribution of nonspecific lipid transfer protein has been reported in 10 rat tissues (Teerlink *et al.*, 1984), but the highest concentration was found in liver and small intestine and the lowest in testis tissue. PC transfer protein also was most abundant in rat liver and small intestine, with lesser amounts in four other tissues; none was detected in brain or heart high-speed supernatants, although unfractionated tissue homogenates contained measurable levels of protein (Teerlink *et al.*, 1982). A radioimmunoassay was used to quantitate PC transfer protein levels in these tissues, and an enzyme immunoassay was used to determine nonspecific lipid transfer proteins levels.

4.3. Regional Distribution of PI Transfer Activity in Rat Brain

Gross dissection of mature rat brain yielded 10 distinct anatomical regions. After homogenization of pooled tissue specimens and isolation of 150,000 × *g* supernatant fractions, PI and PC transfer activities were measured (Venuti and

Table VI
Phospolipid Transfer Activity in Rat Brain

	Phospholipid transfer activity[a] (nmole/h per mg of protein)		
Anatomical region	PI	PC	PC : PI
Forebrain			
Caudate nucleus	31.2 ± 11.2	553.4 ± 46.2	17.7
Cortex	31.0 ± 4.0	550.0 ± 20.9	17.7
Hippocampus	34.5 ± 7.1	400.2 ± 88.9	11.6
Olfactory bulb	32.9 ± 4.3	519.8 ± 85.0	15.8
Thalamus	36.0 ± 10.1	360.0 ± 67.4	10.0
Midbrain			
Colliculi	40.6 ± 10.5	653.6 ± 78.3	16.1
Hindbrain			
Cerebellum	36.8 ± 9.1	461.6 ± 98.8	12.5
Pons-medulla	54.3 ± 7.2	581.2 ± 96.0	10.7
Spinal cord	38.8 ± 4.7	385.0 ± 127.0	7.2
Other			
Pituitary gland	18.6 ± 6.4	172.8 ± 42.0	9.3

[a]Activity measurements were performed on brain tissue preparations (150,000 × g supernatant fractions) by using a vesicle–vesicle assay system; values represent the mean ± standard deviation of 7–15 determinations (Venuti and Helmkamp, 1988a).

Helmkamp, 1988a). PI transfer activity was highest in the pons–medulla oblongata region of the hindbrain, whereas PC transfer activity was highest in the combined superior and inferior colliculi regions of the midbrain (Table VI). The pituitary gland exhibited the lowest specific activities for both PI and PC transfer; its values were about half of those calculated for the central nervous system regions. In all regions of the hindbrain examined, as well as in the hippocampus and thalamus of the forebrain and the pituitary gland, the ratio of PC to PI transfer activity was 10.2 ± 1.9, in excellent agreement with values for a number of other unfractionated rat tissues and purified PI transfer protein. In contrast, the other regions of the forebrain, caudate nucleus, cortex, and olfactory bulb, and the midbrain superior and inferior colliculi exhibited a disproportionately higher ratio of PC to PI transfer activity. One interpretation of this difference would require an unusual localization to specific brain anatomical regions of PC transfer protein, a distinct possibility, since PC transfer protein has been detected by radioimmunoassay in unfractionated rat brain homogenates (Teerlink *et al.*, 1982).

The identification of rat PI transfer activity was first made by Possmayer (1974), who studied preparations of forebrain and used an assay system consisting of radiolabeled rat brain microsome donor membranes and rat brain mitochondrion acceptor membranes. A cytosolic protein-mediated transfer of 17–21% of the donor PI pool could be recovered in the acceptor membranes; under comparable conditions, approximately 2% of the microsomal PC and PA was transferred. Wirtz *et al.* (1976) proceeded to isolate myelin and synaptosome fractions from rat cerebral cortex and to investigate PI and PC transfer activities. Both transfer activities were detected in a 105,000 × *g* supernatant and were modestly enriched (two- to eightfold) in the two membranous fractions. Further localization at the cellular level to synaptic vesicles, soma, or axoplasm was not undertaken, although such information would be invaluable in defining a physiological function for PI transfer protein. The pons–medulla oblongata region contains characteristic nerve bundles in the pyramidal tracts and an unusually abundant proportion of myelinated cell processes. This region also exhibits high specific activities, relative to the other nine regions described earlier, of four enzymes that are known to participate in the *de novo* synthesis of phospholipids (Venuti and Helmkamp, 1988a). Two of the enzymes, CTP: phosphatidate cytidylyltransferase and CDP-diacylglycerol:inositol phosphatidyltransferase, are associated with PI synthesis, and two, CTP:phosphocholine cytidylyltransferase and CDP-choline:diacylglycerol cholinephosphotransferase, are associated with PC synthesis. It is tempting to speculate that in myelinated nerves phospholipid metabolism is particularly active and requires the concerted involvement of synthetic and intermembrane transfer catalytic proteins to maintain the appropriate membrane structure and function necessary for the propagation of action potentials.

4.4. Developmental Patterns of PI Transfer Activity

Pertinent to the regional distribution of PI transfer protein in rat brain are several reports of phospholipid transfer activity in developing tissue. Ruenwongsa *et al.* (1979) quantitated a transfer of PI and PC between radiolabeled rat brain microsome donor membranes and rat brain myelin acceptor membranes which was protein dependent and decreased with increasing temperature (>50°C) and pH (>7). PI transfer was also demonstrated in a myelin donor membrane-microsome acceptor membrane system. The pH 5.1 supernatant fractions from the whole brains of rats of different ages were analyzed for their ability to increase the specific radioactivity of PI in myelin acceptor membranes; a continual increase was seen between the ages of 4 and 21 days, after which no further change was observed. This is the same time frame during which myelination of the rat central nervous system is occurring. In contrast, Brophy and Aitken (1979), using high-speed supernatants of cerebral cortex, observed a slight postnatal decrease in specific PI transfer activity between rat liver microsome donor membranes and small unilamellar vesicle acceptor membranes, which was followed by three to fourfold increase until the age of 18 days and a decline afterwards to an adult level, which was not significantly different from that seen between days 4 and 10. Carey and Foster (1984) made a similar observation with high-speed supernatants of whole rat brain and an assay system composed of radiolabeled rat liver microsome donor membranes and rat liver mitochondrion acceptor membranes: the transfer of PI increased postnatally to a maximum at day 9 and then decreased during the following week to adult levels. In an attempt to resolve this difference in the literature, we have prepared high-speed supernatant fractions of whole rat brain and cerebellum from fetal day 16 to postnatal day 35 and measured PI transfer activity by using the vesicle-vesicle assay system (Nyquist and Helmkamp, 1989). For the whole-brain samples we noted a postnatal decrease at day 2, a maximum level attained by day 10, and a gradual decrease during the following 2 weeks, a developmental profile similar to that described by Brophy and Aitken (1979). For the cerebellum samples the postnatal pattern was nearly identical, with the important difference that the maximum occurred on or near day 19. It can therefore be concluded that PI transfer activity follows a unique developmental pattern: a small, but significant decrease during postnatal days 2–4 is followed by a significant increase to a maximum at days 9–18. The coincidence of the increase in PI transfer activity with the onset of myelination again is noteworthy and suggestive of a strong association between catalytic activity and morphological changes.

4.5. PI Transfer Protein Isoforms in Rat Testis

Analysis of rat testis high-speed supernatants indicated a high-molecular-mass protein, which, after electrophoresis under denaturing conditions on polyacrylamide gels, reacted with antibody to bovine PI transfer protein (Figure 4). The testis-specific band corresponded to a protein with a mass of 41 kDa, clearly larger than the more ubiquitous protein of 36 kDa seen in all other rat tissues tested. Molecular sieve chromatography on a calibrated Sephadex G-100 column could differentiate the two proteins and, more importantly, indicated that the larger protein could also mediate the transfer of phospholipids. Both testis proteins transferred PI and PC between small unilamellar vesicles. The ratio of PI and PC activities for the two proteins after partial purification was not appreciably different, suggesting a preservation of catalytic properties and substrate specificity in the high-molecular-mass isoform.

The normal 36-kDa protein was separated from the 41-kDa protein by DEAE-cellulose column chromatography (Thomas *et al.*, 1989). Again, PI and PC transfer activities were associated with both resolved proteins. Some of these characteristics are presented in Table VII. Under the nondenaturing conditions of the Sephadex molecular sieve column, the testis-specific protein has a mass of 66 kDa, suggesting a homodimeric structure. As discussed in Section 2.2, it is not uncommon for molecular sieve chromatographic procedures to underestimate the apparent size of PI transfer proteins from a number of species and tissues.

Table VII
Comparison of PI Transfer Proteins in Rat Testis

Property	PI transfer protein	Testis-specific PI transfer protein
Phospholipids transferred	PI, PC	PI, PC
Age of animal at appearance in tissue	Birth	Week 4
Electrophoretic migration on polyacrylamide gels (denaturing conditions)	$R_f = 0.49$; $M_r \approx 36,000$	$R_f = 0.41$; $M_r \approx 41,000$
Chromatographic behavior on Sepahdex G-100 (nondenaturing conditions)	$K_{av} = 0.40$; $M_r \approx 22,000$	$K_{av} = 0.14$; $M_r \approx 58,000$
Elution conditions from DEAE-cellulose (pH 7.0)	85 mM NaCl	200 mM NaCl

Therefore, the native protein mass of the testis-specific PI transfer protein can be estimated to be approximately 82 kDa, a dimer of apparently identical subunits.

Up to postnatal week 3 there is no immunological evidence of the 41-kDa isoform; during week 4 its immunoblot intensity begin to increase, and by week 6 it is comparable to that seen in the adult male. A small, but statistically significant ($p < 0.05$) increase in PI transfer activity was observed between weeks 1 and 6. Developmental or hormonal regulation of the testis-specific protein is a possibility (Ewing *et al.*, 1966; Mills *et al.*, 1977; Sharpe, 1986). Increases in testis weight begin to accelerate during week 3, as does the diameter of the seminiferous tubules. These changes are accompanied by increases in the weight ratio of phospholipid to neutral lipid. The amounts of DNA, RNA, and protein in the maturing testis actually decrease between days 5 and 60. Spermatogenesis in the testis requires the hormonal maturation and responsiveness of both Leydig and Sertoli cells; the latter cells attain adult morphology and function during week 5 (Chemes *et al.*, 1979). The influence of pituitary peptide hormones (luteinizing hormone, follicle-stimulating hormone) and testis-derived steroid hormones (testosterone, estradiol) on PI metabolism and PI transfer protein levels may provide additional insight into the two forms of PI transfer protein in mature rat testis.

5. MEMBRANE ORGANIZATION OF PHOSPHATIDYLINOSITOL

5.1. Membrane Levels of Inositol-Containing Lipids

Phosphoinositides are found in virtually all eukaryotic cells. The major species is the glycerolipid, PI. The quantity of PI in mammalian membranes varies between 2 and 15%, on a phospholipid molar basis (White, 1973). In lower eukaryotes, such as *S. cerevisiae,* the proportion of PI may be as high as 25% (Arnold, 1981). Eukaryotic cells also contain two phosphorylated derivatives of PI, namely, PI4P and PI(4,5)P_2, which are usually present in much smaller quantities. For example, in human platelets and erythrocytes the molar distributions of PI, PI4P, and PI(4,5)P_2 are 6.0, 1.0, and 0.4% and 1.2, 0.8, and 1.2%, respectively (Wilson *et al.*, 1985; Ferrell and Huestis, 1984); in rat forebrain the comparable distribution is 1.8, 0.5, and 1.2% (Nishihara and Keenan, 1985). The acute sensitivity of phosphoinositides to agonist-induced metabolic turnover, earlier recognized primarily as chemical instability, prompts necessary caution in interpreting any compositional analysis.

PI is widely distributed among subcellular organelles and membranes. As summarized by McMurray (1973), the proportion of PI in the total phospholipid pool is 8–14% for preparations of rat liver plasma membrane, nuclear membrane, rough endoplasmic reticulum, Golgi apparatus, and inner and outer mitochondrial membranes. When this rather narrow range of PI composition is superimposed on the distribution of membrane surface area in the rat liver cell (Jelsema and Morré, 1978), it is apparent that the largest pool of PI resides in the endoplasmic reticulum (approximately 35%), whereas the plasma membrane contains about 4–10% of the PI of the cell. The consequences of this distribution on PI dynamics within eukaryotic cells are discussed in Section 7.2.

5.2. Bilayer Orientation of PI

The bilayer structure of biological membranes dictates that phospholipids be localized to the cytoplasmic and extracytoplasmic layers. This localization often results in an asymmetric distribution of phospholipids across the bilayer (Op den Kamp, 1979; Houslay and Stanley, 1982). PI is preferentially localized to the cytoplasmic layer in erythrocytes and platelets. In several domains of rat liver plasma membrane, PI has been exclusively identified on the cytoplasmic surface. In rat liver endoplasmic reticulum, however, PI is more equally distributed on both layers or prefers the extracytoplasmic, i.e., luminal surface. The transbilayer distribution of PI follows closely that of PS, the other major anionic phospholipid in most eukaryotic membranes. Unfortunately, little information is available on the distribution of PI4P and PI(4,5)P_2, either among subcellular membranes or across membrane bilayers.

5.3. Covalent PI–Protein Complexes

An unusual class of inositol-containing lipid has been recently described. Proteins may be covalently attached to a complex oligosaccharide which is terminally linked glycosidically to the inositol ring of PI (Low, 1987; Sefton and Buss, 1987; Ferguson and Williams, 1988). Most examples of these modified proteins are found on the external surface of cells. The hydrophobic fatty acyl chains of the PI moiety presumably function to anchor the protein to the bilayer. Since the covalent nature of the lipid–protein complex would most probably preclude the inclusion of the PI in the usual lipid extraction procedures, it is difficult to estimate how much additional cellular PI may be involved in such protein modifications. One must at least entertain the possible role of PI transfer protein in the assembly of the glycosyl–phosphoinositide unit prior to its attachment to a protein.

6. PATHWAYS OF CELLULAR PHOSPHATIDYLINOSITOL METABOLISM

6.1. *De Novo* Synthesis of PI

PI is synthesized in eukaryotic cells from CDP-diacylglycerol (CMP-phosphatidate, phosphatidyl-CMP) through the displacement of CMP by *myo*-inositol, a reaction that is catalyzed by CDP-diacylglycerol:inositol phosphatidyltransferase (Agranoff *et al.*, 1958; Paulus and Kennedy, 1960). The liponucleotide substrate is derived from PA and, in addition to being utilized for PI synthesis, may be partitioned to PS, phosphatidylglycerol phosphate (PGP), and cardiolipin synthesis. Inositol is generated from D-glucose 6-phosphate, with L-*myo*-inositol 1-phosphate as a metabolic intermediate (Parthasarathy and Eisenberg, 1986). Salvage of existing pools of acylglycerols and inositol has also been demonstrated. In the latter case, the inositol is a hydrolysis product of D-*myo*-inositol 1-phosphate (Agranoff, 1986). The phosphatidyltransferase, often referred to as PI synthase, is a membrane-bound protein and exhibits a divalent cation requirement which is usually satisfied *in vitro* by Mn^{2+} or Mg^{2+}. Enzymatic activity has been found in most animal tissues (Abdel-Latif, 1983). Attempts to purify and characterize PI synthase have proven particularly difficult. Values for certain kinetic parameters and information on stabilization by detergents or phospholipid mixtures have been obtained for partially purified proteins from dog pancreas (Parries and Hokin-Neaverson, 1984) and *S. cerevisiae* (Fischl *et al.*, 1986). Both of these groups reported a minimum molecular weight of PI synthase in the range 30,000–34,000. More recently, genetic complementation has permitted the cloning and sequencing of the structural gene that encodes PI synthase in *S. cerevisiae* (Nikawa *et al.*, 1987); the deduced primary structure of the putative enzyme consisted of 220 amino acids, a molecular weight of 24,800, and two potential sites for N-linked glycosylation.

In addition to the *de novo* pathway for PI synthesis, a PI-inositol exchange activity has been reported (Abdel-Latif, 1983). Such an activity, although not generating a net production of PI, could account for the metabolic turnover of the inositol polar headgroup and the incorporation of radiolabeled inositol into a lipid-soluble fraction. The recent findings of Berry *et al.* (1986) would suggest that this base exchange activity is a reflection of the reversible character of PI synthase. Consistent with this suggestion is the observation that inositol base exchange is dependent upon micromolar concentrations of Mn^{2+} and CMP, the latter being a reaction product in the direction of PI formation.

6.2. Determinants of PI Fatty Acyl Composition

One of the hallmarks of mammalian PI is its unusual fatty acyl composition. A single molecular class, 1-stearoyl-2-arachidonoyl PI, predominates in most animal species and tissues and accounts for as much as 80% of the total (Holub and Kuksis, 1978). This skewed distribution toward a (saturated plus tetraenoic) species in the resting cell is at odds with numerous metabolic studies, which suggest that the *de novo* synthetic pathway described above produces a mixture of (monoenoic plus dienoic) and (monoenoic plus trienoic) species. Modification of the acyl residues on the glycerophosphorylinositol backbone most probably occurs by selective deacylation and reacylation steps at both the *sn*-1 and *sn*-2 positions (Holub and Kuksis, 1978; Dawson, 1985). Several recent observations lend support to this pathway. Arachidonoyl-coenzyme A (CoA) synthetase has been identified and partially characterized from a broad distribution of human tissues and cells and shown to be distinct from nonspecific acyl-CoA synthetase (Laposata *et al.*, 1985). Moreover, 1-acyl-lyso-PI, as well as other lysoglycerophospholipids, serves as an acyl acceptor in CoA-mediated arachidonoyl transacylation in human platelets (Kramer *et al.*, 1984). These membrane-bound catalytic activities, acting sequentially, could represent a straightforward mechanism for the enrichment of PI with arachidonic acid.

6.3. Subcellular Location of PI Synthesis

Membrane-associated proteins catalyze the synthesis of PI and its immediate precursor, CDP-diacylglycerol. In their careful and thorough survey of phospholipid synthetic enzymes among subcellular compartments of rat liver, Jelsema and Morré (1978) found that virtually all (>96%) PI synthesis was localized in the endoplasmic reticulum. The sensitivity of PI synthase to proteolytic inactivation and mercury-dextran inhibition has been used to establish the orientation of the enzyme catalytic site in rat liver microsomes (Bell *et al.*, 1981). Once the integrity of the membrane preparation had been verified, 89–99% of the PI synthase could be accounted for on the cytoplasmic surface. Thus, the newly synthesized molecules of PI are topologically oriented within the cytoplasmic monolayer of the endoplasmic reticulum and exposed to the cytosolic compartment.

A two-compartment distribution of PI synthesis has been described for the rat pituitary tumor cell line GH_3 (Imai and Gershengorn, 1987). In addition to the expected endoplasmic reticulum pool of synthetic activity, a second activity was identified in the plasma membrane fraction. Although these two activities differ in kinetic parameters, the relative contribution of each to total cellular PI synthesis had not been determined. Nor has it been demonstrated that *de novo* synthesis, rather than a simple exchange of radiolabeled inositol, has occurred. It is worth noting that no PI synthesis could be quantitated in the plasma membrane in human and rat mammary tumor cell lines (Monaco, 1987).

The only other eukaryotic cell to have been studied in detail with respect to subcellular localization of phospholipid synthesis is *S. cerevisiae*. In this yeast, PI synthase activity was detected in the outer mitochondrial membrane and the microsomal fraction; the distribution of total activity (specific activity times protein) between these two membranes was in the ratio 1.0 : 1.3 (Kuchler *et al.*, 1986). In general, these investigators found a significant contribution by the inner and outer mitochondrial membranes to the synthesis of most classes of phospholipids.

6.4. Ancillary Intermediates in PI Synthesis

In keeping with its utilization in multiple synthetic reactions, the synthesis of CDP-diacylglycerol was associated with the endoplasmic reticulum and the inner mitochondrial membrane in a ratio greater than 4 (Jelsema and Morré, 1978). In the former membrane, CDP-diacylglycerol would be incorporated primarily into PI, while in the mitochondrion the liponucleotide would serve exclusively as a precursor of cardiolipin. Little is known about the molecular properties of the enzyme that catalyzes the synthesis of CDP-diacylglycerol in higher eukaryotes. However, CTP:phosphatidate cytidylyltransferase in *S. cerevisiae* exists as a heterodimer of 56- and 54-kDa subunits in mitochondrial membranes (Kelly and Carman, 1987). The metabolic fate of diacylglycerol is likewise critical to its dual function as an intermediate in glycerolipid synthetic pathways and as an activator of protein kinase C. In the absence of hydrolysis to release fatty acid, diacylglycerol is phosphorylated to PA. This reaction is catalyzed by diacylglycerol kinase, an 80-kDa protein that has been characterized from porcine brain and shown to express activity in the cytosolic, microsomal, and synaptosomal fractions (Kanoh *et al.*, 1986). A similar subcellular distribution has been noted in rat brain (Stubbs *et al.*, 1988). It appears that the localization of diacylglycerol kinase in the cell can be attributed to a translocation between aqueous and membranous compartments (Besterman *et al.*, 1986), a translocation that is dependent upon membrane levels of diacylglycerol and specific phospholipids. Diacylglycerol kinase shares this interesting property of lipid-induced translocation with protein kinase C (Nishizuka, 1986). Not only does the phosphorylation of diacylglycerol make it directly available to the *de novo* synthesis of PI, but the metabolism to PA fully compromises its capacity to activate protein kinase C. Indeed, much recent attention has been focussed on the investigation of physiological and pharmacological modulators of diacylglycerol kinase activity and the subsequent effects on protein kinase C-related phenomena (de Chaffoy de Courcelles *et al.*, 1985; Bishop *et al.*, 1986; Setty *et al.*, 1987).

6.5. Phosphorylation of PI

PI is readily converted to two additional phosphoinositides, PI4P and $PI(4,5)P_2$. Evidence of these phosphorylated derivatives has been observed in most eukaryotic cells, including yeast and human cells. Phosphorylation of the 4-position of the *myo*-inositol ring is catalyzed by PI kinase, and, in a subsequent reaction, phosphorylation of the 5-position is catalyzed by PI4P kinase. For both reactions ATP serves as the phosphoryl group donor.

PI kinase activity has been localized to particulate fractions of various cell and tissue homogenates (Abdel-Latif, 1983; Dawson, 1985). Attempts to localize PI kinase more precisely to specific membranes have yielded a large body of literature and a broad distribution of activity. From a straightforward identification of the plasma membrane in human erythrocytes (Harwood and Hawthorne, 1969), the picture becomes more complex, with multiple sites in other cells. For example, in rat brain the microsomal and myelin-containing fractions, as well as the nonsynaptic plasma membrane fraction, demonstrate high PI phosphorylation capacity (Harwood and Hawthorne, 1969; Deshmukh *et al.*, 1978; Stubbs *et al.*, 1988). In rat liver the plasma membrane, microsome, lysosome, and Golgi fractions have all been implicated (Harwood and Hawthorne, 1969; Collins and Wells, 1983; Jergil and Sundler, 1983). In bovine aortic smooth muscle, PI kinase is associated primarily with the sarcoplasmic reticulum (Doctrow and Lowenstein, 1987).

Bovine cerebral cortex membranes, human erythrocyte ghosts, and murine fibroblasts possess a PI kinase that can be solubilized by neutral or bile acid detergents (Endemann *et al.*, 1987). This apparently ubiquitous kinase exhibits a mass of 55 kDa on sucrose density gradients and phosphorylates PI but not PI4P or diacylglycerol. Of these three tissue sources, bovine brain contained a second PI kinase, larger in protein mass (230 kDa), more relaxed in substrate specificity (PI4P and diacylglycerol were also phosphorylated), and less sensitive by nearly 100-fold to inhibition by adenosine. A 45-kDa protein that phosphorylates PI has been solubilized from bovine brain myelin (Saltiel *et al.*, 1987). The specificity of PI kinase toward PI, but not PI4P, in murine liver has been established independently (Suarez-Quian *et al.*, 1987).

Parenthetically, certain transformed cells express a third class of PI kinase. In polyoma virus-transformed mouse 3T3 fibroblasts, approximately 3% of the total PI phosphorylation activity was distinguished from the more general PI kinase by different kinetic properties and, more interestingly, an exclusive specificity for the 3-position, yielding the PI3P product (Whitman *et al.*, 1988). This unusual phospholipid phosphorylation activity may be similar to one found in human epidermoid carcinoma A431 cells, in which epidermal growth factor stimulated a minor component of total cellular PI kinase activity (Thompson *et al.*, 1985; Walker and Pike, 1987). Thus, there appear to be both general and tissue-specific membrane enzymes that catalyze the phosphorylation of the 3- and 4-positions of the inositol structure.

Phosphorylation of the 5-position of PI4P is facilitated by one or more enzymes that are complex in their cellular localization. Human and rabbit erythrocytes, rat kidney cortex, rat hepatocytes, and rat brain synaptosomes exhibited membrane-associated PI4P kinase activity (Abdel-Latif, 1983; Quist and Barker, 1983; Seyfred and Wells, 1984; Stubbs *et al.*, 1988). In most of these examples it was not established whether PI4P kinase activity was distinct from PI kinase activity. However, rat cerebral cortex cytosol yielded a homogeneous 45-kDa protein with specificity only toward the PI4P substrate molecule (Van Dongen *et al.*, 1984). A protein with similar catalytic properties but a mass of 100–110 kDa has been identified in rat brain extracts (Cochet and Chambaz, 1986). The subsequent observation that the 45-kDa protein could be detected immunologically in other rat tissues (Van Dongen *et al.*, 1986) is consistent with the broad distribution of phosphorylated PI derivatives discussed in Section 5.1. Moreover, it appears likely, on the basis of the physical properties of the few proteins that have been characterized, that different enzymes are responsible for the two phosphorylation steps.

6.6. Hydrolysis of Phosphoinositides by Phosphomonoesterases

Selective hydrolysis of the 4- and 5-phosphate groups from PI4P and $PI(4,5)P_2$ is accomplished by phosphomonoesterases that are distributed widely among animal tissues (Abdel-Latif, 1983). These activities are found in both soluble and membranous fractions and may, in certain cases, reside on the same protein. Human erythrocytes were shown to have a membrane-bound PI4P monoesterase with a pronounced substrate specificity and lack of divalent cation requirement (Mack and Palmer, 1984). By contrast, a highly specific, Mg^{2+}-dependent $PI(4,5)P_2$ monoesterase was partially purified from human erythrocyte cytosol (Roach and Palmer, 1981). A similar distribution was noted for erythrocytes from four other mammals. Nijjar and Hawthorne (1977) have purified to homogeneity a phosphomonoesterase from rat brain cytosol which is equally active toward PI4P and $PI(4,5)P_2$ in the presence of Mg^{2+}. The soluble character and catalytic properties of rat brain $PI(4,5)P_2$ monoesterase have been confirmed (Irvine *et al.*, 1984).

6.7. Hydrolysis of Phosphoinositides by Phosphodiesterases

Cleavage of the phosphodiester bond in phosphoinositides generates diacylglycerol and a family of six inositol phosphates, i.e., the noncyclic and cyclic-1:2 phosphate ester products derived from PI, PI4P, and $PI(4,5)P_2$ (Majerus *et al.*, 1988). Of recent significance are the findings of Hokin and coworkers with a mouse pancreatic tissue explant system: The three cyclic inositol phosphates were formed in measurable quantities under physiological conditions in response to secretagogue stimulation (Dixon and Hokin, 1985; Sekar *et al.*, 1987). Nearly one-third of the 66 possible inositol phosphates have thus far been identified in various eukaryotic tissues and cells, and many of these phosphates are interconvertible by a host of nonspecific and highly specific kinases and phosphomonoesterases (Majerus *et al.*, 1988).

Eukaryotic cells contain a number of phospholipase activities, including both carboxyester and phosphodiester hydrolysis activities. Many of these have broad phospholipid substrate specificity (van den Bosch, 1980; Waite, 1986). One striking exception to this general picture is emerging, i.e., that of a phosphoinositide-specific phospholipase C that attacks PI, PI4P, and $PI(4,5)P_2$. Like PI transfer protein, phosphoinositide-specific phospholipase C is a cytosolic protein whose substrates are strongly membrane bound. The enzyme is activated by Ca^{2+}, a property that dictates the rate of hydrolysis and substrate specificity. $PI(4,5)P_2$ is the preferred substrate at low Ca^{2+} concentrations (ca. 10^{-7} M), and PI is most rapidly degraded at Ca^{2+} concentrations in the range 10^{-4}–10^{-3} M (Majerus *et al.*, 1986; Ryu *et al.*, 1987).

Multiple forms of phosphoinositide-specific phospholipase C have been identified in human, rat, rabbit, and bovine tissues (Low *et al.*, 1984; Majerus *et al.*, 1986; Ryu *et al.*, 1987). Although the enzyme activities differ in size, antigenicity, and tissue distribution, they share low pH optima and the ability to hydrolyze all three phosphoinositides to generate the full spectrum of inositol phosphate products. Catalytic activity is a reflection of substrate availability and Ca^{2+} concentration. Although there is excellent agreement on tissue distribution, there is significant variation in protein size, a point that may reflect species differences or, more simply, differences in experimental methodology. All tissues investigated contain a common phosphoinositide-specific phospholipase C, a 90-kDa protein in rat and rabbit tissue and a 145-kDa protein in bovine tissue. In addition, a tissue-specific, somewhat larger phosphoinositide-specific phospholipase C accounts for the majority of catalytic activity in rat, rabbit, and bovine brain cytosol. Finally, another form of phosphoinositide-specific phospholipase C appears to be unique to human platelet homogenates.

6.8. Stimulated Phosphoinositide Turnover

Phosphoinositide phosphodiester hydrolysis is currently viewed as a critical step in the coupling of cellular responsiveness to environmental stimulation, a phenomenon that is cell surface receptor mediated and proceeds through one or more intracellular messengers. Among these messengers are *sn*-1,2-diacylglycerol, an activator of protein phosphorylation by protein kinase C, and inositol 1,4,5-trisphosphate, which is capable of mobilizing Ca^{2+} from sequestered intracellular stores (Berridge, 1987). Another important consequence of increased diacylglycerol formation from the phosphoinositide pool is the release of arachidonate, which can then be directed into eicosanoid synthesis. The temporal and spatial relationships in agonist-stimulated phosphoinositide turnover constitute a carefully orchestrated series of metabolic events. Although only $PI(4,5)P_2$ can serve as the phosphoinositide-specific phospholipase C substrate for inositol 1,4,5-trisphosphate formation, there is convincing evidence that the other phosphoinositides, and particularly PI, must be hydrolyzed to account for agonist-induced changes in diacylglycerol levels. Long recognized as an important metabolic process in animal cells (Hokin, 1985), a rapid, hormone-induced change in the levels of PI, PI4P, and $PI(4,5)P_2$ has now been described in cultured plant cells (Ettlinger and Lehle, 1988). The current working hypothesis, successfully tested in several cells, is a rapid-onset yet transient hydrolysis of $PI(4,5)P_2$ followed by a delayed but sustained hydrolysis and resynthesis of PI (Majerus *et al.*, 1988).

Metabolically active pools of phosphoinositides, other than that associated with the plasma membrane, may participate in agonist-initiated phosphoinositide turnover. The localization of receptor proteins within the plasma membrane which are sensitive to extracellular ligands is consistent with the degradation of plasma membrane phospholipids. This linkage is further strengthened by the likely participation of one or more guanine nucleotide-binding proteins in the receptor–messenger coupling process, as evidenced by increased cellular phosphoinositide hydrolysis in the presence of GTP, GTP analogues, and AlF_4^-, and, in certain cells, blockage by pertussis toxin (Fain *et al.*, 1988). The characterization of the guanine nucleotide-binding protein and the mechanism by which phosphoinositide-specific phospholipase C is activated remain to be accomplished. What must be kept in mind, however, is the fact the phosphoinositide-specific phospholipase C is a cytosolic protein and, following activation by Ca^{2+} or via a guanine nucleotide-binding protein, could presumably gain access to multiple intracellular pools of PI, PI4P, and $PI(4,5)P_2$.

The concept of a phosphoinositide cycle, first elaborated by Hokin and Hokin (1953), has received considerable attention in the past few years with the degradation of the inositol phosphates, ultimately to inositol, and the conversion of diacylglycerol to PA, which are intermediates in the *de novo* synthesis of PI. A number of metabolic relationships have been described in Sections 6.1, 6.4–6.7. An appreciation of the relative importance of these in a given cell has been made possible by a kinetic model of phosphoinositide turnover under physiological conditions (Chahwala *et al.*, 1987). By using rat kidney fibroblasts, the fluxes of [^{32}P]phosphate and *myo*-[2-^{3}H]inositol were monitored through the three phosphoinositides and their respective noncyclic inositol phosphates. From the pseudo-first-order rate constants calculated for the metabolic interconversions, before the after stimulation with a nonhydrolyzable analog of GTP, a number of conclusions were drawn: (1) the rate of phospholipase C hydrolysis of PI(4,5)P_2 exceeded that of PI4P by an order of magnitude and was increased significantly after stimulation; (2) the rate of phospholipase C hydrolysis of PI remained low, accounting for only 1% of the production of diacylglycerol; (3) the rate of PI4P phosphorylation exceeded that of PI phosphorylation by two orders of magnitude, and both kinase activities increased 10-fold after stimulation; and (4) phosphomonoesterase-catalyzed hydrolysis of PI4P and PI(4,5)P_2 was negligible. In summary, Chahwala *et al.* (1987) argue that agonist-stimulated production of diacylglycerol and inositol 1,4,5-trisphosphate is sensitive to the regulation of PI and PI4P kinase activities.

7. PHOSPHATIDYLINOSITOL TRANSFER PROTEIN AND PHOSPHOLIPID METABOLISM

7.1. Dynamics of Phospholipid Transport

In the eukaryotic cell, phospholipids may be transported between membrane domains by a number of mechanisms. As monomers, phospholipids may move by diffusion from one membrane domain to another through the cytosol, within one-half of the bilayer in a lateral manner, or across the bilayer (Sleight, 1987). Each of these processes may occur spontaneously or be facilitated by proteins. The low aqueous solubility of most phospholipids is a strong argument against significant spontaneous transport through the cytosol. Indeed, lipid transfer proteins in general and PI transfer protein in particular fulfill many of the requirements for increasing the aqueous concentration of phospholipids and for functioning as a catalytic vehicle for the aqueous transport of PI and PC. In contrast, the movement of phospholipids within the bilayer is consistent with a protein-independent, diffusion-controlled reaction governed largely by the viscometric properties of the membrane. For the translocation of a phospholipid molecule from one side of the bilayer to the other, there is increasing evidence to suggest that membrane-associated proteins provide rate enhancement and selectivity (Bishop and Bell, 1985; Zachowski *et al.*, 1986). It is important to note that only by transport through the cytosol can phospholipids migrate between membrane domains that are not physically or topologically continuous, e.g., endoplasmic reticulum and outer mitochondrial membrane.

Vesicular transport is another means by which phospholipids may move between membrane domains in the eukaryotic cell (Morré *et al.*, 1979; Sleight, 1987). The well-established routes of exocytosis and endocytosis involve small organelle-like structures. Thus, the phospholipid molecules residing within the vesicle bilayer membrane become exposed to the other cellular membranes that form the exocytotic or endocytotic axes: endoplasmic reticulum, Golgi elements, secretory vesicles, lysosomes, and plasma membrane. Multiple steps of membrane budding and fusion may be required to complete such transport.

7.2. Membrane Biogenesis and PI Transfer Protein

In a very broad sense, the metabolic turnover of membrane lipids and proteins is a phenomenon that encompasses synthesis, sorting, and degradation. A significant proportion of this trafficking of phospholipid molecules occurs by means of vesicular transport. To cite one example, many different lipids and proteins are involved in the assembly of new plasma membrane domains during cell division. The exocytotic vesicular route might account for the bulk of new phospholipid molecules; protein-mediated transport events could further modify phospholipid composition. Analgous to deacylation–acylation cycles for altering the fatty acyl chains of a phospholipid molecule, phospholipid transfer proteins could exchange phospholipid molecules that differ in polar headgroup, fatty acyl chains, or both, between cytoplasmic domains of intracellular membranes. This would be accomplished without change to the number of phospholipid molecules in the membrane domain.

The substrate specificity of PI transfer protein makes it an ideal catalyst for modifying composition during membrane biogenesis. This and other properties of PI transfer protein are summarized in Table VIII. PC originates in the endoplasmic reticulum and is a major phospholipid in all eukaryotic membranes. One may readily assume that PC would be a significant component of any vesicular structures by which *de novo*-synthesized lipids and proteins are directed to the plasma membrane. If PI were excluded from this vesicular route, PI transfer protein, in a subsequent step, could facilitate the monomeric exchange of "rare" PI molecules in the endoplasmic reticulum with "excess" PC molecules in the plasma membrane. This can be thought of as two opposing net transfers of phospholipid, which would proceed until the ratio of PI and PC in the plasma membrane equaled that in the endoplasmic reticulum (van Paridon *et al.*, 1987a). The intrinsic binding preference for PI should be taken as the principal driving force in the net transfer of that phospholipid. PI transfer-protein-mediated exchange would also take advantage of a number of important observations regarding cellular PI metabolism: (1) *De novo* PI synthesis is unique to the endoplasmic

Table VIII
General Properties of PI Transfer Proteins

Property	Description
Range of species	*S. cerevisiae* to *Homo sapiens*
Substrate specificity	PI > PC ⋙ PG, to CerPCho; no transfer of PI4P or $PI(4,5)P_2$
Molecular mass	34–37 kDa
Isoelectric points	Acidic (pH 4.6–5.6)
Composition	Similar amino acid content; immunologic cross-reactivity
Catalytic forms	Generally two, one containing bound PI and the other, bound PC
Assay systems	A variety of biological and artificial membranes
Transfer mechanism	Bidirectional exchange of PI, PC; unidirectional net transfer of PI or PC

membrane; (2) PI tends to be distributed asymmetrically to the cytoplasmic surfaces of cellular membranes; and (3) further metabolism of PI, in general, involves cell compartments other than the endoplasmic reticulum.

7.3. Phosphoinositide Turnover and PI Transfer Protein

Phospholipids are subjected to degradation and synthesis at rates consistent with appreciable turnover during the life of the cell. This is particularly true for the phosphoinositides, PI, PI4P, and $PI(4,5)P_2$. An agonist-stimulated phosphoinositide-specific phospholipase C hydrolysis, first of $PI(4,5)P_2$ and then of PI and perhaps PI4P, generates lipophilic and hydrophilic mediators of protein phosphorylation and cation mobilization. This initial event is followed in most cells by a recovery phase, during which phosphoinositide metabolism is increased and the membrane concentrations of PI, PI4P, and $PI(4,5)P_2$ return to resting levels. The sequence of metabolic intermediates can be described in a simple manner:

$$\mathrm{PI(4,5)P_2} \rightarrow \text{diacylglycerol} + \text{inositol 1,4,5-trisphosphate} \rightarrow \mathrm{PA}$$
$$\rightarrow \text{CDP-diacylglycerol} \rightarrow \mathrm{PI} \rightarrow \mathrm{PI4P} \rightarrow \mathrm{PI(4,5)P_2}$$

These general aspects of the phosphoinositide cycle are entirely consistent with both the experimental observations in the early 1950s and the more recent advances in chromatographic separations and have been extensively reviewed

(Hokin, 1985; Berridge, 1987; Majerus *et al.*, 1988). It can be anticipated that the turnover of phosphoinositides occurs to a more limited extent in the unstimulated, quiescent cell. What follows is an elaboration of the cellular compartments in which the phosphoinositide cycle operates and the points at which lipid transfer proteins may become involved. Mitchell (1975) first suggested such an involvement.

- *Plasma membrane.* Oriented, at least in part, to the extracytoplasmic surface of the plasma membrane is a *receptor protein* with the ability to bind an extracellular agonist; conversely, oriented to the cytoplasmic surface is a *guanine-nucleotide binding protein.* Also within the cytoplasmic leaflet of the plasma membrane bilayer is a relatively small proportion of the total cellular pool of *PI* and the major cellular pools of *PI4P* and *PI(4,5)P*$_2$. *PI* and a portion of *PI4P kinase* also reside in the plasma membrane. *Protein kinase C,* activated by PS and diacylglycerol, may be associated with the plasma membrane.
- *Endoplasmic reticulum. PI synthase* is localized to cytoplasmic surface of the endoplasmic reticulum. The enzyme catalyzing the preceding step of the *de novo* pathway, *CTP: phosphatidate cytidylyltransferase,* is likewise associated with the endoplasmic reticulum. As a result of this synthetic activity the major cellular pool of *PI* is in this membrane, to a large part oriented on the cytoplasmic leaflet.
- *Cytosol.* The cytosol contains *phopshoinositide-specific phospholipase C,* whose substrates are any of the membrane-bound, cytoplasmically-oriented phosphoinositides. *PI4P kinase* is partially distributed in the cytosol, as are inactive or less active forms of *protein kinase C* and *diacylglycerol kinase.* This aqueous compartment also includes *PI transfer protein* and *nonspecific lipid transfer protein.*

This cast of characters is set to participate in the drama of (stimulated) phosphoinositide turnover:

Act One takes place at the plasma membrane, where an extracellular signal is processed, PI(4,5)P$_2$ is hydrolyzed, and diacylglycerol and inositol 1,4,5-trisphosphate are released as intracellular messengers. A continuing metabolism of PI to PI4P and PI(4,5)P$_2$ is ensured by both plasma membrane-bound and cytosolic kinases. Protein kinase C is activated by membrane PS and the newly generated diacylglycerol; diacylglycerol kinase may undergo similar activation.

Act Two follows at the endoplasmic reticulum, where an already activated phosphoinositide-specific phospholipase C attacks the larger pool of PI, generating additional diacylglycerol, which may be destined to activate cytosolic protein kinase C and diacylglycerol kinase. An increase of metabolic flux through the PI-synthetic pathway occurs, utilizing as substrates some of the salvaged *myo*-inositol and diacylglycerol from the agonist-stimulated hydrolytic events and culminating with the appearance of PI on the cytoplasmic leaflet of the endoplasmic reticulum.

Act Three (or the Interlude between Acts One and Two) occurs in the cytosol and provides essential links between events at the plasma and endoplasmic reticulum membranes. Replenishment of plasma membrane PI and disposal of elevated diacylglycerol levels are required to return the cell to a quiescent, agonist-responsive state. Vesicular transport could accommodate both processes, but a more selective mechanism would be desirable. PI transfer protein transports newly synthesized PI from the endoplasmic reticulum to the plasma membrane, two membranes with which it can interact under *in vitro* conditions, and transports PC in the opposite direction. With stimulated hydrolysis of plasma membrane PI, increased phosphorylation to PI4P and PI(4,5)P_2, and elevated *de novo* synthesis, an even steeper gradient for a unidirectional net transfer of PI to the plasma membrane is generated.

Complementing the concerted flux of PI and PC catalyzed by PI transfer protein is a proposed facilitated transport of diacylglycerol and PA from the plasma membrane to the endoplasmic reticulum. This transport would not only regulate the activity of diacylglycerol as an intracellular mediator by reducing its concentration in the plasma membrane, a step also likely to be shared by the phosphorylation of diacylglycerol, but also deliver diacylglycerol and PA to the cellular site of PI synthesis. Transport of both diacylglycerol and PA would be catalyzed by nonspecific lipid transfer protein. The unusually relaxed substrate specificity of nonspecific lipid transfer protein, already shown to encompass anionic and zwitterionic glycerophospholipids and cholesterol, may extend further to the anionic PA and the monohydroxylated diacylglycerol amphiphilic structures. The broad tissue distribution of nonspecific lipid transfer protein is in keeping with a significant role in cellular metabolism.

Returning to the participation of PI transfer protein in quiescent and agonist-stimulated phosphoinositide metabolism, several unique aspects of its molecular and catalytic properties are worth consideration (Table VIII): (1) PI transfer activity and PI transfer protein have been identified and purified from a wide spectrum of eukaryotic cells, including animals, plants, and fungi, a spectrum coinciding with that of cells which exhibit agonist-stimulated phosphoinoside hydrolysis and resynthesis. No other class of lipid transfer protein studied to date demonstrates this characteristic. (2) The molecular mass, amino acid composition, and antigenic determinants of PI transfer protein are highly conserved, especially among higher eukaryotes, and reflect limited evolutionary divergence and a position as a critical housekeeping genetic translation product. No other class of lipid transfer protein studied to date demonstrates this characteristic. (3) The narrow substrate specificity of PI transfer protein, coupled with an absence of activity with PI4P or PI(4,5)P_2, permits bidirectional net transfer of PI and PC and retention of phosphorylated PI derivatives in the plasma membrane. No other class of lipid transfer protein studied to date demonstrates this characteristic.

7.4. Future Directions

Although an activity has yet to be established unambiguously for PI transfer protein in an intact, viable cell, the evidence is becoming more convincing that an intimate involvement in phosphoinositide metabolism is likely. The case for such an involvement will be strengthened by additional knowledge of the structure and catalytic mechanism of PI transfer protein, details of the transcriptional and translational activity of the genetic elements that encode PI transfer protein, and a more precise description of phosphoinositide metabolism in selected cells and tissues.

8. REFERENCES

Abdel-Latif, A. A., 1983, Metabolism of phosphoinositides, in *Handbook of Neurochemistry* (A. Lajtha, ed.), Vol. 3, pp. 91–131, Plenum Press, New York.

Agranoff, B. W., 1986, Phosphorylated derivatives of *myo*-inositol, *Fed. Proc.* **45:**2629–2633.

Agranoff, B. W., Bradley, R. M., and Brady, R. O., 1958, The enzymatic synthesis of inositol phosphatide, *J. Biol. Chem.* **233:**1077–1083.

Akeroyd, R., Moonen, P., Westerman, J., Puyk, W. C., and Wirtz, K. W. A., 1981, The complete primary structure of the phosphatidylcholine-transfer protein from bovine liver. Isolation and characterization of the cyanogen bromide peptides, *Eur. J. Biochem.* **114:**385–391.

Arnold, W. N., 1981, Lipids, in *Yeast Cell Envelopes: Biochemistry, Biophysics, and Ultrastructure* (W. N. Arnold, ed.), Vol. 1, pp. 97–114, CRC Press, Boca Ration, Fla.

Badger, C. R., and Helmkamp, G. M., Jr., 1982, Modulation of phospholipid transfer protein activity. Inhibition by local anesthetics, *Biochim. Biophys. Acta* **692:**33–40.

Baranska, J., and Grabarek, Z., 1979, Rat liver proteins binding and transferring phosphatidylserine, *FEBS Lett.* **104:**253–257.

Bell, R. M., Ballas, L. M., and Coleman, R. A., 1981, Lipid topogenesis, *J. Lipid Res.* **22:**391–403.

Berridge, M. J., 1987, Inositol trisphosphate and diacylglycerol: Two interacting second messengers, *Annu. Rev. Biochem.* **56:**159–193.

Berry, G., Yandrasitz, J. R., Cipriano, V. M., Hwang, S. M., and Segal, S., 1986, Phosphatidylinositol:*myo*-inositol exchange activity in intact nerve endings: Substrate and cofactor dependence, nucleotide specificity, and effect on synaptosomal handling of *myo*-inositol, *J. Neurochem.* **46:**1073–1080.

Besterman, J. M., Pollenz, R. S., Booker, E. L., Jr., and Cuatrecasas, P., 1986, Diacylglycerol-induced translocation of diacylglycerol kinase: Use of affinity-purified enzyme in a reconstitution system, *Proc. Natl. Acad. Sci. U.S.A.* **83:**9378–9382.

Bishop, W. R., and Bell, R. M., 1985, Assembly of the endoplasmic reticulum phospholipid bilayer: The phosphatidylcholine transporter, *Cell* **42:**51–60.

Bishop, W. R., Ganong, B. R., and Bell, R. M., 1986, Attenuation of *sn*-1,2-diacylglycerol second messengers by diacylglycerol kinase. Inhibition by diacylglycerol analogs *in vitro* and in human platelets, *J. Biol. Chem.* **261:**6993–7000.

Bloj, B., and Zilversmit, D. B., 1977, Rat liver proteins capable of transferring phosphatidylethanolamine. Purification and transfer activity for other phospholipids and cholesterol, *J. Biol. Chem.* **252:**1613–1619.

Bozzato, R. P., and Tinker, D. O., 1982, Effects of membrane fluidity and identification of the rate-limiting step in the protein-mediated phosphatidylcholine exchange reaction, *Can. J. Biochem.* **60:**409–418.

Bozzato, R. P., and Tinker, D. O., 1987, Purification and properties of two phospholipid transfer proteins from yeast, *Biochem. Cell Biol.* **65:**195–202.

Brophy, P. J., and Aitken, J. W., 1979, Phosphatidylinositol transfer activity in rat cerebral hemispheres during development, *J. Neurochem.* **33:**355–356.

Carey, E. M., and Foster, P. C., 1984, Phospholipid exchange activity in developing rat brain, *Biochim. Biophys. Acta* **792:**48–58.

Chahwala, S. B., Fleischman, L. F., and Cantley, L., 1987, Kinetic analysis of guanosine 5′-*O*-(3-thiotriphosphate) effects on phosphatidylinositol turnover in NRK cell homogenates, *Biochemistry* **26:**612–622.

Chemes, H. E., Dym, M., and Raj, H. G. M., 1979, Hormonal control of Sertoli cell differentiation, *Biol. Reprod.* **21:**251–262.

Cochet, C., and Chambaz, E. M., 1986, Catalytic properties of a purified phosphatidylinositol 4-phosphate kinase from rat brain, *Biochem. J.* **237:**25–31.

Collins, C. A., and Wells, W. W., 1983, Identification of phosphatidylinositol kinase in rat liver lysosomal membranes, *J. Biol. Chem.* **258:**2130–2134.

Daum, G., and Paltauf, F., 1984, Phospholipid transfer in yeast. Isolation and partial characterization of a phospholipid transfer protein from yeast cytosol. *Biochim. Biophys. Acta* **794:**385–391.

Dawson, R. M. C., 1985, Enzymic pathways of phospholipid metabolism in the nervous system, in *Phospholipids in Nervous Tissues* (J. Eichberg, ed.), pp. 45–78, John Wiley & Sons, New York.

De Chaffoy de Courcelles, D., Roevens, P., and van Belle, H., 1985, R59022, a diacylglycerol kinase inhibitor. Its effect on diacylglycerol and thrombin-induced C kinase activation in the intact platelet, *J. Biol. Chem.* **260:**15762–15770.

Demel, R. A., Kalsbeek, R., Wirtz, K. W. A., and van Deenen, L. L. M., 1977, The protein-mediated net transfer of phosphatidylinositol in model systems, *Biochim. Biophys. Acta* **466:**10–22.

Demel, R. A., van Bergen, B. G. M., van den Eeden, A. L. G., Zborowski, J., and Defize, L. H. K., 1982, Transfer properties of the bovine brain phospholipid transfer protein. Specificity towards phosphatidylcholine analogs and the inhibitory effect of sphingomyelin, *Biochim. Biophys. Acta* **710:**264–270.

Deshmukh, D. S., Bear, W. D., and Brockerhoff, H., 1978, Phosphoinositide biosynthesis in three subfractions of rat brain myelin, *J. Neurochem.* **30:**1191–1193.

Dickeson, S. K., Lim, C. N., Schuyler, G. T., Dalton, T. P., Helmkamp, G. M., Jr., and Yarbrough, L. R., 1989, Isolation and sequence of cDNA clones encoding rat phosphatidylinositol transfer protein, *J. Biol. Chem.* **264:**16557–16564.

DiCorleto, P. E., Warach, J. B., and Zilversmit, D. B., 1979, Purification and characterization of two phospholipid exchange proteins from bovine heart, *J. Biol. Chem.* **254:**7795–7802.

Dixon, J. F., and Hokin, L. E., 1985, The formation of inositol 1,2-cyclic phosphate on agonist stimulation of phosphoinositide breakdown in mouse pancreatic minilobules, *J. Biol. Chem.* **260:**16068–16071.

Doctrow, S. R., and Lowenstein, J. M., 1987, Inhibition of phosphatidylinositol kinase in vascular smooth muscle membranes, *Biochem. Pharmacol.* **36:**2255–2262.

Dyatlovitskaya, E. V., Timofeeva, N. G., and Bergelson, L. D., 1978, A universal lipid exchange protein from rat hepatoma, *Eur. J. Biochem.* **82**:463–471.

Endemann, G., Dunn, S. N., and Cantley, L. C., 1987, Bovine brain contains two types of phosphatidylinositol kinase, *Biochemistry* **26**:6845–6852.

Ettlinger, C., and Lehle, L., 1988, Auxin induces rapid changes in phosphatidylinositol metabolites, *Nature* **331**:176–178.

Ewing, L. L., Means, A. R., Beames, C. G., Jr., and Montgomery, R. D., 1966, Biochemical changes in rat testis during postnatal maturation, *J. Reprod. Fertil.* **12**:296–307.

Fain, J. N., Wallace, M. A., and Wojcikiewicz, R. J. H., 1988, Evidence for involvement of guanine nucleotide-binding regulatory proteins in the activation of phospholipases by hormones, *FASEB J.* **2**:2569–2574.

Ferguson, M. A. J., and Williams, A. F., 1988, Cell-surface anchoring of proteins via glysocyl-phosphatidylinositol structures, *Annu. Rev. Biochem.* **57**:285–320.

Ferrell, J. E., Jr., and Huestis, W. H., 1984, Phosphoinositide metabolism and the morphology of human erythrocytes, *J. Cell Biol.* **98**:1992–1998.

Fischl, A. S., Homann, M. J., Poole, M. A., and Carman, G. M., 1986, Phosphatidylinositol synthase from *Saccharomyces cerevisiae*. Reconstitution, characterization, and regulation of activity, *J. Biol. Chem.* **261**:3178–3183.

Funkhouser, J. D., 1987, Amino-terminal sequence of a phopsholipid transfer protein from rat lung, *Biochem. Biophys. Res. Commun.* **145**:1310–1314.

George, P. Y., and Helmkamp, G. M., Jr., 1985, Purification and characterization of a phosphatidylinositol transfer protein from human platelets, *Biochim. Biophys. Acta* **836**:176–184.

Harwood, J. L., and Hawthorne, J. N., 1969, The properties and subcellular distribution of phosphatidylinositol kinase in mammalian tissues, *Biochim. Biophys. Acta* **171**:75–88.

Helmkamp, G. M., Jr., 1980, Effects of phospholipid fatty acid composition and membrane fluidity on the activity of bovine brain phospholipid exchange protein, *Biochemistry* **19**:2050–2056.

Helmkamp, G. M., Jr., 1985, Phosphatidylinositol transfer proteins: Structure, catalytic activity, and physiological function, *Chem. Phys. Lipids* **38**:3–16.

Helmkamp, G. M., Jr., 1986, Phospholipid transfer proteins: Mechanism of action, *J. Bioenerg. Biomembr.* **18**:71–91.

Helmkamp, G. M., Jr., Harvey, M. S., Wirtz, K. W. A., and van Deenen, L. L. M., 1974, Phospholipid exchange between membranes. Purification of bovine brain proteins that preferentially catalyze the transfer of phosphatidylinositol, *J. Biol. Chem.* **249**:6382–6389.

Helmkamp, G. M., Jr., Nelemans, S. A., and Wirtz, K. W. A., 1976a, Immunological comparison of phosphatidylinositol and phosphatidylcholine exchange proteins in bovine brain, liver and heart, *Biochim. Biophys. Acta* **424**:168–182.

Helmkamp, G. M., Jr., Wirtz, K. W. A., and van Deenen, L. L. M., 1976b, Phosphatidylinositol exchange protein. Effects of membrane structure on activity and evidence for a ping-pong mechanism, *Arch. Biochem. Biophys.* **174**:592–602.

Hokin, L. E., 1985, Receptors and phosphoinositide-generated second messengers, *Annu. Rev. Biochem.* **54**:205–235.

Hokin, M. R., and Hokin, L. E., 1953, Effects of acetylcholine on phospholipids in the pancreas, *J. Biol. Chem.* **209**:549–558.

Holub, B. J., and Kuksis, A., 1978, Metabolism of diacylglycerophospholipids, *Adv. Lipid Res.* **16**:1–125.

Houslay, M. D., and Stanley, K. K., 1982, *Dynamics of Biological Membranes*, John Wiley & Sons, Chichester, England.

Imai, A., and Gershengorn, M. C., 1987, Independent phosphatidylinositol synthesis in pituitary plasma membrane and endoplasmic reticulum, *Nature* **325**:726–728.

Irvine, R. F., Letcher, A. J., and Dawson, R. M. C., 1984, Phosphatidylinositol-4,5-bisphosphate phosphodiesterase and phosphomonoesterase activities of rat brain, *Biochem, J.* **218:**177–185.

Jelsema, C. L., and Morré, D. J., 1978, Distribution of phospholipid biosynthetic enzymes among cell components of rat liver, *J. Biol. Chem.* **253:**7960–7971.

Jergil, B., and Sundler, R., 1983, Phosphorylation of phosphatidylinositol in rat liver Golgi, *J. Biol. Chem.* **258:**7968–7973.

Kader, J.-C., Douady, D., and Mazliak, P., 1982, Phospholipid transfer proteins, in *Phospholipids* (J. N. Hawthorne and G. B. Ansell, eds.), pp. 279–311, Elsevier Biomedical Press, Amsterdam.

Kanoh, H., Iwata, T., Ono, T., and Suzuki, T., 1986, Immunological characterization of *sn*-1,2-diacylglycerol and *sn*-2-monoacylglycerol kinase from pig brain, *J. Biol. Chem.* **261:**5597–5602.

Kasper, A. M., and Helmkamp, G. M., Jr., 1981, Intermembrane phospholipid fluxes catalyzed by bovine brain phospholipid exchange protein, *Biochim. Biophys. Acta* **664:**22–32.

Kelly, M. J., and Carman, G. M., 1987, Purification and characterization of CDP-diacylglycerol synthase from *Saccharomyces cerevisiae, J. Biol. Chem.* **262:**14563–14570.

Kramer, R. M., Pritzker, C. R., and Deykin, D., 1984, Coenzyme A-mediated arachidonic acid transacylation in human platelets, *J. Biol. Chem.* **259:**2403–2406.

Kuchler, K., Daum, G., and Paltauf, F., 1986, Subcellular and submitochondrial localization of phospholipid-synthesizing enzymes in *Saccharomyces cerevisiae, J. Bacteriol.* **165:**901–910.

Laposata, M., Reich, E. L., and Majerus, P. W., 1985, Arachidonoyl-CoA synthetase. Separation from nonspecific acyl-CoA synthetase and distribution in various cells and tissues, *J. Biol. Chem.* **260:**11016–11020.

Low, M. G., 1987, Biochemistry of the glycosyl-phosphatidylinositol membrane protein anchors, *Biochem. J.* **244:**1–13.

Low, M. G., Carroll, R. C., and Weglicki, W. B., 1984, Multiple forms of phosphoinositide-specific phospholipase C of different relative molecular masses in animal tissues, *Biochem. J.* **221:**813–820.

Lumb, R. H., Kloosterman, A. D., Wirtz, K. W. A., and van Deenen, L. L. M., 1976, Some properties of phospholipid exchange proteins from rat liver, *Eur. J. Biochem.* **69:**15–22.

Lutton, C., and Zilversmit, D. B., 1976, Phospholipid exchange proteins in rat intestine, *Lipids* **11:**16–20.

Mack, S. E., and Palmer, F. B. S. C., 1984, Evidence for a specific phosphatidylinositol 4-phosphate phosphatase in human erythrocyte membranes, *J. Lipid Res.* **25:**75–85.

Majerus, P. W., Connolly, T. M., Deckmyn, H., Ross, T. S., Bross, T. E., Ishii, H., Bansal, V. S., and Wilson, D. B., 1986, The metabolism of phosphoinositide-derived messenger molecules, *Science* **234:**1519–1526.

Majerus, P. W., Connolly, T. M., Bansal, V. S., Inhorn, R. C., Ross, T. S., and Lips, D. L., 1988, Inositol phosphates: Synthesis and degradation, *J. Biol. Chem.* **263:**3051–3054.

McMurray, W. C., 1973, Phospholipids in subcellular organelles and membranes, in *Form and Function of Phospholipids* (G. B. Ansell, J. N. Hawthorne, and R. M. C. Dawson, eds.), 2nd ed., pp. 205–251, Elsevier Scientific Publishing Company, Amsterdam.

Michell, R. H., 1975, Inositol phospholipids and cell surface receptor function, *Biochim. Biophys. Acta* **415:**81–147.

Mills, N. C., Mills, T. M., and Means, A. R., 1977, Morphological and biochemical changes which occur during postnatal development and maturation of the rat testis, *Biol. Reprod.* **16:**124–130.

Monaco, M. E., 1987, Inositol metabolism in WRK-1 cells. Relationship of hormone-sensitive to -insensitive pools of phosphoinositides, *J. Biol. Chem.* **262:**13001–13006.

Morré, D. J., Kartenback, J., and Franke, W. W., 1979, Membrane flow and interconversions among endomembranes, *Biochim. Biophys. Acta* **559:**71–152.

Mullikin, L. J., and Helmkamp, G. M., Jr., 1984, Bovine brain phosphatidylinositol transfer protein. Selective inhibition by chlorpromazine and other amphiphilic amines, *J. Biol. Chem.* **259:**2764–2768.

Nichols, J. W., 1987, Binding of fluorescent-labeled phosphatidylcholine to rat liver nonspecific lipid transfer protein, *J. Biol. Chem.* **262:**14172–14177.

Nichols, J. W., 1988, Kinetics of fluorescent-labeled phosphatidylcholine transfer between nonspecific lipid transfer protein and phospholipid vesicles, *Biochemistry* **27:**1889–1896.

Nichols, J. W., and Pagano, R. E., 1983, Resonance energy transfer assay of protein-mediated lipid transfer between vesicles, *J. Biol. Chem.* **258:**5368–5371.

Nijjar, M. S., and Hawthorne, J. N., 1977, Purification and properties of polyphosphoinositide phosphomonoesterase from rat brain, *Biochim. Biophys. Acta* **480:**390–402.

Nikawa, J., Kodaki, T., and Satoshi, Y., 1987, Primary structure and disruption of the phosphatidylinositol synthase gene of *Saccharomyces cerevisiae, J. Biol. Chem.* **262:**4876–4881.

Nilsson, O. S., and Dallner, G., 1979, Transverse asymmetry of phospholipids in subcellular membranes of rat liver, *Biochim. Biophys. Acta* **464:**453–468.

Nishihara, M., and Keenan, R. W., 1985, Inositol phospholipid levels of rat forebrain obtained by freeze-blowing method, *Biochim. Biophys. Acta* **835:**415–418.

Nishizuka, Y., 1986, Studies and perspectives on protein kinase C, *Science* **233:**305–312.

Nyquist, D. A., and Helmkamp, G. M., Jr., 1989, Developmental patterns in rat brain of phosphatidylinositol synthetic enzymes and phosphatidylinositol transfer protein, *Biochim. Biophys. Acta* **987:**165–170.

Op den Kamp, J. A. F., 1979, Lipid asymmetry in membranes, *Annu. Rev. Biochem.* **48:**47–71.

Parries, G. S., and Hokin-Neaverson, M., 1984, Phosphatidylinositol synthase from canine pancreas: Solubilization by *n*-octyl glucopyranoside and stabilization by manganese, *Biochemistry* **23:**4785–4791.

Parthasarathy, R., and Eisenberg, F., Jr., 1986, The inositol phospholipids: A stereochemical view of biological activity, *Biochem. J.* **235:**313–322.

Pastuszyn, A., Noland, B. J., Bazan, J. F., Fletterick, R. J., and Scallen, T. J., 1987, Primary sequence and structural analysis of sterol carrier protein 2 from rat liver: Homology with immunoglobulins, *J. Biol. Chem.* **262:**13219–13227.

Paulus, H., and Kennedy, E. P., 1960, The enzymatic synthesis of inositol monophosphatide, *J. Biol. Chem.* **235:**1303–1311.

Poorthuis, B. J. H. M., van der Krift, T. P., Teerlink, T., Akeroyd, R., Hostetler, K. Y., and Wirtz, K. W. A., 1980, Phospholipid transfer activities in Morris hepatomas and the specific contribution of the phosphatidylcholine exchange protein, *Biochim. Biophys. Acta* **600:**376–386.

Poorthuis, B. J. H. M., Glatz, J. F. C., Akeroyd, R., and Wirtz, K. W. A., 1981, A new high-yield procedure for the purification of the non-specific phospholipid transfer protein from rat liver, *Biochim. Biophys. Acta* **665:**256–261.

Possmayer, F., 1974, Evidence for a specific phosphatidyl inositol transferring protein in rat brain, *Brain Res.* **74:**167–174.

Quist, E. E., and Barker, R. C., 1983, Properties of phosphatidylinositol kinase activities in rabbit erythrocyte membranes, *Arch. Biochem. Biophys.* **222:**170–178.

Read, R. J., and Funkhouser, J. D., 1983, Properties of a non-specific phospholipid-transfer protein purified from rat lung, *Biochim. Biophys. Acta* **752:**118–126.

Roach, P. D., and Palmer, F. B. S. C., 1981, Human erythrocyte cytosol phosphatidylinositol-bisphosphate phosphatase (E.C. 3.1.3.36), *Biochim. Biophys. Acta* **661:**323–333.

Ruenswongsa, P., Singh, H., and Jungalwala, F. B., 1979, Protein-catalyzed exchange of phosphatidylinositol between rat brain microsomes and myelin, *J. Biol. Chem.* **254:**9385–9393.

Runquist, E. A., and Helmkamp, G. M., Jr., 1988, Effect of acceptor membrane phosphatidylcholine on the catalytic activity of bovine liver phosphatidylcholine transfer protein, *Biochim. Biophys. Acta* **940**:21–32.

Ryu, S. H., Suh, P.-G., Cho, K. S., Lee, K.-Y., and Rhee, S. G., 1987, Bovine brain cytosol contains three immunologically distinct forms of inositolphospholipid-specific phospholipase C, *Proc. Natl. Acad. Sci. U.S.A.* **84**:6649–6653.

Saltiel, A. R., Fox, J. A., Sherline, P., Sahyoun, N., and Cuatrecasas, P., 1987, Purification of phosphatidylinositol kinase from bovine brain myelin, *Biochem. J.* **241**:759–763.

Scallen, T. J., Noland, B. J., Gavey, K. L., Bass, N. M., Ockner, R. K., Chanderbhan, R., and Vahouny, G. V., 1985, Sterol carrier protein 2 and fatty acid-binding protein. Separate and distinct physiological functions, *J. Biol. Chem.* **260**:4733–4739.

Schermoly, M. J., and Helmkamp, G. M., Jr., 1983, The inactivity of brain phospholipid transfer toward phosphatidylinositol 4-phosphate, *Brain Res.* **268**:197–200.

Schroeder, F., 1985, Fluorescence probes unravel asymmetric structure of membranes, in *Subcellular Biochemistry* (D. B. Roodyn, ed.), Vol. 11, pp. 51–101, Plenum Press, New York.

Sefton, B. M., and Buss, J. E., 1987, The covalent modification of eukaryotic proteins with lipid, *J. Cell Biol.* **104**:1449–1453.

Sekar, M. C., Dixon, J. F., and Hokin, L. E., 1987, The formation of inositol 1,2-cyclic 4,5-trisphosphate and inositol 1,2-cyclic 4-bisphosphate on stimulation of mouse pancreatic minilobules with carbamoylcholine, *J. Biol. Chem.* **262**:340–344.

Setty, B. N. Y., Graeber, J. E., and Stuart, M. J., 1987, The mitogenic effect of 15- and 12-hydroxyeicosatetraenoic acid on endothelial cells may be mediated via diacylglycerol kinase inhibition, *J. Biol. Chem.* **262**:17613–17622.

Seyfred, M. A., and Wells, W. W., 1984, Subcellular incorporation of ^{32}P into phosphoinositides and other phospholipids in isolated hepatocytes, *J. Biol. Chem.* **259**:7659–7665.

Sharpe, R. M., 1986, Paracrine control of the testis, *Clin. Endocrin. Metab.* **15**:185–207.

Sleight, R. G., 1987, Intracellular lipid transport in eukaryotes, *Annu. Rev. Physiol.* **49**:193–208.

Somerharju, P., Van Paridon, P., and Wirtz, K. W. A., 1983, Phosphatidylinositol transfer protein from bovine brain. Substrate specificity and membrane binding properties, *Biochim. Biophys. Acta* **731**:186–195.

Stubbs, E. B., Jr., Kelleher, J. A., and Sun, G. Y., 1988, Phosphatidylinositol kinase, phosphatidylinositol-4-phosphate kinase and diacylglycerol kinase activities in rat brain subcellular fractions, *Biochim. Biophys. Acta* **958**:247–254.

Suarez-Quian, C. A., O'Shea, J. J., and Klausner, R. D., 1987, Partially purified phosphatidylinositol kinase does not catalyze the formation of phosphatidylinositol 4,5-bisphosphate, *Biochem. Biophys. Res. Commun.* **143**:512–516.

Teerlink, T., van der Krift, T. P., Post, M., and Wirtz, K. W. A., 1982, Tissue distribution and subcellular localization of phosphatidylcholine transfer protein in rats as determined by radioimmunoassay, *Biochim. Biophys. Acta* **713**:61–67.

Teerlink, T., van der Krift, T. P., van Heusden, G. P. H., and Wirtz, K. W. A., 1984, Determination of nonspecific lipid transfer protein in rat tissues and Morris hepatomas by enzyme immunoassay, *Biochim. Biophys. Acta* **793**:251–259.

Thomas, P. J., Wendelburg, B. E., Venuti, S. E., and Helmkamp, G. M., Jr., 1989, Mature rat testis contains a high molecular weight species of phosphatidylinositol transfer protein, *Biochim. Biophys. Acta* **982**:24–30.

Thompson, D. M., Cochet, C., Chambaz, E. M., and Gill, G. N., 1985, Separation and characterization of a phosphatidylinositol kinase activity that co-purifies with the epidermal growth factor receptor, *J. Biol. Chem.* **260**:8824–8830.

Van den Besselaar, A. M. H. P., Helmkamp, G. M., Jr., and Wirtz, K. W. A., 1975, Kinetic model of the protein-mediated phosphatidylcholine exchange between single bilayer liposomes, *Biochemistry* **14:**1852–1858.

Van den Bosch, H., 1980, Intracellular phospholipases A, *Biochim. Biophys. Acta* **604:**191–246.

Van Dongen, C. J., Zwiers, H., and Gispen, W. H., 1984, Purification and partial characterization of the phosphatidylinositol 4-phosphate kinase from rat brain, *Biochem. J.* **223:**197–203.

Van Dongen, C. J., Kok, J. W., Schrama, L. H., Oestreicher, A. B., and Gispen, W. H., 1986, Immunochemical characterization of phosphatidylinositol 4-phosphate kinase from rat brain, *Biochem. J.* **233:**859–864.

Van Golde, L. M. G., Oldenborg, V., Post, M., Batenburg, J. J., Poorthuis, B. J. H. M., and Wirtz, K. W. A., 1980, Phospholipid transfer proteins in rat lung. Identification of a protein specific for phosphatidylglycerol, *J. Biol. Chem.* **255:**6011–6013.

Van Paridon, P. A., Gadella, T. W. J., Jr., Somerharju, P. J., and Wirtz, K. W. A., 1987a, On the relationship between the dual specificity of the bovine brain phosphatidylinositol transfer protein and membrane phosphatidylinositol levels, *Biochim. Biophys. Acta* **903:**68–77.

Van Paridon, P. A., Visser, A. J. W. G., and Wirtz, K. W. A., 1987b, Binding of phospholipids to the phosphatidylinositol transfer protein from bovine brain as studied by steady-state and time-resolved fluorescence spectroscopy, *Biochim. Biophys. Acta* **898:**172–180.

Van Paridon, P. A., and Gadella, T. W. J., Jr., and Wirtz, K. W. A., 1988, The effect of polyphosphoinositides and phosphatidic acid on the phosphatidylinositol transfer protein from bovine brain: A kinetic study, *Biochim. Biophys. Acta* **943:**76–86.

Venuti, S. E., and Helmkamp, G. M., Jr., 1988a, Regional distribution in rat brain of phosphatidylinositol and phosphatidylcholine synthetic and intermembrane transfer activities, *Neurochem. Int.* **13:**531–541.

Venuti, S. E., and Helmkamp, G. M., Jr., 1988b, Tissue distribution, purification, and characterization of rat phosphatidylinositol transfer protein, *Biochim. Biophys. Acta* **946:**119–128.

Waite, M., 1986, Phospholipases of mammalian cells, in *Lipids and Membranes: Past, Present and Future* (J. A. F. Op den Kamp, B. Roelofsen, and K. W. A. Wirtz, eds.), pp. 207–230, Elsevier Biomedical Press, Amsterdam.

Walker, D. H., and Pike, L. J., 1987, Phosphatidylinositol kinase is activated in membranes derived from cells treated with epidermal growth factor, *Proc. Natl. Acad. Sci. U.S.A.* **84:**7513–7517.

Wetterau, J. R., and Zilversmit, D. B., 1984, Quantitation of lipid transfer activity, *Methods Biochem. Anal.* **30:**199–226.

White, D. A., 1973, The phospholipid composition of mammalian tissues, in *Form and Function of Phospholipids* (G. B. Ansell, J. N. Hawthorne, and R. M. C. Dawson, eds.), 2nd ed., pp. 441–482, Elsevier Biomedical Press, Amsterdam.

Whitman, M., Downes, C. P., Keeler, M., Keller, T., and Cantley, L., 1988, Type I phosphatidylinositol kinase makes a novel phospholipid, phosphatidylinositol 3-phosphate, *Nature* **332:**644–646.

Wilson, D. B., Neufeld, E. J., and Majerus, P. W., 1985, Phosphoinositide interconversion in thrombin-stimulated human platelets, *J. Biol. Chem.* **260:**1049–1051.

Wirtz, K. W. A., 1982, Phospholipid transfer proteins, in *Lipid-Protein Interactions* (P. C. Jost and O. H. Griffith, eds.), Vol. 1, pp. 151–231, John Wiley & Sons, New York.

Wirtz, K. W. A., Jolles, J., Westerman, J., and Neys, F., 1976, Phospholipid exchange proteins in synaptosome and myelin fraction from rat brain, *Nature* **260:**354–355.

Yamada, K., Sasaki, T., and Sakagami, T., 1978, Separation and purification of phospholipid exchange proteins in rat small intestinal mucosa, *J. Biochem.* **84:**855–863.

Yoshimura, T., and Helmkamp, G. M., Jr., 1984, Bovine brain phosphatidylinositol transfer protein. Effects of pH, ionic strength and lipid composition on transfer activity, *Biochim. Biophys. Acta* **793:**463–470.

Yoshimura, T., Welti, R., and Helmkamp, G. M., Jr., 1988, General kinetic model for protein-mediated phospholipid transfer between membranes, *Arch. Biochem. Biophys.* **266**:299–312.

Zachowski, A., Favre, E., Cribier, S., Hervé, P., and Devaux, P. F., 1986, Outside-inside translocation of aminophospholipids in the human erythrocyte membrane is mediated by a specific enzyme, *Biochemistry* **25**:2585–2590.

Zborowski, J., and Demel, R. A., 1982, Transfer properties of the bovine brain phospholipid transfer protein. Effect of charged phospholipids and of phosphatidylcholine fatty acid composition, *Biochim. Biophys. Acta* **688**:381–387.

Chapter 7

Intracellular Fatty-Acid-Binding Proteins

Characteristics and Function

René J. A. Paulussen and Jacques H. Veerkamp

1. EXTRACELLULAR TRANSPORT OF FATTY ACIDS

Fatty acids are distributed by the bloodstream to peripheral tissues that utilize these molecules for energy production, for storage, or in biosynthetic pathways. Fatty acids are transported in plasma in the form of triacylglycerols in chylomicrons and very-low-density lipoproteins, or noncovalently bound to serum albumin. Lipoprotein lipase liberates the ester-bound fatty acids. On their way from the capillary to the intracellular organelles, fatty acids have to pass a series of aqueous spaces and membrane barriers (Figure 1). Fatty acids may be taken up from blood and transferred to the interstitial space by the endothelial cells of the capillary wall. Albumin–fatty acid complexes may also pass the endothelium through clefts or by way of plasmalemmal vesicles that pass through the capillary endothelial cell (Milici *et al.*, 1987). Specific albumin-binding proteins that may be instruments in a carrier-mediated uptake and transcytosis of albumin and its ligands have been identified in the capillary endothelial cells (Ghinea *et al.*,

Abbreviations used in this chapter: PMFABP, plasma membrane fatty-acid-binding protein; FABP, fatty-acid-binding protein; MDGI, mammary gland-derived growth inhibitor; CRBP, cellular retinol-binding protein; ELISA, enzyme-linked immunosorbent assay; K_d, dissociation constant; NMR, nuclear magnetic resonance; ANS, 1-anilino-8-naphthalene sulfonate; DOCA, deoxycorticosterone acetate; CoA, coenzyme A.

René J. A. Paulussen and Jacques H. Veerkamp Department of Biochemistry, University of Nijmegen, 6500 HB Nijmegen, The Netherlands

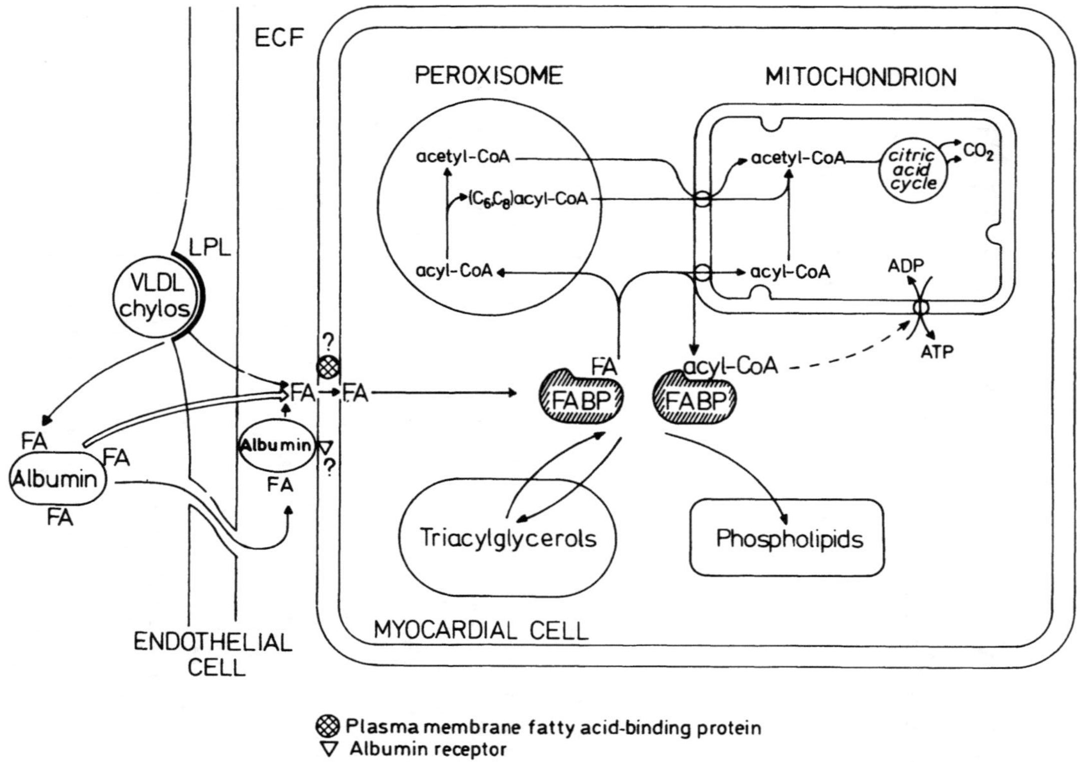

FIGURE 1. Extra- and intracellular transport of fatty acids. Abbreviations: FA, fatty acid; VLDL, very-low-density lipoproteins; chylos, chylomicrons; LPL, lipoprotein lipase; ECF, extracellular fluid.

1988). The presence of cytosolic FABP in endothelial cells of rat heart was demonstrated by immunoelectron microscopy (Fournier and Rahim, 1985) and in cultured endothelial cells by ELISA (Paulussen *et al.*, 1990). This protein may play an important role in the fatty acid transport through the endothelial cells. Albumin probably also plays an important role in fatty acid transport in the interstitial fluid (Julien *et al.*, 1978) and delivers fatty acids directly to the cell. By keeping the extracellular free fatty acid concentration low, albumin may prevent cytotoxic effects of fatty acids.

Quite another mechanism for fatty acid transport from capillaries to the sites of utilization was proposed by Scow and co-workers (Wetzel and Scow, 1984; Scow and Blanchette-Mackie, 1985) on the basis of electron-microscopic studies. In their model, fatty acids are transported by lateral movement in membranes that form a continuum between the capillary endothelium, and parenchymal cells.

2. CELLULAR FATTY ACID UPTAKE

The rate of fatty acid uptake by peripheral tissues depends on the extracellular and the intracellular fatty acid concentrations and the rate of fatty acid metabolism. The free fatty acid concentration in the extracellular fluid is determined by the total fatty acid concentration and the fatty acid : albumin ratio (Spector *et al.*, 1971; Wosilait and Soler-Argilaga, 1977). Fatty acid uptake in cultured and in isolated heart cells (Rauch *et al.*, 1987; Samuel *et al.*, 1976) appears to be related to the total rather than the free fatty acid concentration. The rate of uptake in Ehrlich ascites cells (Spector *et al.*, 1965), perfused rat liver (Soler-Argilaga *et al.*, 1974), rat hepatocytes (Renaud *et al.*, 1985), and cultured human leukemic myeloid cells (Morand *et al.*, 1982) was dependent on the free fatty acid concentration.

Several models have been proposed to explain the mechanism of cellular uptake of fatty acids (Figure 2). The simplest one is passive diffusion through the lipid bilayer of the plasma membrane (Spector *et al.*, 1965; Soler-Argilaga *et al.*, 1974; Dietschy, 1978; Morand *et al.*, 1982; Renaud *et al.*, 1985). For total fatty acids a serum-tissue concentration gradient was found in rat, canine, and human heart tissue (Rose and Goresky, 1977; Van Der Vusse *et al.*, 1983). Fatty acids from the aqueous phase can enter the outer leaflet of the plasma membrane and spontaneously flip-flop to the inner half (Figure 2A). The partition coefficient between unilamellar vesicles and the aqueous phase for palmitic acid is 4×10^6 in favor of the bilayer (Noy and Zakim, 1985) and increases with chain length, together with the passive permeability coefficient (Dietschy, 1978). At physiological pH, the flip-flop rate may be rather high, since approximately 50% of the fatty acids in membranes are un-ionized (Hamilton and Cistola, 1986.) To our knowledge, however, the rate of transbilayer movement of fatty acids has not

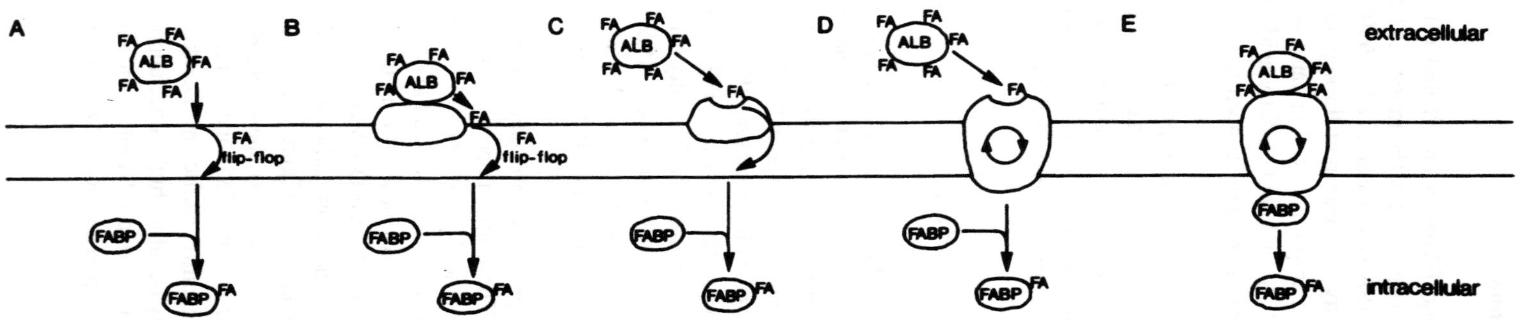

FIGURE 2. Models for the cellular uptake of fatty acids. (A) Passive diffusion of fatty acids through the plasma membrane. (B) Membrane-associated albumin receptor, which facilitates the dissociation of the albumin fatty acid complex. Uptake by passive diffusion. (C) Extrinsic membrane FABP. Uptake by facilitated diffusion. (D) Intrinsic membrane fatty acid translocator. Additional alternatives: (1) extracellular albumin-binding site; (2) intracellular FABP-binding site. (E) Intrinsic membrane fatty acid translocator, with extracellular albumin-binding site and intracellular FABP receptor function.

been determined. It is at least as high as the rate of hydration of fatty acids bound to the bilayer (Noy *et al.*, 1986). For passive diffusion the maximal rate of uptake of a fatty acid will equal the product of its maximal concentration in the extracellular fluid and its passive permeability coefficient (Dietschy, 1978). In this model the rate-limiting step may become either the dissociation of fatty acids from albumin or the transport of fatty acids from the capillary to the plasma membrane. Nevertheless, pure passive diffusion cannot explain the sex differences in fatty acid uptake in perfused rat liver (Kushlan *et al.*, 1981).

Kinetic studies of single-pass hepatic uptake of fatty acids suggested that uptake is mediated primarily by interaction of the albumin–ligand complex with the hepatocyte surface, rather than by the small fraction of unbound fatty acid (Ockner *et al.*, 1983) (Figure 2B). In more recent studies, catalysis of the dissociation of oleic acid from albumin by an albumin receptor could not be established. Dissociation of the bound ligand from albumin could be the rate-limiting step in ligand transport. Therefore, the uptake apparently becomes dependent on the bound ligand (Weisiger and Ma, 1987). The pH and the acyl chain length are important factors in the regulation of fatty acid transfer from albumin to membranes (Spector, 1969; Daniels *et al.*, 1985; Hamilton and Cistola, 1986). According to Noy *et al.* (1986), uptake in the perfused liver is not limited by kinetic factors, but instead is determined by the equilibrium distribution of fatty acids between albumin and the plasma membrane. A change in the molar fatty acid : albumin ratio therefore has a marked influence on fatty acid uptake. The intrahepatic pool sizes for palmitic acid that were predicted on the basis of this model correspond well to the measured values at palmitate : albumin ratios of 2 and 4 (Cooper *et al.*, 1987).

Saturation of uptake of fatty acids cannot be invoked as evidence for specific transport systems (Noy *et al.*, 1986; Cooper *et al.*, 1987). Results of studies of the presence of an albumin receptor in plasma membranes of liver and adipocytes are not equivocal (Weisiger *et al.*, 1981; Brandes *et al.*, 1982; Stremmel *et al.*, 1983b). Influx kinetics indicated a membrane-associated fatty acid transport system in adipocytes (Abumrad *et al.*, 1981, 1984; Schwieterman *et al.*, 1988), hepatocytes (Stremmel and Berk, 1986; Stremmel and Thielmann, 1986; Stremmel *et al.*, 1986), and cardiomyocytes (Sorrentino *et al.*, 1988; Stremmel, 1988). The fatty acid uptake could be inhibited by phloretin and by trypsin, suggesting that a carrier protein was involved (Figure 2C–E). Uptake in rat hepatocytes (Stremmel *et al.*, 1986) and basolateral liver plasma membrane vesicles (Stremmel, 1987) was potential sensitive and sodium coupled and could be inhibited by ouabain and metabolic inhibitors, in contrast to uptake in adipocytes (Abumrad *et al.*, 1981, 1984; Schwieterman *et al.*, 1988) and cardiac myocytes (Sorrentino *et al.*, 1988; Stremmel, 1988).

Heat- and trypsin-sensitive binding of oleic acid was observed with plasma membranes of liver (Stremmel *et al.*, 1983a, 1985b), jejunum (Potter *et al.*,

1987), and heart (Stremmel, 1988). A basic 40-kDa protein was isolated from plasma membranes of rat liver (Stremmel *et al.*, 1985b), from adipocytes (Potter *et al.*, 1987; Schwieterman *et al.*, 1988), and from rat jejunal microvillous membranes (Stremmel et al., 1985a) by procedures including oleate–agarose affinity chromatography. These plasma membrane fatty-acid-binding proteins (PMFABPs) are immunologically related, since all react with a polyclonal antibody raised against the protein from liver (Stremmel *et al.*, 1985a; Potter, *et al.*, 1987; Schwieterman *et al.*, 1988). They differ from the cytosolic FABPs with respect to molecular mass, isoelectric point, and structural and immunochemical properties (Stremmel *et al.*, 1985a,b; Potter *et al.*, 1987). The antibody against the PMFABP from liver can inhibit fatty acid uptake in hepatocytes (Stremmel *et al.*, 1986; Stremmel and Thielmann, 1986), adipocytes (Schwieterman *et al.*, 1988), and cardiac myocytes (Sorrentino *et al.*, 1988). It inhibits fatty acid binding to plasma membranes of liver (Stremmel *et al.*, 1985b) and jejunum (Stremmel *et al.*, 1985a). An antibody against the PMFABP from heart inhibits oleic acid binding to heart plasma membranes and its influx in cardiomyocytes (Stremmel, 1988). The evidence presented for the existence of closely related, if not identical, PMFABPs in various tissues sustains the idea that fatty acid uptake does not occur by diffusion only, but that it also is a facilitated process. The differences among hepatocytes, adipocytes, and cardiomyocytes in the sensitivity of fatty acid uptake to membrane potential, sodium concentration, and energy deprivation may relate to variations in other components or to the microenvironment of the transport system. The liver cell plasma membrane contains multiple transport systems for charged compounds, with overlapping specificity (Berk *et al.*, 1987). Recent studies gave however apparently conflicting data (Potter *et al.*, 1989).

The PMFABP could serve as an acceptor protein that is essential in the sequestration of fatty acids from the extracellular fluid, and/or influences the albumin–fatty acid binding. The fatty acid will flip-flop to the inner side of the bilayer, where translocation to cytosolic FABP takes place (Fig. 2C). Alternatively, other membrane proteins may take over the fatty acid and translocate it into the cell. Also, the protein itself may form the actual transporter, transferring the fatty acid from one side of the plasma membrane to the other. In that case, it could undergo a conformational change (Fig. 2D). The PMFABP may also interact with the albumin–fatty acid complex at the outer surface of the membrane and/or with the FABP at the cytosolic side (Fig. 2E). Little information was available on the primary and secondary structures of the PMFABPs. The protein bears a striking resemblance to mitochondrial glutamate–oxaloacetate transaminase. Their high isoelectric point of 9.0 suggests an electrostatic interaction with a fatty acid anion. The extractability from membranes of the plasma membrane PMFABP by sonication at a high salt concentration (Potter *et al.*, 1987) and the inhibition of fatty acid uptake by antibodies against this protein suggest that it is a superficial protein rather than an integral membrane protein that traverses the entire thickness of the lipid bilayer. Apparently, it is a common fatty acid acceptor in various tissues. Spontaneous transbilayer movement of the fatty acids may occur, or an additional, tissue-specific transfer process may take place after binding of the fatty acid. Such a complex system may be comparable to that of *Escherichia coli* (Nunn *et al.*, 1986). Recent observations on the rapid translocation of oleic acid across the erythrocyte membrane raise doubts on the physiological role of a carrier-mediated translocation of fatty acids across plasma membranes (Bröring et al., 1989). The saturation kinetics of the initial rate of fatty acid uptake may be the result of a simple diffusive equilibration of fatty acids between extracellular albumin and cytosolic FABP (Fig. 2A).

3. INTRACELLULAR TRANSPORT OF FATTY ACIDS

The cellular uptake of fatty acids is also dependent on their metabolism. In this respect, there are large differences among tissues. The intracellular systems utilize fatty acids for oxidation and for synthesis of triacylglycerols, phospholipids, cholesterol, sphingolipids, and eicosanoids, with special preferences for certain fatty acids. Fatty acid synthesis also takes place in various tissues. Therefore, intracellular transport and trafficking of fatty acids are important processes. Transfer of fatty acids through the intracellular water phase can take place as monomers, but the cytosolic FABPs may have an important function in the targeting of fatty acids. They may also play a modulatory role in cellular metabolism. The structural and functional aspects of these cytosolic proteins will be surveyed in the next section.

FABPs form a class of proteins with a molecular mass of 14–15 kDa, which are abundantly present in the cytoplasm of tissues that are involved in the uptake or utilization of fatty acids. Their name was derived from their high affinity for long-chain fatty acids. The FABPs are members of a large group of low-molecular-mass, soluble proteins that are involved in the binding of various hydrophobic substances. They were first discovered when Levi and co-workers fractionated the cytosol (105,000 × *g* supernatant) of liver in their search for proteins that account for the binding of bromosulfophthalein and bilirubin (Levi *et al.*, 1969a). They found two major fractions involved in binding of these ligands; these fractions were designated Y and Z. In the Y class, with molecular masses of 30–50 kDa, binding of organic anions is mainly accounted for by a group of glutathione *S*-transferases. The Z class contains a number of proteins, with molecular masses between 10 and 20 kDa, that each bind a different class of hydrophobic ligands. Among these are the phospholipid transfer proteins, sterol carrier proteins, retinoid-binding proteins, and FABPs. The study of Levi *et al.* (1969a) initiated an extensive research of the proteins involved in binding of organic anions.

FABPs were first recognized in the early 1970s (Ockner *et al.*, 1972; Mishkin *et al.*, 1972; Ketterer *et al.*, 1976). In these studies cytosolic proteins were incubated with [^{14}C]oleic acid and then chromatographed over a gel filtration column. The radioactive label was recovered in two fractions, one with a molecular mass of approximately 68 kDa, containing albumin, and one of about 15 kDa (often given as 12 kDa in the early reports). This fraction was called the fatty-acid-binding protein (FABP). Soon this FABP fraction was demonstrated in a large number of tissues (Table I). Recently the presence of a specific acyl-CoA binding protein was established in various bovine and rat tissues. The protein was isolated from bovine and rat liver and had a molecular mass of 10 kDa. Its amino acid sequence did not show any homology with that of cytosolic FABP. The protein did not bind fatty acids, but it was also mainly located in the cytosol (Mogensen *et al.*, 1987: Mikkelsen *et al.*, 1987: Mikkelsen and Knudsen, 1987; Knudsen *et al.*, 1989). This review will further be focussed on the cytosolic FABPs and not discuss the structure and function of the acyl-CoA binding protein and its relation to the FABPs. In the last few years several reviews have been published on cytosolic FABPs in general (Glatz and Veerkamp, 1985; Sweetser *et al.*, 1987b; Bernier and Jollès, 1987; Bass, 1988), hepatic and intestinal (Bass, 1985) and cardiac (Glatz *et al.*, 1985b) FABPs, and the role of FABPs in muscular fatty acid metabolism (Veerkamp and Paulussen, 1987).

Table I
Occurrence of FABP

	Rat	Human	Pig	Cow	Mouse	Rabbit	Guinea pig	Chicken
Liver	+	+	+	+	+	+	+	+
Heart	+	+	+	+		+	+	+
Intestine	+	+		+	+			+
Skeletal muscle	+	+	+	+		+		+
Kidney	+	+	+					
Spleen	+		+					
Brain	+	+						
Adipose tissue	+	+	+	+				+
Lung	+	+	+					
Adrenal	+							
Mammary gland	+			+				
Placenta	+	+	+			+		
Testes	+							
Retina								+
Aorta	+							
Serum	−	−						−
Erythrocyte	−	−						

4. QUANTITATION OF FATTY-ACID-BINDING PROTEINS

FABPs in cytosols have been quantitated by a variety of assay methods (Table II), which each have their specific advantages and shortcomings. The method of coelution of proteins with radioactive fatty acid, which was used in many studies, is inadequate for the quantitative determination of FABP because of the relatively high affinity of fatty acids for the gel filtration matrix and the fact that this method requires nonequilibrium conditions. These disadvantages also apply to electrophoretic quantitation of FABP in the presence of labeled ligand.

Fatty-acid-binding assays for FABP that enable postcolumn detection (off-line) of the protein are useful for quantitative purposes only in the absence of other proteins with fatty-acid-binding activity, e.g., albumin. Removal of albumin by immunoaffinity chromatography is therefore a necessary step prior to FABP determination.

Assessment of FABP by affinity chromatography on oleoyl-aminohexylamino-Sepharose (Kawashima *et al.*, 1983) involves the use of denaturating conditions for elution of the protein. Thus, this method is not suitable for a study of fatty-acid-binding characteristics of FABPs. For quantitation of FABPs in terms of total binding capacity and investigation of fatty-acid-binding characteristics, assays are available that involve binding of radioactive or fluorescent fatty acids. The radioactive fatty acids may be presented to the protein either free in aqueous solution or incorporated in multilamellar liposomes. After establishment of the binding equilibrium, free and protein-bound ligands can be separated with dextran-coated charcoal (Morrow and Martin, 1983) or Lipidex 1000 (Glatz and Veerkamp, 1983) or by centrifugation in the case of liposomes (Brecher *et al.*, 1984). With fluorescent labels, binding is determined from the enhancement of fluorescence that occurs upon binding of the fatty acid analogue, requiring no

Table II
Assays of FABPs

Coelution on gel filtration
Electrophoretic comigration
Immunological assay with specific antibodies
Affinity chromatography on oleoyl-aminohexylamino-Sepharose
Equilibrium dialysis
Binding of fluorescent fatty acid analog
Transfer of labeled fatty acid from (multilamellar) liposomes
Radiochemical binding assay (charcoal, Lipidex 1000)
Nuclear magnetic resonance
Electron spin resonance

separation of bound and free ligand (Wilkinson and Wilton, 1986; Peeters *et al.*, 1989a). This assay is most sensitive for liver FABPs, owing to their high enhancement of fluorescence as a result of binding of fluorescent fatty acids (Peeters *et al.*, 1989a). Before the determination of its fatty-acid-binding capacity, a protein sample has to be delipidated to prevent competition of endogenously bound ligand, especially at low fatty acid concentrations (Glatz *et al.*, 1984). Lipidex 1000 is preferred both for delipidation and as absorbent for unbound fatty acids, because, unlike charcoal, it does not retain or denature protein (Glatz and Veerkamp, 1983).

The immunochemical detection of FABP depends on the availability of appropriate antibodies and can only provide quantitative data without further information on functional properties. The use of antibodies involves the risk of cross-reactivity with other proteins. These may be other FABP types, or even nonrelated proteins. Therefore, the antibodies that are applied in these assays must be well characterized, e.g., by immunoblotting, with respect to their monospecificity on cytosolic protein preparations and their reactivity with distinct FABP types. Recently two well-documented immunochemical studies of FABPs were presented (Crisman *et al.*, 1987; Paulussen *et al.*, 1989a). Physical methods as nuclear magnetic resonance and electron spin resonance spectroscopy or fluorescence are especially suited for determination of the binding constants and the binding number of fatty acids to FABPs, since no separation of bound and unbound fatty acids is necessary (Cistola *et al.*, 1988, 1989; Fournier and Richard, 1988; Peeters *et al.*, 1989a).

5. ISOLATION OF FATTY-ACID-BINDING PROTEINS

FABPs are relatively abundant in some tissues. They make up 2–6% of total cytosolic protein in heart and liver (Glatz *et al.*, 1984; Paulussen *et al.*, 1986; Wilkinson and Wilton, 1987b). Since their discovery, they have been isolated from a number of tissues from different species. Most isolation procedures are based on a combination of gel filtration and ion exchange chromatography (e.g., Glatz *et al.*, 1985a; Offner *et al.*, 1986; Paulussen *et al.*, 1988; Peeters *et al.*, 1989a). Sometimes affinity chromatography (Ockner *et al.*, 1982), preparative isoelectric focusing (Ketterer *et al.*, 1976; Fournier *et al.*, 1978; Kamisaka *et al.*, 1978; Trulzsch and Arias, 1981; Haq *et al.*, 1982; Takahashi *et al.*, 1983; Lee and Wiggert, 1984; Senjo *et al.*, 1985), heat treatment (Billheimer and Gaylor, 1980; Haunerland *et al.*, 1984; Jagschies *et al.*, 1985; Unterberg *et al.*, 1986) or ammonium sulfate precipitation (Haunerland *et al.*, 1984; Jagschies *et al.*, 1985; Unterberg *et al.*, 1986; Takahashi *et al.*, 1982; Avanzati and Catala, 1983) is incorporated. A detailed comparison of the FABP isolation protocols from different sources reveals clear-cut differences, especially concerning the buffers used in the ion exchange steps. These differences reflect the structural differences of the FABP types isolated from different tissues. It appears that the pure FABP preparations obtained from the liver generally have a higher specific fatty-acid-binding activity than those from the heart and other tissues (Glatz *et al.*, 1985a; Paulussen *et al.*, 1988; Peeters *et al.*, 1989a).

6. STRUCTURAL FEATURES OF FATTY-ACID-BINDING PROTEINS

6.1. Amino Acid Composition and Primary Structure

The existence of distinct FABP types becomes clear from the comparison of their amino acid composition, from amino acid sequencing, and from the nucleotide sequences of cloned cDNA or mRNA. The data reported for the amino acid composition of heart and liver FABPs from cattle, humans, pigs, and rats are summarized, together with values for intestinal FABPs from humans and rats in Table III. Except for some discrepancies (Billheimer and Gaylor, 1980; Alpers *et al.*, 1984; Haunerland *et al.*, 1984; Jagschies *et al.*, 1985; Unterberg *et al.*, 1986), all these studies agree that cysteine is present only in hepatic FABPs, whereas tryptophan is present in all FABP types, except the liver. Intestinal FABPs contain no proline, and their serine content is lower than that of the other FABPs. The threonine and leucine contents of heart FABP are higher. The arginine and asparagine (and aspartic acid) contents are higher in heart than in liver FABP, but they are even markedly more elevated in intestinal FABPs. Isoleucine and glutamine/glutamic acid residues is the highest in liver FABPs. Recently the amino acid composition of a rat kidney-specific FABP was described (Lam *et al.*, 1988). This FABP contains both cysteine and tryptophan residues.

At least two isoforms can be distinguished in human, pig, and rat liver FABPs; they behave differently on DEAE–anion exchange chromatography (Ketterer *et al.*, 1976; Trulzsch and Arias, 1981; Takahashi *et al.*, 1983; Bass, 1985; Peeters *et al.*, 1989a). No significant differences could be detected in the amino acid composition of the isoforms (Ketterer *et al.*, 1976; Trulzsch and Arias, 1981; Takahashi *et al.*, 1983; Peeters *et al.*, 1989a). The isoforms contain different endogenously bound fatty acids (Bass, 1985; Takahashi *et al.*, 1983). Bovine heart and liver FABPs could be separated into two isoforms with distinct pI values by CM-cellulose cation exchange chromatography (Jagschies *et al.*, 1985) and isoelectric focusing (Haunerland *et al.*, 1984), respectively. Only the bovine liver FABP isoforms differ in amino acid composition (Haunerland *et al.*, 1984).

A number of characteristics of FABPs from different tissues are given in Table IV. The amino-terminal amino acid is acetylated (Gordon *et al.*, 1983;

Table III
General Amino Acid Composition of Heart and Liver FABPs from Cattle, Humans, Pigs, and Rats, and Intestinal FABPs from Humans and Rats

Amino acid	Composition[a] (residues per mol of FABP) in: Liver	Heart	Intestine
Asx	11–13	14–17	19–21
Thr	10–12	17–19 (7 for pigs)	10
Ser	6–7	6–9	1–5
Glx	16–20	12–17	15–18
Pro	1–4	1–2 (6 for pigs)	0
Gly	11–13	9–12	10–14
Ala	2–6	5–7 (11 for pigs)	6–7
Cys	1–3	0	0
Val	9–12	8–15	9–11
Met	1–7	2–3	1–3
Ile	6–12	3–7	7
Leu	3–10	9–14	10
Tyr	1–3	2–3	3–4
Phe	6–9	5–6	8–9
Trp	0	1–4	1–2
His	1–2	2–4	1–2
Lys	13–17	10–18	14–16
Arg	2–3	3–5	6–7

[a]Data are derived from Alpers *et al.* (1984), Billich *et al.* (1988), Boguski *et al.* (1984), Chan *et al.* (1985), Gibson *et al.* (1988); Gordon *et al.* (1983), Haunerland *et al.* (1984), Jagschies *et al.* (1985), Lowe *et al.* (1985), Ockner *et al.* (1982), Offner *et al.* (1986, 1988), Paulussen *et al.* (1988), Peeters *et al.* (1989a), Sacchettini *et al.* (1986), Sweetser *et al.* (1987a), Takahashi *et al.* (1983), Unterberg *et al.* (1986).

Alpers *et al.*, 1984). This was thought to be an indication of their cytoplasmic destination (Gordon *et al.*, 1983), but the acetylation appears to be related to the residues adjacent to the amino-terminal amino acid (Flinta *et al.*, 1986).

Molecular masses of liver FABPs are slightly above 14 kDa, in contrast to 15 kDa for all other types. Isoelectric points of the hepatic FABPs are generally higher than those of the other FABP types.

The amino acid sequence of the FABPs appears to have been highly conserved throughout their evolution (Chan *et al.*, 1985). Corresponding FABP types from different species show a remarkably high degree of sequence homology (Table V). Human and rat liver FABPs have an 82% sequence identity (Lowe *et al.*, 1985). A similar sequence identity (82%) was found for human and rat intestinal FABPs (Sweetser *et al.*, 1987a). Bovine, human, and rat heart FABPs

Table IV
Structure of FABPs from Various Tissues

Origin	Molecular mass (Da)	No. of amino acids	IEP	N terminus	No. of residues per molecule: Cys	Trp	Pro	Reference
Rat								
Liver	14,272	127	n.d.[a]	Blocked	1	0	2	213
	14,184	127	5.5	Blocked	1	0	2	75
Intestine	15,062	132	n.d.	Blocked	0	1	0	5
Heart	14,776	133	n.d.	n.d.	0	1	1	43
	14,992	134	5.0	Blocked	0	2	1	192
Kidney	15,500	n.d.	5.3	n.d.	2	1	1	109
Quadriceps muscle	16,000	n.d.	5.2	n.d.	0	n.d.	n.d.	158
Human								
Liver	14,178	127	5.8	Blocked	1	0	1	36
Heart	15,500	n.d.	5.3	Blocked	2	2	2	220
	15,500	n.d.	5.2	n.d.	0	1	3	154
	14,768	132	5.2	Blocked	0	2	1	158
Intestine	15,100	132	n.d.	n.d.	0	1	0	210
Psoas muscle	15,000	n.d.	5.2	n.d.	0	2	3	158
Cattle								
Liver	11,800	n.d.	7.0	n.d.	2	0	1	85
Heart	15,300	n.d.	4.9	Blocked	3	2	1	83
	14,673	132	4.9	Blocked	1	2	2	21
Pig								
Liver	14,500	n.d.	5.8	n.d.	1	0	1	157
Heart	15,000	n.d.	4.9	n.d.	0	5	6	154

[a]n.d., Not determined.

also show a very high degree of homology (Billich *et al.*, 1988; Offner *et al.*, 1988). Rat intestinal FABP has only 31 and 26% homology with heart and liver FABP, respectively (Alpers *et al.*, 1984; Sacchettini *et al.*, 1986). The amino acid sequences of heart FABPs also have a large degree of homology with those of the murine adipocyte P_2 (p422) protein (Bernlohr *et al.*, 1984), the human (Suzuki *et al.*, 1982) and bovine (Kitamura *et al.*, 1980) myelin P_2 proteins, bovine mammary-derived growth inhibitor (MDGI) (Böhmer *et al.*, 1987), and the cellular retinoid-binding proteins (Sundelin *et al.*, 1985a; Li *et al.*, 1986).

No sequence homology between rat liver and intestinal FABPs (Gordon and Lowe, 1985) could be established with serum albumin (McLachlan and Walker, 1977), nonspecific lipid transfer protein (sterol carrier protein 2) (Westerman and

Table V
Amino Acid Sequence Homology of FABPs and Other Lipid-Binding Proteins

Protein	% Homology	Protein	% Homology
Rat heart FABP	100	Rat liver FABP	100
Human heart FABP	90	Human liver FABP	82
Bovine heart FABP	89	Rat intestinal FABP	26
Murine adipocyte P_2 (p422)	67	Rat heart FABP	38
Human myelin P_2	61	Human myelin P_2	27
Rat intestinal FABP	31	Murine adipocyte P_2 (p422)	23
Bovine MDGI	84		

Wirtz, 1985), phosphatidylcholine transfer protein (Akeroyd *et al.*, 1981b), and the apolipoproteins apo A-I and apo A-IV (Lowe *et al.*, 1985; Sweetser *et al.*, 1987a). Statistical analyses of mRNA and protein sequence data (Chan *et al.*, 1985) showed that distinct FABP types must have diverged some 650 million years ago, long before today's species diverged from their common ancestor. Divergence of FABPs and other, closely related proteins such as murine adipocyte P_2 (p422) and the myelin P_2 protein must have taken place several hundred million years earlier, before the invertebrate–vertebrate divergence (Chan *et al.*, 1985).

6.2. Secondary Structure

Circular dichroism measurements and computer-aided predictions of secondary structures (Chou and Fasman, 1979) reveal a relatively high content of α-helical structures in all FABPs (Table VI). Furthermore, a high percentage of β-structures is predicted. The α-helices and β-structures form an alternating pattern of hydrophobic and hydrophilic domains (Takahashi *et al.*, 1983; Offner *et al.*, 1988), which probably yield a protein with a hydrophilic surface for solubilization in the aqueous cytoplasm and a distinct hydrophobic domain for ligand binding. The predicted secondary structure of human liver FABP (Offner *et al.*, 1988) suggests that this protein may orient at a lipid/water interface, a property that may play an important role in the binding of fatty acids from membranes (Chan *et al.*, 1985).

Investigation of the amino acid sequence of rat and human liver FABPs presented evidence for the presence of tandemly arranged repeats that contain a relatively high degree of α-helix structure (Takahashi *et al.*, 1983; Lowe *et al.*, 1985). The existence of two regions of high homology, which contain tandem repeats, has also been found in rat intestinal (Gordon and Lowe, 1985) and rat (Sacchettini *et al.*, 1986) and human (Offner *et al.*, 1988) cardiac FABPs. No

Table VI
Structural Details of FABPs from Various Tissues[a]

<table>
<tr><th rowspan="3">Origin</th><th colspan="4">% of following structure:</th><th rowspan="3">Method</th><th rowspan="3">Reference</th></tr>
<tr><th rowspan="2">α-helix</th><th colspan="2">β-structures</th><th rowspan="2">Random coil</th></tr>
<tr><th>β-sheet</th><th>β-turn</th></tr>
<tr><td>Heart</td><td></td><td></td><td></td><td></td><td></td><td></td></tr>
<tr><td>Rat</td><td>22</td><td colspan="2">45</td><td>33</td><td>CD</td><td>150</td></tr>
<tr><td>Human</td><td>32</td><td colspan="2">44</td><td>24</td><td>CF</td><td>151</td></tr>
<tr><td></td><td>25</td><td colspan="2">45</td><td>30</td><td>CD</td><td>151</td></tr>
<tr><td>Cattle</td><td>15</td><td>51</td><td>8</td><td>26</td><td>CD</td><td>93</td></tr>
<tr><td></td><td>19</td><td>49</td><td>27</td><td>5</td><td>CF</td><td>21</td></tr>
<tr><td>Pig</td><td>19</td><td>48</td><td>8</td><td>25</td><td>CD</td><td>59</td></tr>
<tr><td>Liver</td><td></td><td></td><td></td><td></td><td></td><td></td></tr>
<tr><td>Rat</td><td>14</td><td colspan="2">39</td><td>47</td><td>CD</td><td>96</td></tr>
<tr><td></td><td>29</td><td>30</td><td>22</td><td>19</td><td>CF</td><td>213</td></tr>
<tr><td></td><td>38</td><td>27</td><td>8</td><td>27</td><td>CF</td><td>75</td></tr>
<tr><td>Human</td><td>30</td><td colspan="2">37</td><td>33</td><td>CF</td><td>36</td></tr>
<tr><td>Cattle</td><td>12</td><td>45</td><td>15</td><td>27</td><td>CD</td><td>180</td></tr>
<tr><td>Intestine</td><td></td><td></td><td></td><td></td><td></td><td></td></tr>
<tr><td>Rat</td><td>33</td><td>14</td><td>20</td><td>33</td><td>CF</td><td>5</td></tr>
</table>

[a]Values are given in %, and were obtained by the method of Chou and Fasman (1979) (CF) or by circular dichroism measurement (CD).

significant relationships could be demonstrated among these sequences of the different FABP types. These repeats appear to be the most highly conserved amino acid sequences of the entire protein, together with the amino-terminal part of the molecule (Offner *et al.*, 1988). Despite the high degree of amino acid sequence homology between human and rat heart FABPs, the internally repeating sequences of human heart FABP did not match with the repeats of its rat counterpart (Offner *et al.*, 1988). The occurrence of two small, highly homologous oligopeptides, each consisting of tandemly repeated sequences, in the two halves of rat and human liver FABP molecules led to the assumption that two separate intragenic duplications occurred in the evolution of these proteins (Takahashi *et al.*, 1983; Gordon and Lowe, 1985; Lowe *et al.*, 1985). However, this hypothesis appears unlikely on the basis of a structural analysis of rat liver FABP, which showed that the segments containing the highly homologous polypeptides are not located symmetrically in each half of the FABP molecule (Sweetser *et al.*, 1986).

6.3. cDNA Sequences and FABP Genes

Complementary DNA (cDNA) sequences have been established for rat liver FABP and rat intestinal FABP on the basis of intestinal mRNAs (Gordon *et al.*,

1983; Alpers *et al.*, 1984; Gordon and Lowe, 1985), for murine adipocyte P_2 protein (Bernlohr *et al.*, 1984), and for human liver FABP (Chan *et al.*, 1985; Sweetser *et al.*, 1986). Recently the cDNA structures for rat (Claffey *et al.*, 1987; Heuckeroth *et al.*, 1987) and bovine (Billich *et al.*, 1988) heart FABPs were also described. All cDNAs contain 3′ and 5′ noncoding regions, and the former show more homology between the same FABP types of different species than do the latter (Sweetser *et al.*, 1986; Billich *et al.*, 1988).

The cDNA coding sequences of rat liver and intestinal FABPs have been inserted in an appropriate expression vector (Lowe *et al.*, 1984, 1987) and could be efficiently expressed in *E. coli*. The liver FABP from *E. coli* (Lowe *et al.*, 1984), however, has not got a blocked amino terminus, in contrast to that in liver cytosol (Gordon *et al.*, 1983; Takahashi *et al.*, 1983). A frameshift mutation, introduced *in vitro* in the coding region of liver FABP cDNA, resulted in the production of a less stable FABP, with a lower fatty-acid-binding affinity (Lowe *et al.*, 1984).

Stable L6 myoblast transfectants, made with heart FABP cDNA, contained immunoreactive rat heart FABP, in contrast to transfectants with the cDNA sequence in the antisense orientation (Claffey *et al.*, 1988).

The nucleotide sequences of the genes encoding rat liver FABP (Sweetser *et al.*, 1986) and human intestinal FABP (Sweetser *et al.*, 1987a) have been completely established. These genes contain 3790 and 3382 nucleotides, respectively, and consist of four exons and three introns. The rat intestinal FABP gene has been partially defined (Sweetser *et al.*, 1987a). Also, the genes for murine adipocyte P_2 (p422) protein (Hunt *et al.*, 1986) and cellular retinol-binding protein II (Demmer *et al.*, 1987) have been characterized. The genes encoding these proteins have a similar organization of four exons and three introns. Comparably positioned introns vary considerably in size, but the introns in each of the genes are identically positioned relative to their exons (Demmer *et al.*, 1987; Sweetser *et al.*, 1987a).

The polypeptide sequences encoded by each exon of rat liver FABP were compared with rat intestinal FABP, rat cellular retinol-binding protein (CRBP), bovine cellular retinoic-acid-binding protein, human myelin P_2 protein, and murine adipocyte P_2 protein (p422) (Sweetser *et al.*, 1986). The first exon of liver FABP was significantly related to that of all the other members of this multigene protein family. Only the rat intestinal FABP sequence showed similarity with the sequence encoded by the third exon of liver FABP (Sweetser *et al.*, 1986). The amino-terminal sequence of these proteins (exon I) may define a structural and functional domain that is involved with the common property of these proteins, namely, the binding of hydrophobic ligands (Lowe *et al.*, 1985; Sweetser *et al.*, 1986). Exons III and IV may be related to the differences in ligand specificities of the members of the protein family and/or to differences in the enzymes or cellular membranes with which they interact.

6.4. Tertiary Structure

A highly comparable secondary structure model has been proposed for all proteins of the above-described multigene family, together with specific variations for each separate protein (Sundelin *et al.*, 1985b). Generally, the secondary structure predictions show a high level of β-structures, which can assemble into a β-barrel tertiary structure. A thorough investigation of the tertiary structure by X-ray diffraction at 0.25-nm resolution has been reported for rat intestinal FABP (Sacchettini *et al.*, 1988, 1989) by using the crystallized protein derived from *E. coli* (Sacchettini *et al.*, 1987). The protein is built up of 10 β-strands that are arranged in an antiparallel direction. These β-strands from a shell-like structure or β-barrel, with two α-helices on top in a parallel arrangement. The outer surface of the protein predominantly contains hydrophilic side chains, whereas most nonpolar residues are located at the inner side of the shell, thus creating a hydrophobic cavity where one fatty acid molecule can be bound.

The internal repeat sequences are located in areas of the FABPs with predicted β-structures and decreasing hydrophobicity (Offner *et al.*, 1988), yielding a protein with alternating hydrophobic and hydrophilic regions. Offner *et al.* (1988) predicted a secondary structure for human heart FABP consisting of six regions of β-structure, separated by reverse turns or α-helices. Although slightly different from the 10 β-strands of intestinal FABP (Sacchettini *et al.*, 1987), this secondary structure for heart FABP can allow the protein to be folded into a β-barrel tertiary structure. On the basis of the high degree of sequence homology and the predicted comparable secondary structure (Sundelin *et al.*, 1985b), it is likely that other members of the multigene family of FABPs have a comparable tertiary structure.

A corresponding conformation of a β-barrel, composed of eight antiparallel β-strands, was reported for another group of proteins that carry hydrophobic ligands, such as serum retinol-binding protein (Newcomer *et al.*, 1984), bilin-binding proteins (Holden *et al.*, 1987; Huber *et al.*, 1987), and β-lactoglobulin (Papiz *et al.*, 1986). These proteins show only 25–30% sequence homology and have neither an evolutionary nor a structural relationship to the FABP family (Demmer *et al.*, 1987). Appropriate modification of the primary structure of the residues that build up the central cavity of the proteins may lead to their different ligand specificities (Godovac-Zimmermann, 1988).

7. LIGAND-BINDING SITE OF FATTY-ACID-BINDING PROTEINS

Most of the information concerning the ligand-binding site of FABPs has been derived from comparative studies of amino acid sequences and secondary structure predictions. The remarkably high degree of sequence homology of the amino-terminal segments of FABPs and the closely related cellular retinoid-binding proteins, adipocyte P_2 (p422) protein, and myelin P_2 proteins, indicates an involvement of this part of the protein in the binding of hydrophobic ligands (Gordon and Lowe, 1985; Lowe *et al.*, 1985; Sacchettini *et al.*, 1986; Sweetser *et al.*, 1986; Offner *et al.*, 1988). Other sequences within the proteins may be involved in the specificity of the ligand binding, either separate from or in concert with the amino-terminal domain (Lowe *et al.*, 1985). Domains with amphipathic α-helical structure have been suggested to be involved in the binding of FABPs to hydrophobic ligands, since this was observed for albumin (Brown and Shockley, 1982) and the apolipoproteins apo A-I and apo A-IV (Boguski *et al.*, 1984; Lowe *et al.*, 1985). However, homologous sequences with the latter proteins and such amphipathic structures were not found in FABPs (Gordon and Lowe, 1985; Lowe *et al.*, 1985; Sacchettini *et al.*, 1986).

The hydropathy of FABPs is of great importance for their interaction with ligands. Relative hydropathy plots are clearly different for liver, intestinal, and heart FABPs (Sacchettini *et al.*, 1986), but hydropathy has been highly conserved during evolution. Of the 23 amino acids that are nonidentical in human and rat liver FABP, 74% maintain the same relative hydropathy (Lowe *et al.*, 1985).

Binding of long-chain fatty acids to FABP is noncovalent (Takahashi *et al.*, 1983) and involves both hydrophobic and ionic interactions (Keuper *et al.*, 1985; Schulenberg-Schell *et al.*, 1988). The fatty-acid-binding sites of liver and heart FABPs are different, as can be deduced from their interaction with fluorescent fatty acid analogues (Keuper *et al.*, 1985; Storch *et al.*, 1987; Peeters *et al.*, 1989a). The ionic interactions may be ascribed to arginine in liver, intestinal, and heart FABPs. Modification of arginine residues of human (Peeters *et al.*, 1989a) and bovine (Schulenberg-Schell *et al.*, 1988) liver FABP and of pig heart FABP (Peeters *et al.*, 1989a) strongly reduced the binding of long-chain fatty acids. The remaining binding activity was entirely due to hydrophobic interactions (Schulenberg-Schell *et al.*, 1988). Detailed X-ray diffraction analysis of rat intestinal FABP confirmed the interaction of an arginine residue at position 127 of the protein with the ionized carboxyl moiety of the fatty acid (Sacchettini *et al.*, 1988). The bound fatty acid is located in the interior of the protein and has a bent conformation with several gauche bonds. The arginine involved is the only basic amino acid that extends into the hydrophobic core of the FABP molecule. More support for the involvement of this amino acid in the ligand binding came from comparison of the amino acid sequence of other FABPs and closely related proteins. All, except CRBP, have an arginine residue at the same position (Schulenberg-Schell *et al.*, 1988). The absence of this arginine in a mutant FABP produced in *E. coli* (Lowe *et al.*, 1984) results in a reduction of fatty acid binding, comparable to chemical modifications of this amino acid (Schulenberg-Schell *et al.*, 1988; Peeters *et al.*, 1989a). Arginine residues are also involved in ionic interactions with the polar headgroup in the binding of phospholipids by the phosphatidylcholine transfer protein (Akeroyd *et al.*, 1981a). Studies with 9-anthroyloxy-labeled fatty acids suggest that the carboxyl group of the acyl chain is located near the aqueous surface of rat liver FABP (Storch *et al.*, 1989).

The structural interactions between [carboxyl-^{13}C]oleic acid and rat liver FABP were probed with ^{13}C nuclear magnetic resonance (NMR) spectroscopy (Cistola *et al.*, 1988). Oleic acid bound to FABP appeared to be fully ionized at neutral pH. The anionic carboxyl groups of oleic acid are, however, not involved in electrostatic interactions with cationic residues on FABP. This is in contrast with the important role of those electrostatic interactions in the high-affinity binding of fatty acids to albumin (Cistola *et al.*, 1987). ^{1}H NMR studies of ligand interaction with bovine liver FABP strongly indicated that conformational changes occur upon fatty acid binding (Schulenberg-Schell *et al.*, 1988). Involvement of histidine and lysine residues was excluded by ^{1}H NMR and acetylation studies of bovine liver FABP (Schulenberg-Schell *et al.*, 1988).

It is tempting to suggest that the sulfhydryl group of the cysteine residue is involved in the liver FABP–ligand interaction, since apparently cysteine is present only in this FABP type [and in the kidney type, as was recently observed (Lam *et al.*, 1988)] and its ligand specificity is different from that of the other FABP types. Modification of the sulfhydryl group of human, bovine, and rat liver FABPs, however, did not affect fatty acid binding (Ockner and Manning, 1974; Schulenberg-Schell *et al.*, 1988; Peeters *et al.*, 1989a). Considering the striking conservation of cysteine in human and rat liver FABPs (Gordon and Lowe, 1985) at position 69, this amino acid may be involved in the binding of other hydrophobic ligands or in another function specific for hepatic FABPs. A role for cysteine has been suggested in transient interactions of the nonspecific lipid transfer protein with intracellular membranes (Van Amerongen *et al.*, 1985).

8. TISSUE DISTRIBUTION AND CONTENT OF FATTY-ACID-BINDING PROTEINS

When FABPs were isolated from different rat tissues, it soon became clear that although they are closely related, they are different proteins. They have distinct physicochemical characteristics and ligand specificity. Until now at least four distinct FABP forms have been identified: hepatic, cardiac, intestinal, and renal. Preliminary evidence is available for the existence of a distinct muscular FABP (Veerkamp and Paulussen, 1987). These names are derived from the tissue of their first detection and do not imply a confinement solely to these tissues. Murine adipocytes contain a specific lipid-binding protein (murine adipocyte P_2 or p422) that binds both oleic acid and retinoic acid (Matarese and Bernlohr,

1988). FABPs have been demonstrated in many tissues, mainly with rat and human material, but also in a number of tissues from pigs, cattle, mice, guinea pigs, rabbits, and chickens (Table I). FABPs are not present in detectable amounts in serum (Ockner *et al.*, 1972; Haunerland *et al.*, 1984; Crisman *et al.*, 1987; Paulussen *et al.*, 1989a) or erythrocytes (Ockner *et al.*, 1972; Paulussen *et al.*, 1989a).

The tissue distribution of the different FABP types is rather complex. The tissue distribution of rat, human, and pig cardiac and hepatic FABPs has been investigated by using specific antibodies in immunodiffusion experiments (Ockner and Manning, 1974; Bass *et al.*, 1985b; Bass and Manning, 1986), enzyme-linked immunosorbent assay (ELISA) procedures (Crisman *et al.*, 1987; Paulussen, *et al.*, 1989a,b), and Western immunoblotting (Claffey *et al.*, 1987; Paulussen *et al.*, 1989a,b). The heart and liver FABP types are clearly distinguishable with specific antibodies (Said and Schulz, 1984; Bass *et al.*, 1985b; Jagschies *et al.*, 1985; Offner *et al.*, 1986; Paulussen *et al.*, 1989a). Intestinal FABP is immunochemically distinct from liver FABP (Bass and Manning, 1986). Immunoreactivity with antibodies to rat liver FABP is restricted to liver, small-intestinal mucosa, and adipose tissue (Haq *et al.*, 1982; Ockner *et al.*, 1982; Bass and Manning, 1986; Fukai *et al.*, 1987; Paulussen *et al.*, 1989a), whereas anti-heart FABP antibodies react with a variety of rat tissues (Bass and Manning, 1986; Claffey *et al.*, 1987; Crisman *et al.*, 1987; Lam *et al.*, 1988; Paulussen *et al.*, 1989a,b). Cardiac FABP is found in heart tissue as well as in several skeletal muscles. Reactivity was also found with kidney, aorta, brain, testis, ovary, lung, and adipose tissue. In slow-twitch muscles the content of heart FABP is significantly higher than in the fast-twitch type (Claffey *et al.*, 1987; Crisman *et al.*, 1987; Miller *et al.*, 1988; Paulussen *et al.*, 1989a). Low-molecular-mass proteins reactive with antibodies against intestinal FABP are present in the intestine and stomach (Bass and Manning, 1986). Concomitant expression of two distinct FABP types has been described for two tissues. Intestinal mucosa contains the liver and intestinal FABP types (Ockner and Manning, 1974), and recently heart FABP and a kidney-specific FABP type have been identified in rat kidneys (Lam *et al.*, 1988). The kidney FABP type is present predominantly in the cortex, whereas heart FABP is found in both cortex and medulla.

Indications for the presence of more FABP types than the heart, liver, intestinal, and kidney types can be derived from the variable dissociation constant (K_d) values of dealbuminized cytosolic proteins of various rat tissues (Paulussen *et al.*, 1986) and especially from the differences between the fatty-acid-binding capacity and the content of heart and liver types in these tissues (Paulussen *et al.*, 1989) (Figure 3). Many investigators have determined the FABP content of various tissues with antibodies against liver FABP (Ockner and Manning, 1974; Ketterer *et al.*, 1976; Kamisaka *et al.*, 1981; Rüstow *et al.*, 1982). Comparison of the FABP content of various rat tissues, determined with

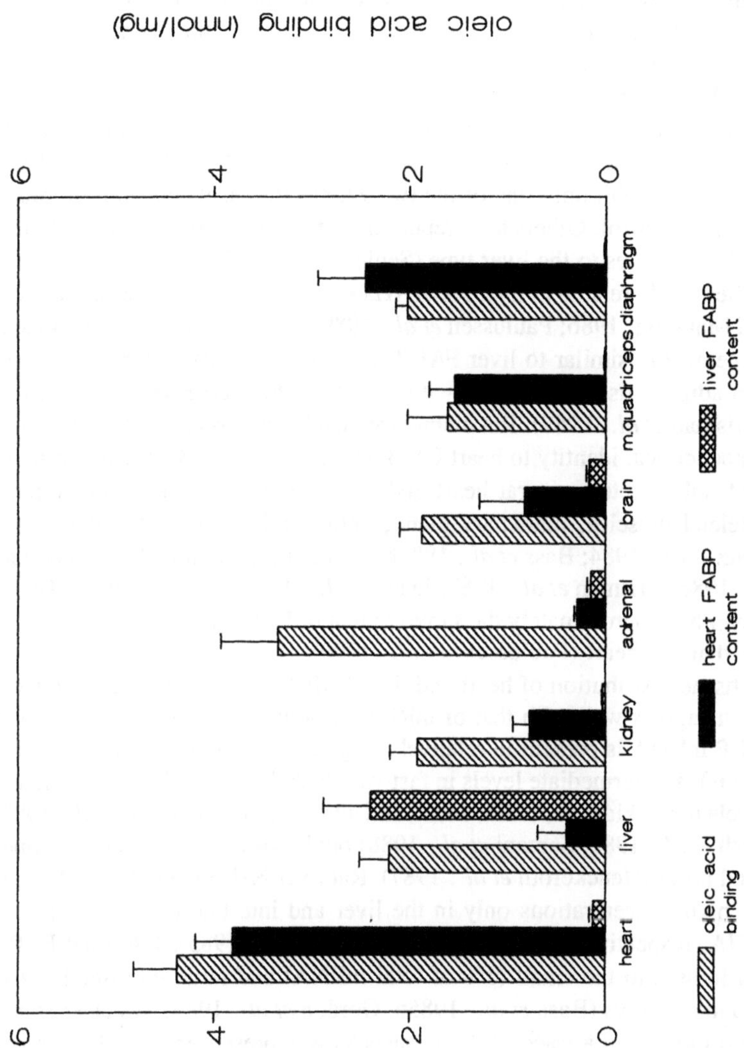

FIGURE 3. Comparison of the fatty-acid-binding capacity and the immunoreactivities with anti-rat heart and liver FABP antibodies of various rat tissues.

anti-liver FABP, with the cytosolic oleic acid-binding capacity clearly demonstrates the inability of anti-liver FABP antibodies to determine the FABP content in many tissues. Similarly, antiserum directed against heart FABP does not detect FABP in the liver, whereas the FABP content of, e.g., the kidneys, adrenal glands, and brain is grossly underestimated (Figure 3).

The relationship of liver FABP to its homologs from brain and lung cytosol still remains unclear. We found minor reactivity of anti-liver FABP antisera with rat brain cytosol (Paulussen *et al.*, 1989a). Bass *et al.* (1984) could detect immunoreactivity with only one of the two FABPs they isolated from rat brain, not with whole cytosol. Others have reported a total absence of reactivity of brain FABP with antibodies to the liver type (Senjo *et al.*, 1985). Lung FABP may be immunochemically related to liver FABP (Haq *et al.*, 1985) or to the cardiac type (Bass and Manning, 1986; Paulussen *et al.*, 1989a). FABP from adipose tissue is immunochemically similar to liver FABP (Haq *et al.*, 1982), but cytosolic proteins from adipose tissue also react with antibodies to heart FABP (Bass *et al.*, 1985b; Crisman *et al.*, 1987). Rat mammary gland contains an FABP that shows an immunochemical identity to heart FABP (Jones *et al.*, 1988). FABPs form 4–6% of cytosolic proteins in rat heart and 2–4% in rat liver, kidneys, adrenal gland, skeletal muscles, mammary gland, brain, and intestine (Ockner *et al.*, 1982; Glatz *et al.*, 1984; Bass *et al.*, 1985b; Fournier and Rahim, 1985; Bass and Manning, 1986; Crisman *et al.*, 1987; Jones *et al.*, 1988; Paulussen *et al.*, 1989). Human liver has approximately the same content of FABP as rat liver, but human heart less than rat heart (Paulussen, *et al.*, 1990).

The tissue distribution of heart and liver FABPs based on immunochemical observations agrees well with that of mRNA of both types in these tissues. Rat heart FABP mRNA is present in relatively large amounts in slow-twitch muscle (e.g., soleus), at intermediate levels in fast-twitch skeletal muscle and testes, and at low levels in the kidneys, brain, aorta, and adrenal gland (Claffey *et al.*, 1987; Heuckeroth *et al.*, 1987; Sarzani *et al.*, 1988) but is absent in the liver, intestine, spleen, and lungs (Heuckeroth *et al.*, 1987). Rat liver FABP mRNA was demonstrated in high concentrations only in the liver and intestine (0.7 and 2.1% of total mRNA, respectively) in male rats (Gordon *et al.*, 1985). Intestinal FABP mRNA is limited to the small (1% of total mRNA) and large intestine in rats, humans, and monkeys (Bass *et al.*, 1985b; Gordon *et al.*, 1985; Sweetser *et al.*, 1987a). No more than a trace of these mRNAs was present in any of the other tissues investigated.

9. CELLULAR AND SUBCELLULAR DISTRIBUTION OF FATTY-ACID-BINDING PROTEINS

In human and rat liver, FABP is present only in hepatocytes (Capron *et al.*, 1979; Kamisaka *et al.*, 1981; Suzuki and Ono, 1987). No other cell type present in the liver reacts with antibodies to hepatic FABP. In fetal human liver, FABP is detectable after 7 weeks of gestation (Suzuki and Ono, 1987). The uniform distribution throughout the fetal liver changes into a primarily periportal localization in adult human liver (Bass and Ockner, 1987; Suzuki and Ono, 1987). The intestinal and hepatic FABP types both decrease in concentration from the proximal to the distal part of the intestine in suckling and adult rats (Ockner and Manning, 1974; Shields *et al.*, 1986), whereas this decline is not found in weanling animals (Shields *et al.*, 1986). Both FABP types are located predominantly in the top of the villi (Ockner and Manning, 1974; Shields *et al.*, 1986).

In the heart, liver, and intestine, FABP is only detectable in the cytoplasm by immunocytochemical staining (Capron *et al.*, 1979; Shields *et al.*, 1986; Iseki *et al.*, 1988) and in the cytosol by immunochemical quantitation of cell fractions (Crisman *et al.*, 1987; Paulussen *et al.*, 1989). In fed rats, reactivity with anti-liver FABP is located mainly around glycogen areas, where smooth endoplasmic reticulum and mitochondria are located (Capron *et al.*, 1979; Iseki *et al.*, 1988). After 2 days of fasting this pattern changed into a diffuse distribution throughout the hepatocyte (Iseki *et al.*, 1988). No FABP was found in association with mitochondria (Capron *et al.*, 1979; Crisman *et al.*, 1987; Paulussen *et al.*, 1989a), smooth endoplasmic reticulum (Capron *et al.*, 1979; Iseki *et al.*, 1988), or lysosomes (Capron *et al.*, 1979). A small fraction of intestinal FABP (16%) was found to be associated with intracellular membranes (Ockner and Manning, 1974). For rat heart, a gradientlike distribution of FABP (with 15% present in mitochondria) was suggested, based on immunoelectron-microscopic observations (Fournier and Rahim, 1985). Binding of FABP to bovine liver and heart mitochondria was shown by using immunogold labeling and immunodiffusion (Rüdel *et al.*, 1985; Bordewick *et al.*, 1986). Recently the same group demonstrated the presence of FABP in association with mitochondria and nuclei of bovine myocardium (Börchers *et al.*, 1989). FABP accounted for 0.02 and 0.005% of total protein of mitochondria and nuclei, respectively.

The subcellular distribution of FABP differs clearly from that of nonspecific lipid transfer protein (sterol carrier protein 2) and phosphatidylcholine transfer protein (Teerlink *et al.*, 1982, 1984; Megli *et al.*, 1986). Only 60–70% of these proteins are located in the cytosol.

10. BINDING CHARACTERISTICS OF FATTY-ACID-BINDING PROTEINS

Up to 85% of the fatty acids endogenously bound to rat liver FABP are unsaturated, with a high percentage (50–75%) of polyunsaturated species (Rüstow *et al.*, 1978; Burnett *et al.*, 1979; Ockner *et al.*, 1982; Bass, 1985; Cistola *et al.*, 1988). For bovine heart FABP similar data were reported (Jagschies, 1984). The different isoforms of rat liver FABP have distinct profiles of endogenously bound fatty acids (Takahashi *et al.*, 1983; Bass, 1985). The molar ratio of endogenously bound fatty acids varied between 0.4 and 1.3 mol/mol of FABP (Burnett *et al.*, 1979; Takahashi *et al.*, 1983; Glatz *et al.*, 1984; Bass, 1985).

FABPs were first recognized by the coelution of a mixture of (labeled) fatty acid and cytosolic protein from a gel filtration column. The affinity for long-chain fatty acids was estimated from these experiments to be in the micromolar range (Levi *et al.*, 1969a; Mishkin *et al.*, 1972; Ockner *et al.*, 1972; Mishkin and Turcotte, 1974b). However, because of the nonequilibrium conditions of the coelution method, accurate determination of dissociation constants (K_d) necessitates another approach. Assays employing an equilibrium between bound and free fatty acids allow Scatchard plot analysis of binding data. Equilibrium dialysis, fluorescence, and radiochemical-binding assays have been used to determine the dissociation constants of FABPs for a variety of ligands (Table VII). K_d values for long-chain fatty acids are in the range of 0.1–1 μM for all FABP types. The K_d values for their CoA and carnitine esters are in the same order of magnitude (Ketterer *et al.*, 1976; Bass, 1985; Burrier *et al.*, 1987; Paulussen *et al.*, 1988).

Both heart and liver FABPs show a preference for the longer, unsaturated fatty acid species (Mishkin *et al.*, 1972; Ockner *et al.*, 1972, 1982; Haunerland *et al.*, 1984; Wilkinson and Wilton, 1987b; Paulussen *et al.*, 1988; Peeters *et al.*, 1989a). Human and rat liver FABP and human heart FABP have a preference for long-chain saturated fatty acids of 16–19 C atoms, but mono- and polyunsaturated fatty acids up to 22 C atoms are quite well bound (Wilkinson and Wilton, 1987b; Paulussen *et al.*, 1988; Peeters *et al.*, 1989a). Bromo-substitution did not affect binding, in contrast to substitution of a hydroxyl group (Peeters *et al.*, 1989a). The fluorescent fatty acids, 11-(dansylamino)undecanoic acid, 16-anthroyloxypalmitic acid and 1-pyrenedodecanoic acid, all show K_d values comparable to those of oleic acid (Peeters *et al.*, 1989a). The K_d values are somewhat higher with the former two acids for the FABPs from heart than from liver (Peeters *et al.*, 1989a). For oleic acid the reverse is the case.

Binding is generally considered to occur noncovalently, at an equimolar ratio (Glatz *et al.*, 1984, 1985a; Bass, 1985; Dutta-Roy *et al.*, 1987; Wilkinson and Wilton, 1987b; Paulussen *et al.*, 1988), but some controversy has arisen concerning the stoichiometry of fatty acid binding to FABP. Haunerland *et al.*, (1984) first reported binding of 2 mol of oleate per mole of FABP from bovine liver. The binding of both fatty acid molecules would occur at a single binding site, whereby the molecules are in parallel orientation (Keuper *et al.*, 1985). The marked difference in K_d values (Table VII) indicates that binding of the first ligand alters the affinity of the binding site for the second molecule (Schulenberg-Schell *et al.*, 1988). Offner *et al.*, (1986, 1988) observed binding of oleic and palmitic acid to rat heart and liver FABP and human heart FABP at a 2-mol/mol ratio, using liposomes as fatty acid donors. Kidney FABP showed only

Table VII
Ligand-Binding Affinities of FABPs

Origin ligand	K_d (μM)	Reference
Rat		
Liver		
Palmitic acid	0.6–1.0	8, 69, 103
Palmitoyl-CoA	0.14–2.8	8, 103, 132
Stearic acid	0.7	8
Oleic acid	0.07–1.5	8, 52, 69, 157, 179, 214, 227
Oleoyl-CoA	1.2–2.0	8, 33
Linoleic acid	0.5	8
Arachidonic acid	0.6; 0.92	8, 69
Erucic acid	0.9	8
Retinol	21	227
Retinyl palmitate	1.4	62
Heme	0.15; 1.5	8, 227
Hematin	0.07	214
Lysophosphatidylcholine	15	32
Prostaglandin E_1	0.037	52
Bromosulfophthalein	0.69	214
Bilirubin	0.27; 1.10	214, 227
11-(Dansylamino)undecanoic acid	0.24; 0.03 + 0.5	157, 227
16-Anthroyloxypalmitic acid	0.72	157
1-Pyrene-dodecanoic acid	0.64	157
Oestrone sulfate	30	103
Dehydroepiandrosterone	20	103
Heart		
Myristic acid	0.95	184
Palmitic acid	0.78; 1.02	69, 154
Palmitoyl-CoA	0.51	154
Palmitoyl-carnitine	0.59	154
Oleic acid	0.38	154, 157
Arachidonic acid	0.16	154
11-(Dansylamino)undecanoic acid	1.84	157
16-Anthroyloxypalmitic acid	1.33	157
1-Pyrene-dodecanoic acid	0.22	157
Kidney		
Palmitic acid	1.0	60
Oleic acid	1.0	60
Arachidonic acid	0.9	60
Human		
Liver		
Hematin	0.97	214
Bromosulfophthalein	0.25	214
Bilirubin	0.78	214
Oleic acid	0.04; 0.62	157, 214

(*continued*)

Table VII
(*Continued*)

Origin ligand	K_d (μM)	References
11-(Dansylamino)undecanoic acid	0.38	157
16-Anthroyloxypalmitic acid	0.28	157
1-Pyrene-dodecanoic acid	0.65	157
Heart		
Myristic acid	0.88	154
Palmitic acid	0.60	154
Palmitoyl-CoA	0.50	154
Palmitoyl-carnitine	0.42	154
Oleic acid	0.20	154
Arachidonic acid	0.17	154
Pig		
Liver		
Oleic acid	0.63	157
11-(Dansylamino)undecanoic acid	0.44	157
16-Anthroyloxypalmitic acid	0.25	157
1-Pyrene-dodecanoic acid	0.39	157
Heart		
Myristic acid	1.81	154
Palmitic acid	1.26	154
Oleic acid	0.43	154, 157
Arachidonic acid	0.24	154
12-Doxylstearic acid	0.85	59
11-(Dansylamino)undecanoic acid	1.62	157
16-Anthroyloxypalmitic acid	0.69	157
1-Pyrene-dodecanoic acid	0.25	157
Cattle		
Liver		
Oleic acid	0.24; 2.15	180
Heart		
Palmitic acid	0.90	92
Oleic acid	0.39	92
Arachidonic acid	0.30	92

one binding site under similar conditions (Lam *et al.*, 1988). In our laboratory this assay produced maximal values of 0.7 mol of oleic acid per mole of human heart FABP (Paulussen *et al.*, 1988). Transfer assays of oleic acid from mitochondria to liver or heart FABPs and Lipidex binding assays showed a comparable binding of stoichiometry, which did not exceed 1 mol of fatty acid per mole of FABP (Peeters *et al.*, 1989a).

In NMR studies on [^{13}C]oleic acid binding to rat liver and intestinal FABPs, Cistola *et al.*, (1988, 1989) could not demonstrate a stoichiometry exceeded 1 mol/mol. Palmitic acid, however, was bound up to 2–3 mol/mol by liver FABP, but maximally in an equimolar ratio by the intestinal type. The measurements of the higher binding ratios were allowed by the preservation of the equilibrium condition with the employed technique, since it did not require separation of bound and free fatty acids. The materials applied for separation of bound and free ligand in other assays may compete with FABP for ligand binding, thus leading to an underestimation of the binding capacity of FABP (Cistola *et al.*, 1988). An increase of the amount of Lipidex does not, however, affect the binding of fatty acid to FABP (Glatz and Veerkamp, 1983). Apparently the stoichiometry of fatty acid binding to FABP is greatly dependent on the assay conditions applied. Especially important may be the solubility and the associative behavior of the fatty acid. The binding ratio of palmitic acid to hepatic FABP was shown to depend on the physical state of the unbound fatty acid, thus on temperature, and on the pH of the assay mixture (Cistola *et al.*, 1989).

Rat liver and intestinal FABP genes have been inserted and expressed in *E. coli* (Lowe *et al.*, 1984, 1987). The bacterially derived FABPs were used to study ligand binding and secondary and tertiary structures. The liver FABP type could bind 2 mol of long-chain fatty acid per mole with a K_d value of about 1.7–3.2 μM, and a higher affinity for unsaturated fatty acids (Lowe *et al.*, 1987). Site-directed mutagenesis of the carboxy-terminal part of liver FABP markedly reduced fatty acid binding (Lowe *et al.*, 1984). Binding properties of *E. coli*-derived intestinal FABP were similar to those of the native intestinal FABP. It bound 1 mol of fatty acid per mole of protein, but the apparent K_d values were also clearly increased (Lowe *et al.*, 1987). Intestinal FABP has an approximately equal affinity for oleic, palmitic, and arachidonic acids (K_d, 2.9–3.7 μM). Analysis of its tertiary structure showed that intestinal FABP noncovalently binds 1 mol of fatty acid per mole of protein (Sacchettini *et al.*, 1988).

Pig heart FABP was reported to self-associate into several forms of higher molecular weight (Fournier *et al.*, 1983). This concentration-dependent self-aggregation was suggested to be an important regulatory step in the control of energy production through the β-oxidative pathway in the heart (Fournier and Rahim, 1985). The self-aggregated heart FABP forms might play different roles in cardiac energy metabolism. Only two forms may act in the translocation of acylcarnitine to mitochondria (Fournier and Richard, 1988). Self-aggregation has not been reported for any other FABP type. Other investigators found no evidence for a concentration-dependent self-association of heart FABP from cattle (Jagschies *et al.*, 1985), rats (Offner *et al.*, 1986), or humans (Offner *et al.*, 1988).

The ligand specificity of liver FABP is clearly distinct from that of the other FABP types. Liver FABPs appear to function as a general carrier of hydrophobic ligands, whereas the other FABP types apparently function more specifically as carriers for long-chain fatty acids and their CoA and carnitine esters. Human heart FABP binds only fatty acids, and long-chain acyl-CoA and acylcarnitine esters (Paulussen *et al.*, 1988). Liver FABP has been reported to bind (*in vitro*) retinyl-palmitate (Fukai *et al.*, 1987), heme (Bass, 1985; Vincent and Muller-Eberhard, 1985; Wilkinson and Wilton, 1987b), prostaglandins (Dutta-Roy *et al.*, 1987), hematin, 1-anilino-8-naphthalene sulfonate (ANS), bilirubin, and bromosulfophthalein (Takikawa and Kaplowitz, 1986) with affinities comparable to those for long-chain fatty acids (Table VII). Binding of lysophosphatidylcholine (Burrier and Brecher, 1986), steroids (Ketterer *et al.*, 1976), and bile acids (Takikawa and Kaplowitz, 1986) to liver FABP has also been demonstrated, but with somewhat lower affinity. No binding of retinoids (Peeters *et al.*, 1989a), clofibric acid (Vincent and Muller-Eberhard, 1985; Peeters *et al.*, 1989a), prostaglandins, long-chain alcohols (Wilkinson and Wilton, 1987b; Peeters *et al.*, 1989a), methyl-palmitate, CoA, malonyl-CoA (Wilkinson and Wilton, 1987b), or cholesterol (Vincent and Muller-Eberhard, 1985; Wilkinson and Wilton, 1987b) to rat and human liver FABP could be detected in studies of competitive displacement of 11-(dansylamino)undecanoic acid or ANS. Palmitoyl-carnitine and palmitoyl-glycerol are bound with only very low affinity (Peeters *et al.*, 1989a). Monohydroxy-bile acids showed a higher affinity to liver FABP than did trihydroxy-bile acids in competition with ANS (Takikawa and Kaplowitz, 1986). Various inhibitors of carnitine-palmitoyl transferase have a high affinity for FABP (Peeters *et al.*, 1989a).

To date, binding of ligands other than fatty acids has been reported only for hepatic FABP, not for any of the other FABP types. Binding characteristics are similar for the isolated isoforms of liver FABP (Bass, 1985; Peeters *et al.*, 1989a). Recently, it was found that the murine adipocyte P_2 (p422) protein is able to bind both oleic and retinoic acids from liposomes (Matarese and Bernlohr, 1988). The K_d value for oleic acid was, however, much lower (3 μM).

11. FUNCTIONAL PROPERTIES OF FATTY-ACID-BINDING PROTEINS

Although the involvement of FABPs in intracellular fatty acid transport and utilization is generally accepted, the massive amount of research into these proteins that has been carried out since the late 1960s has provided much circumstantial evidence for their physiological role(s), but conclusive evidence is not yet available.

Two major functions have been assigned to the FABPs. First, the facilitation of transport, which may include the release of fatty acids and/or movement through the aqueous phase, and second, modulation of enzyme and/or metabolic activities by storage of substrates, sequestration of substrates and/or effectors, and protection against inhibitors. The function may be related to the ligand bound to the protein: fatty acid, acyl-CoA, acylcarnitine, or other hydrophobic ligands.

The major role assigned to FABPs is facilitation of intracellular transport or specific trafficking of fatty acids and acyl derivatives, thereby providing the possibility of targeting of these compounds toward oxidation or esterification. Tipping and Ketterer (1981) concluded on the basis of theoretical considerations that FABP could enhance the rate of intracellular transport of its ligand by an order of magnitude. Cooper *et al.* (1987) presented evidence that the metabolism of palmitate in liver is diffusion limited and that almost all fatty acids in the cytosol are bound to FABP. From the concomitant increase of Z-protein (FABP) content and palmitate uptake by livers isolated from clofibrate-treated rates, Renaud *et al.* (1978) concluded that this protein is involved in fatty acid uptake. Experiments with inhibitors of fatty acid binding to FABP, such as flavaspidic acid or α-bromopalmitic acid, with jejunal preparations, hepatocytes, and perfused liver did not give unequivocal results (Mishkin *et al.*, 1975; Ockner and Manning, 1976; Goresky *et al.*, 1978; Burnett *et al.*, 1979).

The ability of FABPs to release fatty acids from membranes has been demonstrated by many investigators (Catala and Avanzati, 1983; Burrier *et al.*, 1987; McCormack and Brecher, 1987; Peeters *et al.*, 1989a,c). This property can be used for the quantitative determination of these proteins in the transfer assay (Brecher *et al.*, 1984; Peeters *et al.*, 1989a). Liver FABP is, however, less effective than albumin in preventing fatty acid accumulation in membranes (Cistola *et al.*, 1988). Evidence for in vitro fatty acid transfer between membranes by FABP could be obtained for membrane systems separated by a polycarbonate filter (McCormack and Brecher, 1987; Peeters *et al.*, 1989b), and for two separated mono-layers (Peeters *et al.*, 1989c). FABPs are able to deliver long-chain fatty acids to the mitochondrial β-oxidation system for oxidation (Fournier *et al.*, 1978; Fournier and Rahim, 1985; Glatz *et al.*, 1985b; Peeters *et al.*, 1989c) or to microsomes for acyl-CoA synthesis (McCormack and Brecher, 1987; Peeters *et al.*, 1989a). At a 1 : 1 ratio of fatty acid to protein, FABP-bound palmitate proved to be a better substrate for the mitochondria than albumin-bound fatty acid was (Glatz *et al.*, 1985b).

Maximal rates of palmitoylcarnitine oxidation in rat heart mitochondria varied according to an FABP concentration-dependent function (Fournier and Richard, 1988). These investigators postulated a model for a FABP-dependent flux of acylcarnitine between the cytoplasm and the inner mitochondrial membrane was postulated on the basis of electron spin resonance studies of 16-doxylstearoylcarnitine distribution and oxidation. The occurrence of two specific self-aggregated FABP species as acylcarnitine translocators would contribute to the regulation of β-oxidation and energy production in the heart.

Evidence for the physiologically important function of FABPs in the modulation of metabolism has been derived from their influence on the *in vitro* activity of a number of enzymes. Rat liver FABP is able to preclude or reverse the inhibition by fatty acids or their acyl-CoA esters of mitochondrial ATP/ADP translocase (Barbour and Chan, 1979), microsomal 4-methyl sterol oxidase (Billheimer and Gaylor, 1980; Grinstead *et al.*, 1983), acyl-CoA:cholesterol acyltransferase (Grinstead *et al.*, 1983), cytosolic acetyl-CoA carboxylase (Lunzer *et al.*, 1977), and both the mitochondrial (Burnett *et al.*, 1979) and microsomal acyl-CoA synthases (Ockner and Manning, 1976; Wu-Rideout *et al.*, 1976; Haq *et al.*, 1985; Burrier *et al.*, 1987). When liposomes were used as the fatty acid donor instead of salt or detergent suspensions, liver FABP markedly decreased the activity of microsomal acyl-CoA synthase (Noy and Zakim, 1985). Wu-Rideout *et al.* (1976) described an inhibitory effect of liver FABP on mitochondrial long-chain acyl-CoA synthase. Heart FABP does not stimulate microsomal acyl-CoA synthase activity (Burrier *et al.*, 1987). Hepatic and intestinal FABPs stimulate the activity of several microsomal enzymes involved in the synthesis of triacylglycerols and phospholipids, such as acyl-CoA:glycerol-3-phosphate acyltransferase (Mishkin and Turcotte, 1974a; Burnett *et al.*, 1979), diacylglycerol acyltransferase (O'Doherty and Kuksis, 1975; Iritani *et al.*, 1980), and phosphatidate phosphohydrolase (Roncari and Mack, 1981). Lung FABP stimulates the activity of pulmonary microsomal glycerophosphate acyltransferase two- to fourfold (Haq *et al.*, 1987). The effects of FABPs on the activities of a number of enzymes may depend on the system in which the fatty acids or acyl-CoA esters are presented. They may reflect either a specific influence of these proteins on the release of fatty acids from membranes or a site-directed transfer of fatty acids or acyl-CoA esters to the active site of the enzyme. A nonspecific solubilizing effect may be involved in their influence on several enzymes, but the (partial) inability observed in some cases (O'Doherty and Kuksis, 1975; Ockner and Manning, 1976; Burnett *et al.*, 1979; Haq *et al.*, 1987) to duplicate the observed effects with albumin reflects a more direct involvement of FABP. Furthermore, McCormack and Brecher (1987) showed that fatty acids must desorb from liposomal membranes before FABP can have an effect on acyl-CoA formation by acyl-CoA synthase. FABPs apparently penetrate a phospholipid bilayer only to a minor extent (Peeters *et al.*, 1989c).

The relative cytosolic abundance of FABPs (Ockner *et al.*, 1982; Bass *et al.*, 1984, 1985b; Glatz *et al.*, 1984; Fournier and Rahim, 1985; Bass and Manning, 1986; Crisman *et al.*, 1987; Jones *et al.*, 1988; Paulussen *et al.*, 1989, 1990) suggests a role for these proteins in temporary intracellular storage of fatty acids. Binding of fatty acids and their CoA and carnitine esters by FABPs enables the maintenance of a low concentration of the unbound form of these substances within the cell over a large concentration range, which is important because of the damaging effects of high concentrations of these compounds on membranes (Gutknecht, 1988) and the marked influence of acyl-CoA esters on many enzyme activities.

A protective role may be of special importance during ischemia in the heart, when high concentrations of free fatty acids are present (Van Der Vusse *et al.*, 1982). However, recent observations of ischemia in perfused rat hearts showed a marked release of FABP from the heart, together with only minor amounts of fatty acids (Glatz *et al.*, 1988a). This seems to argue against a protective role of FABP in the heart, but fatty acids may be stripped from FABP during membrane passage.

The intracellular storage pool of FABP-bound fatty acids can readily deliver substrate for energy production or biosynthetic processes. Binding to FABP may regulate the intracellular concentrations of fatty acids and acyl-CoA and acyl carnitine esters and thus influence the overall rate of intracellular lipid metabolism. Inversely, the amount of fatty acid stored intracellularly by binding to FABP may be determined by the rates of fatty acid uptake and utilization.

Some recent studies have shown a relationship between FABPs or related proteins and (indirect) regulation of growth and differentiation. The level of liver FABP is positively associated with the growth activity of rat hepatocytes (Bassuk *et al.*, 1987). Ligands of this protein may be involved in normal mitosis and carcinogen-induced cell proliferation of hepatocytes (Bassuk *et al.*, 1987). A polypeptide growth inhibitor purified from bovine mammary gland (MDGI), which shows extensive homology with rat heart FABP, inhibits the proliferation of Ehrlich ascites and mammary carcinoma cells (Böhmer *et al.* 1987). CRBP and cellular retinoic acid-binding protein may mediate the transport of these ligands to their nuclear receptor (Giguere *et al.*, 1987) and, in this way, influence their action on epithelial differentiation and tumorigenesis (Chytil and Ong, 1987). The adipocyte differentiation is accompanied by the transcriptional activation of the gene of the lipid-binding adipocyte P_2 (p422) protein (Bernlohr *et al.*, 1985). Recently the amino acid sequence of porcine gastrotropin appeared to be similar to rat liver FABP, with 44 of 127 residues being identical. This compound is completely relegated to the distal ileum, but it stimulates gastric secretion and growth of cell cultures of gastrointestinal epithelium (Walz *et al.*, 1988).

12. INFLUENCE OF PHYSIOLOGICAL CONDITIONS AND DRUG TREATMENT ON FATTY-ACID-BINDING PROTEINS

Most of the support for the physiological role(s) of FABPs has been gathered from studies of the relation of the FABP content or the fatty-acid-binding activity of cytosolic proteins with the physiological conditions and drug treatments that involve changes in lipid metabolism (Table VIII). Different tissues show clearly different responses of the FABP concentration to physiological changes. The concentration of liver FABP can change markedly, whereas that of heart FABP is relatively stable. Although less information is available for

Table VIII
Response of FABP Content and Fatty-Acid-Binding Capacity to Physiological Conditions and Drugs

		FABP content		Fatty-acid-binding capacity	
Parameter	Tissue	%[a]	Reference	%[a]	Reference
Sex (female versus male)	Liver	128	155	156	153
	Heart	n.s.[b]/128	46, 155	136	153
	Intestine	n.s.	14		
Steroid hormones					
Estrogens	Liver	153	146		
Testosterone	Liver	109	146		
Postnatal age (0 versus 70 days)	Liver	16	155	n.s.	153
	Heart	33	155	n.s.	153
Starvation	Liver	n.s.	155	47	153
	Heart	75	155	n.s.	153
	Intestine	201			
Diet					
High fat	Liver			148	81
	Heart	145	56		
	Intestine	132/144[c]	143		
	Adipose tissue			125	192
	Mammary gland	127	94		
High carbohydrate	Liver			120	81
Diurnal rhythm (dark versus light phase)	Liver	n.s.	155	141	153
	Heart	75	155	59	153
Drugs					
Clofibrate	Liver	138	155	213	153
	Heart	n.s.	155	75	153
	Intestine	128	14		
	Muscle			n.s.	153
	Kidney			165	153
Cholestyramine	Liver	22	155	137	102
Mevinolin	Liver	29	155		
Cholesterol	Liver	41	155		
Cholate	Liver	41	155		
Alcohol	Liver	318	160	217	160
	Liver	100	158,		
Peroxisomal Proliferators[d]	Liver	120–>200	See text	>200	401
				260	234

[a]Values were derived from data in nanomoles per milligram of protein and expressed relative to those of control, adult, male fed animals (100%).
[b]n.s., No significant change.
[c]In the middle and distal parts of the intestine, respectively. No significant effect in proximal section.
[d]Tiadenol, nafenopin, phthalic acid esters, and acetylsalicylic acid.

intestinal FABP, it appears to be closer to the hepatic type in its response to variations. Both the FABP contents of the heart and liver, but not of the intestine, are higher in females (Ockner *et al.*, 1980, 1982; Kawashima *et al.*, 1982; Bass *et al.*, 1985b; Paulussen *et al.*, 1986, 1989). The sex influence in the liver corresponds to the effects of steroid hormones (Ockner *et al.*, 1980).

Starvation (Stein *et al.*, 1976; Brandes and Arad, 1983; Bass *et al.*, 1985a; Paulussen *et al.*, 1986, 1989a) and (hypolipidemic) drugs and peroxisome proliferators (Kawashima *et al.*, 1983; Bass *et al.*, 1985a,b; McTigue *et al.*, 1985; Paulussen *et al.*, 1986, 1989; Wilkinson and Wilton, 1987a) have a marked effect on the FABP content of the liver only. Clofibrate also has an increasing effect on the fatty-acid-binding capacity of kidney cytosol (Paulussen *et al.*, 1986) and on the liver FABP type in rat intestine (Bass *et al.*, 1985b). The capacity of the β-oxidation system in several rat tissues (Veerkamp and Van Moerkerk, 1986) is strongly correlated with the cytosolic fatty-acid-binding capacity in control and clofibrate-treated animals (Glatz *et al.*, 1988b). Brain tissue is an exception in this respect. An increase of the fat content or addition of fat to the diet clearly affects the FABP content of the liver (Rüstow *et al.*, 1982; Haq and Shrago, 1983; St. John *et al.*, 1987) and heart (Fournier and Rahim, 1985). A high-carbohydrate diet also increases the liver FABP content (Haq and Shrago, 1983). In the intestine, a high-fat diet produces a relatively larger induction of the intestinal FABP type (Ockner *et al.*, 1980). Chronic alcohol consumption by rats induced a dramatic increase of hepatic FABP content (Pignon *et al.*, 1987), but this was not always the case (Reyes *et al.*, 1971). A choline-deficient diet increased the FABP concentration of rat liver threefold (Dutta-Roy *et al.*, 1988).

During postnatal development, the FABP content of the heart (Figure 4) and liver increases (Foliot *et al.*, 1973; Ockner *et al.*, 1980, 1982; Crisman *et al.*, 1987, 1988; Sheridan *et al.*, 1987; Paulussen *et al.*, 1989), as do the β-oxidative capacities of these tissues (Veerkamp and Van Moerkerk, 1986). Interestingly, the cytosolic fatty-acid-binding capacity of both tissues is already at the adult level in newborn animals (Paulussen *et al.*, 1986). On the basis of bromosulfophthalein binding, the liver FABP content is already at the adult level at the late fetal stage in guinea pigs (Levi *et al.*, 1970) and monkeys (Levi *et al.*, 1969b). FABP-immunoreactive hepatocytes were found in the human liver as early as week 7 of gestation (Suzuki and Ono, 1987) and remained present at a frequency of approximately 80% of total cell number throughout the gestational period (Suzuki and Ono, 1988). This percentage is lower in the adult liver. In the human intestine, immunoreactive FABP was demonstrated at week 23 of gestation. Changes of the α-fetoprotein content could not explain the differences between fatty-acid-binding capacity and FABP content (Paulussen *et al.*, 1989). The two FABPs in the kidneys show a distinctive developmental pattern (Lam *et al.*, 1988). Whereas the heart type is detectable at comparable levels in neonatal and adult kidneys, the concentration of the kidney FABP type markedly increases

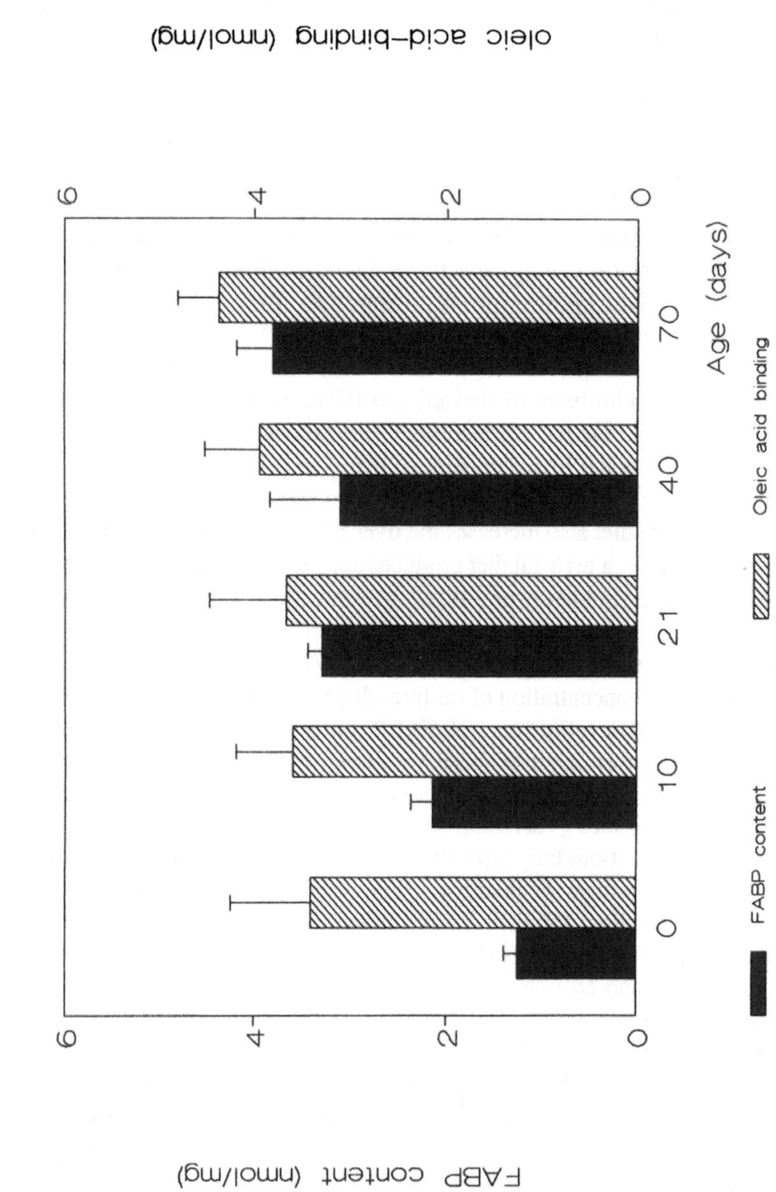

FIGURE 4. Postnatal development of FABP content and oleic acid-binding capacity in rat heart.

during postnatal development (Lam *et al.*, 1988). The diurnal rhythm had only a minor influence on the levels of liver and heart FABP (Bass *et al.*, 1985a,b; Paulussen *et al.*, 1986, 1989a; Wilkinson and Wilton, 1987a).

The FABP concentration varies markedly among various types of skeletal muscles and is higher in slow-twitch than in fast-twitch muscles (Claffey *et al.*, 1987; Crisman *et al.*, 1987; Miller *et al.*, 1988; Paulussen *et al.*, 1989). The FABP content and fatty acid oxidation capacity of rat and human skeletal muscles show a good correlation (Peeters *et al.*, 1989b). Chronic electrostimulation of rat tibialis anterior (a fast-twitch muscle) caused a marked rise of oxidative metabolism which is accompanied by an increase of the FABP content to a level comparable to that of soleus muscle and close to that of heart (Kaufmann *et al.*, 1989).

13. STABILITY AND TURNOVER

Most investigations concerning stability and turnover rates of FABPs have been carried out with the hepatic and intestinal types. FABPs are very stable proteins, with a high resistance to enzymic digestion (Ockner and Manning, 1974; Takahashi *et al.*, 1983; Sacchettini *et al.*, 1986; Peeters *et al.*, 1989a) and radiation (Sacchettini *et al.*, 1987). The binding of fatty acids to liver (Takahashi *et al.*, 1982) and heart (Sacchettini *et al.*, 1986) FABPs actually appears to increase this resistance. The liver FABP produced in *E. coli* had a comparable stability (Lowe *et al.*, 1984). However, a mutant liver FABP, produced by site-directed mutagenesis of an FABP gene introduced in *E. coli,* had a clearance rate that was 10-fold higher than that of the wild-type FABP (Lowe *et al.*, 1984). The mutant protein had a modified carboxy-terminal sequence, resulting in an altered secondary structure with a decreased hydrophobicity and a markedly reduced binding of oleic acid.

During isolation of bovine FABPs, Spener and co-workers applied temperatures up to 50°C and a pH of 4.0, without significant effects on the activity of the proteins (Haunerland *et al.*, 1984; Jagschies *et al.*, 1985). Ion exchange chromatography and delipidation at an early stage of purification, however, led to loss of binding activity (Jagschies *et al.*, 1985).

Bass *et al.* (1985a) determined the rates of degradation of FABP and total cytosolic protein after pulse-labeling with $NaH^{14}CO_3$ and calculated half-lives of 3.1 and 2.9 days for FABP and total protein, respectively. Apparently the degradation rate of liver FABP is comparable to that of an average cytosolic protein. Compared with the rate of fluctuations of fatty acid metabolism, FABP concentrations change only relatively slowly, based on the observed half-life. This is in agreement with the slight diurnal rhythm of the protein (Bass *et al.*, 1985a,b; Paulussen *et al.*, 1986, 1989a; Wilkinson and Wilton, 1987a). The isoforms of rat liver FABP have a similar degradation rate in vivo (Bass *et al.*, 1985a), although their endogenous ligands differ (Takahashi *et al.*, 1983; Bass, 1985).

The half-life of [^{35}S]methionine-labeled FABP in cultured neonatal rat cardiomyocytes is 2.5 days (Crisman *et al.*, 1988). The ability of hepatocytes to modulate the concentration of a protein which appeared to be identical to liver FABP (Bassuk *et al.*, 1987) within the relatively short period spanned by mitosis (Custer and Sorof, 1984) is not compatible with a half-life of 2–3 days. The half-life of 2 h reported for sterol carrier protein and hepatic FABP (McGuire *et al.*, 1984) markedly contrasts the findings of Bass *et al.* (1985a). However, despite their substantial similarity, they are now considered to be different proteins (Scallen *et al.*, 1985; Vahouny *et al.*, 1987).

The higher levels of liver FABP in female rats, as well as the increased content of this protein in the livers of male animals on clofibrate treatment, are not related to differences in the turnover rate of the protein. They appear to be correlated with an increase of the specific FABP mRNA content of the liver (Bass *et al.*, 1985b).

The liver and intestinal FABP types exhibit different regulatory responses in intestinal mucosa (Bass *et al.*, 1985b). Although the mRNA concentration of the liver FABP type is approximately 50% higher in the intestines of female rats than of males, there is no sex-related difference in the steady-state level of this protein, owing to a 1.35-fold-higher turnover rate in the female jejunum (Bass *et al.*, 1985b). No sex-related differences are observed for the intestinal FABP type or its mRNA.

A differential regulatory response is also present in the effect of clofibrate treatment on both types of FABP in the intestinal mucosa of male rats. Whereas the liver type is induced up to twofold by clofibrate, the level of the intestinal FABP type is only 25% higher. Clofibrate treatment doubled the concentration of liver FABP mRNA in the intestine, but had no effect on the intestinal FABP mRNA (Bass *et al.*, 1985b). The influence of sex difference and of clofibrate indicate that the concentrations of both proteins are modulated via changes in their rates of synthesis (Bass, 1985; Bass *et al.*, 1985b).

Postnatal development of mRNA levels for the different FABP types appears to be quite different in various rat tissues. In small intestinal epithelium, both mRNAs for liver and intestinal types of FABP are first detectable between days 19 and 21 of gestation (Gordon *et al.*, 1985). They undergo a coordinated three- to fourfold increase during the first postnatal day. Thereafter, the level of intestinal FABP mRNA remains constant, while the level of liver FABP mRNA increases further in the intestine. Levels of both mRNAs increase in this tissue during growth from 120 to 400 g of body weight.

The liver FABP mRNA concentration increases about twofold in liver within the first 24 h after birth, but it does not change further until 35 days (Gordon *et al.*, 1985). During growth from 120 to 400 g, the FABP mRNA concentration doubles again.

Distinct patterns of developmental change in heart FABP mRNA concentration are observed in the heart, placenta, brain, and testes of the rat (Heuckeroth *et al.*, 1987). This specific mRNA sequence was already detectable in myocardium at day 19 of gestation. Its level rises rapidly during the first 48 h after birth and reaches the maximal (adult) value at approximately 14 days. The brain and testes also show a postnatal increase of cardiac FABP mRNA, but later than heart. Maximal values are reached at 24 and 70 days, respectively, in these tissues. Changes in the heart FABP mRNA concentration in the kidneys form a contrasting pattern to that observed in the heart and brain (Heuckeroth *et al.*, 1987). The renal concentration of heart FABP mRNA is maximal during the fetal period, then falls rapidly during the first 2 days after birth, peaks again around day 8, and finally declines to a very low adult level. The significance of this aberrant developmental pattern is still unclear. Possibly, a quantitative switch between the two FABP types present in the kidneys (Lam *et al.*, 1988) is related to this phenomenon. Rat placenta shows a rapid increase in the heart FABP mRNA concentration in the late gestational period (Heuckeroth *et al.*, 1987).

14. FATTY-ACID-BINDING PROTEINS AND PATHOLOGY

Structural mutations or deficiencies of FABP, due to genetic aberrations, have not been observed in humans or animals. Some pathological or experimental conditions show changes in the content or appearance of FABPs (Table IX). In diabetes a decline of liver FABP content was observed, together with the appearance of a FABP form with a molecular mass of 400 kDa (Brandes and Arad, 1983). The appearance of a high-molecular-mass form was also observed under diabetic conditions in human and rabbit placenta (Thomas *et al.*, 1986). Hypertension, induced by deoxycorticosterone acetate (DOCA)-NaCl treatment or infusion of angiotensin II, resulted in the disappearance of the kidney FABP type from kidney cortex, whereas the heart FABP type remained unchanged in cortex and medulla (Lam *et al.*, 1988). The same models of experimental hypertension induced a marked decrease of the heart FABP type and its mRNA in rat aorta, but not in heart, skeletal muscle, or kidneys (Sarzani *et al.*, 1988).

FABP content and intracellular fatty acid metabolism may be closely related. In Morris hepatoma a nutritional regulation of fatty acid synthesis was absent, together with a decrease of the FABP content to 20% of that in the liver (Mishkin and Halperin, 1977). The FABP level appears to be associated with normal mitosis and carcinogen-induced proliferation of hepatocytes (Bassuk *et al.*, 1987). This suggests that liver FABP may carry not only ligands that pro-

Table IX
FABP and Pathology

Condition	Reported changes	Reference
Diabetes		
Rat liver	Decreased FABP content, appearance of high-molecular-mass form	26
Rabbit placenta	Appearance of different form	217
Human placenta	Appearance of different form	217
Hypertension		
Spontaneous		
Rat kidney	Increased FABP content in medulla	61
Induced by DOCA-NaCl or angiotensin II		
Rat kidney	Disappearance of kidney FABP from cortex, no effect on heart FABP type	109
Rat aorta	Marked reduction of heart FABP type	178
Morris hepatoma		
Rat	Low FABP content	130
Hepatitis, cirrhosis, Gilbert's syndrome, Dubin-Johnson syndrome		
Human liver	Decreased FABP content	
Human serum	Elevated FABP levels	98
Obesity (Zucker rat)		
Rat liver	Increased FABP content	138
Ischemia/Ca paradox		
Perfused rat heart	Release of FABP	71, 106

mote hepatocyte division, but also certain chemically activated carcinogens in their passage from cytoplasm to chromatin (Bassuk *et al.*, 1987). The covalent binding of metabolites of carcinogens to this protein was already detected before its fatty-acid-binding capacity was known (Ketterer *et al.*, 1976). The liver FABP content is markedly decreased in patients with various liver diseases, whereas their serum FABP concentration is higher than in normal subjects (Kamisaka *et al.*, 1981). A release of heart FABP in the perfusate was found in experimental ischemia and Ca paradox in perfused rat hearts (Van Der Vusse *et al.*, 1982; Knowlton *et al.*, 1989).

15. CHROMOSOMAL LOCALIZATION OF FATTY-ACID-BINDING PROTEIN GENES

Localization of the genes encoding FABPs and other hydrophobic ligand-binding proteins may provide new insights into the physiological roles and reg-

Table X
Chromosomal Localization of Genes Encoding FABP and Other Proteins with High Affinity for Hydrophobic Ligands

	Chromosome region	
Protein	Human	Mouse
Liver FABP	2 (p12–q12)	6
Intestinal FABP	4 (q28–q31)	3
Heart FABP	?	4, 8 and 10 or 15
Albumin	4 (q11–q22)	?
CRBP	3	9
CRBP-II (intestine)	3	9
Adipocyte P_2 (p422)	?	3

ulation of these proteins. Useful information may be derived from the comparison of inbred strains with known mutations or deletions and from in situ hybridization to metaphase chromosomes and somatic cell hybrid clones. Table X summarizes the available data concerning chromosomal localization of the genes encoding heart, liver, and intestinal FABP, CRBPs, adipocyte P_2 protein, and albumin in humans and mice. No data are available for rats. The mouse genes for liver (Sweetser *et al.*, 1987a), intestinal (Sweetser *et al.*, 1987a), and heart (Heuckeroth *et al.*, 1987) FABPs are located on different chromosomes, as are the human intestinal and liver FABP genes (Chen *et al.*, 1986; Sweetser *et al.*, 1987a). Human heart FABP-encoding DNA sequences have not yet been assigned to a specific chromosome, but, interestingly, a cDNA probe to its mouse counterpart hybridized with sequences on at least three chromosomes (Heuckeroth *et al.*, 1987). These may represent pseudogenes, or actively transcribed genes that encode other, closely related proteins that have not yet been identified (Heuckeroth *et al.*, 1987). These proteins may be other FABP types that are closely related to heart FABP, such as the FABP from skeletal muscle that is also detected by antibodies to heart FABP (Claffey *et al.*, 1987; Crisman *et al.*, 1987; Paulussen *et al.*, 1989).

Although a number of hydrophobic ligand-binding proteins are closely related, and probably originate from a common ancestor gene (Chan *et al.*, 1985), the genes encoding these proteins are dispersed throughout the genome in both mice and humans. Only the CRBP and CRBP-II genes are closely linked on the same chromosome. The genes for human intestinal FABP (Chen *et al.*, 1986) and albumin (Minghetti *et al.*, 1986) are both located on the long arm of chromosome 4.

16. CONCLUSIONS

The abundance of the FABPs and their mRNAs in the cytosol of various cell types suggests an important role for these low-molecular-weight proteins. The occurrence of various types of FABPs, their tissue distribution, and the differences in their ligand specificity, postnatal development, and response to drugs, physiological changes, and experimental pathological conditions indicate an adaptation to distinct cellular environments and/or functions. This may relate to specific metabolic needs of the cell: targeting of the bound ligand (fatty acid, acyl-CoA, acylcarnitine, or other) to specific subcellular systems, interaction with and protection of cellular membranes or enzyme systems, or modulation of lipid metabolism. FABPs may also indirectly be involved in the regulation of growth, differentiation, and transmembrane signaling.

Structural defects or deficiencies of FABPs have not yet been described for human or other mammalian tissues. The availability of cDNA probes and gene transfer techniques may, however, allow expression studies in various cell types and site-directed mutagenesis in order to gain more insight into the functions of the different FABP types. The regulation and expression of the various FABP genes under normal physiological and pathological conditions, and the relation of the FABPs to other members of their multigene family and to other carriers of hydrophobic ligands, remain other important topics for future research.

17. REFERENCES

1. Abumrad, N. A., Perkins, R. C., Park, J. H., and Park, C. R., 1981, Mechanism of long-chain fatty acid permeation in the isolated adipocyte, *J. Biol. Chem.* **256**:9183–9191.
2. Abumrad, N. A., Park, J. H., and Park, C. R., 1984, Permeation of long-chain fatty acid into adipocytes. Kinetics, specificity, and evidence for involvement of a membrane protein, *J. Biol. Chem.* **259**:8945–8953.
3. Akeroyd, R., Lange, L. G., Westerman, L., and Wirtz, K. W. A., 1981a, Modification of phosphatidylcholine-transfer protein from bovine liver with butanedione and phenylglyoxal, *Eur. J. Biochem.* **121**:77–81.
4. Akeroyd, R., Moonen, P., Westerman, J., Puyk, C., and Wirtz, K. W. A., 1981b, The complete primary structure of the phosphatidylcholine-transfer protein from bovine liver, *Eur. J. Biochem.* **114**:385–391.
5. Alpers, D. H., Strauss, A. W., Ockner, R. K., Bass, N. M., and Gordon, J. I., 1984, Cloning of a cDNA encoding rat intestinal fatty acid-binding protein, *Proc. Natl. Acad. Sci. USA* **81**:313–317.
6. Avanzati, B., and Catala, A., 1983, Partial purification of fatty acid-binding protein by ammonium sulphate precipitation, *Arch. Int. Physiol. Biochem.* **91**:103–108.
7. Barbour, R. L., and Chan, S. H. P., 1979, Regulation of palmitoyl-CoA inhibition of mitochondrial adenine nucleotide transport byu cytosolic fatty acid-binding protein, *Biochem. Biophys. Res. Commun.* **89**:1168–1177.
8. Bass, N. M., 1985, Function and regulation of hepatic and intestinal fatty acid-binding proteins, *Chem. Phys. Lipids* **38**:95–114.

9. Bass, N. M., 1988, The cellular fatty acid binding proteins: Aspects of structure, regulation and function, *Int. Rev. Cytol.* **111**:143–183.
10. Bass, N. M., and Manning, J. A., 1986, Tissue expression of the three structurally different fatty acid-binding proteins from rat heart, muscle, liver, and intestine, *Biochem. Biophys. Res. Commun.* **137**:929–935.
11. Bass, N. M., and Ockner, R. K., 1987, Interaction of lipid-lowering drugs with fatty acid-binding proteins, in *Drugs Affecting Lipid Metabolism* (R. Paoletti, D. Kritchevsky, and W. L. Holmes, eds.), pp. 105–110, Springer Verlag, New York.
12. Bass, N. M., Raghupathy, E., Rhoads, D. E., Manning, J. A., and Ockner, R. K., 1984, Partial purification of molecular weight 12 000 fatty acid-binding proteins from rat brain and their effect on synaptosomal Na^+-dependent amino acid uptake, *Biochemistry* **23**:6539–6544.
13. Bass, N. M., Manning, J. A., and Ockner, R. K., 1985a, Turnover and short-term regulation of fatty acid-binding protein in liver, *J. Biol. Chem.* **260**:9603–9607.
14. Bass, N. M., Manning, J. A., Ockner, R. K., Gordon, J. I., Seetharam, S., and Alpers, D. H., 1985b, Regulation of the biosynthesis of two distinct fatty acid-binding proteins in rat liver and intestine. Influence of sex difference and of clofibrate, *J. Biol. Chem.* **260**:1432–1436.
15. Bassuk, J. A., Tsichlis, P. N., and Sorof, S., 1987, Liver fatty acid-binding protein is the mitosis-associated polypeptide target of a carcinogen in rat hepatocytes, *Proc. Natl. Acad. Sci. USA* **84**:7547–7551.
16. Berk, P. D., Potter, B. J., and Stremmel, W., 1987, Role of plasma membrane ligand-binding proteins in the hepatocellular uptake of albumin-bound organic anions, *Hepatology* **7**:165–176.
17. Bernier, I., and Jollès, P., 1987, A survey on cytosolic nonenzymic proteins involved in the metabolism of lipophilic compounds: From organic anion binders to new protein families, *Biochimie* **69**:1127–1152.
18. Bernlohr, D. A., Angus, C. W., Lane, M. D., Bolanowski, M. A., and Kelly, T. J., 1984, Expression of specific mRNAs during adipose differentiation: Identification of an mRNA encoding a homologue of myelin P2 protein, *Proc. Natl. Acad. Sci. USA* **81**:5468–5472.
19. Bernlohr, D. A., Bolanowski, M. A., Kelly, T. A., Jr., and Lane, M. D., 1985, Evidence for an increase in transcription of specific mRNAs during differentiation of 3T3-L1 adipocytes, *J. Biol. Chem.* **260**:5563–5567.
20. Billheimer, J. T., and Gaylor, J. L., 1980, Cytosolic modulators of activities of microsomal enzymes of cholesterol biosynthesis. Role of a cytosolic protein with properties similar to Z protein (fatty acid-binding protein), *J. Biol. Chem.* **255**:8128–8135.
21. Billich, S., Wissel, T., Kratzin, H., Hahn, U., Hagendorff, B., Lezius, A. G., and Spener, F., 1988, Cloning of a full length complementary DNA for fatty acid-binding protein from bovine heart, *Eur. J. Biochem.* **175**:549–556.
22. Boguski, M. S., Elshourbagy, N., Taylor, J. M., and Gordon, J. I., 1984, Rat apolipoprotein A-IV contains 13 tandem repetitions of a 22-amino acid segment with amphipathic helical potential, *Proc. Natl. Acad. Sci. U.S.A.* **81**:5021–5025.
23. Böhmer, F. D., Kraft, R., Otto, A., Wernstedt, C., Hellman, U., Kurtz, A., Müller, T., Rohde, K., Etzold, G., Lehmann, W., Langen, P., Heldin, C.-H., and Grosse, R., 1987, Identification of a polypeptide growth inhibitor from bovine mammary gland, *J. Biol. Chem.* **262**:15137–15143.
24. Börchers, T., Unterberg, C., Rüdel, H., Robenek, H., and Spener, F., 1989, Subcellular distribution of cardiac fatty acid-binding proteins in bovine heart muscle and quantitation with an enzyme-linked immunosorbent assay, *Biochim. Biophys. Acta* **1002**: 54–61.
25. Bordewick, U., Schulenberg, H., Schäfer, P., and Spener, F., 1986, Recognition of fatty acids by hepatic fatty acid-binding protein, *Fette Seifen Anstrichm.* **88**:588–592.
26. Brandes, R., and Arad, R., 1983, Liver cytosolic fatty acid-binding proteins. Effect of diabetes and starvation, *Biochim. Biophys. Acta* **750**:334–339.

27. Brandes, R., Ockner, R. K., Weisiger, R. A., and Lysenko, N., 1982, Specific and saturable binding of albumin to rat adipocytes: Modulation by epinephrine and possible role in free fatty acid transfer, *Biochem. Biophys. Res. Commun.* **105**:821–827.
28. Brecher, P., Saouaf, R., Sugarman, J. M., Eisenberg, D., and LaRosa, K., 1984, Fatty acid transfer between multilamellar liposomes and fatty acid-binding proteins, *J. Biol. Chem.* **259**:13395–13401.
29. Bröring, K., Haest, C. W. M., and Deuticke, B., 1989, Translocation of oleic acid across the erythrocyte membrane. Evidence for a fast process, *Biochim. Biophys. Acta* **986**:321–331.
30. Brown, J. R., and Shockley, P., 1982, Serum albumin: Structure and characterization of its ligand binding sites, in *Lipid-Protein Interactions* (P. C. Jost, and O. H. Griffith, eds.), Vol. 1, pp. 25–68, John Wiley & Sons, New York.
31. Burnett, D. A., Lysenko, N., Manning, J. A., and Ockner, R. K., 1979, Utilization of long-chain fatty acids by rat liver: Studies of the role of fatty acid-binding protein, *Gastro enterology* **77**:247–249.
32. Burrier, R. E., and Brecher, P., 1986, Binding of lysophosphatidylcholine to the rat liver fatty acid-binding protein, *Biochim. Biophys. Acta* **879**:229–239.
33. Burrier, R. E., Mansson, C. R., and Brecher, P., 1987, Binding of acyl-CoA to liver fatty acid-binding protein: Effect on acyl-CoA synthesis, *Biochim. Biophys. Acta* **919**:221–230.
34. Capron, F., Coltoff-Schiller, B., Johnson, A. B., Fleischner, G. M., and Goldfischer, S., 1979, Immunocytochemical localization of hepatic ligandin and Z protein utilizing frozen sections for light and electron microscopy, *J. Histochem. Cytochem.* **27**:961–966.
35. Catala, A., and Avanzati, B., 1983, Oleic acid transfer from microsomes to egg lecithin liposomes: Participation of fatty acid-binding protein, *Lipids* **18**:803–807.
36. Chan, L., Wei, C.-F., Li, W.-H., Yang, C.-Y., Ratner, P., Pownall, H., Gotto, A. M., Jr., and Smith, L. C., 1985, Human liver fatty acid-binding protein cDNA and amino acid sequence. Functional and evolutionary implications, *J. Biol. Chem.* **260**:2629–2632.
37. Chen, S. H., VanTuinen, P., Ledbetter, D. H., Smith, L. C., and Chan, L., 1986, Human liver fatty acid-binding protein gene is located on chromosome 2, *Somatic Cell Mol. Genet.* **12**:303–306.
38. Chou, P. Y., and Fasman, G. D., 1979, Prediction of β-turns, *Biophys. J.* **26**:367–384.
39. Chytil, F., and Ong, D. E., 1987, Intracellular vitamin A-binding proteins, *Annu. Rev. Nutr.* **7**:321–335.
40. Cistola, D. P., Sacchettini, J. C., Banaszak, L. J., Walsh, M. T., and Gordon, J. I., 1989, Fatty acid interactions with rat intestinal and liver fatty acid-binding proteins expressed in Escherichia coli. A comparative ^{13}C NMR study, *J. Biol. Chem.* **264**:22700–2710.
41. Cistola, D. P., Small, D. M., and Hamilton, J. A., 1987, Carbon 13 NMR studies of saturated fatty acids bound to bovine serum albumin. II. Electrostatic interactions in individual fatty acid binding sites, *J. Biol. Chem.* **262**:10980–10985.
42. Cistola, D. P., Walsh, M. T., Corey, R. P., Hamilton, J. A., and Brecher, P., 1988, Interactions of oleic acid with liver fatty acid-binding protein: A carbon-13 NMR study, *Biochemistry* **27**:711–717.
43. Claffey, K. P., Herrera, V. L., Brecher, P., and Ruiz-Opazo, N., 1987, Cloning and tissue distribution of rat heart fatty acid-binding protein mRNA: Identical forms in heart and skeletal muscle, *Biochemistry* **26**:7900–7904.
44. Claffey, K. P., Crisman, T. S., Ruiz-Opazo, N., and Brecher, P., 1988, Expression of rat heart fatty acid-binding protein cDNA in rat L6 myoblasts, *FASEB J.* **2**:A1783.
45. Cooper, R., Noy, N., and Zakim, D., 1987, A physical-chemical model for cellular uptake of fatty acids: Prediction of intracellular pool, *Biochemistry* **26**:5890–5896.
46. Crisman, T. S., Claffey, K. P., Saouaf, R., Hanspal, J., and Brecher, P., 1987, Measurement of rat heart fatty acid-binding protein by ELISA. Tissue distribution, developmental changes and subcellular distribution, *J. Mol. Cell. Cardiol.* **19**:423–431.

47. Crisman, T. S., Claffey, K. P., Labrie, E. D., and Brecher, P., 1988, Rat heart fatty cid-binding protein in cultured neonatal myocytes, *FASEB J.* **2:**A1037.
48. Custer, R. P., and Sorof, S., 1984, Target polypeptide of a carcinogen is associated with normal mitosis and carcinogen-induced hyperplasias in adult hepatocytes, *Proc. Natl. Acad. Sci. USA* **81:**6738–6742.
49. Daniels, C., Noy, N., and Zakim, D., 1985, Rates of hydration of fatty acids bound to unilamellar vesicles of phosphatidylcholine or to albumin, *Biochemistry* **24:**3286–3292.
50. Demmer, L. A., Birkenmeier, E. H., Sweetser, D. A., Levin, M. S., Zollman, S., Sparkes, R. S., Mohandas, T., Lusis, A. J., and Gordon, J. I., 1987, The cellular retinol binding protein II gene. Sequence analysis of the rat gene, chromosomal localization in mice and humans, and documentation of its close linkage to the cellular retinol binding protein gene, *J. Biol. Chem.* **262:**2458–2467.
51. Dietschy, J. M., 1978, General principles governing movement of lipids across biological membranes, in *Disturbances in Lipid and Lipoprotein Metabolism* (J. M. Dietschy, A. M. Gotto, and S. A. Ontko, eds.), pp. 1–28, The Williams & Wilkins Co., Baltimore.
52. Dutta-Roy, A. K., Gopalswamy, N., and Trulzsch, D. V., 1987, Prostaglandin E_1 binds to Z protein of rat liver, *Eur. J. Biochem.* **162:**615–619.
53. Dutta-Roy, A. K., Trinh, M. V., Sullivan, T. F., and Trulzsch, D. V., 1988, Choline-deficient diet increases Z protein concentration in rat liver, *J. Nutr.* **118:**1116–1119.
54. Flinta, C., Persson, B., Jörnvall, H., and Von Heijne, G., 1986, Sequence determinants of cytosolic N-terminal protein processing, *Eur. J. Biochem.* **154:**193–196.
55. Foliot, A., Housset, E., Ploussard, J. P., Petite, J. P., and Infante, R., 1973, Study of Y and Z, two cytoplasmic bilirubin binding proteins. Their development in the liver of foetal and newborn Wistar and Gunn rats, *Biomedicine* **19:**488–491.
56. Fournier, N. C., and Rahim, M., 1985, Control of energy production in the heart, A new function for fatty acid-binding protein, *Biochemistry* **24:**2387–2396.
57. Fournier, N. C., and Richard, M. A., 1988, Fatty acid-binding protein, a potential regulator of energy production in the heart. Investigation of mechanisms by electron spin resonance, *J. Biol. Chem.* **263:**14471–14479.
58. Fournier, N. C., Geoffroy, M., and Deshusses, J., 1978, Purification and characterization of a long-chain fatty acid-binding protein supplying the mitochondrial β-oxidative system in the heart, *Biochim. Biophys. Acta* **533:**457–464.
59. Fournier, N. C., Zuker, M., Williams, R. E., and Smith, I. C. P., 1983, Self-association of the cardiac fatty acid-binding protein. Influence on membrane-bound, fatty acid dependent enzymes, *Biochemistry* **22:**1863–1872.
60. Fujii, S., Kawaguchi, H., and Yasuda, H., 1987a, Purification and characterization of fatty acid-binding protein from rat kidney, *Arch. Biochem. Biophys.* **254:**552–558.
61. Fujii, S., Kawaguchi, H., and Yasuda, H., 1987b, Fatty acid-binding protein in kidney of normotensive and genetically hypertensive rats, *Hypertension* **10:**93–99.
62. Fukai, F., Kase, T., Shidotani, T., Nagai, T., and Katayama, T., 1987, A novel role of fatty acid-binding protein as a vehicle for retinoids, *Biochem. Biophys. Res. Commun.* **147:**899–903.
63. Ghinea, N., Fixman, A., Alexandru, D., Popov, D., Hasu, M., Ghitescu, L., Eskenasy, Simionescu, M., and Simionescu, N., 1988, Identification of albumin-binding proteins in capillary endothelial cells, *J. Cell Biol.* **107:**231–239.
64. Gibson, B. W., Yu, Z., Aberth, W., Burlingame, A. L., and Bass, N. M., 1988, Revision of the blocked N terminus of rat heart fatty acid-binding protein by liquid secondary ion mass spectrometry, *J. Biol. Chem.* **263:**4182–4185.
65. Giguere, V., Ong, E. S., Segui, P., and Evans, R. M., 1987, Identification of a receptor for the morphogen retinoic acid, *Nature* **330:**624–629.
66. Glatz, J. F. C., and Veerkamp, J. H., 1983, A radiochemical procedure for the assay of fatty acid binding by proteins, *Anal. Biochem.* **132:**89–95.

67. Glatz, J. F. C., and Veerkamp, J. H., 1985, Intracellular fatty acid-binding proteins, *Int. J. Biochem.* **17**:13–22.
68. Glatz, J. F. C., Baerwaldt, C. C. F., Veerkamp, J. H., and Kempen, H. J. M., 1984, Diurnal variation of cytosolic fatty acid-binding protein content and of palmitate oxidation in rat liver and heart, *J. Biol. Chem.* **259**:4295–4300.
69. Glatz, J. F. C., Janssen, A. M., Baerwaldt, C. C. F., and Veerkamp, J. H., 1985a, Purification and characterization of fatty acid-binding proteins from rat heart and liver, *Biochim. Biophys. Acta* **837**:57–66.
70. Glatz, J. F. C., Paulussen, R. J. A., and Veerkamp, J. H., 1985b, Fatty acid-binding proteins from heart, *Chem. Phys. Lipids* **38**:115–129.
71. Glatz, J. F. C., Van Bilsen, M., Paulussen, R. J. A., Veerkamp, J. H., Van Der Vusse, G. J., and Reneman, R. S., 1988a, Release of fatty acid-binding protein from isolated rat heart subjected to ischemia and reperfusion or to the calcium paradox, *Biochim. Biophys. Acta* **961**:148–152.
72. Glatz, J. F. C., Van Der Vusse, G. J., and Veerkamp, J. H., 1988b, Fatty acid-binding proteins and their physiological significance, *News Physiol. Sci.* **3**:41–43.
73. Godovac-Zimmermann, J., 1988, The structural motif of β-lactoglobulin and retinol-binding protein: A basic framework for binding and transport of small hydrophobic biomolecules?, *Trends Biochem. Sci.* **13**:64–66.
74. Gordon, J. I., and Lowe, J. B., 1985, Analyzing the structures, functions and evolution of two abundant gastrointestinal fatty acid-binding proteins with recombinant DNA and computational techniques, *Chem. Phys. Lipids* **38**:137–158.
75. Gordon, J. I., Alpers, D. H., Ockner, R. K., and Strauss, A. W., 1983, The nucleotide sequence of rat liver fatty acid-binding protein mRNA, *J. Biol. Chem.* **258**:3356–3363.
76. Gordon, J. I. Elshourbagy, N., Lowe, J. B., Liao, S., Alpers, D. H., and Taylor, J. M., 1985, Tissue specific expression and developmental regulation of two genes coding for rat fatty acid-binding proteins, *J. Biol. Chem.* **260**:1995–1998.
77. Goresky, C. A., Daly, D. S., Mishkin, S., and Arias, I. M., 1978, Uptake of labelled palmitate by the intact liver: Role of intracellular binding sites, *Am. J. Physiol.* **234**:E542–E553.
78. Grinstead, G. F., Trzaskos, J. M., Billheimer, J. T., and Gaylor, J. L., 1983, Cytosolic modulators of activities of microsomal enzymes of cholesterol biosynthesis. Effects of acyl-CoA inhibition and cytosolic Z protein, *Biochim. Biophys. Acta* **751**:41–51.
79. Gutknecht, J., 1988, Proton conductance caused by long-chain fatty acids in phospholipid bilayer membranes, *J. Membr. Biol.* **106**:83–93.
80. Hamilton, J. A., and Cistola, D. P., 1986, Transfer of oleic acid between albumin and phospholipid vesicles, *Proc. Natl. Acad. Sci. U.S.A.* **83**:82–86.
81. Haq, R. U., and Shrago, E., 1983, Dietary induction of rat liver fatty acid-binding protein, *Nutr. Res.* **3**:329–333.
82. Haq, R. U., Christodoulides, L., Ketterer, B., and Shrago, E., 1982, Characterization and purification of fatty acid-binding protein in rat and human adipose tissue, *Biochim. Biophys. Acta* **713**:193–198.
83. Haq, R. U., Shrago, E., Christodoulides, L., and Ketterer, B., 1985, Purification and characterization of fatty acid-binding protein in mammalian lung, *Exp. Lung Res.* **9**:43–55.
84. Haq, R. U., Tsao, F., and Shrago, E., 1987, Relation of lung fatty acid-binding protein to the biosynthesis of pulmonary phosphatidic acid and phosphatidylcholine, *J. Lipid Res.* **28**:216–220.
85. Haunerland, N., Jagschies, G., Schulenberg, H., and Spener, F., 1984, Occurrence of two fatty acid-binding proteins in bovine liver cytosol and their binding of fatty acids, cholesterol, and other lipophilic ligands, *Hoppe-Seyler's Z. Physiol. Chem.* **365**:365–376.
86. Heuckeroth, R. O., Birkenmeier, E. H., Levine, M. S., and Gordon, J. I., 1987, Analysis of the tissue-specific expression, developmental regulation, and linkage relationships of a rodent gene encoding heart fatty acid-binding protein, *J. Biol. Chem.* **262**:9709–9717.

87. Holden, H. M., Rypniewski, W. R., Law, J. H., and Rayment, I., 1987, The molecular structure of insecticyanin from the tobacco hornworm Manduca sexta L. at 2.6 Å resolution, *EMBO J.* **6:**1565–1570.
88. Huber, R., Schneider, M., Epp, O., Mayr, I., Messerschmidt, A., Pflugrath, J., and Kayser, H., 1987, Crystallization, crystal structure analysis and preliminary molecular model of the bilin binding protein from the insect Pieris brassicae, *J. Mol. Biol.* **195:**423–434.
89. Hunt, C. R., Ro, J. H.-S., Dobson, D. E., Min, H. Y., and Spiegelman, B. M., 1986, Adipocyte P2 gene: Developmental expression and homology of 5′-flanking sequences among fat cell-specific genes, *Proc. Natl. Acad. Sci. U.S.A.* **83:**3786–3790.
90. Iritani, N., Fukuda, E., and Inoguchi, K., 1980, A possible role of Z protein in dietary control of hepatic triacylglycerol synthesis, *J. Nutr. Sci. Vitaminol.* **26:**271–277.
91. Iseki, S., Kondo, H., Hitomi, M., and Ono, M., 1988, Immunocytochemical localization of hepatic fatty acid-binding protein in the liver of fed and fasted rats, *Histochemistry* **89:**317–322.
92. Jagschies, G., 1984, Fettsäurebindungsproteine aus Rinderherz, Ph.D. Thesis, University of Münster, Münster, Federal Republic of Germany.
93. Jagschies, G., Reers, M., Unterberg, C., and Spener, F., 1985, Bovine fatty acid-binding proteins. Isolation and characterization of two cardiac fatty acid-binding proteins that are distinct from corresponding hepatic proteins, *Eur. J. Biochem.* **152:**537–545.
94. Jones, P. D., Caine, A., Bass, N. M., and Grigor, M. R., 1988, Isolation and characterization of fatty acid-binding proteins from mammary tissue of lactating rats, *Biochem. J.* **251:**919–925.
95. Julien, P., Dagenais, G. R., Gailis, J., and Roy, P. E., 1978, Role of cardiac lymph and interstitial fluid in lipid metabolism of canine heart, *Can. J. Physiol. Pharmacol.* **56:**1041–1046.
96. Kamisaka, K., Listowsky, I., Gatmaitan, Z., and Arias, I. M., 1975, Circular dichroism analysis of the secondary structure of Z protein and its complexes with bilirubin and other organic anions, *Biochim. Biophys. Acta* **393:** 24–30.
97. Kamisaka, K., Hirano, M., and Tsuru, M., 1978, Purification of acidic Z protein from human liver, *Experientia* **34:**1265–1266.
98. Kamisaka, K., Maezawa, H., Inagaki, T., and Okano, K., 1981, A low molecular weight binding protein for organic anions (Z protein) from human hepatic cytosol: Purification and quantitation, Hepatology **1:**221–227.
99. Kaufmann, M., Simoneau, J.-A., Veerkamp, J. H., and Pette, D., 1989, Electrostimulation-induced increases in fatty acid-binding protein and myoglobin in rat fast-twitch muscle and comparison with tissue levels in heart, *FEBS Lett.* **245:**181–184.
100. Kawashima, Y., Nakagawa, S., and Kozuka, H., 1982, Effects of some hypolipidemic drugs and phthalic acid esters on fatty acid-binding protein in rat liver, *J. Pharmacobio-Dyn.* **5:**771–779.
101. Kawashima, Y., Nakagawa, S., Tachibana, Y., and Kozuka, H., 1983, Effects of peroxisome proliferators on fatty acid-binding protein in rat liver, *Biochim. Biophys. Acta* **754:**21–27.
102. Kempen, H. J. M., Glatz, J. F. C., De Lange, J., and Veerkamp, J. H., 1983, Concomitant increase in hepatic triacylglycerol biosynthesis and cytosolic fatty acid-binding protein content after feeding rats with a cholestyramine-containing diet, *Biochem. J.* **216:**511–514.
103. Ketterer, B., Tipping, E., Hackney, J. F., and Beale, D., 1976, A low-molecular-weight protein from rat liver that resembles ligandin in its binding properties, *Biochem. J.* **155:**511–521.
104. Keuper, H. J. K., Klein, R. A., and Spener, F., 1985, Spectroscopic investigations on the binding site of bovine hepatic fatty acid-binding protein. Evidence for the existence of a single binding site for two fatty acid molecules, *Chem. Phys. Lipids* **38:**159–173.
105. Kitamura, K., Suzuki, M., Suzuki, A., and Uyemura, K., 1980, The complete amino acid sequence of the P2 protein in bovine peripheral nerve myelin, *FEBS Lett.* **115:**27–30.

106. Knolton, A. A., Apstein, C. S., Saouaf, R., and Brecher, P., 1989, Leakage of heart fatty acid-binding protein with ischemia and reperfusion in the rat, *J. Mol. Cell. Cardiol.* **21**:577–583.
107. Knudsen, J., Højrup, P., Hansen, H. O., Hansen, H. F., and Roepstorff, P., 1989, Acyl-CoA-binding protein in the rat. Purification, binding characteristics, tissue concentrations and amino acid sequence, *Biochem. J.* **262**:513–519.
108. Kushlan, M. C., Gollan, J. L., Ma, W.-L., and Ockner, R. K., 1981, Sex differences in hepatic uptake of long-chain fatty acids in single-pass perfused rat liver, *J. Lipid Res.* **22**:431–436.
109. Lam, K. T., Borkan, S., Claffey, K. P., Schwartz, J. H., Chobanian, A. V., and Brecher, P., 1988, Properties and differential regulation of two fatty acid-binding proteins in the rat kidney, *J. Biol. Chem.* **263**:15762–15768.
110. Lee, L., and Wiggert, B., 1984, Isolation and characterization of an unsaturated fatty acid-binding protein from developing chick neutral retina, *J. Neurochem.* **42**:47–53.
111. Levi, A. J., Gatmaitan, Z., and Arias, I. M., 1969a, Two hepatic cytoplasmic protein fractions, Y and Z, and their possible role in the hepatic uptake of bilirubin, sulfobromophthalein, and other anions, *J. Clin. Invest.* **48**:2156–2167.
112. Levi, A. J., Gatmaitan, Z., and Arias, I. M., 1969b, Deficiency of hepatic organic anion-binding protein as a possible cause of non-haemolytic unconjugated hyperbilirubinaemia in the newborn, *Lancet* **II**:139–140.
113. Levi, A. J., Gatmaitan, Z., and Arias, I. M., 1970, Deficiency of hepatic organic anion-binding protein, impaired organic anion uptake by liver and "physiologic" jaundice in newborn monkeys, *N. Engl. J. Med.* **283**:1136–1139.
114. Li, E., Demmer, L. A., Sweetser, D. A, Ong, D. E., and Gordon, J. I., 1986, Rat cellular retinol-binding protein II: Use of a cloned cDNA to define its primary structure, tissue specific expression, and developmental regulation, *Proc. Natl. Acad. Sci. U.S.A.* **83**:5779–5783.
115. Lowe, J. B., Strauss, A. W., and Gordon, J. I., 1984, Expression of a mammalian fatty acid-binding protein in Escherichia coli, *J. Biol. Chem.* **259**:12696–12704.
116. Lowe, J. B., Boguski, M. S., Sweetser, D. A., Elshourbagy, N. A., Taylor, J. M., and Gordon, J. I., 1985, Human liver fatty acid-binding protein. Isolation of a full length cDNA and comparative sequence analyses of orthologous and paralogous proteins, *J. Biol. Chem.* **260**:3413–3417.
117. Lowe, J. B., Sacchettini, J. C., Laposata, M., McQuillan, J. J., and Gordon, J. I., 1987, Expression of rat intestinal fatty acid-binding protein in Escherichia coli. Purification and comparison of ligand binding characteristics with that of Escherichia coli-derived rat liver fatty acid-binding protein, *J. Biol. Chem.* **262**:5931–5937.
118. Lunzer, M. A., Manning, J. A., and Ockner, R. K., 1977, Inhibition of rat liver acetyl coenzyme A carboxylase by long-chain acyl coenzyme A and fatty acid. Modulation by fatty acid-binding protein, *J. Biol. Chem.* **252**:5483–5487.
119. Matarese, V., and Bernlohr, D. A., 1988, Purification of murine adipocyte lipid-binding protein. Characterization as a fatty acid- and retinoic acid-binding protein, J. Biol. Chem. **263**:14544–14551.
120. McCormack, M., and Brecher, P., 1987, Effect of liver fatty acid-binding protein on fatty acid movement between liposomes and rat liver microsomes, *Biochem. J.* **244**:717–723.
121. McGuire, D. M., Olson, C. D., Towle, H. C., and Dempsey, M. E., 1984, Translational control of the circadian rhythm of liver sterol carrier protein, *J. Biol. Chem.* **259**:5368–5371.
122. McLachlan, A. D., and Walker, J. E., 1977, Evolution of serum albumin, *J. Mol. Biol.* **112**:543–558.
123. McTigue, J., Taylor, J. B., Craig, R. K., Christodoulides, L., and Ketterer, B., 1985, Effect of the peroxisome proliferator tiadenol on levels of mRNA for fatty acid-binding protein, *Biochem. Soc. Trans.* **13**:896–897.
124. Megli, F. M., De Lisi, A., Van Amerongen, A., Wirtz, K. W. A., and Quagliariello, E., 1986, Nonspecific lipid transfer protein (sterol carrier protein 2) is bound to rat liver mitochondria: Its

role in spontaneous intermembrane phospholipid transfer, *Biochim. Biophys. Acta* **861**:463–470.

125. Mikkelsen, J., and Knudsen, J., 1987, Acyl-CoA-binding protein from cow. Binding characteristics and cellular and tissue distribution, *Biochem. J.* **248**:709–714.
126. Mikkelsen, J., Hojrup, P., Nielsen, P. F., Roepstorff, P., and Knudsen, J., 1987, Amino acid sequence of acyl-CoA binding protein from cow liver, *Biochem. J.* **245**:857–861.
127. Milici, A. J., Watrous, N. E., Stukenbrok, H., and Palade, G. E., 1987, Transcytosis of albumin in capillary endothelium, *J. Cell Biol.* **105**:2603–2612.
128. Miller, W. C., Hickson, R. C., and Bass, N. M., 1988, Fatty acid-binding proteins in the three types of rat skeletal muscle, *Proc. Soc. Exp. Biol. Med.* **189**:183–188.
129. Minghetti, P. P., Ruffner, D. E. Kuang, W. J., Dennison, O. E., Hawkins, J. W., Beattie, W. G., and Dugaiczyk, A., 1986, Molecular structure of the human albumin gene is revealed by nucleotide sequence within q11-22 of chromosome 4, *J. Biol. Chem.* **261**:6747–6757.
130. Mishkin, S., and Halperin, M. L., 1977, Loss of regulation of lipid metabolism in Morris hepatomas: A potential role for cytoplasmic binding proteins, *Adv. Exp. Med. Biol.* **92**:583–605.
131. Mishkin, S., and Turcotte, R., 1974a, Stimulation of monoacylglycerolphosphate formation by Z protein, *Biochem. Biophys. Res. Commun.* **60**:376–381.
132. Mishkin, S., and Turcotte, R., 1974b, The binding of long-chain fatty acid CoA to Z, a cytoplasmic protein present in liver and other tissues of the rat, *Biochem. Biophys. Res. Commun.* **57**:918–926.
133. Mishkin, S., Stein, L., Gatmaitan, Z., and Arias, I. M., 1972, The binding of fatty acids to cytoplasmic proteins: Binding to Z protein in liver and other tissues of the rat, *Biochem. Biophys. Res. Commun.* **47**:997–1003.
134. Mishkin, S., Stein, L., Fleischner, G., Gatmaitan, Z., and Arias, I. M., 1975, Z protein in hepatic uptake and esterification of long-chain fatty acids, *Am. J. Physiol.* **228**:1634–1640.
135. Mogensen, I. B., Schulenberg, H., Hansen, H. O., Spener, F., and Knudsen, J., 1987, A novel acyl-CoA-binding protein from bovine liver. Effect on fatty acid synthesis, *Biochem. J.* **241**:189–192.
136. Morand, O., Fibach, E., and Gatt, S., 1982, Effect of albumin, low temperature and metabolic inhibitors on transport of fatty acids into cultured human leukemic myeloid cells, *Biochim. Biophys. Acta* **693**:143–150.
137. Morrow, F. D., and Martin, R. J., 1983, Quantitation of hepatic fatty acid-binding proteins by post-chromatographic ligand binding assay, *J. Lipid Res.* **24**:324–331.
138. Morrow, F. D., Allen, C. E., and Martin, R. J., 1979, Intracellular fatty acid-binding protein: Hepatic levels in lean and obese rats, *Fed. Proc.* **38**:280.
139. Newcomer, M. E., Jones, T. A., Åqvist, J., Sundelin, J., Eriksson, U., Rask, L., and Peterson, P. A., 1984, The three dimensional structure of retinol binding protein, *EMBO J.* **3**:1451–1454.
140. Noy, N., and Zakim, D., 1985, Fatty acids bound to unilamellar lipid vesicles as substrates for microsomal acyl-CoA ligase, *Biochemistry* **24**:3521–3525.
141. Noy, N., Donnelly, T. M., and Zakim, D., 1986, Physical-chemical model for the entry of water-insoluble compounds into cells. Studies of fatty acid uptake by the liver, *Biochemistry* **25**:2013–2021.
142. Nunn, W. D., Colburn, R. W., and Black, P. N., 1986, Transport of long-chain fatty acids in Escherichia coli. Evidence for role of fadL gene product as long-chain fatty acid receptor, *J. Biol. Chem.* **261**:167–171.
143. Ockner, R. K., and Manning, J. A., 1974, Fatty acid-binding protein in small intestine. Identification, isolation and evidence for its role in cellular fatty acid transport, *J. Clin. Invest.* **54**:326–338.
144. Ockner, R. K., and Manning, J. A., 1976, Fatty acid-binding protein. Role in esterification of absorbed long-chain fatty acid in rat intestine, *J. Clin. Invest.* **58**:632–641.

145. Ockner, R. K., Manning, J. A., Poppenhausen, R. B., and Ho, W. K. L., 1972, A binding protein for fatty acids in cytosol of intestinal mucosa, liver, myocardium, and other tissues, *Science* **177**:56–58.
146. Ockner, R. K., Lysenko, N., Manning, J. A., and Monroe, S. E., 1980, Sex steroid modulation of fatty acid utilization and fatty acid-binding protein concentration in rat liver, *J. Clin. Invest.* **65**:1013–1023.
147. Ockner, R. K., Manning, J. A., and Kane, J. P., 1982, Fatty acid-binding protein. Isolation from rat liver, characterization, and immunochemical quantification, *J. Biol. Chem.* **257**:7872–7878.
148. Ockner, R. K., Weisiger, R. A., and Gollan, J. L., 1983, Hepatic uptake of albumin-bound substances: Albumin receptor concept, *Am. J. Physiol.* **245**:G13–G18.
149. O'Doherty, P. J. A., and Kuksis, A., 1975, Stimulation of triacylglycerol synthesis by Z protein in rat liver and intestinal mucosa, *FEBS Lett.* **60**:256–258.
150. Offner, G. D., Troxler, R. F., and Brecher, P., 1986, Characterization of a fatty acid-binding protein from rat heart, *J. Biol. Chem.* **261**:5584–5589.
151. Offner, G. D., Brecher, P., Sawlivich, W. B., Costello, C. E., and Troxler, R. F., 1988, Characterization and amino acid sequence of a fatty acid-binding protein from human heart, *Biochem. J.* **252**:191–198.
152. Papiz, M. Z., Sawyer, L., Eliopoulos, E. E., North, A. C. T., Findlay, J. B. C., Sivaprasadarao, R., Jones, T. A., Newcomer, M. E., and Kraulis, P. J., 1986, The structure of β-lactoglobulin and its similarity to plasma retinol-binding protein, *Nature* **324**:383–385.
153. Paulussen, R. J. A., Jansen, G. P. M., and Veerkamp, J. H., 1986, Fatty acid-binding capacity of cytosolic proteins of various rat tissues: Effect of postnatal development, starvation, sex, clofibrate feeding and light cycle, *Biochim. Biophys. Acta* **877**:342–349.
154. Paulussen, R. J. A., Van Der Logt, C. P. E., and Veerkamp, J. H., 1988, Characterization and binding properties of fatty acid-binding proteins from human, pig and rat heart, *Arch. Biochem. Biophys.* 264:533–545.
155. Paulussen, R. J. A., Geelen, M. J. H., Beynen, A. C., and Veerkamp, J. H., 1989, Immunochemical quantitation of fatty acid-binding proteins. I. Tissue and intracellular distribution, postnatal development and influence of physiological conditions on rat heart and liver FABP, *Biochim. Biophys. Acta* **1001**:201–209.
156. Paulussen, R. J. A., Van Moerkerk, H. T. B., and Veerkamp, J. H., 1990, Immunochemical quantitation of fatty acid binding proteins II: Tissue distribution of liver and heart FABP types in human and porcine tissues, *Int. J. Biochem.*, in press.
157. Peeters, R. A., In't Groen, M. A. P. M., De Moel, M. P., Van Moerkerk, H. T. B., and Veerkamp, J. H., 1989a, The binding affinity of fatty acid-binding proteins from human, pig, and rat liver for different fluorescent fatty acids and other ligands, *Int. J. Biochem.* **21**:407–418.
158. Peeters, R. A., In 't Groen, M. A., and Veerkamp, J. H., 1989b, The fatty acid-binding protein from human skeletal muscle, *Arch. Biochem. Biophys.* **274**:556–563.
159. Peeters, R. A., Veerkamp, J. H., and Demel, R. A., 1989c, Are fatty acid-binding proteins involved in fatty acid transfer? *Biochim. Biophys. Acta* **1002**:8–13.
160. Pignon, J. P., Bailey, N. C., Baraona, E., and Lieber, C. S., 1987, Fatty acid-binding protein: A major contributor to the ethanol-induced increase in liver cytosolic proteins in the rat, *Hepatology* **7**:865–871.
161. Potter, B. J., Sorrentino, D., and Berk, P. D., 1989, Mechanisms of cellular uptake of free fatty acids, *Ann. Rev. Nutr.* **9**:253–270.
162. Potter, B. J., Stump, D., Schwieterman, W., Sorrentino, D., Jacobs, L. N., Kiang, C.-L., Rand, J. H., and Berk, P. D., 1987, Isolation and partial characterization of plasma membrane fatty acid-binding proteins from myocardium and adipose tissue and their relationship to analogous proteins in liver and gut, *Biochem. Biophys. Res. Commun.* **148**:1370–1376.

163. Rauch, B., Bode, C., Piper, H. M., Hütter, J. F., Zimmermann, R., Braunwell, E., Hasselbach, W., and Kübler, W., 1987, Palmitate uptake in calcium tolerant, adult rat myocardial single cells—evidence for an albumin mediated transport across sarcolemma, *J. Mol. Cell. Cardiol.* **19**:159–166.
164. Renaud, G., Foliot, A., and Infante, R., 1978, Increased uptake of fatty acids by the isolated rat liver after raising the fatty acid-binding protein concentration with clofibrate, *Biochem. Biophys. Res. Commun.* **80**:327–334.
165. Renaud, G., Bouma, M. E., Foliot, A., and Infante, R., 1985, Free fatty acid uptake by isolated rat hepatocytes, *Arch. Int. Physiol. Biochim.* **93**:313–319.
166. Reyes, H., Levi, A. J., Gatmaitan, Z., and Arias, I. M., 1971, Studies of Y and Z, two hepatic cytoplasmic anion-binding proteins: Effect of drugs, chemicals, hormones, and cholestasis, *J. Clin. Invest.* **50**:2242–2252.
167. Roncari, D. A. K., and Mack, E. Y. W., 1981, Purification of liver cytosolic proteins that stimulate triacylglycerol synthesis, *Can. J. Biochem.* **59**:944–950.
168. Rose, C. P., and Goresky, C. A., 1977, Constraints on the uptake of labelled palmitate by the heart. The barriers at the capillary and sarcolemmal surfaces and the control of intracellular sequestration, *Circ. Res.* **41**:534–545.
169. Rüdel, H., Unterberg, C., and Spener, F., 1985, Wechselwirkungen kardialer Fettsäurebindungsproteine mit Fettsäuren und intracellulären Membranen, *Fette Seifen Anstrichm.* **87**:561–567.
170. Rüstow, B., Hodi, J., Kunze, D., Reichmann, G., and Egger, E., 1978, Specific binding of saturated and unsaturated fatty acids on the Z protein of rat liver cytosol, *FEBS Lett.* **95**:225–228.
171. Rüstow, B., Risse, S., and Kunze, D., 1982, Endogenes Lipidmuster, Organverteilung und Diätbeeinflussung einer fettsäurebindenden Proteinfraktion des Leberzytosols der Ratte, *Acta Biol. Med. Ger.* **41**:439–445.
172. Sacchettini, J. C., Said, B., Schulz, H., and Gordon, J. I., 1986, Rat heart fatty acid-binding protein is highly homologous to the murine adipocyte 422 protein and the P2 protein of peripheral nerve myelin, *J. Biol. Chem.* **261**:8218–8223.
173. Sacchettini, J. C., Meininger, T. A., Lowe, J. B., Gordon, J. I., and Banaszak, L. J., 1987, Crystallization of rat intestinal fatty acid-binding protein. Preliminary X-ray data obtained from protein expressed in Escherichia coli, *J. Biol. Chem.* **262**:5428–5430.
174. Sacchettini, J. C., Gordon, J. I., and Banaszak, L. J., 1988, The structure of crystalline Escherichia coli-derived rat intestinal fatty acid-binding protein at 2.5-Å resolution, *J. Biol. Chem.* **263**:5815–5819.
175. Sacchettini, J. C., Gorton, J. I., and Banaszak, J. L., 1989, Crystal structure of rat intestinal fatty acid-binding. Refinement and analysis of the Escherichia coli-derived protein with bound palmitate, *J. Mol. Biol.* **208**:327–339.
176. Said, B., and Schulz, H., 1984, Fatty acid-binding protein from rat heart. The fatty acid-binding proteins from rat heart and liver are different proteins, *J. Biol. Chem.* **259**:1155–1159.
177. Samuel, D., Paris, S., and Alihaud, G., 1976, Uptake and metabolism of fatty acids and analogues by cultured cardiac cells from chick embryo, *Eur. J. Biochem.* **64**:583–595.
178. Sarzani, R., Claffey, K. P., Chobanian, A. V., and Brecher, P., 1988, Hypertension induces tissue-specific gene suppression of a fatty acid-binding protein in rat aorta, *Proc. Natl. Acad. Sci. U.S.A.* **85**:7777–7781.
179. Scallen, T. J., Noland, B. J., Gavey, K. L., Bass, N. M., Ockner, R. K., Chanderbhan, R., and Vahouny, G. V., 1985, Sterol carrier protein 2 and fatty acid-binding protein. Separate and distinct physiological functions, *J. Biol. Chem.* **260**:4733–4739.
180. Schulenberg-Schell, H., Schäfer, P., Keuper, H. J., Stanislawski, B., Hoffmann, E., Rüterjans, H., and Spener, F., 1988, Interactions of fatty acids with neutral fatty acid binding protein from bovine liver, Eur. J. Biochem. **170**:565–574.

181. Schwieterman, W., Sorrentino, D., Potter, B. J., Rand, J., Kiang, C.-L., Stump, D., and Berk, P. D., 1988, Uptake of oleate by isolated rat adipocytes is mediated by a 40 kDa plasma membrane fatty acid-binding protein closely related to that in liver and gut, *Proc. Natl. Acad. Sci. U.S.A.* **85**:359–363.
182. Scow, R. O., and Blanchette-Mackie, E. J., 1985, Why fatty acids flow in cell membranes, *Prog. Lipid Res.* **24**:197–241.
183. Senjo, M., Ishibashi, T., Imai, Y., Takahashi, K., and Ono, T., 1985, Isolation and characterization of fatty acid-binding protein from rat brain, *Arch. Biochem. Biophys.* **236**:662–668.
184. Sheridan, M., Wilkinson, T. C. I., and Wilton, D. C., 1987, Studies on fatty acid-binding proteins. Changes in the concentration of hepatic fatty acid-binding protein during development of the rat, *Biochem. J.* **242**:919–922.
185. Shields, H. M., Bates, M. L., Bass, N. M., Best, C. J., Alpers, D. H., and Ockner, R. K., 1986, Light microscopic immunocytochemical localization of hepatic and intestinal types of fatty acid-binding proteins in rat small intestine, *J. Lipid Res.* **27**:549–557.
186. Soler-Argilaga, C., Infante, R., Renaud, G., and Polonovski, J., 1974, Factors influencing free fatty acid uptake by isolated perfused rat liver, *Biochimie* **56**:757–761.
187. Sorrentino, D., Stump, D., Potter, B. J., Robinson, R. B., White, R., Kiang, C.-L., and Berk, P. D., 1988, Oleate uptake by cardiac myocytes is carrier mediated and involves a 40 kD plasma membrane fatty acid-binding protein similar to that in liver, adipose tissue, and gut, *J. Clin. Invest.* **82**:928–935.
188. Spector, A. A., 1969, Influence of pH of the medium on free fatty acid utilization by isolated mammalian cells, *J. Lipid Res.* **10**:207–215.
189. Spector, A. A., Steinberg, D., and Tanaka, A., 1965, Uptake of free fatty acids by Ehrlich ascites tumor cells, *J. Biol. Chem.* **240**:3747–3753.
190. Spector, A. A., Fletcher, J. E., and Ashbrook, J. D., 1971, Analysis of long-chain free fatty acid binding to bovine serum albumin by determination of stepwise equilibrium constants, *Biochemistry* **10**:3229–3232.
191. Stein, L. B., Mishkin, S., Fleischner, G., Gatmaitan, Z., and Arias, I. M., 1976, Effect of fasting on hepatic ligandin, Z protein, and organic anion transfer from plasma in rats, *Am. J. Physiol.* **231**:1371–1376.
192. St. John, L. C., Rule, D. C., Knabe, D. A., Mersmann, H. J., and Smith, S. B., 1987, Fatty acid-binding protein activity in tissues from pigs fed diets containing 0 and 20% high oleate oil, *J. Nutr.* **117**:2021–2026.
193. Storch, J., Bass, N. M., and Kleinfeld, A. M., 1989, Studies of the fatty acid-binding site of rat liver fatty acid-binding protein using fluorescent fatty acids, *J. Biol. Chem.* **264**:8708–8713.
194. Storch, J., Kleinfeld, A., and Bass, N. M., 1987, Comparison of cardiac fatty acid-binding proteins from rat: Physical interaction with long-chain fluorescent fatty acids, *Fed. Proc.* **46**:1417.
195. Stremmel, W., 1987, Translocation of fatty acids across the basolateral rat liver plasma membrane is driven by an active potential-sensitive sodium-dependent transport system, *J. Biol. Chem.* **262**:6284–6289.
196. Stremmel, W., 1988, Fatty acid uptake by isolated rat heart myocytes represents a carrier-mediated transport process, *J. Clin. Invest.* **81**:844–852.
197. Stremmel, W., and Berk, P. D., 1986, Hepatocellular influx of [^{14}C]oleate reflects membrane transport rather than intracellular metabolism or binding, *Proc. Natl. Acad. Sci. USA* **83**:3086–3090.
198. Stremmel, W., and Thielmann, L., 1986, Selective inhibition of long-chain fatty acid uptake in short-term cultured rat hepatocytes by an antibody to the rat liver plasma membrane fatty acid-binding protein, *Biochim. Biophys. Acta* **877**:191–197.
199. Stremmel, W., Kochwa, S., and Berk, P. D., 1983a, Studies of oleate binding to rat liver plasma membranes, *Biochem. Biophys. Res. Commun.* **112**:88–95.

200. Stremmel, W., Potter, B. J., and Berk, P. D., 1983b, Studies of albumin binding to rat liver plasma membranes: Implications for the albumin receptor hypothesis, *Biochim. Biophys. Acta* **756:**20–27.
201. Stremmel, W., Lotz, G., Strohmeyer, G., and Berk, P. D., 1985a, Identification, isolation, and partial characterization of a fatty acid-binding protein from rat jejunal microvillous membranes, *J. Clin. Invest.* **75:**1026–1076.
202. Stremmel, W., Strohmeyer, D., Borchard, F., Kochwa, S., and Berk, P. D., 1985b, Isolation and partial characterization of a fatty acid-binding protein rat liver plasma membranes, *Proc. Natl. Acad. Sci. U.S.A.* **82:**4–8.
203. Stremmel, W., Strohmeyer, G., and Berk, P. D., 1986, Hepatocellular uptake of oleate is energy dependent, sodium linked, and inhibited by an antibody to a hepatocyte plasma membrane fatty acid-binding protein, *Proc. Natl. Acad. Sci. U.S.A.* **83:**3584–3588.
204. Sundelin, J., Anundi, H., Tragårdh, L., Eriksson, U., Lind, P., Ronne, H., Peterson, P. A., and Rask, L., 1985a, The primary structure of rat liver cellular retinol-binding protein, *J. Biol. Chem.* **260:**6488–6493.
205. Sundelin, J., Eriksson, U., Melhus, H., Nilsson, M., Lundvall, J., Bavik, C. O., Hansson, E., Laurent, B., and Peterson, P. A., 1985b, Cellular retinoid-binding proteins, *Chem. Phys. Lipids* **38:**175–185.
206. Suzuki, M., Kitamura, K., Sakamoto, Y., and Uyemura, K., 1982, The complete amino acid sequence of human P2 protein, *J. Neurochem.* **39:**1759–1762.
207. Suzuki, T., and Ono, T., 1987, Immunohistochemical studies on the distribution and frequency of fatty acid-binding protein positive cells in human fetal, newborn and adult liver tissues, *J. Pathol.* **153:**385–394.
208. Suzuki, T., and Ono, T., 1988, Ontogeny of hepatic fatty acid-binding protein immunoreactivity in human liver and intestinal tract, *Acta Pathol. Jpn.* **38:**979–987.
209. Sweetser, D. A., Lowe, J. B., and Gordon, J. I., 1986, The nucleotide sequence of the rat liver fatty acid-binding protein gene. Evidence that exon I encodes an oligopeptide domain shared by a family of proteins which bind hydrophobic ligands, *J. Biol. Chem.* **261:**5553–5561.
210. Sweetser, D. A., Birkenmeier, E. H., Klisak, I. J., Zollman, S., Sparkes, R. S., Mohandas, T., Lusis, A. J., and Gordon, J. I., 1987a, The human and rodent intestinal fatty acid-binding protein genes. A comparative analysis of their structure, expression, and linkage relationships, *J. Biol. Chem.* **262:**16060–16071.
211. Sweetser, D. A., Heuckeroth, R. O., and Gordon, J. I., 1987b, The metabolic significance of mammalian fatty acid-binding proteins: Abundant proteins in search of a function, *Annu. Rev. Nutr.* **7:**337–359.
212. Takahashi, K., Odani, S., and Ono, T., 1982, Primary structure of rat liver Z protein. A low Mr cytosol protein that binds sterols, fatty acids and other small molecules, *FEBS Lett.* **140:**63–66.
213. Takahashi, K., Odani, S., and Ono, T., 1983, Isolation and characterization of the three fractions (DE-I, DE-II and DE-III) of rat liver Z protein and the complete primary structure of DE-I, *Eur. J. Biochem.* **136:**589–601.
214. Takikawa, H., and Kaplowitz, N., 1986, Binding of bile acids, oleic acid, and organic anions by rat and human hepatic Z protein, *Arch. Biochem. Biophys.* **251:**385–392.
215. Teerlink, T., Van Der Krift, T. P., Post, M., and Wirtz, K. W. A., 1982, Tissue distribution and subcellular localization of phosphatidylcholine transfer protein in rats as determined by radioimmunoassay, *Biochim. Biophys. Acta* **713:**61–67.
216. Teerlink, T., Van Der Krift, T. P., Van Heusden, G. P. H., and Wirtz, K. W. A., 1984, Determination of nonspecific lipid transfer protein in rat tissues and Morris hepatomas by enzyme immunoassay, *Biochim. Biophys. Acta* **793:**251–259.
217. Thomas, C. R., Lowy, C., and Evans, J. L., 1986, The occurrence of different forms of fatty acid-binding protein in diabetic and normal placentae of both humans and rabbits may explain altered lipid transfer in diabetic pregnancy, *Diabetologia* **29:**600A–601A.

218. Tipping, E., and Ketterer, B., 1981, The influence of soluble binding proteins on lipophile transport and metabolism in hepatocytes, *Biochem. J.* **195**:441–452.
219. Trulzsch, D., and Arias, I. M., 1981, Z protein: Isolation and characterization of multiple forms in rat liver cytosol, *Arch. Biochem. Biophys.* **209**:433–440.
220. Unterberg, C., Heidl, G., Von Bassewitz, D.-B., and Spener, F., 1986, Isolation and characterization of the fatty acid-binding protein from human heart, *J. Lipid Res.* **27**:1287–1293.
221. Vahouny, G. V., Chanderbhan, R., Kharroubi, A., Noland, B. J., Pastuszyn, A., and Scallen, T. J., 1987, Sterol carrier and lipid transfer proteins, *Adv. Lipid Res.* **22**:83–113.
222. Van Amerongen, A., Teerlink, T., Van Heusden, G. P. H., and Wirtz, K. W. A., 1985, The non-specific lipid transfer protein (Sterol carrier protein-2) from rat and bovine liver, *Chem. Phys. Lipids* **38**:195–204.
223. Van Der Vusse, G. J., Roemen, T. H. M., Prinzen, F. W., Coumans, W. A., and Reneman, R. S., 1982, Uptake and tissue content of fatty acids in dog myocardium under normal and ischemic conditions, *Circ. Res.* **50**:538–546.
224. Van Der Vusse, G. J., Roemen, T. H. M., Flameng, W., and Reneman, R. S., 1983, Serum-myocardium gradients of non-esterified fatty acids in dog, rat and man, *Biochim. Biophys. Acta* **752**:361–370.
225. Veerkamp, J. H., and Paulussen, R. J. A., 1987, Fatty acid transport in muscle: The role of fatty acid-binding proteins, *Biochem. Soc. Trans.* **15**:331–336.
226. Veerkamp, J. H., and Van Moerkerk, H. T. B., 1986, Peroxisomal fatty acid oxidation in rat and human tissues. Effect of nutritional state, clofibrate treatment and postnatal development with the rat, *Biochim. Biophys. Acta* **875**:301–310.
227. Vincent, S., and Muller-Eberhard, U., 1985, A protein of the Z class of liver cytosolic proteins in the rat that preferentially binds heme, *J. Biol. Chem.* **260**:14521–14528.
228. Walz, D. A., Wider, M. D., Snow, J. W., Dass, C., and Desiderio, D. M., 1988, The complete amino acid sequence of porcine gastrotropin, an ileal protein which stimulates gastric acid and pepsinogen secretion, *J. Biol. Chem.* **263**:14189–14195.
229. Weisiger, R. A., and Ma, W.-L., 1987, Uptake of oleate from albumin solutions by rat liver. Failure to detect catalysis from the dissociation of oleate from albumin by an albumin receptor, *J. Clin. Invest.* **79**:1070–1077.
230. Weisiger, R. A., Gollan, J. L., and Ockner, R. K., 1981, Receptor for albumin on the liver cell surface may mediate uptake of fatty acids and other albumin-bound substances, *Science* **211**:1048–1051.
231. Westerman, J., and Wirtz, K. W. A., 1985, The primary structure of the nonspecific lipid transfer protein (sterol carrier protein 2) from bovine liver, *Biochem. Biophys. Res. Commun.* **127**:333–338.
232. Wetzel, M. G., and Scow, R. O., 1984, Lipolysis and fatty acid transport in rat heart: Electron microscopic study, *Am. J. Physiol.* **246**:C467–C485.
233. Wilkinson, T. C. I., and Wilton, D. C., 1986, Studies on fatty acid-binding proteins. The detection and quantitation of the protein from rat liver using a fluorescent fatty acid analogue, *Biochem. J.* **238**:419–424.
234. Wilkinson, T. C. I., and Wilton, D. C., 1987a, Studies on fatty acid-binding proteins. The diurnal variation shown by rat liver fatty acid-binding protein, *Biochem. J.* **242**:913–917.
235. Wilkinson, T. C. I., and Wilton, D. C. 1987b, Studies on fatty acid-binding proteins. The binding properties of rat liver fatty acid-binding proteins, *Biochem. J.* **247**:485–488.
236. Wosilait, W. D., and Soler-Argilaga, C., 1977, A comparative analysis of the binding of different long-chain free fatty acids by human serum albumin, *FEBS Lett.* **73**:72–76.
237. Wu-Rideout, M. Y. C., Elson, C., and Shrago, E., 1976, The role of fatty acid-binding protein on the metabolism of fatty acids in isolated rat hepatocytes, *Biochem. Biophys. Res. Commun.* **71**:809–816.

Chapter 8

Intracellular and Extracellular Flow of Dolichol

G. Van Dessel, M. De Wolf, H. J. Hilderson, A. Lagrou, and W. Dierick

1. INTRODUCTION

Dolichols belong to a family of polymeric lipids with the isoprene unit as the repeating building block. This lipid class represents a relative large group of natural products displaying a wide variety of biological functions (Table I) (Rip *et al.*, 1985). The existence of dolichols was first reported by Hemming and coworkers (Pennock *et al.*, 1960; Burgos *et al.*, 1963). The isolation of this new lipid from the nonsaponifiable fraction of pig liver was not met with a wave of

Abbreviations used in this chapter: d, density; DMP, dolichyl monophosphate; DSC, differential scanning calorimetry; EM, electron microscopy; ER, endoplasmic reticulum; ESR, electron spin resonance; GlcNac, *N*-acetylglucosamine; HDL, high-density lipoproteins; HPLC, high-performance liquid chromatography; LCAT, lecithin-cholesterol acyltransferase; LDL, low-density lipoproteins; L, light mitochrondrial fraction; LUV, large unilamellar vesicles; M, mitochondrial fraction; NMR, nuclear magnetic resonance; N, nuclear fraction; ns-LTP, nonspecific lipid transfer protein; PC, phosphatidylcholine; PE, phosphatidylethanolamine; PI, phosphatidylinositol; P, microsomal fraction; RER, rough endoplasmic reticulum; RSA, relative specific activity; S, supernatant fraction; SUV, small unilamellar vesicles; TLC, thin layer chromatography; VLDL, very-low-density lipoproteins.

G. Van Dessel UIA-Laboratory for Pathological Biochemistry, University of Antwerp, B2610 Antwerp, Belgium. **M. De Wolf, H. J. Hilderson, A. Lagrou, and W. Dierick** RUCA-Laboratory for Human Biochemistry, University of Antwerp, B2020 Antwerp, Belgium.

Table I.
Biological Functions of Isoprene-Derived Compounds

Compound	Function
Vitamin K	Blood clotting
Coenzyme Q	Respiration chain
Retinol	Vision process
Cholesterol	Membrane structure
Chlorophyls	Light energy conversion
Polyprenols	Synthesis of cell wall peptidoglycans
Dolichols	N-glycoprotein synthesis

enthusiasm among lipidologists (Keller, 1987). Indeed, more than 10 years e-lapsed before this class of polyisoprenoic alcohols could gain "full membership" in the lipid family. The acceptance was due solely to the discovery in the early 1970s that phosphorylated dolichols act as the coenzyme in the *N*-glycosylation of proteins (Behrens and Leloir, 1970).

Previous reviews on dolichol and related compounds (Hemming, 1974, 1983, 1985; Lennarz, 1975, 1983, 1985; Lucas and Waechter, 1976; Waechter and Lennarz, 1976; Elbein, 1979, 1984, 1987; Dallner and Hemming, 1981; Hanover and Lennarz, 1981; Staneloni and Leloir, 1982; Jamieson, 1983; Rip *et al.*, 1985; Hirschberg and Snider, 1987; Keller, 1987; Kukururinska *et al.*, 1987; Chojnacki and Dallner, 1988) have dealt mainly with their metabolism and physiological significance, particularly their involvement in the assembly of glycoconjugates. From recent experimental evidence it appears that dolichol and derivatives are subject to an intensive intracellular flow and redistribution (Rip *et al.*, 1985; Chojnacki and Dallner, 1988). Therefore, in this chapter we focus on the intracellular traffic of these isoprenoids. The extra- and intracellular traces of exogenously administered as well as *in vivo*-synthesized dolichol will be described. Furthermore, we will tackle the problem of whether lipid-specific or non-lipid-specific binding/transfer proteins are implicated in the mechanisms of dolichol movement.

2. CHEMISTRY

Dolichols are member of the group of poly-*cis*-isoprenoic alcohols. They differ from the fully unsaturated polyprenols in having a saturated α-residue, displaying the *S*-configuration (Adair and Robertson, 1980; Suzuki *et al.*, 1983; Chojnacki *et al.*, 1984), and bearing the single alcoholic function, next to a variable number of built-in isoprene monomers (16–21 in higher eukaryotes) (Figure 1). All but the three isoprene units at the ω end are in the *cis* configuration. The presence of the 2,3 double bond strongly affects the stability of the phosphorylated derivatives toward acid hydrolysis (Hemming, 1974, 1985). The most prominent molecular features of dolichols are the pronounced hydrophobicity and their extreme length (ca. 10 nm), by virtue of which they are able to overspan the thickness of a membrane bilayer (Hanover and Lennarz, 1981).

In many tissues dolichols are found as free alcohol. Appreciable amounts also occur esterified with long-chain fatty acids. A minor, although variable, portion of this lipid is (pyro)phosphorylated and additionally conjugated with carbohydrate components. Recently a novel ester derivative, dolichyl dolichoate, has been identified in bovine thyroid (Steen *et al.*, 1984), resulting from a random combination of free dolichol and dolichoic acid homologs. This waxlike substance also seems to be present in other organs and tissues (G. Van Dessel *et al.*, unpublished results).

In nature, dolichols are found not as a single molecular entity with a definite molecular weight, but as a family of homologs, with a difference in size in each tissue of four or five isoprene units (Rip *et al.*, 1983a) (Figure 2). High-performance liquid chromatography (HPLC) has been proven to be the ideal tool for establishing the isoprenolog pattern as well as for quantitating dolichol (Keenan *et al.*, 1977c; Tavares *et al.*, 1977; Freeman *et al.*, 1980; Keller *et al.*, 1982b; Eggens *et al.*, 1983; Palmer *et al.*, 1984; Steen, 1985). Dolichol levels in tissues have previously been assayed by a number of less specific, time-consuming, rather elaborate analytical methods (Dallner *et al.*, 1972; Keller and Adair, 1977; Van Dessel *et al.*, 1979; Coolbear and Mookerjea, 1981; Keller *et al.*, 1981; Hilderson *et al.*, 1984). The homolog patterns can vary appreciably from one

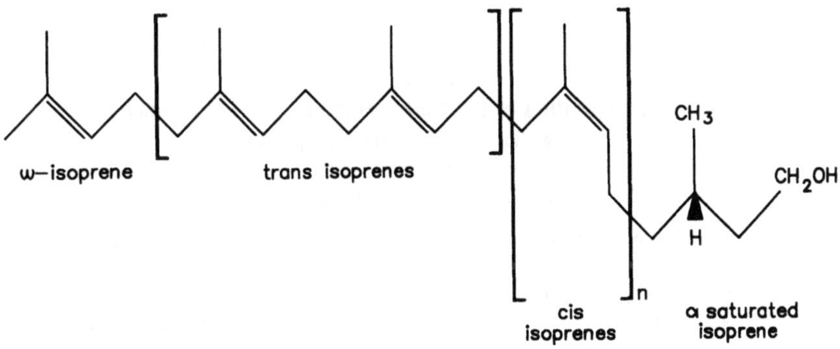

FIGURE 1. Structure of dolichol. The number of *cis*-isoprene units (n) varies between 13 and 18 in higher eukaryotes. The primary function can be phosphorylated or esterified with long-chain fatty acids.

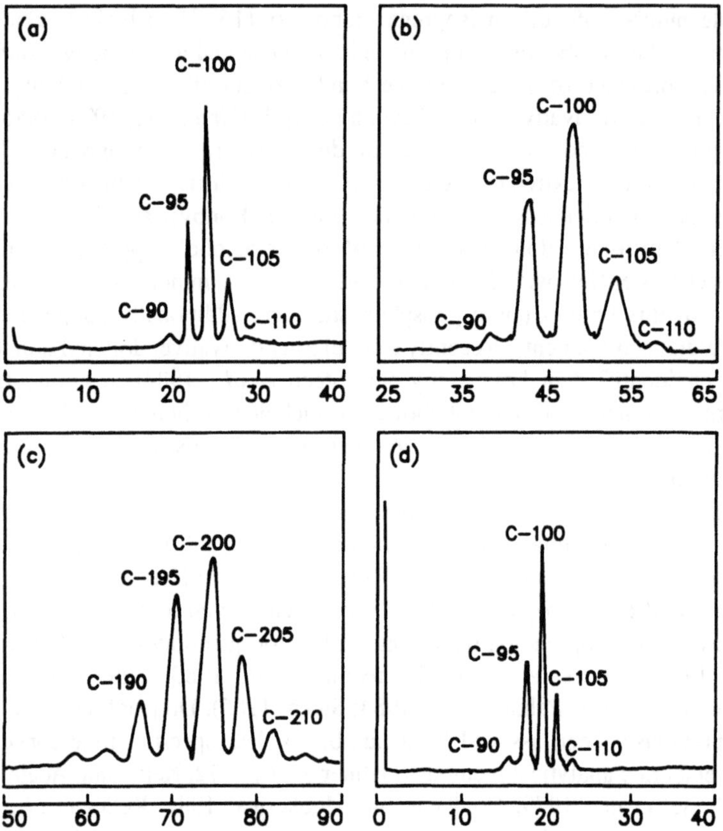

FIGURE 2. HPLC pattern of free dolichol (a), dolichyl fatty acid esters (mainly esterified with palmitic and oleic acid) (b), dolichyl dolichoate (c), and DMP (d) in bovine thyroid (Steen *et al.*, 1984, 1985). C-100, etc., indicate the number of carbon atoms in the polyisoprene chain.

Table II
Percent Distribution of the Major Homologs of Free Dolichol in the Livers of Different Species

	Percent distribution of following dolichol type:						
Species	C-85	C-90	C-95	C-100	C-105	C-110	Reference
Cattle	2.2	8.1	32.2	38.7	15.0	3.8	Steen (1985)
Human	4	12	49	27	8	—	Tollbom and Dallner (1986)
Mouse	9.8	40.6	38.4	8.9	—	—	Pullarkat *et al.* (1984)
Rat	11	42	32	10	5	—	Eggens *et al.* (1983)
Trout	4	28	47	17	4	—	Bizri *et al.* (1986)

Table III
Percent Distribution of the Major Homologs of Free Dolichol in Various Bovine Tissues[a]

Tissue	Percent distribution of following dolichol type: C-85	C-90	C-95	C-100	C-105	C-110
Brain	0.7	4.0	23.8	47.9	19.7	3.9
Heart	—	8.2	54.5	34.3	3.0	—
Kidney	0.5	4.9	36.8	45.8	10.6	1.3
Liver	2.2	8.1	32.2	38.7	15.0	3.8
Pancreas	1.4	14.6	48.8	30.3	4.9	—
Spleen	0.5	3.4	24.5	45.4	20.9	5.3
Testes	—	1.7	16.9	52.2	25.1	4.1
Thyroid	1.0	3.9	24.1	53.0	15.6	2.4

[a]Steen (1985).

organism to another, whereas tissue specificity has also become evident (Table II). Generally, within a single species the same homologs are found, although the percent distribution oscillates from one organ to another, e.g., bovine heart and pancreas show a tendency toward shorter-chain dolichols (Table III) (Steen, 1985). On the other hand, the dolichyl derivatives in the respective tissues display the same isoprenolog pattern as observed for the maternal dolichol (Eggens *et al.*, 1983; Tollbom and Dallner, 1986).

With regard to the total dolichol concentration marked differences are found between vertebrate species (Hemming, 1985) (Table IV); human tissues seem to

Table IV
Dolichol and DMP Content in Livers of Different Species

Species	Amount (μg/g wet wt) of: Dolichol	DMP	References
Cattle	56	—	Steen (1985)
Chicken	182	—	Keller and Adair (1977)
Human	452–1226	6.1	Rupar and Carroll (1978); Tollbom and Dallner (1986)
Pig	21.5–129	—	Keller and Adair (1977); Tavares *et al.* (1977)
Rat	17.1–51	2.4–17.5	Keller *et al.* (1981); Chaudhary *et al.* (1982); Eggens *et al.* (1983); Yamada *et al.* (1985); Bizri *et al.* (1986); Elmberger *et al.* (1987)
Sheep	108	—	Palmer *et al.* (1984)
Trout	15.7	8.9	Bizri *et al.* (1986)

be exceptionally enriched; e.g., the human pituitary gland has been reported to contain even more dolichol (±7 mg/g of wet tissue) than phospholipids (Carroll *et al.*, 1973; Tollbom and Dallner, 1986). Additionally, the dolichol content also displays prominent organ specificity, endocrine organs having considerably higher dolichol levels than other tissues examined (Rupar and Carroll, 1978; Tollbom and Dallner, 1986). Another interesting feature of cellular dolichol levels is the increase upon aging; e.g., in human cerebral cortex, the dolichol level at the age of 20 years, (±100 μg/g of tissue) is nearly tripled by the age of 80 years (Pullarkat and Reha, 1982; Ng *et al.*, 1983; Pullarkat *et al.*, 1984; Sakakihara and Volpe, 1984; Keller and Nellis, 1986; Andersson *et al.*, 1987). In pathological conditions (e.g., drug treatment, alcohol abuse, induced and spontaneous cancers, liver cirrhosis, ceroid lipofuscinosis, Alzheimer's disease) considerable changes in concentrations and/or homolog profiles have been noted (Rip *et al.*, 1986; Chojnacki and Dallner, 1988).

3. FUNCTION

The only function for dolichol so far established in animal tissues is the role of essential coenzyme and glycosyl carrier attributed to the phosphorylated derivative in the synthesis of N-linked glycoproteins. Primary N-glycosylation involves the assembly of the so-called G-oligosaccharide by stepwise addition of simple sugars (Sharon and Lis, 1981). This G-oligosaccharide contains two *N*-acetylglucosamine (GlcNAc), nine mannose, and three glucose residues linked to dolichol via a pyrophosphate bridge (dolichol-P-P GlcNAc$_2$-Man$_9$-Glc$_3$) (Hemming, 1985). In this process dolichol plays a dual role (Figure 3). First, it serves

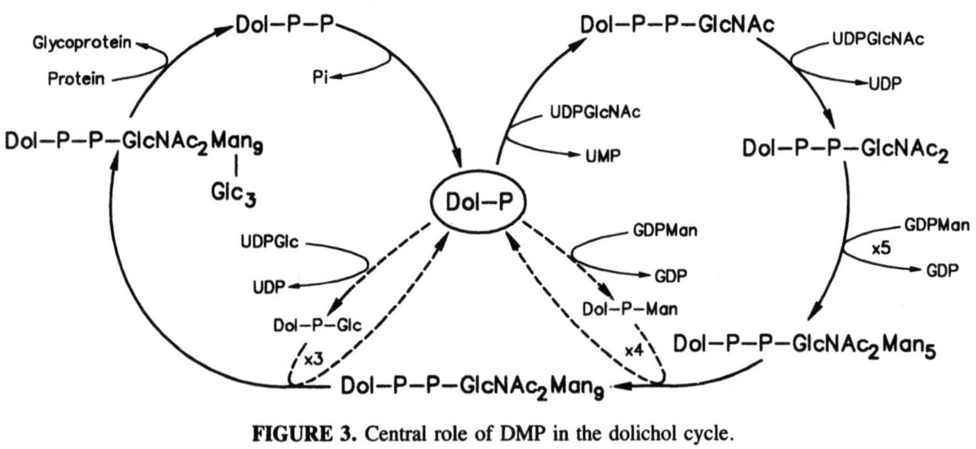

FIGURE 3. Central role of DMP in the dolichol cycle.

as a site of initiation and elongation, representing a kind of anchor function (Czichi and Lennarz, 1977). Second, the last seven sugar residues are not attached as nucleotide sugars, but are in the form of dolichylphosphate sugars, representing a sort of carrier function (Elbein, 1979; Parodi and Leloir, 1979; Staneloni and Leloir, 1982). At the end, the fully assembled G-oligosaccharide is transferred *en bloc* to a nascent polypeptide (Hubbard and Ivatt, 1981; Mills and Adamany, 1981; Hemming, 1982; Presper and Heath, 1983). The asparagine residue can be glycosylated only if it is part of the "canonical triplet" Asn-X-Thr/Ser, where X represents any amino acid except proline. Concomitant with the oligosaccharide transfer dolichyl pyrophosphate is generated, which, after dephosphorylation, is theoretically at hand for reutilization (this has not yet been experimentally demonstrated). The same holds for dolichyl monophosphate (DMP) regenerated following transfer of monosaccharides from DMP sugars.

The complex chemical machinery of N-glycosylation is associated with a number of topographical problems, as can be concluded from a series of experimental data. A great deal of evidence has been obtained in support of the rough endoplasmic reticulum (RER) lumen as the stage for the ultimate G-oligosaccharide transfer to protein acceptors (Hirschberg and Snider, 1987; Roth, 1987). On the other hand, intermediates have been localized at both sides of the RER membrane, suggesting that the lipid–oligosaccharide assembly occurs on both cytoplasmic and luminal faces (Hanover and Lennarz, 1981, 1982; Haselbeck and Tanner, 1982, 1984; Snider and Robbins, 1982; Snider and Rogers, 1984). Additionally, a number of transferases involved have protease-sensitive sites (Hanover and Lennarz, 1982; Spiro and Spiro, 1985) or are sensitive to the non-penetrating stilbene derivative 4,4′-diisothiocyanatostilbene-2,2′-disulfonic acid (DIDS) in intact microsomal vesicles (Spiro and Spiro, 1985). Furthermore, all nucleotide sugar substrates are of cytoplasmic origin. For UDP-GlcNAc and UDP-Glc, specific uptake systems for entering the RER lumen are available; for GDP-mannose, however, no such translocation process has yet been detected (Perez and Hirschberg, 1985, 1986). Finally, DMP seems to display a conformation in which the polar portion of the molecule bearing the phosphoryl group is exposed at the bilayer/water interface (Valtersson *et al.*, 1985). From these data a hypothetical model for the glycosylation process has been proposed. The assembly of the lipid-linked G-oligosaccharide starts with the translocation of UDP-GlcNAc from the cytoplasmic side of the RER to the luminal side and subsequent reaction with DMP. The Dol-P-P-$GlcNAc_2$ formed would then be transferred back to the cytoplasmic side to be processed to Dol-P-P-$GlcNAc_2$-Man_5. In its turn, this G-oligosaccharide precursor is flipped, together with Dol-P-Man and UDP-Glc, to the luminal side of the RER, whether the mature G-oligosaccharide-P-P-Dol is produced and becomes available for N-glycosylation. This topographical model has the attractive feature of having individual sugar transfer reactions compartmentalized, possibly providing multiple sites for control of assembly of oligosaccharide chains (Hirschberg and Snider, 1987; Lennarz, 1987; Roth, 1987). An early postulate was that dolichol participated only in the active translocation. It was suggested that the carbohydrate units esterified to the phosphorylated dolichol were transported from one bilayer surface to the other, while the nonpolar ω-terminus probably remained burried in the hydrophobic interior of the membrane. This hypothesis has entered the literature to a surprising extent, despite the lack of biophysical data to support it (Kanegasaki and Wright, 1970; Johnston and Neuhaus, 1976; Weppner and Neuhaus, 1978; Robyt, 1979). Recent studies estimating the transverse diffusion rate of spin-labeled polyprenols in model membrane systems have shown that such an unfacilitated flip-flop mechanism is unlikely to occur. The actual rate ($t_{1/2} > 5$ h) of the unassisted flip-flop is much too low (it should be less than 1 sec) to account for a role in polysaccharide and glycoprotein synthesis (McCloskey and Troy, 1980b). Therefore, one may wonder whether dolichyl phosphate molecules are asymmetrically distributed in the lipid bilayer. So far, no experiments to test this possibility in the RER have been reported (Lennarz, 1987). Also, Hanover and Lennarz (1981, 1982), using [^{14}C]chitobiosyl lipid, could not detect any significant mobility in the transverse phase of the membrane. Finally from chemical intuition one would argue that the motion of such a bulky polar headgroup across a nonpolar membrane region is energetically unfavorable. Therefore, a protein-mediated flip-flop mechanism looks more attractive, perhaps as a result of the specific transmembrane orientation in the RER membranes of the glycosyltransferases involved. In this respect, Haselbeck and Tanner (1982, 1984) have tested this possibility with a solubilized, partially purified yeast mannosyltransferase catalyzing the formation of DMP-mannose. The enzyme, together with DMP, was incorporated into soybean lecithin liposomes. When these liposomes were preloaded with GDP, the transfer of radioactive mannosyl residues from GDP-[^{14}C]mannose outside to GDP inside could be demonstrated. The presence of both DMP and mannosyltransferase in the liposomes was obligatory. A GDP/GDP-mannose exchange, although not to be excluded *per se*, seems unlikely because the obligatory role of DMP in the system would otherwise be puzzling. Although the mechanism of the "transmembrane flow" of DMP and corresponding saccharide derivatives have not yet been elucidated (one of the more challenging aspects of transmembrane movement of polar compounds), recent studies indicate that dolichol and derivatives by themselves are able to locally induce nonbilayer phases in artificial membranes (Jensen and Schutzbach, 1984; Vigo *et al.*, 1984; de Ropp and Troy, 1985; Valtersson *et al.*, 1985).

In recent years, the scientific community has also been actively searching for other putative functions of dolichol. Most of these studies have been conducted with model membrane systems using sophisticated instrumental techniques such as ^{31}P nuclear magnetic resonance (NMR) spectroscopy, electron spin resonance (ESR) spectroscopy, differential scanning calorimetry (DSC), electron microscopy (EM), steady-state and dynamic fluorescence, stop-flow spectrophotometry, and small-angle X-ray scattering. These experimental approaches have provided more insight in the submembrane localization and spatial distribution of dolichols, the effect of dolichol incorporation on the macroscopic organization of the phospholipid packing (lipid polymorphism and thermotropic behavior), and on a number of membrane characteristics (fluidity, permeability, fusion capacity, and modulation of membrane-associated enzyme activities).

From the vast collection of data, several conclusions can be drawn. Free and esterified dolichols occupy sites near the hydrophobic center of phospholipid bilayers, whereas the phosphate moiety of DMP is located near the membrane/water interface (McCloskey and Troy, 1980a,b; de Ropp and Troy, 1984, 1985; Valtersson *et al.*, 1985; Lai and Schutzbach, 1986). Recent investigations (Murgolo *et al.*, 1989) using the small-angle X-ray scattering method, indicate that the solution conformation of dolichol is comprised of a central coiled region flanked by two, approximately diametrically opposed, arms. These authors contend that although esterified dolichols may adopt an unconventional orientation, the free dolichol and its phosphorylated derivatives should be expected to adopt a head group at interface orientation. Indeed, the model proposed by Murgolo *et al.* (1989) is compact enough to exist entirely within the membrane bilayer, without the unfavorable hydrophobic–hydrophilic interactions associated with the submembranal localization proposed by others. Moreover, these findings coupled with the determined "corkscrew-like" conformation of dolichol would represent a plausible mechanism for membrane transport of sugars in *N*-linked protein glycosylation.

The limited solubility of dolichol and its derivatives depends on the nature of the derivative as well as the type of phospholipid and can lead to situations with either a homogeneous distribution or segregation phenomena (Valtersson *et al.*, 1985). Dolichol and dolichol analogues greatly influence the macroscopic organization and phospholipid packing in model membranes in a way that is dependent on both the phospholipid and polyprenol type (Wood *et al.*, 1986). Dolichol and its derivatives promote the formation of the hexagonal H_{II} phase of phosphatidylethanolamine (PE) and PE-phosphatidylcholine (PC)-containing membranes (Jensen and Schutzbach, 1984; Gruner, 1985; Kirk and Gruner, 1985; Valtersson *et al.*, 1985). The addition of dolichols to phospholipid membranes seems to increase the fluidity of the acyl chains of both PE and PC (disruption of the chain packing), although the mobility of spin-labeled fatty acids is restricted by the presence of dolichols (Vigo *et al.*, 1984). Dolichol and its derivatives have been shown to induce vesicle fusion. Moreover, DMP can facilitate the transbilayer movement of PC during the fusion process. The membrane fusions are paralleled by the appearance of intermediate nonbilayer structures (Sunamoto *et al.*, 1983; Van Duijn *et al.*, 1986). The incorporation of dolichol in phospholipid bilayers affects the permeability properties of membranes, which diverge depending on the phospholipid species (PC versus PE)(Lai and Schutzbach, 1984; Boscoboinik *et al.*, 1985; Monti *et al.*, 1987; Van Dessel *et al.*, 1987a; Jonas and Tien, 1988). The effect of dolichol incorporation upon bilayer stabilization and fluidization may modulate certain enzyme activities (Jensen and Schutzbach, 1985). Finally, diverging results have been reported (Vigo *et al.*, 1984; Depauw *et al.*, 1988). Variations may result from the use of different experimental techniques (time course of the measurement; the probe or method applied), the choice of the phospholipid and dolichol species tested, the structural makeup of the artificial membranes [large unilamellar vesicles (LUV), small unilamellar vesicles (SUV), multilamellar] and experimental conditions (e.g., experimental temperature versus transition temperature). By and large, caution is always required when extrapolation data obtained with artificial membranes to native membranes (Schroeder *et al.*, 1987; Depauw *et al.*, 1988).

4. METABOLISM

In view of the regulatory role of DMP on N-protein glycosylation under certain conditions, considerable interest has been developed in identifying the exact biosynthetic pathway, the chemical nature of the end product, and the regulation of dolichol production, with special emphasis upon the interrelation between dolichol and cholesterol biosynthesis. Many tissues incorporate appropriate precursors into dolichyl compounds, although the synthesis capacity can vary appreciably (Elmberger *et al.*, 1987). Synthesis is confined to the endoplasmic reticulum (ER) (Adair and Keller, 1982; Wong and Lennarz, 1982; Edlund *et al.*, 1983), although earlier work by Daleo *et al.* (1977) pointed to a localization in the outer mitochrondrial membrane. Peroxisomes also seem able to synthesize these isoprenoids (Appelkvist and Dallner, 1987). Three lines of evidence suggest that tissue levels can be maintained by *de novo* synthesis: (i) many tissues (liver, brain, testes, aorta, smooth muscle, and thyroid) incorporate precursors into dolichyl compounds (Van Dessel *et al.*, 1986c; Elmberger *et al.*, 1987); (ii) the absolute rate of dolichol synthesis by a 230-g rat has been estimated at 25 nmol/day per liver (Keller, 1987), far exceeding the absorption efficiency from the diet (Keller *et al.*, 1982b); and (iii) pigs maintained on a dolichol-free diet have normal hepatic levels of dolichol (Butterworth *et al.*, 1963). This finding was recently confirmed for rats by Connelly and Keller (1984).

The biosynthesis of dolichol starts at acetyl-CoA (Figure 4). The initial steps proceed along the same path (mevalonate route) as for cholesterol and ubiquinone up to all-*trans*-farnesyl pyrophosphate (Dallner and Hemming, 1981; Hemming, 1983). At this point there is a bifurcation (Wong and Lennarz, 1982). The bulk of the farnesyl pyrophosphate precursor is harnessed (500 : 1 in liver)

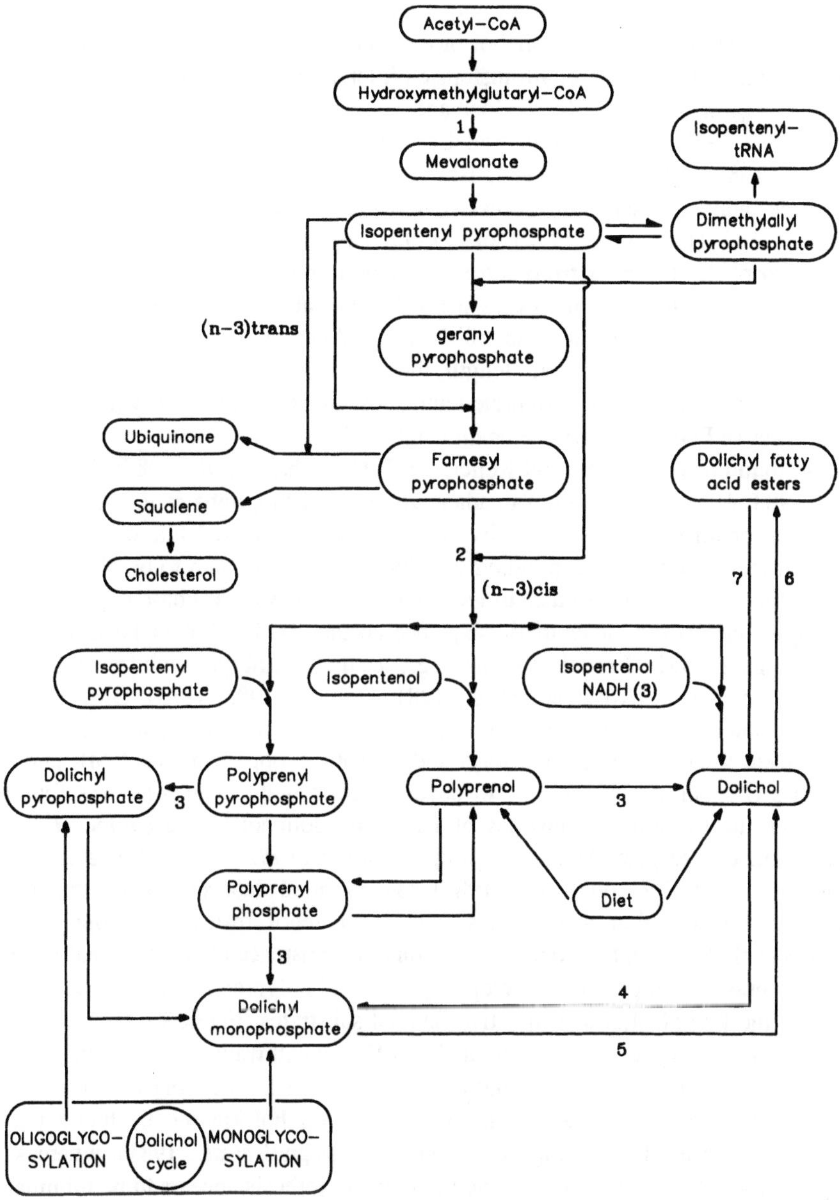

FIGURE 4. Biosynthesis and interconversion of dolichol and derivatives. Numbers indicate reactions catalyzed by enzymes referred to in the manuscript: 1, β-hydroxymethylglutaryl-CoA reductase; 2, long-chain prenyltransferase; 3, α-saturase; 4, dolichol kinase; 5, DMP phosphatase; 6, dolichol acyltransferase; 7, dolichyl fatty acid ester hydrolase. Not all interconversions depicted in the figure have been demonstrated experimentally.

for cholesterol production (Adair and Keller, 1982; Keller, 1986). A small portion is directed toward the dolichol biosynthetic branch through the action of a microsomal long-chain prenyltransferase (Allen *et al.*, 1976; Grange and Adair, 1977; Sagami *et al.*, 1978; Wellner and Lucas, 1979; Baba and Allen, 1980; Adair and Keller, 1982; Adair and Cafmeyer, 1983, 1987a,b; Takahashi and Ogura, 1982; Adair et al., 1984; Ishii *et al.*, 1986; Baba *et al.*, 1987). This enzyme catalyzes a series of successive head-to-tail condensations (*cis*-additions; Gough and Hemming, 1970) of 14–17 isoprene units from isopentenyl-pyrophosphate to *trans,trans*-farnesyl pyrophosphate (Struck and Lennarz, 1980). The polymerization process seems to be catalyzed by a single enzyme, at least in bacteria (molecular mass, 60–70 kDa) (Allen *et al.*, 1976; Sagami *et al.*, 1978). At the appropriate chain length, the primary product, α-unsaturated polyprenyl pyrophosphate, is α-saturated and subsequently dephosphorylated, generating active DMP. *In vitro,* the chain lengths of the end products are shorter than those found *in vivo* (Adair and Keller, 1982; Chaudhary *et al.*, 1982; Eggens *et al.*, 1983; Adair *et al.*, 1984; Yamada *et al.*, 1985, 1986; Baba *et al.*, 1987). The exact mechanism and sequence by which the α-unsaturated intermediates are transformed into the corresponding dolichyl derivative remain to be elucidated.

Other biosynthetic studies *in vivo* and *in vitro* as well as chase experiments strongly suggest the biosynthetic sequence acetate → dolichol → DMP (Wong and Lennarz, 1982). In such systems, e.g., sea urchin embryos with free dolichol as the end product (Rossignol *et al.*, 1981; Lennarz, 1983, 1985), DMP levels are regulated by the interplay of two ER enzymes: a dolichol-specific CTP-dependent kinase (salvage pathway) and its natural opponent, the DMP phosphatase (Lennarz, 1983, 1985). In developing rat brain, a parallel was found between the evolutionary profiles of DMP and dolichol kinase (Volpe *et al.*, 1987). In this respect, Dallner's group (Ekström *et al.*, 1987; Chojnacki and Dallner, 1988; Ericsson *et al.*, 1988) has provided data suggesting that the ultimate condensation step occurs with isopentenol instead of isopentenyl pyrophosphate. From this, one could assume the existence of two different types of α-saturases, one preferring pyrophosphorylated polyprenol and the other preferring the free alcohol as respective favored substrates (Figure 4).

Summarizing, we may conclude that different pathways for the synthesis of DMP and perhaps other functional dolichol may exist depending on the metabolic or developmental state of the organism. Dolichol kinases are thought to play an important role during development (Rossignol *et al.*, 1981), whereas, e.g., in liver its contribution relative to *de novo* synthesis appears to be minimal (Astrand *et al.*, 1986).

A major problem in studying dolichol metabolism is that the radio-labeled precursors used are subject to variable dilution by intracellular pool and therefore can yield misleading data. Measurements of absolute rates are best obtained by quantitating the rate of ^3H_2O incorporation (Lakshmanan and Veech, 1977;

Anderson and Dietschy, 1979). As ^3H_2O is not a practical precursor for dolichol synthesis, experimentally absolute rates of cholesterol synthesis found by using ^3H_2O are compared with the relative rates of cholesterol and dolichol synthesis from [^{14}C]acetate (Adair and Keller, 1982).

Many studies have been devoted to the regulation of dolichol biosynthesis, with—in view of their partial common pathway—special attention to the interrelation between cholesterol and dolichol metabolism (Keller *et al.*, 1979; James and Kandutsch, 1979, 1980a,b; Potter *et al.*, 1981; Tavares *et al.*, 1981; White *et al.*, 1981; Keller, 1986; Kabakoff and Kandutsch, 1987). The data are very convincing in showing that the two pathways are controlled independently in a number of systems. This raises the question of how a system with a branching biosynthetic pathway sharing common intermediates and a common regulatory enzyme can maintain a high degree of regulatory independance. One explanation hypothesizes that in the dolichol pathway a second rate-limiting enzyme downstream of hydroxymethylglutaryl-CoA reductase is saturated at a much lower level of intermediates. Recent data by Keller (1986) do not support this theory. As an alternative explanation, the flux diversion hypothesis as advanced by Brown and Goldstein (1980) should be considered. A coordinate regulation of two (perhaps more) enzymes is supposed to maintain the levels of mevalonate-derived intermediates for nonsterol pathways. This apparent secondary regulation site for dolichol products still remains to be identified.

Concerning the catabolism and/or excretion of dolichol theoretically, three mechanisms can be considered: (1) oxidative degradation to shorter-chain-length metabolites or CO_2; (2) minor chemical modification whether or not coupled with conjugation, to yield more polar secretion products; and (3) secretion of dolichol with an intact backbone. Respiration seems not to be a major route for elimination: after intravenous injection of [1-^{14}C]dolichol, $^{14}CO_2$ is evolved at a very low rate (Rip and Carroll, 1985). Moreover, many experiments have shown that the dolichol backbone is chemically extremely stable. On the other hand, a limited number of data (Connely and Keller, 1984; Rip and Carroll, 1985, 1986), such as the finding of radioactivity in the methanolic-acqueous phase after a Folch distribution following intubation of labeled dolichol (Chojnacki and Dallner, 1983), the existence of dolichyl dolichoate (Steen *et al.*, 1984), and the presence of a not yet identified dolichyl derivative more polar than dolichol itself in urine (Rip and Carroll, 1985), sustain the possibility of a minor chemical modification solely or in combination with some type of conjugation. No such conversion has yet been shown experimentally. It appears that the major route for dolichol outflow passes via bile, feces and, to a much lesser extent, urine (Connelly and Keller, 1984; Pullarkat and Raguthu, 1985; Rip and Carroll, 1985; Turpeinen, 1986; Patton and Poulos, 1987). The secretion occurs for the most part as intact dolichol or esterified to fatty acids. Taking into account this unresolved problem of dolichol elimination, the increase of the dolichol content upon aging, and the changes observed under pathological conditions, one may say that the enigma of dolichol homeostatis still represents a challenge for further research.

5. INFLOW AND FATE OF EXOGENOUS DOLICHOL

Dolichol in an organism can originate either from an exogenous source, i.e., the diet, or from *de novo* synthesis of endogenous dolichol (Figure 4). A restricted number of studies have focused on the absorption efficiency from the natural diet by comparing daily consumption versus excretion via the feces. The fate of intubated dolichol has been monitored with regard to its time-dependent distribution and redistribution in the different organs and tissues, and even at subcellular level. Experiments allied to intravenous injection of exogenous radiolabeled dolichol have been conducted as well. Finally, the uptake of exogenous dolichol by isolated cells or cells in culture has also been investigated. In addition, in collecting information on the destination of administered exogenous dolichol, answers to the following questions were sought: What is the relative contribution of dietary dolichol with respect to *de novo* synthesis? Are ingested shorter-chain polyprenols subjected to further elongation and α-saturation? Can exogenous polyprenols and dolichols be converted into biologically active intermediates through subsequent phosphorylation and glycosylation? Is there a translocation toward the site of action (RER membranes)?

5.1. Dietary Dolichol

In contrast to plants, bacteria, and several chow diets, mammalian tissues possess mainly longer-chain and α-saturated compounds (85–115 carbon atoms), although shorter-chain and fully unsaturated polyprenols are present as well but in much smaller quantities. Since rats ingest ca. 15 g of chow per day, equivalent to ca. 125 μg of dolichol and ca. 30 μg of polyprenol, it is very attractive to speculate that the diet, at least partially, could serve as a supply of dolichol. The first suggestion in this direction was made by Keenan *et al.*, (1977a), who found that dolichyl palmitate could be absorbed when intubated in an olive oil suspension. On the other hand, Keller *et al.* (1982a) arrived at a conclusion that there is no substantial contribution from the diet to total rat liver dolichol (rat liver dolichol amounts to ca. 40 μg/g of tissue) from the following data. The polyprenol : dolichol ratio and the homolog pattern in rat feces were indisguishable from that found for the chow prenols. Quantitation of the prenols in the rat feces indicated that the rat excretes about the same amount of total prenols as is consumed on a daily basis. Intubation experiments with [^{3}H]dolichol and [^{3}H]polyprenol by using [^{14}C] sitosterol, a poorly absorbed cholesterol analog as an internal standard, showed the $^3H : ^{14}C$ ratios in the fecal lipid extracts of the animals to be very similar to the starting ratios in the intubation suspensions (aqueous emulsion in triolein). Lipid analysis 30 h after intubation revealed that only 0.05% (0.06μg) of the total dolichol dose administered (120μg) (in parallel experiments an average value of 5.5% was found for cholesterol) had reached the liver (Keller *et al.*, 1982). In comparison, the *de novo* rate of hepatic dolichol synthesis has been estimated at 2–3 μg/day [in this respect, diverging data have been reported in the literature, varying from 2 (Adair and Keller, 1982) up to 25 nmol/liver per day (Elmberger *et al.* 1987; Keller, 1987)].

In other similar intubation experiments however, appreciable amounts of radioactivity could be recovered from the liver, but also from other organs. In their approach, Chojnacki and Dallner (1983) presented the radiolabeled polyprenols and dolichols to the alimentary tract as a sonicated aqueous lecithin emulsion (liposomes). A preferential uptake of shorter-chain compounds was observed, although this could have resulted from the experimental differences in preparing a proper emulsion with long-chain polyprenols. Moreover, these authors were able to demonstrate that, e.g., undecaprenol is effectively transformed to dolichol by reduction of the double bond in the terminal α-isoprene unit, which, upon phosphorylation, yields a metabolically active lipid intermediate. Because of the low specific activities, the authors did not find whether any chain elongation occurred. Dolichol, taken up by the liver cells from the gastrointestinal tract, displayed the highest specific activity in the mitochrondrial fraction, with a preference for the outer mitochrondrial membrane (ratio of specific activity, in mitochrondria : specific activity in microsomes ≈ 6–20; specific activity of outer mitochrondrial membrane : specific activity of inner mitochrondrial membrane ≈ 100). Upon subcellular fractionation of the liver homogenate, nonnegligible radioactivity could also be recovered from the particulate-free supernatant, one of the first pieces of data pointing to the occurrence of nonmembrane-associated dolichol. Using the intestinal loop and everted sac methods, Kimura *et al.* (1989) studied the intestinal absorption of dolichol. The highest uptake (39.1%) was found when a liposomal preparation (egg yolk or soybean lecithin) was used. From kinetic experiments they suggest a mechanism in which dolichol is released from the liposomes into the aqueous phase and subsequently partitioned into the lipid phase of the intestinal tissue.

Recently it has been shown by Jakobsson *et al.* (1989) that after incubation of dolichols and polyprenols (incorporated in liposomes), these lipids were taken up to a small extent by the different organs. Significant differences were found between uptake of free and esterified dolichol. Liver takes up preferentially free α-unsaturated prenols, while for example in stomach and intestine no significant difference between uptake of alcohols and their esterified derivatives was noted. Most of the administered lipids were recovered in the lysosomes. The exogenous isoprenols were, at least in the liver both partly esterified and phosphorylated; a portion of the polyprenols was also α-saturated.

5.2. Intravenously Injected Dolichol

Uptake of dietary dolichol by different organs is in a second phase, followed by redistribution among organs, which can only occur via the blood circulation. Second, considering the high biosynthetic capacity of the liver, it is possible that part is reserved for transport to other extrahepatic tissues. Such an export still has to proceed through the bloodstream. In this respect, a number of experiments have been performed to study the fate of radiolabeled dolichol after intravenous injection.

Following intravenous injection of [^{3}H]dolichol, suspended in a steroid-suspending vehicle, Keenan *et al.* (1977a) calculated a plasma half-life of 9–13 h (a relatively long $t_{1/2}$ compared with many other lipids such as retinol and cholesterol), where it remained as free dolichol. The decline of radioactivity in the plasma was paralleled by a progressive labeling of other tissues (lungs, heart, pancreas, kidneys, and brain), especially the liver (63% after 24 h) and the spleen. In the liver, the mitochrondrial fraction contained the bulk of radioactivity, where it was found concentrated within the outer mitochrondrial membrane. Vigo and Adair (1982), when injecting liposomes loaded with (pyro)phosphorylated and free polyprenols into the portal vein, found a preferential uptake of the free polyprenols by the liver.

Recently, intravenous injection experiments have been repeated by Rip and Carroll (1985,1986) with [1-^{14}C]dolichol. In their approach, dolichol was administered after *in vitro* mixing with rat serum. The clearance of dolichol from the blood displayed a biphasic profile consistent with a two-compartment model ($t_{1/2}$ of ca. 1 h and ca. 1 day). It was suggested that the rapid phase reflected fast uptake by the liver, whereas the second, slower phase may represent organ redistribution among organs. The lungs, e.g., take up considerable amounts of radioactivity, but do not retain it for any length of time. Moreover, the dolichol uptake differs considerably from one tissue to another, liver being the most active. After 1 day the liver contained more than 60% of the total dose administered. Some tissues (lung, spleen, and gastrointestinal tract) also took up significant quantities, whereas others (kidney, testes, heart, and especially brain) showed minimal uptake. Dolichol accounted for at least 98% of the radioactivity present after 24 h and did not change significantly after 4–21 days. In the liver 40–80% was esterified. Outflow of dolichol from the tissues also differed appreciably ($t_{1/2}$ for total rat and rat liver was ca. 16 days). The spleen actually accumulated radioactivity for the duration of the experiment. In longer-term studies (up to 130 days) the specific radioactivity in the spleen was shown to be higher than in the liver. Whether dolichol in the spleen serves a unique function or whether the spleen is unable to synthesize sufficient dolichol for its own requirement is not yet known.

Subcellular studies indicated that in the liver, intravenously injected dolichol preferentially associates with the so-called R_2 fraction (Rip and Carroll,

1985, 1986), i.e., a lysosome-enriched fraction. The specific activities in the R_2 fraction, but also in the microsomal fraction increased for a period (~ 30 h) after injection. At later times all specific activities declined, but at all times the highest specific activity was always confined to the lysosome-enriched fraction. The initial increase and the subsequent decline of the specific radioactivities suggest the occurrence of both inflow and outflow of dolichol in the different subcellular compartments.

5.3. Uptake by Isolated Cells

Incorporation of dolichol and polyprenols into egg lecithin liposomes proved to be an effective way of transferring individual polyisoprenoids into hepatocytes (Chojnacki *et al.*, 1980; Ekström *et al.*, 1982a,b). Moreover, this procedure looks very promising for studying the intracellular movement of these lipids between membrane compartments. The incorporation resulted in a net increase of the cellular dolichol content, the 95C homolog being more efficiently resorbed than the 55C compound. Part of the lipid was recovered in the phosphorylated form and was available for further glycosylation. Subcellular-localization experiments showed that the highest concentration of the exogenous dolichol was associated with the microsomal fraction. Dolichol and DMP was also observed in the cytosol. The interaction of liposomes with hepatocytes is an interesting phenomenon, but the mode of interaction (e.g., endocytosis or simple dissolution in the plasma membrane) is not quite clear.

Palamarczyk and Butters (1982) also have treated a battery of isolated cells or cells in culture with a source of exogenous dolichol (as a suspension in serum-free medium). All cell lines tested scavenged dolichol from the culture medium, but to significantly different extents. Serum exerted an inhibitory effect. Uptake by bulk-phase endocytosis or other nonspecific mechanisms was contradicted by the poor absorption of [^{14}C]sorbitol, a nonmetabolizable label. Excess cholesterol or α-ecdysone did not inhibit the dolichol uptake, suggesting that two independent resorption mechanisms are operating. In this approach, the highest [^{3}H]dolichol specific radioactivities were found in the plasma membrane enriched fractions. This could reflect either cell surface membranes coated with dolichol–receptor complexes or internalized plasma membranes (receptosomes), or both. Other intracellular membranes and microsomes were also shown to contain radioactivity, although with much lower specific activity. A small but detectable amount of dolichol was recovered modified.

5.4. Conclusions

(1) The contribution of dietary dolichol to total cellular dolichol has not yet been unambiguously clarified. Nevertheless, from a number of data, the *in vivo* biosynthetic pathways seem to be the primary source of the hepatic dolichol pool.

(2) From the different subcellular localizations mentioned above, one could conclude that dolichol and derivatives are subject to an intensive intracellular flow.

(3) Cells, organs, and organisms all tend to respond differently toward very different physiological conditions, as cited here (intubation, intravenous injections, and uptake by isolated cells or cells in culture). As a consequence, one could expect exogenous dolichol to be recovered in different subcellular components (diet: outer mitochondrial membrane of the liver; intravenous injection: scavenging toward liver lysosomes; uptake by isolated cells: recovered from microsomal fraction; uptake by cells in culture: recovered from plasma membranes). In addition, the timescales in these experimental approaches differ appreciably. For example, the timescale in intubation and intravenous injection experiments diverges to a large extent from that in experiments with isolated cells (days or hours versus minutes). Dietary uptake involves a long transport of dolichol before it enters the liver. If the uptake of dolichol by cells occurs by phagocytosis followed by a release from the lysosomes toward other cellular compartments (e.g., ER, mitochrondria), the time course may be an important factor.

(4) The method of uptake and further processing (subcellular fate or chemical modification) of exogenous dolichol probably strongly depends on the physicochemical conditions (oily or aqueous emulsions, liposomes, micelles) presented to the resorbing target by dolichol.

(5) Finally, subcellular studies are very vulnerable with regard to artifacts if the necessary controls [i.e., assaying a whole series of markers and marker enzymes allowing calculation of the relative specific activities (RSA) of "all" fractions isolated as well as mass balances and recoveries] are omitted. Even when the necessary precautions are taken, peculiar results can emerge; e.g., in our laboratory we found the highest-RSA DMP phosphatase to be associated with a 15–20-fold-enriched plasma membrane fraction (see Table X), whereas all other "subcellular arguments" suggested that the enzyme was an intrinsic ER enzyme (Steen *et al.*, 1988).

6. EXTRACELLULAR FLOW OF DOLICHOL

In plasma the majority of lipids are found in association with circulating lipoproteins. Moreover, it has been demonstrated that all lipoprotein classes can exchange lipids with each other and with cells (Pownall *et al.*, 1982). From this, and referring to the arguments mentioned in Section 5.2, one can expect similar phenomena to occur for extracellular dolichol and derivatives. In this respect, other nonpolar polyisoprenoids such as β-carotene, lutein, and lycopene have been reported to bind to serum lipoproteins (Bjornson *et al.*, 1976).

Indirect evidence of dolichol binding by serum lipoproteins came from the experiments of Palamarczyk and Butters (1982), who stated that uptake of dolichol by cells diminished when serum was added to the culture medium. Another argument in favor of serum-mediated dolichol transfer came from an observation of Elmberger and Engfeldt (1985): [^{3}H]dolichol, synthesized in the liver from [^{3}H]mevalonate injected in the portal vein, reappeared in the blood circulation after a lag period of about 25 min.

The dolichol level in blood is rather low (ca. 0.2–1 μg/ml), differing by about a factor of 10^3 from to the cholesterol concentration (millimolar versus micromolar) (Elmberger and Engfeldt, 1985; Rip and Carroll, 1985; Yamada *et al.*, 1985; Morris and Pullarkat, 1987). The serum dolichol content is of the same order of magnitude as found for vitamin A (1.7 μM) (Goodman, 1984). In serum, dolichol is present mainly in the unesterified form (Elmberger and Engfeldt, 1985; Elmberger *et al.*, 1988).

6.1. Lipoprotein-Associated Dolichol

After intravenous injection of [^{3}H]dolichol in the tail vein of rats, Keenan *et al.* (1977b) found that almost all of the radioactivity was rapidly taken up by the high-density lipoprotein (HDL) fraction. At any given time, little radioactivity was associated with erythrocytes or other plasma components (Keenan *et al.*, 1977b; Elmberger and Engfeldt, 1985; Elmberger *et al.*, 1988). *In vitro* incubation of serum with radiolabeled substrate resulted in the binding of dolichol to a plasma fraction with properties corresponding to those found in the plasma of intravenously injected animals (it should be stressed that in these experiments the binding lipoprotein fraction was not identified unambiguously as HDL) (Keenan *et al.*, 1977b). The *in vitro* studies showed that the binding of the plasma protein factor is limited. Dolichyl palmitate, even in a 10^3-fold excess, was ineffective in interfering with the binding of free dolichol. Although other nonpolar isoprenoids have been found associated with HDL, these compounds are present in much higher concentrations in low-density lipoproteins (LDL) and very-low-density lipoproteins (VLDL). On the other hand, the so-called HDL fraction was able to bind 3–6 times more dolichol per milligram of protein than the other lipoproteins did. The authors suggested that HDL would serve as a transport vehicle direction liver → tissue.

Opposite to the "*in vivo*" data of Keenan *et al.* (1977b), in the experiments of Rip and Carroll (1985, 1986), when *in vitro*-labeled dolichol was mixed with serum, the radioactivity initially appeared mostly in the VLDL fraction. However, 4 days after intravenous injection, only 5% of the residual radioactivity (3% of the total dose administered) could be recovered from the VLDL fraction, whereas nearly 50% was associated with HDL.

Recently the kinetics of "*in vitro*" dolichol biding by human serum has been studied in our laboratory by using the dextran-coated charcoal assay (Van Dessel *et al.*, 1986a). The dolichol-binding capacity of native serum is relatively low (3 pmol/mg of protein; equivalent to 0.3 μg/ml of serum) compared with the total amount present in serum. After delipidation of serum (Cham and Knowles, 1976; Assmann and Kladetzky, 1984), the binding capacity increases up to 0.9 μg/ml of serum, close to the dolichol concentration in native human plasma. From kinetic studies it became evident that the binding occurs at a high rate (saturation being reached within 5 min) and results in a rather stable complex. Optimal binding is achieved at physiological pH, and an incubation temperature below room temperature is required. Preincubation of serum at 50°C does not prevent binding. Chase experiments fail to remove bound radiolabeled ligand. In this respect it should be recalled that other ligand–protein interactions also have often been performed at 5°C (Kandutsch *et al.*, 1977; Inoue *et al.*, 1983; Love *et al.*, 1983) and that chase experiments are not always reported in the literature. From the Scatchard plot an apparent dissociation constant (K_d) of 6.9×10^{-6} M and a β_{max} of 432 pmol/mg of protein were calculated. The K_d value is much higher than found for steroids, retinol and a number of drugs (10^{-8}–10^{-9} M) (Scanu *et al.*, 1982), but, on the other hand, it is much lower than the literature values for tricyclic antidepressants (Javaid *et al.*, 1983), tetracycline (Powers, 1974), quinidine (Nilsen and Jacobson, 1975) and clioquinol (Hobara and Takata, 1976). The specificity of the binding process was deduced from competition studies showing that only cold dolichol and dolichyl derivatives compete equally for binding sites (Table V). The inhibition by dolichyl fatty acid esters is in contradiction with the data of Keenan *et al.* (1977b). From our experimental setup, it is hard to conclude whether the inhibition is due to free dolichol generated through the action of ester hydrolases on dolichyl fatty acid esters, as reported for the inhibition of retinol binding by retinyl esters (Chytil and Ong, 1984).

Diverging from the retinol-binding system in serum, the serum-bound dolichol is recovered as a high-molecular-weight complex, as evidenced by Sephacryl S-300 and Sepharose 2B column chromatography. From density gradient centrifugation followed by gel chromatography on Bio-Gel A-5m, as well as by specific precipitation methods, it became obvious that after binding the bulk of [^{3}H]dolichol is likely to be associated with VLDL, as found by Rip and Carroll (1985, 1986) (Figure 5) but in disagreement with the results of Keenan *et al.* (1977b). Whether the binding process reflects nonspecific dissolution of dolichol into the hydrophobic core of the VLDL complex, as found for β-carotene (Bjornson *et al.*, 1976), or whether it results from specific dolichol–VLDL apoprotein interactions remains to be defined. β-Carotene does not contain any polar group; in dolichol the presence of a primary alcoholic function must be taken into account.

One could expect serum albumin and probably many other proteins with hydrophobic binding sites to bind dolichol. Our data clearly indicated that al-

Table V
Specificity of the Dolichol-Binding Activity in Human Serum[a]

Effector	% Binding
None	100
Palmitic acid	102
Retinol	96
Cholesterol	120
Triolein	108
PC	116
PE	90
PI	100
Farnesol	120
Dolichol	23
DMP	14
Dolichyl fatty acid ester	17

[a]All incubations were performed in the presence of at least 125-fold molar excess of unlabeled ligand. Reprinted with permission from *Biochim. Biophys. Acta* (Van Dessel *et al.*, 1986).

bumin, in comparison with VLDL, provides only a minor contribution to dolichol transport, as was confirmed by Affi-Gel Blue chromatography. A similar observation was made for vitamin A, also able to associate with many proteins, which, however, do not participate in the retinol transport under physiological conditions (Chytil and Ong, 1984).

6.2. Dolichol Exchange between Lipoproteins

Several lipids having no measurable or little solubility in water are transferred by specific transfer proteins (Pownall *et al.*, 1982). *In vitro* and *in vivo* studies have demonstrated that, e.g., phospholipid exchange can occur among lipoproteins (Tall, 1986). Plasma lipid transfer proteins, which facilitate the transfer of cholesteryl esters (Pattnaik *et al.*, 1978; Tall, 1986; Hesler *et al.*, 1987), triglycerides (Rajaram *et al.*, 1980; Abbey *et al.*, 1985; Quig and Zilversmit, 1988), PC (Chajek and Eisenberg, 1978; Tall, 1986), and glycolipids (Kwok *et al.*, 1981; Via *et al.*, 1985), have been purified and characterized.

Contrary to the situation after incubating radiolabeled dolichol with serum *in vitro*, the HDL fraction *in vivo* seems to contain the bulk of plasma dolichol (Keenan *et al.*, 1977b; Elmberger and Engfeldt, 1985). This phenomenon may reflect a possible mechanism of transporting dolichol from extrahepatic tissues to the liver (elimination). As already cited, 4 days after intravenous injection, with

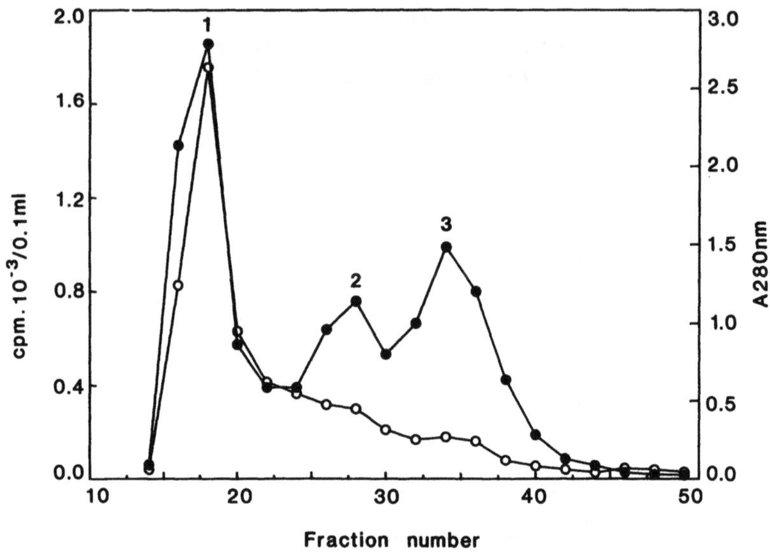

FIGURE 5. Gel filtration of serum lipoproteins on Bio-Gel A-5m. After incubation and dextran-treated charcoal treatment, lipoproteins (80 mg of protein) were applied to a Bio-Gel A-5m column (1 × 100 cm) and eluted with 0.15 M NaCl–0.01% EDTA. The flow rate was about 7 ml/hr, and 2-ml fractions were collected. Symbols = ○, [^{3}H]dolichol; ●, A_{280}. Protein peaks 1, 2, and 3 represent VLDL, LDL, and HDL, respectively. Yield: protein, 81%; [^{3}H]dolichol, 79%. Reprinted with permission from *Biochim. Biophys. Acta* (Van Dessel *et al.*, 1986).

serum VLDL as a transporting vehicle, about half of the remaining dolichol is found within the HDL fraction (Rip and Carroll, 1985).

From these data, Van Dessel *et al.* (1986a) have postulated the occurrence of a dolichol-transferring system in human blood, which should shuttle dolichol from VLDL to HDL as already reported for triglycerides (Rajaram *et al.*, 1980) and esterified cholesterol (Chajek and Fielding, 1978; Barter and Jones, 1979; Ha and Barter, 1985; Quig and Zilversmit, 1986). Considering the length of the aliphatic chain in the dolichol molecule, this transfer process is most probably protein mediated, as hypothesized for glucocerebrosides (Via *et al.*, 1985). Hitherto, most lipid-transferring and lipid-binding proteins in serum have been recovered in the lipoprotein-free fraction, with molecular masses between 21 and 150 kDa (van Jaarsveld *et al.*, 1973; Ihm *et al.*, 1980; Tall, 1986). Therefore, this fraction of human serum was monitored for dolichol transfer activity. For this purpose, the rapid and simple procedure for determining phospholipid transfer activity as described by Damen *et al.* (1982) was adapted, with PC liposomes as donor vesicles and HDL as acceptor (VLDL could serve a donor function). By using this assay, Van Dessel *et al.* (1987b) have demonstrated the presence in serum of a protein-facilitated dolichol transfer between lipoproteins together with limited spontaneous exchange. The transfer activity was exclusively confined to the $d > 1.19$ lipoprotein-free serum fraction. Partial purification could be achieved by hydrophobic interaction chromatography on phenyl-Sepharose followed by heparin-Sepharose column chromatography and an ultimate separation step on SE-Sephadex. The final protein preparation was able to realize a flux rate of 214 pmol of dolichol/3 hr per mg of protein under the assay conditions. The time course of transfer was linear up to 15 min but continued for at least 1 hr. The dolichol exchange was shown to be dose-response related to the transfer protein concentration. The kinetics are summarized in Figure 6.

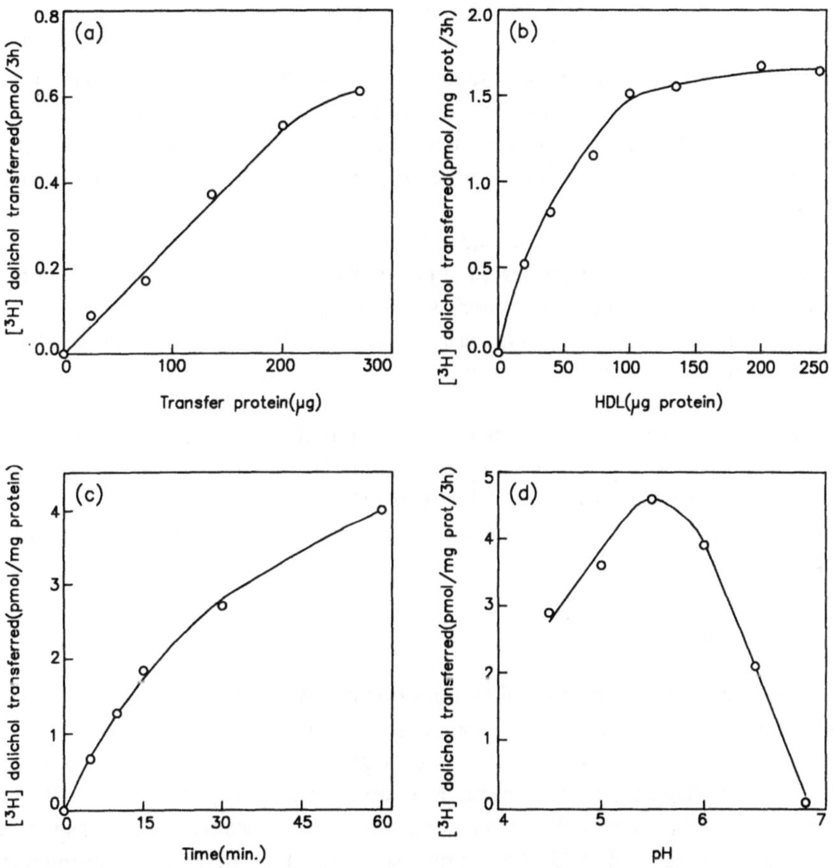

FIGURE 6. Kinetics of [^{3}H]dolichol transfer in human serum. Liposomes (6 pmol of [^{3}H]dolichol per 125 nmol of PC) and HDL (equivalent with 25 μg of protein) were incubated with the fraction to be assayed in a final volume of 400 μl containing 150 mM NaCl and 10 mM Tris hydrochloride (pH 7.4) at room temperature. The reaction was stopped by addition of 300 μl of 500 mM NaCl and 215 mM $MnCl_2$ containing 140 U of heparin. After standing for 10 min at room temperature, the precipitate was removed by centrifugation (12,000 × *g* for 10 min), and the radioactivity in the supernatant was counted (Van Dessel *et al.*, 1987c).

The identity of this dolichol transfer protein in the lipoprotein-free serum fraction, as well as the specificity of the transfer, remains to be elucidated. The observation that the lipid transfer activity is retained on phenyl-Sepharose at salt concentrations of 0.15 M or higher suggests that this transfer protein possesses hydrophobic regions and/or is complexed with lipids as reported for other plasma proteins (Ihm *et al.*, 1982). In the presence of *N*-ethylmaleimide, a thiol residue blocker, the dolichol transfer activity is markedly depleted. Mercurial thiol group inhibitors are known to reduce also the transfer of triglycerides between VLDL and HDL (Hopkins and Barter, 1980), whereas they exert virtually no effect on the transfer of esterified cholesterol or PC (Zilversmit *et al.*, 1980; Ihm *et al.*, 1982). Another difference is that the cholesterol exchange occurs as a free and esterified form where no derivatization seems to precede the dolichol transfer. Indeed, after incubation and extraction of the acceptor fraction, no esterified dolichol could be traced upon inspection of the thin-layer chromatography (TLC) plates. Furthermore, heat treatment of serum to destroy lecithin-cholesterol acyltransferase (LCAT) activity (Ihm *et al.*, 1982) did not influence the dolichol transfer activity.

As serum HDLs represent a heterogeneous class of lipoproteins, the ability of the various-HDL subclasses to serve as the dolichol acceptor from donor liposomes was screened. To this end, serum HDLs were subfractionated according to their apo E content on a heparin-Sepharose affinity column (Marcel *et al.*, 1981). Subsequent exchange experiments revealed that the HDL subfraction containing most of the apo E, all of the apo B, and the majority of the sterols did not act as an acceptor for dolichol, whereas the HDL fraction that eluted with $MnCl_2$ scored the highest as acceptor (Table VI). Finally, the incorporation of dolichol (diameter, 10 nm) in HDL (diameter, 8 nm) poses some structural problems with respect to the orientation of this lipid in HDL particles (Rip *et al.*, 1985).

6.3. Conclusions

Several homeostatic control systems operate by intermediate transport along the blood circulation. Triglycerides are channeled as VLDL from the liver toward adipose tissue for storage; free fatty acids generated upon mobilization of fat depots are transported in the plasma as albumin complexes; cholesterol metabolism is kept in balance by intervening extracellular vehicles (LDL for transport to extrahepatic tissues and HDL for the reverse process). From the data presented in this section, one might speculate that for dolichol homeostatis, serum binding and transfer factors may also be involved. Indeed, dolichol is found in association with VLDL or HDL (also partly with LDL) depending on the experimental conditions (*in vitro* or *in vivo*). At present the interrelation between these associations and the dolichol transfer protein is not quite clear. In this respect, one

Table VI
Dolichol Acceptor Activity of HDL-Subfractions[a]

Fraction	Eluent	Apoproteins[b] E	Apoproteins[b] B	[^{3}H]dolichol transferred[c]
Not retained	25 mM $MnCl_2$ +50 mM NaCl	+		3.2
Fraction I	70 mM NaCl	++		0
Fraction II	300 mM NaCl	+++	+	0

[a]HDLs were subfractionated as described by Marcel *et al.* (1981) on heparin-Sepharose.
[b]The content of apoproteins in the different subfractions according to Marcel *et al.* (1981).
[c]PC-liposomes-HDL assay; picomoles transferred/3 hr/milligram of transfer protein with 1 mg of protein HDL subfraction as acceptor.

should keep in mind that the VLDL fraction is subject to rapid clearance in comparison with HDL and LDL, suggesting that *in vivo* the different dolichol pools in the distinct extracellular compartments are in a dynamic equilibrium (Via *et al.*, 1985; Rip *et al.*, 1985); e.g., the dolichol contents of plasma and erythrocytes also differ appreciably.

The physiological relevance of the protein-mediated shift between lipoprotein classes, as well as the identity of the transfer protein factor involved, awaits further exploration. Finally, the apparent lack of balance between intra- and extracellular dolichol contents remains intriguing.

7. SUBCELLULAR ASPECTS OF DOLICHOL METABOLISM

7.1. Intracellular Traffic of *De Novo* Synthesized Dolichol

As an alternative to an exogenous source, cellular dolichol may arise from *in situ* production. The chemistry and regulation of the biosynthetic pathway have been dealt with in Section 4. As already cited, the primary dolichol manufacture occurs within the RER membranes (Ekström *et al.*, 1982; Ekström and Chojnacki, 1982; Rip and Carroll, 1982; Rip *et al.*, 1985), where this polyisoprenoid is supposed to exert its biological function as a cofactor in the N-glycosylation of proteins. Studying the subcellular fate of radiolabeled dolichol generated *in vivo* from [^{3}H]mevalonate as a precursor, an intracellular flow of newly synthesized dolichol was found to occur. A kinetic study (Wong and Lennarz, 1982) monitoring the distribution of the [^{3}H]dolichol over a 2-hr period in the subcellular fractions from liver slices, revealed that the newly synthesized dolichol first appeared in the microsomal fraction, the site of biosynthesis. However, after a lag period, appreciable radioactivity could be recovered from the mitochrondrial plus light mitochrondrial (M+L) fraction. Furthermore, even after the level of [^{3}H]dolichol in the microsomal fraction had reached a plateau, the radioactivity in the M+L fraction continued to increase. Further subfractionation of this fraction to partially separate mitochondria and lysosomes pointed to the latter as the principal site of [^{3}H]dolichol deposition. Edlund *et al.* (1983) also have observed a similar intracellular shift after biosynthesis of dolichol from *in vivo*-injected [^{3}H]mevalonate, although to a much lesser extent. These results are in agreement with recent subcellular data indicating the existence of an intracellular dolichol concentration gradient increasing from the RER over the Golgi to the lysosomes (Rip *et al.*, 1981; Eggens *et al.*, 1983; Keller *et al.*, 1986; Chojnacki and Dallner, 1988). All these studies have suggested that for some reason, after synthesis in the RER, most of dolichol (with or without having performed its function in the N-glycosylation process) is directed to the lysosomal compartment. The targeting mechanism is still unknown, although several possibilities can be put forward. First, like phospholipids, dolichol may be transferred through the cytoplasm by the aid of lipid-specific or nonspecific carrier proteins. Otherwise, membrane flow (vesicle transport) may be involved, as this process is also associated with the production of primary lysosomes. Third, non-protein-mediated luminal transport could be considered, e.g., with dolichol in micellar form. Recently, in analogy with the signal peptide hypothesis (Walter *et al.*, 1984), a role for fatty acids as a signal for targeting phospholipids to definite subcellular compartments has been proposed (Valtersson *et al.*, 1986; Chojnacki and Dallner, 1988). According to Tolbom *et al.* (1988), luminal transfer of dolichol to the lysosomes can occur only if the lipid is esterified with fatty acids. The logical sequence in the generation of dolichyl fatty acyl ester is dephosphorylation of the biosynthetic endproduct followed by esterification. In this respect the subcellular localization of the two intervening enzymes, i.e., the DMP phosphatase and the acyl-CoA:dolichol acyltransferase, poses some problems. The specific activities of the phosphatase, an integral membrane protein, have been shown to be four- to nine-fold higher in the smooth ER and Golgi, respectively, than in the RER (Rip *et al.*, 1983a,b), while the acyltransferase has been localized at the outer cytoplasmic surface of the ER (Chojnacki and Dallner, 1988). Finally, theoretically one might envisage the possibility that newly synthesized polyprenols can be transported out of the cell and subsequently taken up again by a receptor-mediated process to reach the lysosomal compartments. This could be supported by the fact that dolichol micelles, injected in the portal vein, are taken up (Vigo and Adair, 1982; Edlund *et al.*, 1983). The very low polyprenol blood levels would then reflect a low steady-state concentration.

Besides the chemical nature of lysosomal dolichol, several other questions remain to be answered. (1) Does dolichol upon arrival integrates within the lysosomal membranes, or are we dealing, as suggested by Chojnacki and Dallner (1988), with luminal non-protein-associated dolichol? (2) What is the metabolic fate of this dolichol? (3) Can it become mobilized again for reutilization by a transfer process in the opposite direction (lysosomes to ER), or is it another functional form of dolichol? In unfertilized sea urchin eggs these dolichol reserves are bypassed in favor of *de novo* synthesis (Carson and Lennarz, 1981). Alternatively, lysosomal dolichol could represent "transit" dolichol, a first step in the degradation or elimination of cellular dolichol.

As summarized recently by Chojnacki and Dallner (1988), dolichols can undergo extensive intracellular movements in various directions: (1) ER → cytoplasm (carrier) → organelles; (2) ER → ER (lumen) → Golgi (lumen) → lysosome (lumen); (3) ER → vesicle transport → organelles. In the liver two additional transfer systems must be considered: (4) ER → Golgi (lumen) → blood (this transport should provide a way for introducing dolichol into the blood circulation, probably associated with apoproteins in analogy with other lipids), and (5) ER (?) → bile → intestine (the mechanism of dolichol transfer to the bile has not yet been elucidated, but bile cholesterol also have been suggested to be of lysosomal origin) (Coleman, 1987). Probably this cholesterol is to be considered extrahepatic (uptake via LDL receptors). Therefore, it is possible that extrahepatic dolichol is ingested by hepatocytes, stored in lysosomes, possibly transformed, and released in the bile canaliculi for removal through the intestine.

7.2. Subcellular Localization of Dolichol

Contrary to the information gained from the *in vivo* experiment with labeled precursors, as discussed in Section 7.1, "classical" subcellular studies provide only a steady-state picture, which does not allow us to discriminate between liver dolichol pools of different origin (dolichol obtained from *de novo* synthesis, from extrahepatic tissues, or from the diet).

Several subcellular studies, mostly on the liver, have been performed to define the intracellular localization of dolichol and derivatives, by using either differential pelleting (Wong and Lennarz, 1982), enriched fractions (Rip and Carroll, 1982; Rip *et al.*, 1983a,b), or isopycnic gradient centrifugation (Wong *et al.*, 1982). Different subcellular localizations have been reported.

Initially, the scientific community inclined toward a "mitochrondrial" localization (Burgos and Morton, 1962). Indeed, in the first studies dolichol was recovered in both the 1000 x *g* and 10,000 x *g* sediments, esterified dolichol mainly (53%) in the nuclear pellet and free dolichol for the greater part (77%) in the mitochrondrial fraction (Butterworth and Hemming, 1968). From subsequent studies on a number of fungi (Butterworth and Hemming, 1968; Barr and Hemming, 1972) and holotricha (Adrian and Keenan, 1981), a mitochrondrial localization was also suggested. Incorporation studies with radiolabeled precursors (Daleo and PontLezica, 1970; Daleo *et al.*, 1977; Martin and Thorne, 1974) were in favour of mitochondria as the main intracellular residence of dolichol, as were the data from intubation experiments (Chojnacki and Dallner, 1983) and intravenous injection (Keenan *et al,* 1977).

In the early 1980s a change in opinion emerged, suggesting lysosomes as the primary location. One of the first indications came from an elegant study by Rip *et al.* (1981), who found that the mitochrondrial fraction (31 ng of dolichol per mg of protein) contained only 1⁄10 of the dolichol concentration in the starting homogenate and the plasma membrane fraction (d ca. 300 ng of dolichol per mg of protein). The same line of evidence came from work by Wong *et al.* (1982), who, upon subfractionation of an L fraction on sucrose or metrizamide gradients, stated that the distribution profiles for dolichol (4000 ng of dolichol per mg of protein in the peak fraction) and the lysosomal marker enzyme hexosaminidase coincide. Rip *et al.* (1983a) also reported their so-called R_2 fraction to contain about half of the liver dolichol. This R_2 fraction has a similar composition to the L fraction, which displays the same enrichment for dolichol as for β-D-*N*-acetyl-glucosaminidase (Wong *et al.*, 1982). In these experiments (Rip *et al.*, 1983a), the enrichment in the R_2 fraction was far superior to that in the corresponding R_1 fraction (microsomal fraction). Moreover, in lysosomal preparations the highest specific amounts for dolichol up to ca. 15,000 ng/mg of protein have been found (Wong *et al.*, 1982). On the other hand, relatively high values also have been reported from the Golgi, ranging from ca. 1900 (Rip and Carroll, 1982b) up to 4800 ng/mg protein (Ekström *et al.*, 1982; Eggens *et al.*, 1983), which recently have been confirmed for rat testes (Nyquist and Holt, 1985).

In submicrosomal fractions from rat liver the occurrence of a gradient distribution of dolichol increasing from the RER (90 ng/mg of protein) to the SER (140 ng/mg of protein) to the Golgi (1940 ng/mg of protein) was shown. For phosphorylated dolichol this concentration gradient runs in the opposite direction, the RER displaying the highest specific amounts (107 ng/mg of protein versus about 55 ng/mg in the Golgi) (Rip and Carroll, 1982; Rip *et al.*, 1983a).

Recently, in our laboratory the subcellular localization of dolichol in bovine thyroid was investigated (Steen *et al.*, 1986, 1988) by using three different approaches, i.e., differential pelleting, buoyant density equilibration in a zonal rotor, and isolation of enriched nuclear and plasma membrane fractions. Table VII shows that the percentages of constituents and markers recovered from the supernatant can vary considerably (protein: 90%; 5′-nucleotidase, alkaline phosphatase, hexosaminidase, and NADPH-cytochrome c reductase, 60–70%; cholesterol, phospholipids and glucose-6-phosphatase, 40–50%; mitochrondrial markers, ca. 20%). Dolichol and its derivatives are also found in the supernatant to an appreciable extent (40–50%), whereas the remainder cosediments mainly with the M and L fractions. Upon prolonged centrifugation, it became evident that the supernatant fraction still contains membranous fragments, mostly confined to the so-called S_b and S_c subfractions (Table VIII). Sepharose 2B column

Table VII

Absolute Amounts, Relative Amounts, RSA, and Yield in Subcellular Fractions of Bovine Thyroid after Differential Pelleting[a]

		N fraction		M fraction		L fraction		P fraction		S fraction			
Enzyme or constituent	Amount (μg/mg of protein) in homogenate	Relative amount (%)	RSA	Relative amount (%)	RSA	Relative amount (%)	RSA	Relative amount (%)	RSA	Relative amount (%)	RSA	Yield (% vs. homogenate)	Number of experiments
Proteins	62	1.0 ± 0.1		4.9 ± 0.3		1.3 ± 0.2		2.1 ± 0.2		90.7 ± 3.0		94 ± 5	10
Cholesterol	5.2 ± 0.3	0.8 ± 0.1	0.8 ± 0.3	37.8 ± 0.4	7.7 ± 0.6	14.4 ± 0.5	11.1 ± 2.4	8.6 ± 0.7	4.1 ± 0.6	38.4 ± 1.1	0.4 ± 0.2	96 ± 4	5
Phospholipids	30.9 ± 3.1	0.7 ± 0.1	0.7 ± 0.3	34.2 ± 0.6	7.0 ± 0.5	12.2 ± 0.3	9.4 ± 1.9	7.6 ± 0.3	3.6 ± 0.5	45.3 ± 1.4	0.5 ± 0.1	93 ± 3	5
Dolichol	2.1 ± 0.9	0.8 ± 0.2	0.8 ± 0.2	32.8 ± 7.0	6.7 ± 1.9	11.8 ± 2.4	9.1 ± 2.8	3.8 ± 1.2	1.8 ± 0.6	50.8 ± 9.0	0.6 ± 0.1	90 ± 9	5
Dolichyl fatty acid ester	265 ± 80[b]	0.7 ± 0.1	0.7 ± 0.2	43.8 ± 4.6	8.9 ± 1.6	7.3 ± 2.1	5.6 ± 0.9	4.2 ± 2.0	2.0 ± 1.0	44.0 ± 6.0	0.5 ± 0.1	92 ± 13	5
Dolichyl dolichoate	197 ± 70[b]	0.6 ± 0.1	0.6 ± 0.2	31.1 ± 5.1	6.3 ± 1.5	11.2 ± 2.2	8.6 ± 2.3	5.9 ± 2.0	2.8 ± 1.1	51.2 ± 6.5	0.6 ± 0.1	89 ± 14	5
5′-Nucleotidase		0.3 ± 0.1	0.3 ± 0.1	14.2 ± 0.3	2.9 ± 0.2	9.0 ± 0.6	6.9 ± 1.2	5.9 ± 0.7	2.8 ± 0.6	70.6 ± 2.3	9.8 ± 0.1	100 ± 5	8
Hexosaminidase		0.9 ± 0.1	0.9 ± 0.1	32.8 ± 0.4	6.7 ± 0.5	1.9 ± 0.2	1.5 ± 0.3	1.3 ± 0.3	0.6 ± 0.2	63.1 ± 2.6	0.7 ± 0.1	100 ± 6	8
Glucose-6-phosphatase		1.0 ± 0.1	1.0 ± 0.1	32.6 ± 0.6	6.7 ± 0.5	9.4 ± 0.9	7.2 ± 1.3	6.5 ± 0.8	3.1 ± 0.6	50.5 ± 3.6	0.6 ± 0.1	93 ± 15	8
Alkaline phenylphosphatase		0.5 ± 0.1	0.5 ± 0.1	22.9 ± 0.5	4.7 ± 0.3	8.6 ± 0.3	6.6 ± 1.2	5.8 ± 0.7	2.8 ± 0.4	62.2 ± 4.1	0.7 ± 0.1	85 ± 6	8
Cytochrome c oxidase		1.3 ± 0.1	1.3 ± 0.2	70.4 ± 2.8	14.4 ± 1.5	4.3 ± 0.2	3.3 ± 0.9	4.6 ± 0.3	2.2 ± 0.2	19.4 ± 0.9	0.2 ± 0.1	90 ± 18	8
NADPH cytochrome c reductase		1.1 ± 0.1	1.1 ± 0.1	18.2 ± 0.7	3.7 ± 0.4	7.9 ± 0.4	6.0 ± 0.8	8.6 ± 0.8	4.1 ± 0.5	64.2 ± 3.6	0.7 ± 0.1	102 ± 3	8
Monoamine oxidase		1.5 ± 0.2	1.5 ± 0.1	63.2 ± 0.9	12.9 ± 1.0	7.2 ± 0.3	5.5 ± 0.5	5.5 ± 0.9	2.6 ± 0.6	22.6 ± 1.9	0.2 ± 0.1	97 ± 1	3
[1-3H]dolichol kinase		0.8 ± 0.1	0.8 ± 0.2	29.9 ± 6.1	6.1 ± 1.1	10.5 ± 2.6	8.1 ± 1.5	22.5 ± 3.1	10.7 ± 1.3	36.3 ± 5.1	0.4 ± 0.1	96 ± 10	8
[γ-32P]CTP dolichol kinase		1.4 ± 0.1	1.4 ± 0.1	53.9 ± 4.8	11.0 ± 1.2	14.0 ± 1.8	10.8 ± 1.1	15.1 ± 1.2	7.2 ± 1.0	15.6 ± 2.7	0.2 ± 0.1	89 ± 15	6
[1-3H]dolichylmonophosphate phosphatase		1.2 ± 0.1	1.2 ± 0.1	42.1 ± 4.3	8.6 ± 0.8	13.8 ± 1.7	10.6 ± 0.8	9.7 ± 0.8	4.6 ± 0.9	33.2 ± 2.4	2.4 ± 0.1	94 ± 8	8

[a]All values are mean ± standard error of the mean.

[b]The concentrations of dolichyl fatty acid ester and dolichyl dolichoate are given in nanograms per milligram of protein.

chromatography revealed that nearly 70–80% of phospholipids and cholesterol were eluted with the void volume, together with a number of membrane-bound enzymes (Van Dessel *et al.*, 1977). In contrast, dolichol and its derivatives seem to be enriched in the S_d fraction, the top fraction essentially devoid of membranous material (Table VIII). Whether this "supernatant" dolichol is present as a water cytosolic-soluble complex (e.g., through association with polar lipids and/or proteins), as recently reported for rat liver (Appelkvist *et al.*, 1985), or whether it results from luminal dolichol (e.g., from ER, Golgi, or lysosomes) being released by mechanical forces during homogenization and centrifugation remains to be established. It should be noted that the S_d fraction contains only a small percentage of total supernatant proteins. At any rate, the data strongly suggest the existence of intracellular non-membrane-associated dolichol. With regard to sedimentable dolichol, the distribution profiles of free and conjugated dolichol mutually differ; neither have the distribution patterns coincident with those of the markers and compounds assayed. As a consequence, dolichol and its derivatives could not be localized unambiguously by the differential pelleting procedure.

Upon buoyant density equilibration a localization in mitochondria becomes questionable (Figure 7), and Table IX shows that a nuclear localization seems rather unlikely. The same conclusion can be drawn from Table X as far as plasma membranes are concerned. A partial localization in lysosomes can be considered from Figure 7, because a dolichol peak in the high-density region coincides with a peak of the lysosomal markers in the same density regions. On the other hand, much larger amounts are recovered from the top of the gradient than for the lysosomal hydrolases. However, there is no coincidence, and a separation is achieved, the proteins and acid hydrolases equilibrating at a somewhat higher density level than dolichol. Again, this could be ascribed to the presence of dolichol and derivatives in the form of a "soluble" complex. In addition, this complex should not consist predominantly of proteins; they should be present to a much lesser degree than in biological membranes.

Our results (Steen *et al.*, 1988) revealed that upon subcellular fractionation, dolichol and its derivatives behave totally differently from the classical membrane lipids such as phospholipids and cholesterol. This unique behavior of dolichol and its derivatives might explain the divergence reported in the literature with regard to its subcellular localization. Moreover, the bulk of intracellular dolichol seems to be present in a "soluble" form. Our data do not allow us to conclude whether this finding reflects an *in vivo* situation or originates from the fact that dolichol is easily detached from membranous structures under the rather traumatizing conditions of the homogenization procedures. Luminal and/or cytosolic dolichol could represent a kind of salvage pathway to protect membrane integrity. Recently it has been shown that dolichol can destabilize artificial membranes (see Section 3). It cannot be excluded that the cellular dolichol content is

Table VIII
Absolute Amounts, Relative Amounts, RSA, and Yield of Enzymes and Constituents in the Subfractions After Subfractionation of the Supernatant[a]

Enzyme or constituent	Amount (μg/mg of protein) in S fraction	S_a fraction		S_b fraction		S_c fraction		S_d fraction		Yield (% vs. homogenate)	Number of experiments
		Relative amount (%)	RSA	Relative amount (%)	RSA	Relative amount (%)	RSA	Relative amount (%)	RSA		
Proteins	31.6	30.4 ± 2.8		44.4 ± 4.4		14.4 ± 0.8		1.5 ± 0.2		93 ± 12	8
Cholesterol	2.3 ± 0.2	3.2 ± 0.2	0.11 ± 0.01	10.2 ± 1.8	0.23 ± 0.04	22.5 ± 3.0	1.54 ± 0.06	2.5 ± 0.4	1.67 ± 0.01	84 ± 6	5
Phospholipids	14.3 ± 2.0	4.6 ± 1.0	0.15 ± 0.03	15.8 ± 3.6	0.36 ± 0.01	23.7 ± 3.2	1.65 ± 0.08	1.2 ± 0.2	0.80 ± 0.02	90 ± 11	5
Dolichol	1.1 ± 0.5	6.3 ± 1.6	0.21 ± 0.03	7.4 ± 3.2	0.17 ± 0.01	10.6 ± 1.7	0.74 ± 0.10	26.5 ± 5.1	17.67 ± 0.88	84 ± 12	5
Dolichyl fatty acid ester	0.20 ± 0.08	5.8 ± 1.8	0.19 ± 0.02	10.7 ± 4.4	0.24 ± 0.08	7.2 ± 2.4	0.50 ± 0.11	19.3 ± 5.0	12.87 ± 1.30	79 ± 15	2
Dolichyl dolichoate	0.14 ± 0.07	8.1 ± 2.4	0.27 ± 0.01	16.8 ± 4.9	0.38 ± 0.04	5.4 ± 2.1	0.38 ± 0.08	20.9 ± 1.6	13.93 ± 0.70	81 ± 16	2
5′-Nucleotidase		7.1 ± 2.0	0.23 ± 0.07	40.5 ± 4.0	0.91 ± 0.01	22.5 ± 2.1	1.54 ± 0.01	0.5 ± 0.1	0.33 ± 0.05	87 ± 6	6
Alkaline phenylphosphatase		7.1 ± 0.9	0.23 ± 0.04	33.0 ± 3.1	0.74 ± 0.01	21.6 ± 2.9	1.50 ± 0.01	0.5 ± 0.1	0.33 ± 0.05	86 ± 4	3
[1-3H]dolichol kinase		16.8 ± 3.1	0.55 ± 0.01	24.6 ± 2.6	0.55 ± 0.05	27.9 ± 2.1	1.94 ± 0.06	30.7 ± 2.8	20.47 ± 0.80	74 ± 14	2
[1-3H]dolichylmonophosphate phosphatase		9.1 ± 0.9	0.30 ± 0.02	52.2 ± 6.4	1.18 ± 0.01	38.1 ± 3.8	2.64 ± 0.14	0.6 ± 0.1	0.40 ± 0.01	82 ± 13	2

[a]All values are mean ± standard error of the mean.

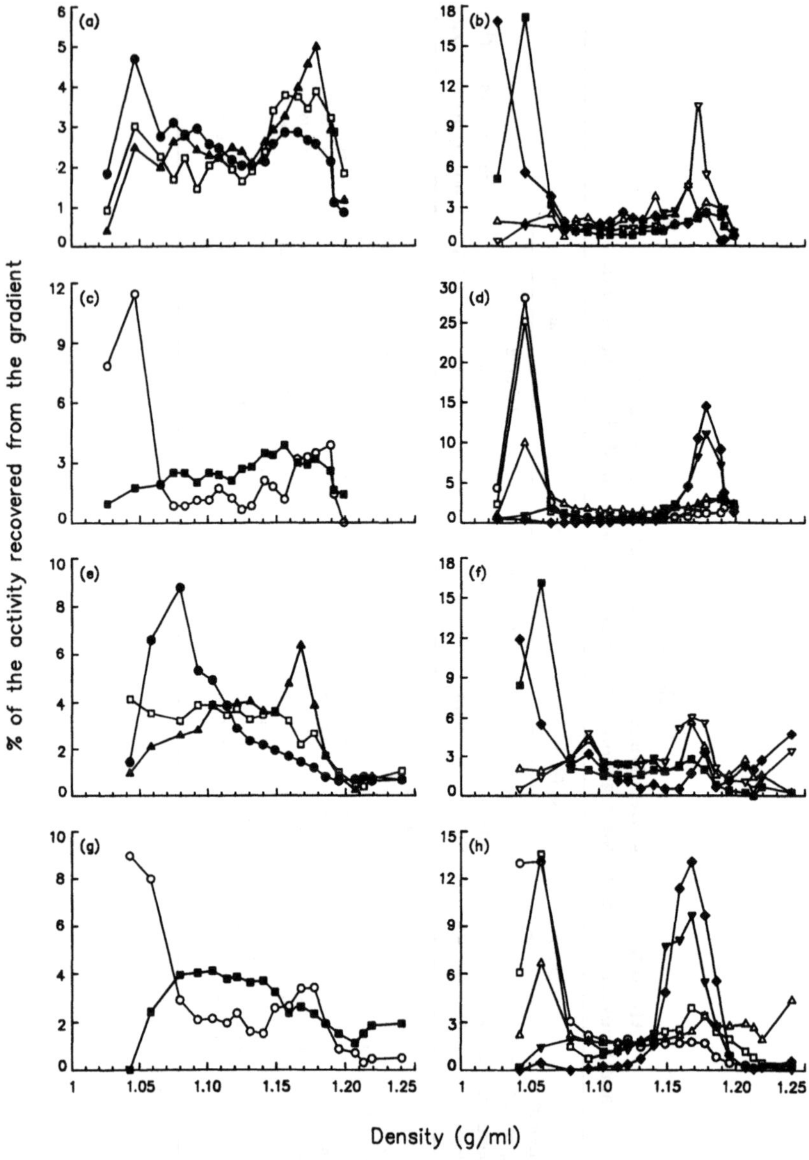

FIGURE 7. Buoyant density equilibration of an M+L fraction from bovine thyroid in a HS zonal rotor. (a–d) 0.25 M sucrose homogenate; (e–h) $NaHCO_3$ homogenate. Panels a and e: ●, 5′-nucleotidase; ▲, alkaline phosphatase; □, glucose-6-phosphatase. Panels b and f: ■, proteins; △, cholesterol; ▽, phospholipids; ◆, dolichol. Panels c and g: ○, [1-^{3}H]dolichol kinase; ■, [1-^{3}H]DMP phosphatase. Panels d and h: ○, acid β-glucosidase; □, β-D-hexosaminidase; △, acid phenylphosphatase; ▼, monoamine oxidase; ◆, cytochrome c oxidase.

TABLE IX
Enrichment Factors of Dolichol and Dolichol-Metabolizing Enzymes in a Purified Nuclear Fraction of Bovine Thyroid

Enzyme or constituent	Enrichment factor[a]
Dolichol	0.8
Dolichyl fatty acid esters	1.1
Dolichyl dolichoate	0.8
[1-^{3}H]dolichol kinase	1.9
[1-^{3}H]dolichyl monophosphate phosphatase	1.9
DNA	41
Cholesterol	1.7
Phospholipids	2.3
5′-Nucleotidase	0.1
Alkaline phosphatase	0.2
NADPH-cytochrome c reductase	1.6
Peroxidase	2.4
Glucose-6-phosphatase	3.2
β-Glucosidase	0.03
β-D-hexosaminidase	0.2
Cytocrome c oxidase	0.3
Monoamine oxidase	0.3

[a]Enrichment factors are calculated with respect to homogenate (Steen *et al.*, 1987, 1988).

out of balance with the incorporation capacity of the disposable intracellular membranes.

Compiling all the foregoing data, there seems to be a growing consensus about the subcellular localization of dolichol. However, some pieces of this "biochemical jigsaw" are still missing, and a number of conflicting findings are hard to reconcile. As already stated in this section, some of the diverging results may arise from an underestimation of the complexity of subcellular methodology [incomplete balances not allowing the calculation of the respective recoveries; single-step procedures; sometimes presentation only of data for a subfractionation of a subcellular fraction; too few marker enzymes tested; the expectation that M+L fractions isolated by different investigators will display different lysosomal enrichment factors, e.g., the incorrectly low dolichol levels reported by Adair and Keller (1982) for an M+L preparation probably may be ascribed to a very limited presence of intact lysosomes]. Furthermore, these subcellar studies have long been hampered by the lack of accurate and sensitive analysis techniques,

Table X
Enrichment of Dolichol and Dolichol-Metabolizing Enzymes in Bovine Thyroid Plasma Membranes

	Enrichment factor[a]		
Enzyme or constituent	L+P	P_2	P_3
[1-^{3}H]dolichol kinase	12.2	14.0	26
[1-^{3}H]dolichyl monophosphate phosphatase	5.5	63.7	50.6
5′-Nucleotidase	5.0	24.2	25.6
Alkaline phosphatase	6.0	27.6	27.4
Glucose-6-phosphatase	5.2	14.6	17.3
Phospholipids[b]		20.8	
Cholesterol[b]		23.2	
Dolichol[b]		1.2	

[a]Enrichment factors are calculated with respect to homogenate.
[b]Cholesterol, phospholipid and dolichol analyses were performed on the pooled P_2 + P_3 fractions (Steen *et al.*, 1987, 1988).

especially for phosphorylated dolichol (Carson and Lennarz, 1981; Chaudhary *et al.*, 1982).

At present, organelle dolichol appears to be mainly confined to the Golgi–ER–lysosomal system. In these subcellular structures dolichol displays a concentration gradient, the lowest values being found in the RER, intermediate values in the Golgi apparatus, and the highest score being reached in the lysosomes. For the phosphorylated derivative, a concentration gradient has been observed, but in the opposite direction. The targeting toward the lysosomal compartment might be dictated by the chemical nature of the fatty acid load. The picture is in agreement with the intracellular flow accompanying the *in vivo* biosynthesis. The presence of dolichol in other subcellular organelles seems to be restricted. Outer and inner mitochondrial dolichol might be related to local N-glycosylation processes (Gateau *et al.*, 1980a,b; Ardail *et al.*, 1985).

A second emerging trend is that dolichol does not behave as the classical membrane lipids. As already referred to, only a limited amount of cellular dolichol seems to be membrane associated. A considerable part is found to be luminal and/or cytosolic. Lipid particles isolated from the lysosomal content contained phospholipids, cholesterol, ubiquinone, and dolichol, but no proteins (Chojnacki and Dallner, 1988). In our hands, upon buoyant density centrifugation of a combined M+L fraction, solubilized dolichol equilibrated at a density differing from that of the protein band at the top of the gradient (Steen *et al.*, 1988). Even so, subfractionation of a bovine thyroid supernatant yielded a so-called S_d fraction enriched in dolichol, but with a rather low protein content. These data strongly suggest the existence of luminal dolichol associated with few or no proteins. On the other hand, the occurrence of supernatant dolichol associated with much higher protein levels has been reported on several occasions (Appelkvist *et al.*, 1985; Van Dessel *et al.*, 1986a). Because of osmotic phenomena and mechanical damage during fractionation procedures, the relative distribution between luminal and cytosolic dolichol cannot be quantitated at present.

7.3. Characterization of Supernatant Dolichol

In the past a number of authors have casually referred to the occurrence of cytosolic, non-membrane-associated supernatant dolichol. In their intubation experiment Chojnacki and Dallner (1983) recovered radioactivity in the particle-free supernatant upon subcellular fractionation. After uptake of the exogenous dolichol by isolated hepatocytes, part of the resorbed dolichol was present in the soluble cytoplasm (Ekström *et al.*, 1982a). On several occasions Carroll's group (Rip and Carroll, 1986; Rip *et al.*, 1985) reported on "cytosolic" dolichol. In rat liver supernatant Appelkvist *et al.* (1985) found dolichol to be present in two cytoplasmic forms, one as a large floating vesicle (top of the gradient) and the other as a complex lipid structure containing phospholipid and cholesterol (midlayer). In the course of their experiment, the top-layer dolichol displayed a continuous increase, whereas the midlayer form showed a transient appearance. Subcellular fractionation of bovine thyroid also points to the existence of supernatant dolichol (Steen *et al.*, 1986, 1988).

In this respect, recently Van Dessel *et al.* (1986b) have studied the dolichol-binding capacity of a bovine liver pH 5.1 supernatant by using the same dextran-coated charcoal assay as applied for the serum dolichol-binding system (Van Dessel *et al.*, 1986a). The binding characteristics were very similar to these found for the human serum binding factor. In serum, however, the binding affinity is appreciably lower ($K_{d\ (serum)} = 6.9 \times 10^{-6}$ M; $K_{d\ (liver)} = 1.8 \times 10^{-11}$), the binding capacity otherwise being higher (serum $\beta_{max} = 432$ pmol/mg of protein; liver $\beta_{max} = 3.6$ pmol/mg protein). The competitive inhibition by dolichyl fatty acyl esters and phosphorylated dolichol suggests that the primary alcoholic function at the α end is not essential for binding in agreement with the results of Futterman *et al.* (1976) on [^{3}H]retinol binding in the presence of retinal or retinoic acid. From the lack of inhibition, it became apparent that the binding of other lipids, e.g., phospholipids and sterols, must proceed at other loci or through other mechanisms than the dolichol binding. The failure of palmitic acid to curtail the [^{3}H]dolichol binding indicated that fatty-acid binding proteins were not involved. This was further sustained by the different elution behavior of labeled [^{3}H]dolichol and [^{14}C]palmitic acid on gel permeation column chromatography after incubation (Figure 8). Trypsin proteolysis and ammonium sulfate precipitation experiments indicated that the proteins were in-

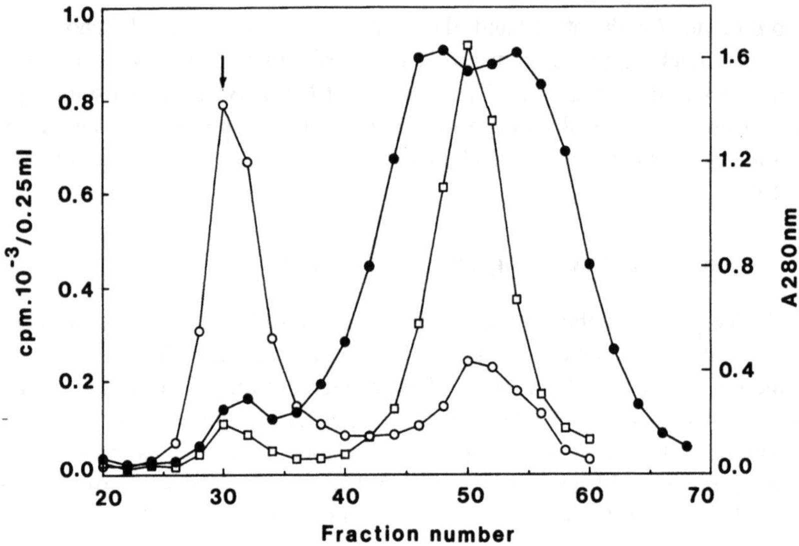

FIGURE 8. Gel filtration of [^{3}H]dolichol- and [^{14}C]palmitate-labeled bovine liver pH 5.1 supernatant. Cytosolic protein (100 mg) was incubated with [^{3}H]dolichol and [^{14}C]palmitate following the standard incubation procedure. The dextran-coated charcoal supernatant was loaded on a Sephacryl S-300 column (2.5 × 60 cm) and eluted with 50 mM Tris hydrochloride (pH 7.4). The flow rate was about 15 ml/hr, and 2-ml fractions were collected. Symbols: ○, [^{3}H]dolichol; □, [^{14}C]palmitate; ●, A_{280}. Arrow indicates the fraction with maximal [^{14}C]cholesterol as deduced from a separate experiment. The elution patterns of blank incubations (incubations without proteins) showed that [^{3}H]dolichol is eluted at fraction 70–75. Yield: proteins, 86%; [^{3}H]dolichol, 91%. Reprinted with permission from *Biochim. Biophys. Acta* (Van Dessel *et al.*, 1986b).

volved in the binding phenomenon. Most of the binding activity (77%) resided in the 40% saturation pellet containing 29% of total proteins in the starting supernatant. Upon chemical analysis of the binding complex, a protein : phospholipid : cholesterol (w/w/w) ratio of 13 : 0.1 : 1 was found, distinctly different from VLDL (0.6 : 1 : 1) but nevertheless endorsing the lipoproteinlike nature of this complex. VLDL and the dolichol-binding complex also displayed a different chromatographic behavior on Bio-Gel A-5m (Figure 9). Exclusion chromatography on Sephacryl S-300 (Figure 8) and gel electrophoresis under nondenaturing conditions revealed a high molecular weight for the binding complex; this was further sustained by dialysis experiments. A high molecular weight has also been reported for the retinol-binding protein in rat liver (Sklan *et al.*, 1982), equine liver (Sklan and Donoghue, 1982a) and chicken liver (Sklan and Donoghue, 1982b) as well as in TM-4 cells (Carson *et al.*, 1984). It was shown that in TM-4 cells the apparent high molecular weight was caused by protein aggregation. In our hands, gel chromatography of the dolichol binding factor at low ionic

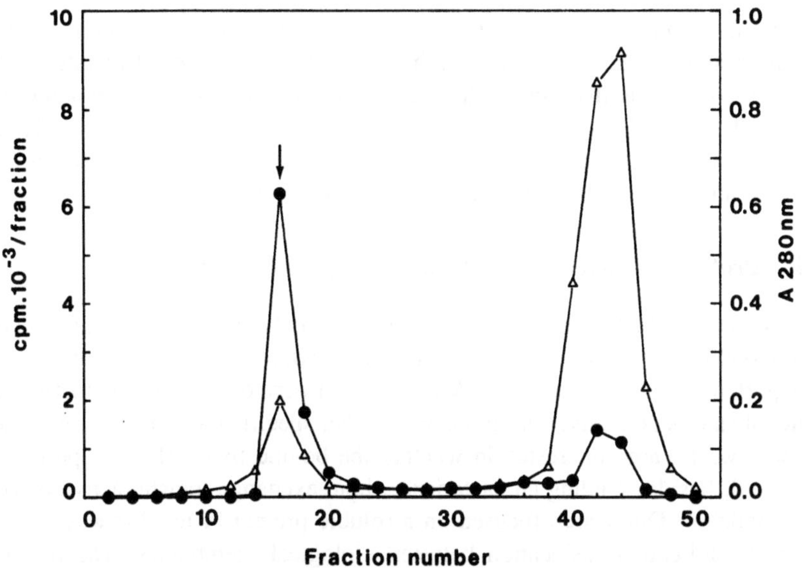

FIGURE 9. Gel filtration on Bio-Gel A-5m. Sephacryl S-300 peak fractions (see Figure 8) were concentrated (Amicon; Minicon macrosolute concentrator), applied to a Bio-Gel A-5m column (1.5 × 100 cm), and eluted with 50 mM Tris hydrochloride (pH 7.4). The flow rate was about 7 ml/hr, and 1-ml fractions were collected. Symbols: Δ, [^{3}H]dolichol; ●, A_{280}. The arrow indicates the fraction where VLDL is eluted. Yield: proteins, 43%; [^{3}H]dolichol, 55%. Reprinted with permission from *Biochim. Biophys. Acta* (Van Dessel *et al.*, 1986b).

strength conditions did not lead to lower-molecular-weight entities. When the binding factor collected from a Bio-Gel A-5m was subjected to sodium dodecyl sulfate (SDS)-gel electrophoresis, the [^{3}H]dolichol radioactivity comigrated with a protein band, which displays a molecular weight around 25,000. At present this low-molecular-weight protein awaits identification and characterization.

The biological significance of supernatant dolichol (luminal or cytosolic) cannot be evaluated at present. First, it could be a storage form of dolichol or a transport form for redistributing dolichol between the site of biosynthesis and the different intracellular membranes and organelles. Second, it could be some sort of salvage dolichol for protecting membrane integrity. Finally, liver luminal dolichol could represent export dolichol either for other nonhepatic tissues or for secretion out of the organism in the bile. With regard to a storage function, it should be recalled that dolichol and cholesterol biosynthesis proceed through pathways with common precursors. One of the key reactions catalyzed by hydroxymethylglutaryl-CoA reductase is "blocked" by dietary cholesterol. Therefore, from simple reasoning one could argue that when cholesterol is supplied in the diet in excess, dolichol synthesis would be severely impaired (Keller *et al.*,

1979; James and Kandutsch, 1980a,b; Tavares *et al.*, 1981; White *et al.*, 1982; Kabakoff and Kandutsch, 1987). On the other hand, a number of literature data are very convincing that both pathways, despite this common branch point, are regulated independently, at least up to a certain level, which remains to be clearly established (Keller *et al.*, 1979; James and Kandutsch, 1980a,b; Potter *et al.*, 1981; White *et al.*, 1981; Dull *et al.*, 1983; Keller, 1986).

7.4. Protein-Mediated Intracellular Transport of Dolichol

In the literature (see Chapters 1–7, 9–10, and 12, in this volume) a platoon of intracellular lipid-specific or nonspecific carrier proteins in charge of net transport or exchange has been described. Furthermore, from the foregoing, dolichol and its derivatives are lipids with rather specific mobilities. As a consequence, we became interested in whether the bovine liver pH 5.1 supernatant displayed, besides binding characteristics, lipid exchange-transfer activities *vis-a-vis* dolichol. Our search focused on a soluble protein factor that was able to stimulate dolichol translocation between biological membranes. The transfer activity, which was monitored by measuring the transfer of [^{3}H]dolichol from small unilamellar PC vesicles to mitochondria, could be partially purified by a four-step procedure including ammonium sulfate precipitation, heat treatment, cation exchange chromatography on CM-Sephadex, and, finally, hydrophobic column chromatography on phenyl-Sepharose (Van Dessel *et al.*, 1987b). With respect to dolichol transfer capacity, a 4000-fold enrichment could be achieved with a protein yield of 12%. Linearity of dolichol transfer was found up to at least 250 μg of protein. Maximal transfer was reached by 15 min, when the transfer activity leveled off. The rate of facilitated dolichol transfer did not vary markedly over the pH range 7–9, whereas virtually no transfer was detectable below pH 5. Including other lipids in the liposomal donor preparations affected the velocity of the protein-mediated transfer of dolichol. Incorporation of negatively charged phospholipids (10 mol% dicetylphosphate or phosphatidic acid) was stimulatory (40–50%). On the other hand, cationic amphiphilic amines caused a decrease in transfer activity (propanolol, 100%; tetracaine, 30%; chloroquine, 10%). Cholesterol (10–15 mol%) had no effect, whereas sphingomyelin (5 mol%) was highly inhibitory. The latter inhibition may result either from a decrease in membrane fluidity or from a straightforward lipid–transfer protein interaction as described by DiCorleto *et al.* (1979) for the phospholipid exchange protein from bovine heart. Other proteins, functioning at the water/lipid interphase, are also affected by the presence of sphingomyelin (Verkleij *et al.*, 1973; Demel *et al.*, 1975). At high salt concentrations (1 M) a decrease in transfer activity was noted, but no special effects of divalent cations relative to monovalent cations were observed. At lower concentrations (0.1 M) there was no salt effect, as further sustained by EDTA treatment. The sulfhydryl reagent *N*-

ethylmaleimide depressed the dolichol transfer activity almost completely, as reported for the human serum dolichol transfer protein. On the other hand, reducing conditions (β-mercaptoethanol, dithiothreitol) had no effect, ruling out essential disulfide bridges.

The identity of the dolichol transfer protein in the bovine liver pH 5.1 supernatant remains to be elucidated. On one hand, the isolated protein preparation capable of dolichol transfer showed no narrow specificity, as other lipids, e.g., phosphatidylinositol (PI), PE, PC, and cholesterol were also transported at respective rates of 2.1, 1.3, 1.2, and 4.2 with respect to dolichol. These transfer rates, however, are different from those reported by Crain and Zilversmit (1980) for the nonspecific lipid transfer protein of liver (ns-LTP). One always must be cautious when comparing mutual transfer activities of protein samples isolated in different laboratories, because of differences in isolation procedures and incubation conditions, e.g., chemical composition and ratio of donor and acceptor membranes. On the other hand, several characteristics of the dolichol transfer activity, e.g., its heat stability (it withstands heating at 90°C), lack of lipid specificity, absence of glycosylation, and action of inhibitors, are very alike, as found for the ns-LTP (Crain and Zilversmit, 1980; Wirtz, 1982) from bovine liver. Moreover, recently Depauw *et al.* (1988) in our laboratory has purified this ns-LTP by following an identical procedure to that described by Crain and Zilversmit (1980) to study the effect of dolichol incorporation on membrane fluidity by measuring steady-state fluorescence anisotropies. Depauw *et al.* (1988) demonstrated that this ns-LTP indeed was able to enhance the dolichol level in native thyroidal membranes by dolichol transport from donor SUV. As such, the transfer action radius of this ns-LTP must be extended to dolichol.

It has been suggested that the different lipids associating with the liver ns-LTP bind to the same recognition site, probably via a common denominator. Therefore, one could imagine that a highly hydrophobic lipid, such as dolichol, also might fit in the lipophilic pocket of this protein(s). On the other hand, one should keep in mind that the molecular dimensions of dolichols and phospholipids differ appreciably.

Some experimental data present evidence that the dolichol molecules within biological membranes are sandwiched between the two monolayers (Valtersson *et al.*, 1985). Such a topographic orientation imposes a "technical" problem with regard to the action mechanism of the transfer protein. In this context, one could assume that the transfer protein must penetrate rather deeply into the hydrophobic core of the membrane to pick up the dolichol molecule. Otherwise, one could hypothesize that the initial interaction of the transfer protein with the membrane induces changes in the membrane characteristics, leading to an exposure of the dolichol molecule at its surface.

The physiological role of lipid transfer proteins has not yet been unambiguously established. An interesting feature of the dolichol transfer factor is that in its presence the activity of the dolichol kinase is increased up to 30% (Van Dessel *et al.*, unpublished results). Such an amplifying effect has also been reported for other transfer proteins and enzyme systems (Poorthuis and Wirtz, 1982; Trzakos and Gaylor, 1983; Kadowaki *et al.*, 1988).

8. CLOSING REMARKS

Dolichol and its derivatives are polyisoprene lipids, which have joined the lipid family rather recently. The most prominent feature of the hydrophobic dolichol molecule is its extreme length (ca. 10 nm), from which the name "dolichol" has been derived (the Greek word δολιχοσ means long). The only function so far established, i.e., a coenzyme role of phosphorylated dolichol in the N-glycosylation process (RER), is somewhat contradictory with the relatively high concentrations present within the cell. Therefore, this isoprenoid and its derivatives are suspected to exert other not yet defined biological functions.

The biosynthesis of dolichol starts at acetyl-CoA and proceeds via an identical pathway to that for cholesterol up to farnesyl pyrophosphate, after which a long-chain polyprenyltransferase catalyzes a variable number (14–17) of *cis* additions from isopentenyl pyrophosphate. The exact chemical nature of the end product, as well as the sequence of the final steps (dephosphorylation/α-saturation), is subject to discussion. Experimental evidence points to at least two different biosynthetic routes for DMP production. Despite the initial common reactions of dolichol and cholesterol biosynthesis, both metabolic pathways seem to be regulated independently. An interesting feature of dolichol is the translocation toward the lysosomal compartments after biosynthesis in the RER. The reasons for this intracellular shift, as well as the exact mechanism(s) of this transfer, are still unknown. Several hypotheses have been proposed but are not yet supported by substantial experimental evidence.

Dolichol and its derivatives do not behave as the classical membrane lipids, i.e., phospholipids and cholesterol. Dolichol changes upon aging and under pathological conditions are not paralleled by similar alterations of phospholipids and cholesterol. Moreover, an appreciable amount of intracellular dolichol is recovered in a non membrane-associated form. The origin and physiological significance of this soluble dolichol is not yet understood. Localization to both lumina and cytosol is likely to occur. The luminal dolichol is found in association with amounts of protein and could result from luminal transport of dolichol to the lysosomes after synthesis in the RER. The cytosolic dolichol, on the other hand, is associated with much higher protein levels. The protein(s) in cytosolic dolichol display(s) transfer activity as well as binding activity. The chemical identity and lipid specificity of this protein, which displays some resemblance with the ns-LTP of bovine liver, remain to be defined. Nonmembranous dolichol may represent either a kind of storage form, a transport form for intracellular distribution from the site of biosynthesis to other organelles and membranes, or alternatively, an export form for distribution among organs or secretion in the bile.

Cells may acquire dolichol from either an exogenous source in the diet or by *de novo* synthesis. The contribution of dietary dolichol is still a matter of dispute. Liver with a high biosynthetic capacity seems to be able to provide the necessary dolichol, which could be even made available for other nonhepatic tissues. Uptake of dietary dolichol or interorgan redistribution both imply that dolichol is transported via the blood circulation. In comparison with cholesterol, dolichol levels in serum have been reported to be very low. Nevertheless, serum lipoproteins have been shown to bind and transport dolichol extracellularly. Dolichol is found in association with VLDL or HDL depending on the experimental conditions. Moreover, dolichol transfer from VLDL to HDL has been demonstrated *in vitro*. The identity of a protein transfer factor, which was recovered from the lipoprotein-free serum fraction, is still under investigation.

Elimination of dolichol and intimately related regulation of "dolichol homeostatis" are unresolved problems. No sound indications are at hand pointing to a drastic degradative pathway. Dolichol seems to be partially excreted via urine and preponderantly via the bile–feces route. Whether these processes are paralleled by chemical conjugation or modification awaits further exploration.

Finally, one can conclude that dolichol metabolism is accompanied by an intensive extracellular and intracellular transport. A number of experimental data support the involvement of transfer and/or exchange proteins. Further research is required to unveil the identity, lipid specificity, and action mechanisms of these vector proteins.

ACKNOWLEDGMENT. The authors are indebted to S. Plompen for technical assistance, to J. Van Sonhoven for producing the photographic illustrations, and to P. Van Dyck for word-processing skills. The original work by the authors was supported by the Belgian NFWO (grant 3.0083.87) and by the Belgian Lotto (grant 9.0013.88).

9. REFERENCES

Abbey, M., Bastiras, S., and Calvert, G., 1985, Immunoprecipitation of lipid transfer protein activity by an antibody against human plasma lipid transfer protein-I, *Biochim. Biophys. Acta* **833**:25–33.

Adair, W., and Cafmeyer, N., 1983, Topography of dolichyl phosphate synthesis in rat liver microsomes. Transbilayer arrangements of dolichol kinase and long chain prenyl transferase, *Biochim. Biophys. Acta* **751**:21–26.

Adair, W., and Cafmeyer, N., 1987a, Cell-cycle dependence of dolichyl phosphate biosynthesis, *Arch. Biochem. Biophys.* **258:**491–497.

Adair, W., and Cafmeyer, N., 1987b, Characterization of the *Saccharomyces cerevisiae* cis-prenyltransferase required for dolichyl phosphate biosynthesis, *Arch. Biochem. Biophys.* **259:**589–596.

Adair, W., and Keller, R., 1982, Dolichol metabolism in rat liver, *J. Biol. Chem.* **257:**8990–8996.

Adair, W., and Robertson, S., 1980, Absolute configuration of dolichol, *Biochem. J.* **189:**441–445.

Adair, W., Cafmeyer, N., and Keller, R., 1984, Solubilization and characterization of the long chain prenyltransferase involved in dolichyl phosphate biosynthesis, *J. Biol. Chem.* **259:**4441–4446.

Adrian, G., and Keenan, R., 1981, Dolichyl phosphate phosphatase from *Tetrahymena pyriformis, Biochim. Biophys. Acta* **663:**637–644.

Allen, C., Keenan, M., and Sack, J., 1976, *Lactobacillus plantarum* undecaprenyl pyrophosphate synthetase : Purification and reaction requirements, *Arch. Biochem. Biophys.* **175:**236–248.

Andersen, J., and Dietschy, J., 1979, Absolute rates of cholesterol synthesis in extrahepatic tissues measured with ^{3}H-labelled water and ^{14}C-labelled substrates, *J. Lipid Res.* **20:**740–752.

Andersson, M., Appelkvist, E., Kristensson, K., and Dallner, G., 1987, Distribution of dolichol and dolichyl phosphate in human brain, *J. Neurochem.* **49:**685–691.

Appelkvist, E., and Dallner, G., 1987, Dolichol biosynthesis in rat liver peroxisomes, *Acta Chem. Scand.* **B41:**73–75.

Appelkvist, E., Chojnacki, T., and Dallner, G., 1985, The presence of dolichol in liver supernatant, *Acta Chem. Scand.* **B39:**72–74.

Ardail, D., Gateau, O., Morelis, R., and Louisot, P., 1985, Glycosyltransferase activities in rat liver mitochondria, *Eur. J. Biochem.* **149:**497–502.

Assmann, G., and Kladetzky, R., 1984, Plasma lipoproteins, in *CRC Handbook of Chromatography: Lipids* (H. Mangold, ed.), Vol. 2, pp. 241–308, CRC Press, Boca Raton, Fla.

Astrand, I.-M., Fries, E., Chojnacki, T., and Dallner, G., 1986, Inhibition of dolichyl phosphate biosynthesis by compactin in cultured rat hepatocytes, *Eur. J. Biochem.* **155:**447–452.

Baba, T., and Allen, C., 1980, Prenyl transferases from *Micrococcus luteus:* Characterization of undecaprenyl pyrophosphate synthetase, *Arch. Biochem. Biophys.* **200:**474–484.

Baba, T., Morris, C., and Allen, C., 1987, Dehydrodolichyl diphosphate synthetase from rat seminiferous tubules, *Arch. Biochem. Biophys.* **252:**440–450.

Barr, R., and Hemming, F., 1972, Polyprenols of *Aspergillus niger, Biochem. J.* **126:**1193–1202.

Barter, P., and Jones, M., 1979, Kinetic studies of the transfer of esterified cholesterol between human low and high density lipoproteins, *Atherosclerosis* **34:**67–74.

Behrens, N., and Leloir, L., 1970, Dolichol monophosphate glucose: An intermediate in glucose transfer, *Proc. Natl. Acad. Sci. U.S.A.,* **66:**153–159.

Bizri, M., Bouhours, J.-F., Bizri, Y., and Got, R., 1986, Separation and quantitative determination of dolichol and dolichyl phosphate in rat and trout liver, *Comp. Biochem. Physiol.* **84B:**565–569.

Bjornson, L., Kayden, H., Miller, E., and Moshell, A., 1976, The transport of α-tocopherol and β-carotene in human blood, *J. Lipid Res.* **17:**343–352.

Boscoboinik, D., Feliz, M., Disalvo, E., and Belocopitow, E., 1985, The effect of dolichol on the permeability properties of phosphatidylcholine bilayers, *Chem. Phys. Lipids* **38:**343–352.

Brown, M., and Goldstein, J., 1980, Multivalent feedback regulation of HMG-CoA reductase, a control mechanism coordinating isoprenoid synthesis and cell growth, *J. Lipid Res.* **21:**505–517.

Burgos, J., and Morton, A., 1962, Intracellular distribution of dolichol in pig liver, *Biochem. J.* **82:**454–456.

Burgos, J., Hemming, F., Pennock, J., and Morton, R., 1963, Dolichol: A naturally-occurring C100 isoprenoid alcohol, *Biochem. J.* **88:**470–482.

Butterworth, P., and Hemming, F., 1968, Intracellular distribution of free and esterified forms of dolichol in pig liver, *Arch. Biochem. Biophys.* **128:**503–508.

Butterworth, P., Draper, H., Hemming, F., Pennock, J., and Morton, R., 1963, Origin of pig liver dolichol, *Biochem. J.* **89:**32P.

Carroll, K. K., Velim, A., and Wood, M. E., 1973, Polyprenols of beef and human pituitary glands, *Lipids* **8:**246–248.

Carson, D., and Lennarz, W., 1981, Relationship of dolichol synthesis to glycoprotein synthesis during embryonic development, *J. Biol. Chem.* **256:**4679–4686.

Carson, D., Rosenberg, L., Blaner, W., Kato, M., and Lennarz, W., 1984, Synthesis and secretion of a novel binding protein for retinol by a cell line derived from Sertoli cells, *J. Biol. Chem.* **259:**3117–3123.

Chajek, T., and Eisenberg, S., 1978, Very low density lipoproteins. Metabolism of phospholipids, cholesterol and apolipoprotein C in the isolated perfused rat heart, *J. Clin. Invest.* **62:**1654–1665.

Chajek, T., and Fielding, C., 1978, Isolation and characterization of a human serum cholesterol ester transfer protein, *Proc. Natl. Acad. Sci. U.S.A.* **75:**3445–3449.

Cham, B., and Knowles, B., 1976, A solvent system for delipidation of plasma and serum without protein precipitation, *J. Lipid Res.* **17:**176–181.

Chaudhary, N., Freeman, D., Rip, J., and Carroll, K., 1982, Direct estimation of dolichyl phosphate in rat liver by high pressure liquid chromatography, *Lipids* **17:**558–560.

Chojnacki, T., and Dallner, G., 1983, The uptake of dietary polyprenols and their modification to active dolichols by the rat liver, *J. Biol. Chem.* **258:**916–922.

Chojnacki, T., and Dallner, G., 1988, The biological role of dolichol, *Biochem. J.* **251:**1–9.

Chojnacki, T., Ekström, T., and Dallner, G., 1980, Phosphorylation and mannosylation of dolichol (C55) in microsomes from hepatocytes preincubated with dolichol, *FEBS Lett.* **113:**218–220.

Chojnacki, T., Palamarczyk, G., Jankowski, W., Krajewka-Rycklik, L., Szkopinska, A., and Vogtman, T., 1984, The enzymic formation of dolichyl phosphate mannose from C-3 enantiomeric dolichyl phosphate, *Biochim. Biophys. Acta* **793:**187–192.

Chytil, F., and Ong, D., 1984, Cellular retinoid-binding proteins, in *The Retinoids* (M. Sporn, A. Roberts, and D. S. Goodman, eds.) Vol. 2, pp. 89–123, Academic Press, New York.

Coleman, R., 1987, Biochemistry of bile secretion, *Biochem. J.* **244:**249–261.

Connelly, L., and Keller, R., 1984, Quantitation of excretion of dolichol and dolichyl phosphate, *Biosci. Rep.* **4:**771–776.

Coolbear, T., and Mookerjea, S., 1981, Effects of turpentine-induced inflammation on the synthesis of dolichol-linked intermediates of N-glycosylation and the phosphorylation of dolichol by CTP-dependent dolichol kinase, *J. Biol. Chem.* **256:**4529–4535.

Crain, R., and Zilversmit, D., 1980, Two nonspecific phospholipid exchange proteins from beef liver. I. Purification and characterization, *Biochemistry* **19:**1433–1439.

Czichi, V., and Lennarz, W., 1977, Localization of the enzyme system for glycosylation of proteins via the lipid-linked pathway in rough endoplasmic reticulum, *J. Biol. Chem.* **252:**7901–7904.

Daleo, G., and PontLezica, R., 1970, Synthesis of dolichol phosphate by a cell-free extract from pea, *FEBS Lett.* **74:**247–250.

Daleo, G., Hopp, H., Romero, P., and PontLezica, R., 1977, Biosynthesis of dolichol phosphate by subcellular fractions from liver, *FEBS Lett.* **81:**411–414.

Dallner, G., and Hemming, F., 1981, Lipid carriers in microsomal membranes, in *Mitochondria and Microsomes* (C. Lee, G. Schatz, and G. Dallner, eds.), pp. 655–680, Addison-Wesley, Reading, Mass.

Dallner, G., Behrens, N., Parodi, A., and Leloir, L., 1972, Subcellular distribution of dolichol phosphate, *FEBS Lett.* **24:**315–317.

Damen, J., Regts, J., and Scherphof, G., 1982, Transfer of ^{14}C phosphatidylcholine between liposomes and human high density lipoprotein, *Biochim. Biophys. Acta* **712:**444–452.

Demel, R., Geurtsma, W., Zwaal, R., and Van Deenen, L. L. M., 1975, Relation between various phospholipase actions on human red cell membranes and interfacial phospholipid pressure in monolayers, *Biochim. Biophys. Acta* **406:**97–107.

Depauw, H., De Wolf, M., Van Dessel, G., Hilderson, H. J., Lagrou, A., and Dierick, W., 1988, Fluidity of thyroid plasma membranes, *Subcell. Biochem.* **13:**193–240.

de Ropp, J., and Troy, F., 1984, Chemical synthesis and ^{2}H NMR investigations of polyisoprenols: Dynamics in model membranes, *Biochemistry* **23:**2691–2695.

de Ropp, J., and Troy, F., 1985, ^{2}H NMR investigation of the organization and dynamics of polyisoprenols in membranes, *J. Biol. Chem.* **260:**15669–15674.

DiCorletto, P., Warach, J., and Zilversmit, D., 1979, Purification and characterization of two phospholipid exchange proteins from bovine heart, *J. Biol. Chem.* **254:**7795–7802.

Dull, B., McCarthy, R., and Kilara, A., 1983, The modulating effect of an inhibitor of cholesterolgenesis present in bovine milk upon the synthesis of cholesterol, dolichol and ubiquinone, *Atherosclerosis* **49:**231–239.

Edlund, C., Chojnacki, T., and Dallner, G., 1983, Separation of dolichol and dolichyl-P in microsomal and lysosomal compartments in rat liver, *Biochem. Biophys. Res. Commun.* **115:**917–923.

Eggens, I., Chojnacki, T., Kenne, L. and Dallner, G., 1983, Separation, quantitation and distribution of dolichol and dolichyl phosphate in rat and human tissues, *Biochim. Biophys. Acta* **751:**355–368.

Ekström, T., and Chojnacki, T., 1982, Dolichol distribution and biosynthesis in hepatocytes, *Acta Chem. Scand.* **36:**411–412.

Ekström, T., Chojnacki, T., and Dallner, G., 1982a, Enrichment of the intracellular dolichol pool in isolated liver cells, *J. Lipid Res.* **23:**972–983.

Ekström, T., Eggens, I., and Dallner, G., 1982b, Biosynthesis of dolichol and dolichyl phosphate in rat hepatocytes *in vivo, FEBS Lett.* **150:**133–136.

Ekström, T., Chojnacki, T., and Dallner, G., 1987, The α-saturation and terminal events in dolichol biosynthesis, *J. Biol. Chem.* **262:**4090–4097.

Elbein, A., 1979, The role of lipid-linked saccharides in the biosynthesis of complex carbohydrates, *Annu. Rev. Plant Physiol.* **30:**239–272.

Elbein, A., 1984, Inhibitors of the biosynthesis and processing of N-linked oligosaccharides, *Crit. Rev. Biochem.* **16:**21–49.

Elbein, A., 1987, Inhibitors of the biosynthesis and processing of N-linked oligosaccharide chains, *Annu. Rev. Biochem.* **56:**497–534.

Elmberger, P., and Engfeldt, P., 1985, Distribution of dolichol in human and rabbit blood, *Acta Chem. Scand.* **39:**323–325.

Elmberger, P., Kalen, A., Appelkvist, E., and Dallner, G., 1987, *In vitro* and *in vivo* synthesis of dolichol and other main mevalonate products in various organs of the rat, *Eur. J. Biochem.* **168:**1–11.

Elmberger, P. G., Engfeldt, P., and Dallner, G., 1988, Presence of dolichol and its derivatives in human blood, *J. Lipid Res.* **29:**1651–1662.

Ericsson, J., Ekström, T., Chojnacki, T., and Dallner, G., 1988, The terminal reactions in polyisoprenoid biosynthesis, *Acta Chem. Scand.* **B42:**202–205.

Freeman, D., Rupar, C., and Carroll, K., 1980, Analysis of dolichol in human tissues by high pressure liquid chromatography, *Lipids* **15:**191–193.

Futterman, S., Saari, J., and Swanson, D., 1976, Retinol and retinoic acid binding proteins in bovine retina: Aspects of binding specificity, *Exp. Eye Res.* **22:**419–424.

Gateau, O., Morelis, R., and Louisot, P., 1980a, Etude des N-acétylglucosaminyl-transférase de la membrane mitochrondriale externe, *Biochimie* **62:**695–703.

Gateau, O., Morelis, R., and Louisot, P., 1980b, Glucosyltransferase activities in liver mitochondria, *Eur. J. Biochem.* **112:**193–201.

Goodman, D. S., 1984, Plasma retinol binding protein, in *The Retinoids* (M. Sporn, A. Roberts, and D. S. Goodman, eds.), Vol. 2, pp. 41–88, Academic Press, New York.

Gough, D., and Hemming, F., 1970, The stereochemistry of betulaprenol biosynthesis, *Biochem. J.* **117:**309–317.

Grange, D., and Adair, W., 1977, Studies on the biosynthesis of dolichyl phosphate: Evidence for the *in vitro* formation of 2,3-dehydrodolichyl phosphate, *Biochem. Biophys. Res. Commun.* **79:**734–739.

Gruner, S., 1985, Intrinsic curvature hypothesis for biomembrane lipid composition: A role for non bilayer lipids, *Proc. Natl. Acad. Sci. U.S.A.* **82:**3665–3669.

Ha, Y., and Barter, P., 1985, Effects of injecting exogenous lipid transfer protein into rats, *Biochim. Biophys. Acta* **833:**203–210.

Hanover, J., and Lennarz, W., 1981, Transmembrane assembly of membrane and secretory glycoproteins, *Arch. Biochem. Biophys.* **211:**1–19.

Hanover, J., and Lennarz, W., 1982, Transmembrane assembly of N-linked glycoproteins, *J. Biol. Chem.* **257:**2787–2794.

Haselbeck, A., and Tanner, W., 1982, Dolichol phosphate-mediated mannosyl transfer through liposomal membranes, *Proc. Natl. Acad. Sci. U.S.A.* **79:**1520–1524.

Haselbeck, A., and Tanner, W., 1984, Further evidence for dolichyl phosphate-mediated glycosyl translocation through membranes, *FEMS Microbiol. Lett.* **21:**305–308.

Hemming, F., 1974, Formation of sugar derivatives of dolichyl phosphate in mammalian systems, in *Biochemistry* (T. Goodwin, ed.), Series 1, Vol. 4, pp. 39–97, Butterworths, London.

Hemming, F., 1982, Control and manipulation of the phosphodolichol pathway of protein N-glycosylation, *Biosci. Rep.* **2:**203–221.

Hemming, F., 1983, Biosynthesis of dolichols and related compounds, in *Biosynthesis of Isoprenoid Compounds* (J. Porter and S. Spurgeon, eds.), pp. 305–354, John Wiley & Sons, Chichester, England.

Hemming, F., 1985, Glycosyl phosphopolyprenols, in *Glycolipids* (H. Wiegandt, ed.), pp. 261–305, Elsevier Biomedical Press, Amsterdam.

Hesler, C., Swenson, T., and Tall, A., 1987, Purification and characterization of a human plasma cholesterol ester transfer protein, *J. Biol. Chem.* **262:**2275–2282.

Hilderson, H. E. L., Steen, L., De Wolf, M., Lagrou, A., Hilderson, H. J., Van Dessel, G., and Dierick, W., 1984, Spectrophotometric determination of dolichol and dolichyl derivatives using the Chugaev color reaction, *Anal. Biochem.* **141:**116–120.

Hirschberg, C., and Snider, M., 1987, Topography of glycosylation in the rough endoplasmic reticulum and Golgi apparatus, *Annu. Rev. Biochem.* **56:**63–87.

Hobara, M., and Takata, K., 1976, Electrophoretic studies of clioquinol binding to human serum proteins, *Biochem. Pharmacol.* **26:**1601–1606.

Hopkins, G., and Barter, P., 1980, Transfer of esterified cholesterol and triglyceride between high density and very low density lipoproteins in *in vitro* studies of rabbits and humans, *Metabolism* **29:**546–550.

Hubbard, S., and Ivatt, R., 1981, Synthesis and processing of asparagine-linked oligosaccharides, *Annu. Rev. Biochem.* **50:**555–583.

Ihm, J., Harmony, J., Ellsworth, J., and Jackson, R., 1980, Simultaneous transfer of cholesteryl ester and phospholipid by protein(s) isolated from human lipoprotein-free plasma, *Biochem. Biophys. Res. Comm.* **93:**1114–1120.

Ihm, J., Ellsworth, J., Chataing, B., and Harmony, J., 1982, Plasma protein-facilitated coupled exchange of phosphatidylcholine and cholesteryl esters in the absence of cholesterol esterification, *J. Biol. Chem.* **257:**4818–4827.

Inoue, A., Yamakawa, J., Yukioka, M., and Morisawa, S., 1983, Filter-binding assay procedure for thyroid hormone receptors, *Anal. Biochem.* **134:**176–183.

Ishii, K., Sagami, H., and Ogura, K., 1986, A novel prenyltransferase from *Paracoccus denitrificans, Biochem. J.* **233:**773–777.

Jakobsson, A., Swiezewska, E., Chojnacki, T., and Dallner, G., 1989, Uptake and modification of dietary polyprenols and dolichols in rat liver, *FEBS Lett.* **255:**32–36.

James, M., and Kandutsch, A., 1979, Inter-relationships between dolichol and sterol synthesis in mammalian cell cultures, *J. Biol. Chem.* **254:**2844–2846.

James, M., and Kandutsch, A., 1980a, Evidence for independent regulation of dolichol and cholesterol synthesis in developing mouse brain, *Biochim. Biophys. Acta* **619:**432–435.

James, M., and Kandutsch, A., 1980b, Regulation of hepatic dolichol synthesis by β-hydroxy-β-methylglutaryl coenzyme A reductase, J. Biol. Chem. 255:8618–8622.

Jamieson, J., 1983, Glycosylation and migration of membrane and secretory proteins, in *Plasma Protein Secretion by the Liver* (H. Glaumann, I. Peters, and C. Redman, eds.), pp. 257–284, Academic Press, New York.

Janas, T., and Tien, H., 1988, Influence of dolichyl phosphate on permeability and stability of bilayer lipid membranes, *Biochim. Biophys. Acta* **939:**624–629.

Javaid, J., Hendricks, K., and Davis, J., 1983, High-affinity binding of tricyclic antidepressants to human plasma involves alpha 1-acid glycoprotein, *Biochem. Pharmacol.* **32:**1149–1153.

Jensen, J., and Schutzbach, J., 1984, Activation of mannosyltransferase II by nonbilayer phospholipids, *Biochemistry* **23:**1115–1119.

Jensen, J., and Schutzbach, J., 1985, Activation of dolichyl-phosphomannose synthase by phospholipids, *Eur. J. Biochem.* **153:**41–48.

Johnston, L., and Neuhaus, F., 1975, Initial membrane reactions in the biosynthesis of peptidoglycan, *Biochemistry* **14:**2754–2759.

Kabakoff, B., and Kandutsch, A., 1987, Depression of hepatic dolichol levels by cholesterol feeding, *J. Lipid Res.* **28:**305–310.

Kadowaki, H., Symanski, L., and Koff, R., 1988, Nonspecific transfer protein in the assay of a membrane-bound enzyme CMP-N-acetylneuraminate: lactosylceramide sialyl transferase, *J. Lipid Res.* **29:**52–62.

Kandutsch, A., Chen, H., and Shown, E., 1977, Binding of 25-hydroxycholesterol and cholesterol to different cytoplasmic proteins, *Proc. Natl. Acad. Sci. U.S.A.* **74:**2500–2503.

Kanegasaki, S., and Wright, A., 1970, Mechanism of polymerization of the *Salmonella* O-antigen: Utilization of lipid-linked intermediates, *Proc. Natl. Acad. Sci. U.S.A.* **67:**951–958.

Keenan, R., Fischer, J., and Kruczek, M., 1977a, The tissue and subcellular distribution of [^{3}H]-dolichol in the rat, *Arch. Biochem. Biophsy.* **179:**634–642.

Keenan, R., Kruczek, M., and Fischer, J., 1977b, The binding of [^{3}H]-dolichol by plasma high density lipoprotein, *Biochim. Biophys. Acta* **486:**1–9.

Keenan, R., Rice, N., and Quock, R., 1977c, The high-performance liquid-chromatrographic analysis of ficaprenol and dolichol, *Biochem. J. 165:*405–408.

Keller, R., 1986, The mechanism and regulation of dolichyl phosphate biosynthesis in rat liver, *J. Biol. Chem.* **261:**12053–12059.

Keller, R., 1987, Dolichol metabolism in the rat, *Trends Biochem. Sci.* **12:**443–445.

Keller, R., and Adair, W., 1977, Microdetermination of dolichol in tissues, *Biochim. Biophys. Acta* **489:**330–336.

Keller, R., Adair, W., and Ness, G., 1979, Studies on the regulation of glycoprotein synthesis, *J. Biol. Chem.* **254:**9966–9969.

Keller, R., Tamkun, J., and Adair, W., 1981, Development of a quantitative assay for tissue levels of dolichyl phosphate, *Biochemistry* **20:**5831–5836.

Keller, R., Jehle, E., and Adair, W., 1982a, The origin of dolichol in the liver of the rat, *J. Biol. Chem.* **257:**8985–8989.

Keller, R., Rottler, G., and Adair, W., 1982b, Separation of dolichols and polyprenols by straight-phase high-performance liquid chromatography, *J. Chromatogr.* **236:**230–233.

Keller, R., Adair, W., Cafmeyer, N., Simion, F., and Fleischer, B., and Fleischer, S., 1986, Characterization of polyisoprenyl phosphatase activity in rat liver, *Arch. Biochem. Biophys.* **249:**207–214.

Keller, R., and Nellis, S., 1986, Quantitation of dolichyl phosphate and dolichol in major organs of the rat as a function of age, *Lipids* **21:**353–355.

Kimura, T., Takeda, K., Kageyu, A., Toda, M., Kurosaki, Y., and Nakayama, T., 1989, Intestinal absorption of dolichol from emulsions and liposomes in rats, *Chem. Pharm. Bull.* **37:**463–466.

Kirk, G., and Gruner, S., 1985, Lyotropic effects of alkanes and headgroup composition on the L_a-H_{II} lipid liquid crystal phase transition: Hydrocarbon packing versus intrinsic curvature, *J. Phys.* **46:**51–59.

Kukuruzinska, M., Bergh, M., and Jackson, B., 1987, Protein glycosylation in yeast, *Annu. Rev. Biochem.* **56:**915–944.

Kwok, B., Shen, B., and Dawson, G., 1981, Glycosphingolipid-high density lipoprotein-3 interaction. I. Transfer of glycosphingolipid from phosphatidylcholine vesicles to high density lipoprotein-3. *J. Biol. Chem.* **256:**9698–9704.

Lai, C., and Schutzbach, J., 1984, Dolichol induces membrane leakage of liposomes composed of phosphatidylethanolamine and phosphatidylcholine, *FEBS Lett.* **169:**279–282.

Lai, C., and Schutzbach, J., 1986, Localization of dolichol in phospholipid membranes, *FEBS Lett.* **203:**153–156.

Lakshmanan, M., and Veech, R., 1977, Measurements of rate of rat liver sterol synthesis *in vivo* using tritiated water, *J. Biol. Chem.* **252:**4667–4673.

Lennarz, W., 1975, Lipid linked sugars in glycoprotein synthesis, *Science* **138:**986–991.

Lennarz, W., 1983, Glycoprotein synthesis and embryonic development, *Crit. Rev. Biochem.* **14:**257–272.

Lennarz, W., 1985, Regulation of glycoprotein synthesis in the developing sea-urchin, *Trends Biochem. Sci.* **10:**248–251.

Lennarz, W., 1987, Protein glycosylation in the endoplasmic reticulum: Current topological issues, *Biochemistry* **26:**7205–7210.

Love, C., and Cowan, S., Laing, L., and Leake, R., 1983, Stability of the human nuclear estrogen receptor: Influence of the temperature and ion strength, *J. Endocrinol.* **99:**423–433.

Löw, P., Peterson, E., Mizuno, M., Tagikawa, T., Chojnacki, T., and Dallner, G., 1985, Reaction of optically active S- and R-forms of dolichyl phosphates with activated sugars, *Biochem. Biophys. Res. Commun.* **130:**460–466.

Lucas, J., and Waechter, C., 1976, Polyisoprenoid glycolipids involved in glycoprotein synthesis, *Mol. Cell. Biochem.* **11:**67–77.

Marcel, Y., Vezina, C., Emond, D., Verdery, R., and Milne, R., 1981, High density lipoprotein subfractions isolated by heparin-sepharose affinity chromatography and their role in cholesteryl ester transfer to very low density lipoproteins, *J. Lipid Res.* **22:**1198–1205.

Martin, H., and Thorne, K., 1974, Synthesis of radioactive dolichol from [4S-^{3}H]mevalonate in the regenerating rat liver, *Biochem. J.* **138:**277–280.

McCloskey, M., and Troy, F., 1980a, Paramagnetic isoprenoid carrier lipids. 1. Chemical synthesis and incorporation into model membranes, *Biochemistry,* **19:**2056–2060.

McCloskey, M., and Troy, F., 1980b, Paramagnetic isoprenoid carrier lipids. 2. Dispersion and dynamics in lipid membranes, *Biochemistry* **19:**2061–2066.

Mills, J., and Adamany, A., 1981, The role of phosphorylated dolichols in membrane glycoprotein biosynthesis: Relation to cholesterol biosynthesis, *Int. Rev. Cytol.* **73:**103–147.

Monti, J., Christian, S., and Schutzbach, J., 1987, Effects of dolichol on membrane permeability, *Biochim. Biophys. Acta* **905:**133–142.

Morris, G., and Pullarkat, R., 1987, A micro method for the estimation of blood dolichol, *Lipids* **22:**58–60.

Murgolo, N., Patel, A., Stivala, S., and Wong, T., 1989, The conformation of dolichol, *Biochemistry* **28:**253–260.

Ng, Y., Palo, N., Haltia, M., and Wolfe, L., 1983, High levels of brain dolichols in neuronal ceroid-lipofuscinosis, *J. Neurochem.* **40:**1465–1473.

Nilsen, O., and Jacobson, S., 1975, The influence of salts and differences in protein isolated procedure on the binding of quinidine to human serum albumin, *Biochem. Pharmacol.* **24:**995–998.

Nyquist, S., and Holt, S., 1985, Rat testicular dolichols: Distribution and accumulation with age, *Biol. Reprod.* **34:**89–95.

Palamarczyk, G., and Butters, T., 1982, Uptake of dolichol into cultured cells, *FEBS Lett.* **143:**241–246.

Palmer, D., Andersson, M., and Jolly, R., 1984, Separation of some neutral lipids by normal phase high performance liquid-chromatography on a cyanopropyl column: Ubiquinone, dolichol and cholesterol levels in sheep liver, *Anal. Biochem.* **140:**315–319.

Parodi, A., and Leloir, L., 1979, The role of lipid intermediates in the glycosylation of proteins in the eukaryotic cell, *Biochim. Biophys. Acta* **599:**1–37.

Pattnaik, N., Montes, A., Hughes, L., and Zilversmit, D., 1978, Cholesteryl ester exchange protein in human plasma. Isolation and characterization, *Biochim. Biophys. Acta* **530:**428–438.

Patton, B., and Poulos, A., 1987, Normal dolichol concentration in urine sediments from four patients with neuronal ceroid lipofuscinosis (Batten's disease), *J. Inherited Metab. Dis.* **10:**28–32.

Pennock, J., Hemming, F., and Morton, R., 1960, Dolichol: A naturally occurring isoprenoid alcohol, *Nature* **186:**470–472.

Perez, M., and Hirschberg, C., 1985, Translocation of UDP-N-acetylglucosamine into vesicles derived from rat liver rough endoplasmic reticulum and Golgi apparatus, *J. Biol. Chem.* **260:**4671–4678.

Perez, M., and Hirschberg, C., 1986, Topography of glycosylation reactions in the rough endoplasmic reticulum membrane, *J. Biol. Chem.* **261:**6822–6830.

Poorthuis, B., and Wirtz, K., 1982, Increased cholesterol esterification in rat liver microsomes by purified non-specific phospholipid transfer protein, *Biochim. Biophys. Acta* **710:**99–105.

Potter, J., Milette, C., James, M., and Kandutsch, A., 1981, Elevated cholesterol and dolichol synthesis in mouse pachytene spermatocytes, *J. Biol. Chem.* **256:**7150–7154.

Powis, G., 1974, A study of the interaction of tetracycline with human serum lipoproteins and albumin, *J. Pharm. Pharmacol.* **26:**113–118.

Pownall, H., Hickson, D., Gotto, A., Jr., and Massey, J., 1982, Kinetics of spontaneous and plasma-stimulated sphingomyelin transfer, *Biochim. Biophys. Acta* **712:**169–176.

Presper, K., and Heath, G., 1983, Glycosylated lipid intermediates involved in glycoprotein biosynthesis, in *The Enzymes* Vol. XVI (P. Boyer, ed.), pp. 449–488, Academic Press, New York.

Pullarkat, R., and Raguthu, S., 1985, Elevated urinary dolichol levels in chronic alcoholics, *Clin. Exp. Res.* **9:**28–30.

Pullarkat, R., and Reha, H., 1982, Accumulation of dolichols in brains of elderly, *J. Biol. Chem.* **257:**5991–5993.

Pullarkat, R., Reha, H., and Pullarkat, P., 1984, Age-associated increase in free dolichol levels in mice, *Biochim. Biophys. Acta* **793:**494–496.

Quig, D., and Zilversmit, D., 1986, Disappearance and effects of exogenous lipid transfer activity in rats, *Biochim. Biophys. Acta* **879:**171–178.

Quig, D., and Zilversmit, D., 1988, Dissociation between cholesterol secretion and plasma lipid transfer activity in rabbits, *FASEB J.* **2:**2712–2716.

Rajaram, O., White, G., and Barter, P., 1980, Partial purification and characterization of a triacylglycerol transfer protein from rabbit serum, *Biochim. Biophys. Acta* **617:**383–392.

Rip, J., and Carroll, K., 1982, The submicrosomal distribution of dolichol and dolichol phosphokinase activity in rat liver, *Can. J. Biochem.* **60:**36–41.

Rip, J., and Carroll, K., 1985, Distribution, metabolism and excretion of [1-^{14}C]dolichol injected intravenously into rats, *Biochem. J.* **227:**705–710.

Rip, J., and Carroll, K., 1986, Long-term distribution and metabolism of [1-^{14}C]dolichol injected into rats, *Biochem. Cell Biol.* **64:**886–890.

Rip, J., Rupar, C., Chaudhary, N., and Carroll, K., 1981, Localization of a dolichyl phosphate phosphatase in plasma membranes of rat liver, *J. Biol. Chem.* **256:**1929–1934.

Rip, J., Chaudhary, N., and Carroll, K., 1983a, Distribution and metabolism of dolichol and dolichyl phosphate in rat liver, *Can. J. Biochem. Cell Biol.* **61:**1025–1031.

Rip, J., Chaudhary, N., and Carroll, K., 1983b, The submicrosomal distribution of dolichyl phosphate and dolichyl phosphate phosphatase in rat liver, *J. Biol. Chem.* **258:**14926–14930.

Rip, J., Rupar, C., Ravi, K., and Carroll, K., 1985, Distribution, metabolism and function of dolichol and polyprenols, *Prog. Lipid Res.* **24:**269–309.

Robyt, J., 1979, Mechanism involved in the biosynthesis of polysaccharides, *Trends Biochem. Sci.* **4:**47–49.

Rossignol, D., Lennarz, W., and Waechter, C., 1981, Induction of phosphorylation of dolichol during embryonic development of the sea urchin, *J. Biol. Chem.* **256:**10538–10542.

Roth, J., 1987, Subcellular organization of glycosylation in mammalian cells, *Biochim. Biophys. Acta* **906:**405–436.

Rupar, C., and Carroll, K., 1978, Occurrence of dolichol in human tissues, *Lipids* **13:**291–293.

Sagami, H., Ogura, K., and Seto, S., 1978, A new prenyltranferase from *Micrococcus lysodeikticus, Biochem. Biophys. Res. Commun.* **85:**572–578.

Sakahihara, Y., and Volpe, J., 1984, Dolichol deposition in developing mammalian brain: Content of free and fatty-acylated dolichol and proportion of specific isoprenologues, *Dev. Brain Res.* **14:**255–262.

Scanu, A., Edelstein, C., and Shen, B., 1982, Lipid-protein interactions in plasma lipoproteins. Model: High density lipoproteins, in *Lipid-protein interactions* (P. Jost and O. Griffith, eds.), pp. 259–316, John Wiley & Sons, New York.

Schroeder, F., Gorka, C., Williamson, L., and Wood, W., 1987, The influence of dolichols on the fluidity of mouse synaptic plasma membranes, *Biochim. Biophys. Acta* **902:**385–393.

Sharon, N., and Lis, H., 1981, Glycoproteins: Research booming on long-ignored ubiquitous compounds, *Chem. Eng. News* **59:**21–44.

Sklan, D., and Donoghue, S., 1982a, Association of acylglyceride and retinylpalmitate hydrolase activities with zinc and copper metalloproteins in a high molecular weight lipid-protein aggregate fraction from chicken cytosol, *Biochim. Biophys. Acta* **711:**532–538.

Sklan, D., and Donoghue, S., 1982b, Serum and intracellular retinol transport, *Br. J. Nutr.* **47:**273–280.

Sklan, D., Blaner, W., Adachi, N., Smith, J., and Goodman, D. S., 1982, Association of a cellular retinol-binding protein and several lipid hydrolase activities with a vitamin A-containing high molecular weight lipid-protein aggregate from rat liver cytosol, *Arch. Biochem. Biophys.* **214:**35–44.

Snider, M., and Robbins, P., 1982, Transmembrane organisation of protein glycosylation, *J. Biol. Chem.* **257:**6796–6801.

Snider, M., and Rogers, O., 1984, Transmembrane movement of oligosaccharide-lipids during glycoprotein synthesis, *Cell* **36:**753–761.

Spiro, M., and Spiro, G., 1985, Effect of anion-specific inhibitors on the utilization of sugar nucleotides for N-linked carbohydrate unit assembly by the thyroid endoplasmic reticulum vesicles, *J. Biol. Chem.* **260:**5808–5815.

Staneloni, R., and Leloir, L., 1982, The biosynthetic pathway of the asparagine-linked oligosaccharides of glycoproteins, *Crit. Rev. Biochem.* **12:**289–326.

Steen, L., 1985, Dolichol, dolichyl derivatives, dolichol kinase and dolichyl monophosphate phosphohydrolase in bovine thyroid, Ph.D. Thesis (in Dutch), University of Antwerp, Antwerp, Belgium.

Steen, L., Van Dessel, G., De Wolf, M., Lagrou, A., Hilderson, H. J., De Keukeleire, D., Pinkse, F., Fokkens, R., and Dierick, W., 1984, Identification and characterization of dolichyldolichoate, a novel isoprenoic derivative in bovine thyroid, *Biochim. Biophys. Acta* **796:**294–303.

Steen, L., Van Dessel, G., De Wolf, M., Hilderson, H. J., Lagrou, A., and Dierick, W., 1986, On the subcellular localization of dolichol, dolichol kinase and dolichyl monophosphate phosphohydrolase in bovine thyroid, *Arch. Int. Physiol. Biochim.* **94:**B93.

Steen, L., Van Dessel, G., De Wolf, M., Lagrou, A., Hilderson, H. J., and Dierick, W., 1988, Dolichol and dolichol metabolizing enzymes in bovine thyroid: Subcellular distribution and occurrence of non-membrane associated dolichol, *Life Sci. Adv.*, in press.

Struck, D., and Lennarz, W., 1980, The function of saccharide lipids in synthesis of glycoproteins, in *The Biochemistry of Glycoproteins and Proteoglycans* (W. Lennarz, ed.), pp. 35–83, Plenum Press, New York.

Sunamoto, J., Iwamoto, K., Tezuka, T., Kadosaki, K., and Kondo, H., 1983, Liposomal membranes. XIV. Fusion of liposomal membranes induced by polyisoprenoids as monitored by fluorescence quenching methods, *Chem. Phys. Lipids* **33:**127–133.

Suzuki, S., Mori, F., Tagikawa, T., Ibata, K., Ninagawa, Y., Nishida, T., Mizuno, M., and Tanaka, Y., 1983, Synthesis of mammalian dolichols from plant polyprenols, *Tetrahedron Lett.* **24:**5103–5106.

Takahashi, I., and Ogura, K., 1982, Prenyltransferases of *Bacillus subtilis:* Undecaprenyl pyrophosphate synthetase and geranylgeranyl pyrophosphate synthetase, *J. Biochem.* **92:**1527–1537.

Tall, A., 1986, Plasma lipid transfer proteins, *J. Lipid Res.* **27:**361–367.

Tavares, I., Johnson, N., and Hemming, F., 1977, A sensitive quantitative assay method for dolichols, cholesterol and ubiquinone using high-pressure liquid chromatography, *Biochem. Soc. Trans.* **5:**1771–1773.

Tavares, I., Coolbear, R., and Hemming, F., 1981, Increased hepatic dolichol and dolichyl phosphate-mediated glycosylation in rats fed cholesterol, *Arch. Biochem. Biophys.* **207:**427–437.

Tollbom, O., and Dallner, G., 1986, Dolichol and dolichyl phosphate in human tissues, *Br. J. Exp. Pathol.* **67:**757–764.

Tollbom, O., Valtersson, C., Chojnacki, T., and Dallner, G., 1988, Esterification of dolichol in rat liver, *J. Biol. Chem.* **263:**1347–1352.

Trzakos, J., and Gaylor, J., 1983, Cytosolic modulations of activities of microsomal enzymes of cholesterol biosynthesis. Purification and characterization of a non-specific lipid-transfer protein, *Biochim. Biophys. Acta* **751:**52–65.

Turpeinen, U., 1986, Liquid-chromatographic determination of dolichols in urine, *Clin. Chem.* **32:**2026–2029.

Valtersson, C., Van Duijn, G., Verkleij, A., Chojnacki, T., de Kruijff, B., and Dallner, G., 1985, The influence of dolichol, dolichyl esters and dolichyl phosphate on phospholipid polymorphism and fluidity in model membranes, *J. Biol. Chem.* **260:**2742–2751.

Valtersson, C., Filipsson, L., and Dallner, G., 1986, Localization of phosphatidylethanolamine in microsomal membranes and regulation of its distribution by the fatty acid composition, *J. Lipid Res.* **27**:731–741.

Van Dessel, G., Lagrou, A., Hilderson, H. J., and Dierick, W., 1977, Subfractionation of a supernatant from bovine thyroid, *Arch. Int. Physiol. Biochim.* **85**:1022–1023.

Van Dessel, G., Lagrou, A., Hilderson, H. J. Dommisse, R., Esmans, E., and Dierick, W., 1979, Isolation and identification of polyprenols from bovine thyroid gland, *Biochim. Biophys. Acta* **573**:296–300.

Van Dessel, G., De Wolf, M., Lagrou, A., Hilderson, H. J., and Dierick, W., 1986a, On the binding of dolichol by human serum, *Biochim. Biophys. Acta* **787**:346–352.

Van Dessel, G., De Wolf, M., Lagrou, A., Hilderson, H. J., and Dierick, W., 1986b, On the binding of dolichol by bovine liver supernatant, *Biochim. Biophys. Acta* **878**:338–345.

Van Dessel, G., De Wolf, M., Lagrou, A., Hilderson, H. J., and Dierick, W., 1986c, Dolichol biosynthesis in bovine thyroid, *Arch. Int. Physiol. Biochim.* **94**:B55.

Van Dessel, G., De Wolf, M., Lagrou, A., Hilderson, H. J., and Dierick, W., 1987a, Effect of dolichol and dolichyl monophosphate on the stability of small unilamellar vesicles, *Proc. IX Int. Symp. Glycoconj.* (Abstr. G141) Lille, France.

Van Dessel, G., De Wolf, M., Lagrou, A., Hilderson, H. J., and Dierick, W., 1987b, On the occurrence of a dolichol transferring protein in human serum, *Proceedings 28th ICBL,* (Abstr. C31), ICBL Aberdeen, Scotland.

Van Dessel, G., Depauw, H., De Wolf, M., Lagrou, A., Hilderson, H. J., and Dierick, W., 1987c, Protein-mediated dolichol transfer in bovine liver, *Arch. Int. Physiol. Biochim.* **95**:B52.

Van Duijn, G., Valtersson, C., Chojnacki, T., Verkleij, A., Dallner, G., and de Kruijff, B., 1986, Dolichyl phosphate induces non-bilayer structures, vesicles fusion and transbilayer movements of lipids: A model membrane study, *Biochim. Biophys. Acta* **861**:211–223.

van Jaarsveld, P., Edelhoch, H., Goodman, D. S., and Robbins, H., 1973, The interaction of human plasma retinol binding protein with pre-albumin, *J. Biol. Chem.* **248**:4698–4705.

Verkleij, A., Zwaal, R., Roelofson, B., Comfurius, P., Kastelyn, D., and Van Deenen, L. L. M., 1973, Asymmetric distribution of phospholipids in human red-cell membranes, *Biochim. Biophys. Acta* **323**:178–186.

Via, D., Massey, J., Vignale, S., Kundu, S., Marcus, D., Pownall, H., and Gotto, Jr., A, 1985, Spontaneous and plasma factor-mediated transfer of pyrenyl cerebrosides between model and native lipoproteins, *Biochim. Biophys. Acta* **837**:27–34.

Vigo, C., and Adair, W., 1982, *In vivo* biosynthesis of the saturated isoprene unit of dolichyl phosphate, *Biosci. Rep.* **2**:835–840.

Vigo, C., Grossman, S., and Drost-Hansen, W., 1984, Interaction of dolichol and dolichyl phosphate with phospholipids, *Biochim. Biophys. Acta* **774**:221–226.

Volpe, J., Sakakihara, Y., and Rust, R., 1987, Dolicholkinase and the regulation of dolichyl phosphate levels in developing brain, *Dev. Brain Res.* **31**:193–200.

Waechter, C., and Lennarz, W., 1976, The role of polyprenol-linked sugars in glycoprotein synthesis, *Annu. Rev. Biochem.* **45**:95–112.

Walter, P., Gilmore, R., and Blobel, G., 1984, Protein translocation across the endoplasmic reticulum, *Cell* **38**:5–8.

Wellner, R., and Lucas, J., 1979, Evidence for a compound with the properties of 2,3-dehydrodolichyl phosphate, *FEBS Lett.* **104**:379–383.

Weppner, W., and Neuhaus, F., 1978, Biosynthesis of peptidoglycan, *J. Biol. Chem.* **253**:472–478.

White, D., Middleton, B., Dawson, S., Bradshaw, J., Clegg, R., Hemming, F., and Bell, G., 1982, Regulation of dolichol and of cholesterol biosynthesis in cholesterol-fed rabbits, *Arch. Biochem. Biophys.* **208**:30–36.

Wirtz, K., 1982, Phospholipid transfer proteins, in *Lipid-Protein Interactions* (P. Jost and O. Griffith, eds.), pp. 151–231, John Wiley & Sons, New York.

Wong, T., and Lennarz, W., 1982, The site of biosynthesis and intracellular deposit of dolichol in rat liver, *J. Biol. Chem.* **257**:6619–6624.

Wong, T., Decker, G., and Lennarz, W., 1982, Localization of dolichol in the lysosomal fraction of rat liver, *J. Biol. Chem.* **257**:6614–6618.

Wood, W., Gorka, C., Williamson, L., Strong, R., Sun, A., Sun, G., and Schroeder, F., 1986, Dolichol alters dynamic and static properties of mouse synaptosomal plasma membranes, *FEBS Lett.* **205**:25–28.

Yamada, K., Yokohama, H., Abe, S., Katayama, K., and Sato, T., 1985, High-performance liquid chromatographic methods for the determination of dolichols in tissues and plasma, *Anal. Biochem.* **150**:26–31.

Yamada, K., Abe, S., Suzuki, T., Katayama, K., and Sato, T., 1986, A high-performance liquid chromatographic method for the determination of dolichyl phosphate in tissues, *Anal. Biochem.* **156**:380–385.

Zilversmit, D., Hughes, L., and Balmer, J., 1980, Stimulation of cholesterol ester exchange by lipoprotein-free rabbit plasma, *Biochim. Biophys. Acta* **409**:393–398.

Chapter 9

Phosphoplipid Transport in Microorganisms

F. Paltauf and G. Daum

1. INTRODUCTION

Efforts to elucidate mechanisms of intracellular lipid traffic have concentrated mainly on the isolation and characterization of phospholipid transfer proteins from mammalian tissues. These proteins supposedly participate in the transfer of phospholipids from their sites of synthesis to those subcellular membranes that are incapable of synthesizing all constituents of their lipid matrix. The existence of other mechanisms of intermembrane phospholipid transport is conceivable, but cannot easily be tested in complex biological systems such as experimental animals. As an alternative, unicellular organisms can be used to investigate problems of intracellular lipid transfer *in vivo* or in cell-free preparations. Prokaryotes offer the advantage of a relatively simple organization of subcellular membrane structures. Typically, bacteria have only two types of membranes between which lipid transfer can occur: the cytoplasmic membrane and the outer membrane. Photosynthetic bacteria, e.g., *Rhodopseudomonas sphaeroides,* when grown under light, contain an additional inducible membrane system, the intracytoplasmic membrane, where the photosynthetic machinery is located.

Eukaryotic microorganisms, such as yeasts or, to a lesser extent, filamentous fungi, have been well characterized with respect to the genetics and bio-

F. Paltauf and G. Daum Institut für Biochemie und Lebensmittelchemie, Technische Universität Graz, A-8010 Graz, Austria.

chemistry of lipid synthesis and membrane biogenesis. Their subcellular structure is similar to that of higher eukaryotes, which makes them ideal objects for unraveling mechanisms of phospholipid transport, including phospholipid transfer protein function, which are probably common to all eukaryotic cells. The advantage of studying these processes in microorganisms lies in the ease of physiological and genetic manipulation of membrane structure, composition, and biogenesis.

In this review the present knowledge about intermembrane phospholipid transfer in prokaryotes and in the yeast *Saccharomyces cerevisiae* will be summarized. The isolation and characterization of phospholipid transfer proteins from prokaryotic and eukaryotic microorganisms will be described, and the relevance of the findings to an understanding of the physiological role of phospholipid transfer proteins will be discussed. A review on phospholipid transfer proteins in microorganisms has been published by Tai and Kaplan (1985b).

2. PHOSPHOLIPID TRANSPORT IN PROKARYOTES

2.1. Sites of Phospholipid Biosynthesis

In contrast to eukaryotic cells, prokaryotes are characterized by their paucity of subcellular membranes. The only membrane structures are the inner (cytoplasmic) and outer membranes and the intracytoplasmic membrane in photoheterotrophic bacteria, e.g., *Rps. sphaeroides*.

Most phospholipid synthesizing enzymes of prokaryotes are located in the cytoplasmic membrane. CDP-diacylglycerol synthase, phosphatidylglycerolphosphate synthase, and also phosphatidylserine decarboxylase of *Rps. sphaeroides* grown heterotrophically cofractionate with this membrane (Cain *et al.*, 1984). During heterotrophic growth, substantial quantities of phosphatidylserine decarboxylase are also found in the intracytoplasmic membrane. Artificial distribution of phosphatidylserine decarboxylase to this membrane as a result of self-aggregation of the enzyme (Dowhan *et al.*, 1974) could not be excluded.

Surprisingly phosphatidylserine synthase and phosphatidylethanolamine methyltransferase activity were found in the soluble fraction of *Rps. sphaeroides*. These enzymes were obviously not tightly membrane associated and were therefore released during cell disruption and fractionation. The authors suggested that availability of the substrate of phosphatidylserine synthase, CDP-diacylglycerol, might be the limiting factor for phosphatidylserine synthase activity in vivo. Since CDP-diacylglycerol is synthesized in the cytoplasmic membrane and free pools of this intermediate do not exist in *Rps. sphaeroides,* the only site of phosphatidylserine synthesis can be the cytoplasmic membrane.

In vivo studies had shown that phosphatidylcholine synthesis during the cell cycle occurred in the intracytoplasmic membrane (Cain *et al.*, 1981). This observation and the finding that phosphatidylethanolamine methyltransferase was not associated with any membrane fraction following cell disruption led to the conclusion that this enzyme, too, may associate with membranes through association with its substrate.

In *Escherichia coli* phosphatidylserine synthase was found to be associated with ribosomes (Raetz and Kennedy, 1972), although redistribution of this enzyme during disruption of the cells could not be ruled out. The bacterial outer membrane is not a site of phospholipid synthesis, and the capacity of the intracytoplasmic membrane to synthesize its own lipids is very limited. As a result of this distribution of phospholipid-synthesizing enzymes, these two membranes rely to a large extent on the supply of membrane phospholipids that are synthesized by enzymes associated with the cytoplasmic membrane.

2.2. Phospholipid Transport *In Vivo*

Jones and Osborn (1977a) developed a system to study the uptake of lipids from extracellular vesicles into bacterial membranes. This translocation seems to be vectorial rather than an exchange and requires Ca^{2+} or spermine. The characteristics of this process are consistent with a direct fusion of vesicles with the outer cellular membrane. The translocation process appears to be nonspecific with respect to the type of phospholipid: All *Salmonella* phospholipids as well as the foreign phospholipid, phosphatidylcholine, were transferred to the inner membrane to a similar extent. By using this system, the bidirectional translocation of phospholipids between the outer and the inner membrane of *Salmonella typhimurium* was studied (Jones and Osborn, 1977b). Decarboxylation of vesicle-derived radioactive phosphatidylserine by inner-membrane-located phosphatidylserine decarboxylase served as an indication of phosphatidylserine transport across the outer membrane. Phospholipid translocation was rapid and extensive and led to an equilibrium of phospholipids between the two membranes. Jones and Osborn (1977b) suggested that diffusional flow across membrane contact sites might be the mechanism of this intermembrane translocation. This notion was supported by Langley *et al.* (1982), who observed that in *E. coli* movement of phosphatidylserine from the cytoplasmic membrane, where it is synthesized, to the outer membrane occurs with a half-life of 12–13 min. Complete conversion of radioactively labeled phosphatidylserine to phosphatidylethanolamine by inner-membrane phosphatidylserine decarboxylase indicates that intermembrane movement of phosphatidylserine is bidirectional. Lloyd (1987) studied the translocation of phospholipids between membranes of *Micrococcus cryophilus*. When cells were labeled with [^{14}C]acetate, a time-dependent shift of radioactivity from the cytoplasmic membrane to the outer membrane was observed. In contrast, transfer of lipids labeled with [^{14}C] palmitic acid could not be measured. This was attributed to the loss of outer membrane lipids, either by blebbing of outer membrane vesicles or by degradation processes catalyzed by phospholipases of the outer membrane followed by reutilization of ^{14}C-fatty acid in the cytoplasmic membrane. Palmitic acid may enter a pool of phospholipids which is subject to rapid phospholipid degradation. [^{14}C]Acetate-labeled phospholipids and wax esters were transferred between membranes to a similar extent. Lloyd (1987) concluded that a rather nonspecific process should govern this translocation. Migration of lipids through zones of adhesion between the outer membrane and the cytoplasmic membrane was suggested as a possible mechanism, although direct evidence could not be presented.

In their studies of intracytoplasmic membrane biogenesis, Tai *et al.* (1986) used synchronous cultures of *Rhodopseudemonas sphaeroides*. Translocation of phospholipids from their site of synthesis, the cytoplasmic membrane, into the growing intracytoplasmic membrane was shown to be concurrent with cell division. Measurements of the specific activity of a protein catalyzing phospholipid transfer in cell lysates showed a cyclical oscillation with its highest value at the completion of the cell division, although the absolute amount of phospholipid transfer activity per cell remained constant. Isolated intracytoplasmic membrane with a low phospholipid : protein ratio (prior to cell division) served as a better phospholipid acceptor than intracytoplasmic membrane with a higher phospholipid to protein ratio (after cell division). These data support the view that protein catalysis is a mechanism for the transfer of phospholipids between bacterial membranes in vivo. Further experimental evidence is necessary to substantiate this hypothesis.

The process of transmembrane movement of newly synthesized phospholipids was investigated by Rothman and Kennedy (1977). Phosphatidethanolamine synthesized on the cytoplasmic side of the plasma membrane of *Bacillus megaterium* is rapidly translocated to the external leaflet, indicating that transmembrane flip-flop must be a catalyzed process in membranes of living cells. The existence of flippases has since been demonstrated in higher eukaryotic cells (Backer and Dawidowicz, 1987; Devaux, 1988).

In a series of papers, Bogdanov *et al.* (1984, 1985, 1986) reported that protein secretion in *E. coli* is coupled to the translocation and turnover of the acidic phospholipids phosphatidylglycerol and cardiolipin.

2.3. Phospholipid Transport Proteins in Bacteria

In analogy to eukaryotic systems (Helmkamp, 1986) the cytosol of prokaryotic cells contains proteins that are able to catalyze phospholipid transfer between membranes (Table I). Transfer assays were adapted to prokaryotic membrane organization. Cohen *et al.* (1979) developed a system with vesicles (iso-

Table I
Phospholipid Transfer Proteins of Microorganisms

Organism[a]	Subcellular localization	Molecular weight	PI	Specificity[a]	Assay system[a]	Reference
M. mucedo	Cytosol	ca. 20,000	–	PC	ULV-mitochondria	Chavant and Kader (1982)
B. subtilis						
PT I	Soluble	18,000	–	PE > PG > CL	Mesosomes-protoplasts	Lemaresquier *et al.* (1982)
PT II		15,000	–	PG > CL > PE	Mesosomes-protoplasts	Lemaresquier *et al.* (1982)
Rps. sphaeroides	Cytoplasm	27,000	5.2	PG > PE > PC	Chromatophores-ULV	Tai and Kaplan (1984)
		–	4.6	–		
	Periplasm	56,000	–	PE > PG > PC	Chromatophores-ULV	Tai and Kaplan (1985)
S. cerenisiae	Cytosol	35,000	5.2	PI > PC	ULV-mitochondria; Fluorescently labeled ULV-unlabeled ULV	Daum and Paltauf (1984); Szolderits *et al.* (1989)
	Cytosol	33,400	6.3; 6.1[b]	PC > PE > PI > PS	ULV-MLV	Bozzato and Tinker (1987)

[a]Abbreviations: PT I, PT II, phospholipid transfer proteins I and II; PC, phosphatidylcholine; PE, phosphatidylethanolamine; PG, phosphatidylglycerol; CL, cardiolipin; PI, phosphatidylinositol; PS, phosphatidylserine; ULV, unilamellar vesicles; MLV, multilamellar vesicles.
[b]The pI 6.1 protein is most probably a product of proteolytic modification of the pI 6.3 protein.

lated from cells incubated in the presence of [^{3}H]acetate) of purified intracytoplasmic membranes as the donor and unilamellar phospholipid vesicles containing the nontransferable marker [^{14}C]triolein as the acceptor. Donor and acceptor vesicles were separated from each other by using an antiserum directed against the intracytoplasmic membrane. Using this assay system, the authors were able to detect phosphatidylglycerol, phosphatidylethanolamine, and phosphatidylcholine transfer activity in the soluble fraction of *Rps. sphaeroides*. The most abundantly transferred phospholipid was phosphatidylglycerol, whereas phosphatidylcholine constituted the smallest percentage of the transferred phospholipids. The relative transfer rates do not simply reflect the phospholipid composition of the donor membrane, since phosphatidylethanolamine is the major phospholipid component of the intracytoplasmic membrane. Tai and Kaplan (1984) purified a phospholipid transfer protein from extracts of photoheterotrophically grown *Rps. sphaeroides*. Conventional protein purification techniques (ammonium sulfate precipitation; DEAE-cellulose, Sephadex G-75, and hydroxyapatite chromatography) and vesicle association combined with molecular sieve chromatography led to a 1400-fold-enriched protein, with a recovery of 12.5%. The order of transfer efficiency was phosphatidylglycerol > phosphatidylcholine > phosphatidylethanolamine. Under denaturing conditions the molecular weight of the purified transfer protein as determined by sodium dodecyl sulfate-polyacrylamide gel electrophoresis was 27,000. During the course of their purification procedure, the authors obtained evidence for the existence of another phospholipid transfer protein in *Rps. sphaeroides*. Following hydroxyapatite chromatography, two peaks with phospholipid transfer activity were detected. They were different with respect to their pI (4.6 and 5.2) and their binding properties to phenyl-Sepharose. The pI 4.6 species was not further characterized. The relationship between intracytoplasmic membrane biogenesis and phospholipid transfer activity led Tai and Kaplan (1985a) to study the influence of environmental factors on phospholipid transfer activity. When the facultative photosynthetic bacterium *Rps. sphaeroides* was grown under strong light (100 W/m^2) the cellular level of phospholipid transfer activity was markedly higher than when weak light conditions (3 W/m^2) were used or when chemoheterotrophic growth took place. The highest level of membrane-bound phospholipid transfer activity was found in cells grown under weak light conditions. The authors argued that this binding is not simply nonspecific. They suggested that phospholipid transfer proteins might have a transient membrane-binding state depending on the physiological state of the cell. *Rps. sphaeroides* grown under weak light conditions has the highest content of intracytoplasmic membrane; the large amount of membrane-associated transfer activity might reflect the increased membrane surface available for interaction. These data suggest that membrane association of phospholipid transfer activity may be involved in intracytoplasmic membrane biosynthesis in photosynthetic cells. Soluble transfer activity was localized in both periplasmic and cytoplasmic fractions. Transfer proteins from these two fractions are different with respect to their molecular weights and their substrate specificities. The molecular weight of the cytoplasmic transfer protein that has been purified to homogeneity (see above) is 27,000; it has a preference for phosphatidylglycerol. The periplasmic space protein has an apparent molecular weight of 56,000 and transfers phospholipid classes corresponding to the bulk composition of the donor membrane. The involvement of the two proteins in the translocation of phospholipids between outer and cytoplasmic membranes, or their role in the biosynthesis of the intracytoplasmic membrane, was suggested but needs further elucidation.

Proteins contained in the soluble fraction of *B. subtilis* homogenates were found to accelerate the transfer of phospholipids from isolated mesosomes to protoplasts. Size fractionation on Sephadex G-100 resulted in two peaks corresponding to molecular weights of 18,000 and 15,000. The larger protein showed a preference for phosphatidylethanolamine transfer, while the smaller protein was more specific for the acidic phospholipids phosphatidylglycerol and cardiolipin (Lemaresquier *et al.*, 1982).

3. PHOSPHOLIPID TRANSPORT IN EUKARYOTIC MICROORGANISMS

3.1. Sites of Phospholipid Biosynthesis in *S. cerevisiae*

Several studies have confirmed that in the yeast *S. cerevisiae,* both the endoplasmic reticulum and mitochondria have the capacity to synthesize glycerophospholipids (Cobon *et al.*, 1974; Kuchler *et al.*, 1986). Some of the enzymes involved cofractionate with both compartments: glycerophosphate acyltransferase, CDP-diacylglycerol synthase, phosphatidylinositol synthase, phosphatidylserine synthase, and choline phosphotransferase occur in the endoplasmic reticulum and in mitochondria. On the other hand, key enzymes of phospholipid biosynthesis are located in only a single membrane compartment. Phosphatidylserine decarboxylase and phosphatidylglycerophosphate synthase are restricted to the inner mitochondrial membrane, and phosphatidylethanolamine methyltransferases were found exclusively in the endoplasmic reticulum. Thus, the quantitatively predominant phospholipids, phosphatidylcholine and phosphatidylethanolamine, must be moved from the membrane where they are synthesized to all other cellular membrane structures. In *S. cerevisiae,* vacuoles and the plasma membrane are apparently devoid of glycerophospholipid-synthesizing enzymes; nuclear membranes have not been investigated in this respect.

3.2. Phospholipid Transport *In Vivo*

Most experiments aimed at understanding phospholipid transport mechanisms in eukaryotic cells have been carried out *in vitro* by using isolated subcellular membranes or artificial lipid vesicles in combination with phospholipid transfer proteins. Jungalwala and Dawson (1970) were among the first to demonstrate that phospholipids synthesized in the endoplasmic reticulum in isolated rat liver cells were transported to mitochondria and to the nuclear and the plasma membranes. Yaffe and Kennedy (1983) compared the transport of a phospholipid analog *in vitro* and *in vivo*. They observed that phosphatidyl-*N*-propyl-*N*,*N*-dimethylethanolamine had no affinity to a partially purified phospholipid transfer protein isolated from baby hamster kidney (BHK-21) cells, but was nevertheless efficiently transported from the endoplasmic reticulum to mitochondria in living BHK cells. This demonstrates that interpretation of data derived only from studies *in vitro* might be misleading. On the other hand, the complexity of the *in vivo* system might result in a misinterpretation, too. In this particular case, e.g., the possible presence of additional phospholipid transfer proteins in BHK-21 cells was not considered. The use of eukaryotic microorganisms such as *S. cerevisiae* to study phospholipid transport in vivo offers the advantage of ease of manipulation of membrane composition and assembly by the alteration of growth conditions or by the use of mutants.

Intracellular phospholipid transport can occur by at least three different mechanisms: (1) spontaneous, energy-independent translocation either by diffusion of free phospholipids or catalyzed by more or less specific lipid transfer proteins; (2) energy-dependent vesicle flux between membranes; and (3) lipid translocation via membrane contact sites.

Experiments were designed in our laboratory (Daum *et al.*, 1986a; G. Daum *et al.*, unpublished data) to investigate which of these mechanisms is operating in *S. cerevisiae* cells. Movement of lipids was traced in two different ways. Cells were pulse-labeled with the water-soluble phospholipid precursor [^{3}H]serine or [*methyl*-^{3}H]methionine, and subcellular fractions were isolated and analyzed with respect to the presence of radioactive phospholipids at certain time intervals during the chase. As an alternative labeling method, spheroplasts were incubated with vesicles composed of radioactively labeled phospholipids in the presence of a phospholipid transfer protein. At certain time points, organelle membranes were isolated and analyzed as described above. A serious drawback of this procedure is the possibility of cross-contamination between membranes or scrambling of membrane-associated radioactive phospholipids during cell fractionation. More reliable data are obtained when translocation of lipids results in their metabolic modification. For example, phosphatidylserine synthesized from radioactive serine in the endoplasmic reticulum and transported to the inner mitochondrial membrane will be decarboxylated to radioactive phosphatidylethanolamine by phosphatidylserine decarboxylase, which is found exclusively at this subcellular site. The radioactive phosphatidylethanolamine will be methylated to radioactive phosphatidylcholine only after translocation to the endoplasmic reticulum, where the respective methyltransferases are located. Similarly, [^{3}H]phosphatidylserine added to cells in the form of vesicles will be converted to [^{3}H]phosphatidylethanolamine only after it has reached the inner mitochondrial membrane.

Intracellular phospholipid translocation is a very fast process (Daum *et al.*, 1986a). After a few minutes of labeling yeast cells with [^{3}H]serine, [^{3}H]phosphatidylethanolamine and [^{3}H]phosphatidylcholine (derived from [^{3}H]phosphatidylethanolamine) were evently distributed over mitochondria and the endoplasmic reticulum. The total radioactivity associated with phosphatidylethanolamine and phosphatidylcholine in the endoplasmic reticulum and with phosphatidylcholine in mitochodria was higher than the radioactivity associated with phosphatidylethanolamine in mitochondria, indicating rapid transport of newly synthesized phosphatidylethanolamine from the inner mitochondrial membrane to the endoplasmic reticulum, where most of it is converted to phosphatidylcholine; part of the phosphatidylcholine then moves to mitochondria. In cells depleted of metabolic energy by poisoning with CN^- and F^- after the pulse with [^{3}H]serine, equilibration of [^{3}H]phosphatidylcholine between the endoplasmic reticulum and mitochondria could be observed. In the same type of experiment, migration of [^{3}H]phosphatidylserine from the endoplasmic reticulum to mitochondria continued after cells were poisoned with CN^- and F^-, although at a reduced rate. Taken together, these data suggest that phosphatidylcholine and phosphatidylserine can be translocated between membranes by an energy-independent process catalyzed by the respective phospholipid transfer proteins present in yeast cytosol (see below).

Energy-independent intracellular movement of phosphatidylserine in *S. cerevisiae* was also demonstrated by the conversion of exogenously added [^{3}H]phosphatidylserine to [^{3}H]phosphatidylethanolamine. When yeast spheroplasts were incubated with [^{3}H]phosphatidylserine vesicles in the presence of a nonspecific lipid transfer protein isolated from bovine liver (Crain and Zilversmit, 1980), the extent of [^{3}H]phosphatidylethanolamine formation catalyzed by mitochondrial phosphatidylserine decarboxylase was essentially the same in energized and deenergized cells. This result is at variance with results reported by Voelker (1985), who found that transfer of phosphatidylserine from the endoplasmic reticulum to mitochondria in hamster kidney cells was significantly accelerated by metabolic energy. It is conceivable, however, that different transport mechanisms are involved in the translocation of exogenously added phosphatidylserine to mitochondria and in the intermembrane transfer of newly synthesized phosphatidylserine. It is tempting to speculate that in *S. cerevisiae* a phospholipid transfer protein with specificity for phosphatidylserine and phosphatidylethanolamine participates in a reciprocal movement of the two phospholipids between the endoplasmic reticulum and mitochondria. Phosphatidylserine synthesized in the endoplasmic reticulum could be exchanged for mitochondrially synthesized phosphatidylethanolamine, which is required as a substrate for phosphatidylcholine synthesis in the endoplasmic reticulum.

Vesicles involved in protein secretion (Holcomb *et al.*, 1988; Walworth and Novick, 1987) migrate from internal membranes to the plasma membrane. Since the plasma membrane is devoid of phospholipid-synthesizing enzymes, it relies on a supply of all the constituent phospholipids. We used temperature-sensitive secretory mutants (*sec* mutants) (Schekman, 1982) to investigate whether secretory vesicles transport a quantitatively significant proportion of phosphatidylcholine and phosphatidylinositol to the plasma membrane. The experimental design was such that we measured the release of the water-soluble products of phospholipid deacylation, glycerophosphocholine and glycerophosphoinositol, into the growth medium. Deacylation of phosphatidylinositol and phosphatidylcholine is catalyzed by phospholipase B attached to the outer side of the plasma membrane (Witt *et al.*, 1982, 1984). The phospholipids hydrolyzed must therefore be part of the plasma membrane. Using this approach, we found that release of radioactive glycerophosphodiesters from cells prelabeled with [^{3}H]inositol or [^{3}H]choline did not cease after a shift from the permissive (24°C) to the nonpermissive (37°C) temperature. The total amount of radioactivity secreted during several hours by far exceeded the amount of radioactively labeled phosphatidylcholine or phosphatidylinositol present initially in the plasma membrane. Therefore, transfer of the two phospholipids from internal membranes to the plasma membrane must have occurred, but this transport was obviously independent of the flow of vesicles competent for protein secretion. This result confirms the postulate of Atkinson (1983) that different types of vesicles are responsible for protein export and membrane growth. However, the existence of specific phospholipid-transporting vesicles must still be verified experimentally. On the other hand, phospholipid transfer proteins could be involved in the transport of phospholipids to the plasma membrane.

The existence of reciprocal phospholipid transport between the plasma membrane and internal membranes was demonstrated by the experiment described above, in which cellular uptake and metabolism of exogenous [^{3}H]phosphatidylserine was studied (G. Daum *et al.*, unpublished data). Conversion of exogenous [^{3}H]phosphatidylserine to [^{3}H]phosphatidylethanolamine occurred only when a phosphatidylserine transfer protein was added to the medium during incubation of spheroplasts with [^{3}H]phosphatidylserine. This requirement for a phospholipid transfer protein suggests that phosphatidylserine has to be integrated into the plasma membrane before it can be transported to mitochondria. Endocytotic vesicles are apparently not involved in this process, because it is independent of metabolic energy and is not inhibited by the cytoskeleton inhibitor nocodazol (Makarow, 1988).

Translocation of proteins across membranes requires energy and—at least for import into mitochondria—a membrane potential (Pfanner and Neupert, 1986; Eilers *et al.*, 1987) in addition to soluble and membrane-bound protein factors that are involved at several stages of this complex process. Does phospholipid transport across internal membranes have similar requirements? Phosphatidylethanolamine synthesized in the inner mitochondrial membrane by decarboxylation of phosphatidylserine has to cross the outer mitochondrial membrane on its way to the endoplasmic reticulum, where it is required for the synthesis of phosphatidylcholine via stepwise methylation by *S*-adenosylmethionine-dependent methyltransferases. Pulse-chase experiments in vivo with [^{3}H]serine clearly showed that the extent of [^{3}H]phosphatidylcholine synthesis was essentially the same in the presence or in the absence of the uncoupler carbonyl cyanide m-chlorophenylhydrazone. This result allows the conclusion that transport of phosphatidylserine from the outside of and across the outer mitochondrial membrane to the inner membrane and transport of phosphatidylethanolamine from the inner membrane across the outer mitochondrial membrane to the endoplasmic reticulum do not require a membrane potential. In Section 3.3, possible mechanisms for phospholipid translocation across the outer mitochondrial membrane will be discussed.

3.3. Intramitochondrial Transfer of Phospholipids

In growing yeast cells a continous flow of phospholipids traverses the outer mitochondrial membrane in both directions. The bulk of phospholipids transported consist of phosphatidylethanolamine on its way from the inner mitochondrial membrane to the endoplasmic reticulum, phosphatidylcholine derived from the endoplasmic reticulum and destined for the inner mitochondrial membrane, and phosphatidylserine synthesized in the endoplasmic reticulum and required in the inner mitochondriel membrane as a substrate for phosphatidylserine decarboxylase. In addition, phosphatidic acid, a precursor of cardiolipin synthesis, and phosphatidylinositol are made both outside of mitochondria and have to be translocated to the inner mitochondrial membrane. Transport of phospholipids from outside the mitochondrion into the inner mitochondrial membrane and vice versa involves two translocation events: movement of phospholipids between the outer and the inner mitochondrial membrane, and translocation across the outer mitochondrial membrane. The latter process might be triggered by a flippase similar to that identified in rat liver microsomes (Backer and Dawidowicz, 1987). In isolated right-side-out outer mitochondrial membrane vesicles (Riezman *et al.*, 1983), only approximately 50% of phosphatidylcholine and phosphatidylinositol and 25% of phosphatidylethanolamine are freely exchangeable during incubation in the presence of artificial phospholipid vesicles and phospholipid transfer proteins specific for these lipids (Sperka-Gottlieb *et al.*, 1988). Thus, in isolated outer membrane vesicles, flip-flop of these phospholipids either does not occur or is too slow to account for the rapid phospholipid translocation observed in vivo.

Phospholipid transfer between the yeast outer and inner mitochondrial membranes does not involve phospholipid transfer protein(s) of the intermembrane space; no such protein could be detected when the intermembrane space contents were tested in appropriate phospholipid transfer systems in vitro (R. Simbeni *et al.*, 1990 data). This finding confirms previous results by Blok *et al.* (1971), who detected no phospholipid transfer activity in rat liver mitochondria.

In intact mitochondria, the outer and inner membranes are linked by stable contact sites (junctions), which are most probably involved in the transport of proteins into the inner membrane and the matrix space (Schleyer and Neupert, 1985; Schwaiger *et al.*, 1987). These contact sites are also good candidates for the transfer of phospholipids between the two mitochondrial membranes. Thus, phospholipid transfer via membrane contact sites seems to be operating in this particular case of intermembrane phospholipid movement. We propose that the junctions are the zones of transmembrane translocation of phospholipids across the outer membrane, as well as of their intermembrane transport between outer and inner mitochondrial membranes. In *S. cerevisiae* mitoplasts (prepared by subjecting isolated mitochondria to a mild hypotonic shock), the outer mitochondrial membrane is disrupted but still adheres to the inner membrane, most probably through the original junctions. Preparations of intact mitochondria and mitoplasts incorporate [^{3}H]serine into phosphatidylserine at the same rate. Moreover, conversion of [^{3}H]phosphatidylserine to [^{3}H]phosphatidylethanolamine occurs at the same rate with both preparations. From this result, we conclude that phosphatidylserine, which is synthesized by phosphatidylserine synthase in the outer mitochondrial membrane (or in residual microsomal membranes attached to the mitochondrial surface), reaches the inner membrane via membrane contact sites and is subsequently decarboxylated by the inner-mitochondrial-membrane-located phosphatidylserine decarboxylase. Control experiments with mitoplasts from a mutant strain deficient in phosphatidylserine synthase (*cho*1 null mutant) and isolated outer mitochondrial membrane vesicles from wild-type cells showed that nonspecific membrane contact between the inner and outer mitochondrial membranes resulted in only marginal rates of conversion of phosphatidylserine to phosphatidylethanolamine. More direct evidence for the involvement of mitochondrial junctions in phospholipid transport might come from experiments with isolated junctions reconstituted into artificial phospholipid membranes.

Evidence for rapid translocation of phospholipids across the outer mitochondrial membrane was obtained with whole cells (see Section 3.2) and could be confirmed by the following experiments with isolated mitochondria (Simbeni *et al.*, 1990). Cells were prelabeled to equilibrium with [^{3}H]inositol; then mitochondria were isolated by a gentle procedure (Daum *et al.*, 1982) after spheroplasting and incubated with [^{14}C]inositol under conditions of active phosphatidylinositol synthesis. After a 10-min incubation, phosphatidylinositol synthesis was stopped by the addition of EDTA, and mitochondria were then incubated in the presence of phosphatidylcholine vesicles and a purified phosphatidylinositol/phosphatidylcholine-specific transfer protein from *S. cerevisiae* (Daum and Paltauf, 1984). Analysis of radioactivity in acceptor vesicles revealed that approximately 30% of newly synthesized [^{14}C]phosphatidylinositol, but only 4–5% of the "old" [^{3}H]phosphatidylinositol, was accessible for transfer. In another experiment, unlabeled mitochondria were incubated with [^{3}H]inositol, phosphatidylinositol synthesis was stopped at time intervals between 1 and 20 min, and the percentage of [^{3}H]phosphatidylinositol accessible for protein-catalyzed exchange was measured. It was found to be close to 80% at 1 min, but declined to approximately 10–20% after 30 min. When phosphatidylinositol synthesis was stopped after 1 min, but incubation was continued for another 19 min, the percentage of [^{3}H]phosphatidylinositol in the outer leaflet of the outer mitochondrial membrane remained constant (approximately 70%). These results led to the following conclusions. (1) Phosphatidylinositol is synthesized at the mitochondrial surface and rapidly transported to the inner membrane; this was confirmed by measuring the distribution of [^{3}H]phopshatidylinositol in isolated outer and inner mitochondrial membranes. (2) Translocation of newly synthesized phosphatidylinositol from the outer aspect of the outer mitochondrial membrane to the inner membrane is a vectorial process linked to phosphatidylinositol synthesis. When isolated mitochondria were incubated with unilamellar vesicles consisting of ^{3}H-phosphatidylinositol in the presence of a phosphatidylinositol transfer protein, the outer, but not the inner mitochondrial membrane became radioactively labeled (R. Simbeni *et al.*, 1990). This result confirms the notion that phosphatidylinositol is translocated between membranes only if a "pressure" is exerted, e.g. by net synthesis, and that free exchange of phosphatidylinositol between the mitochondrial membranes does not occur.

Vectorial transport between mitochondrial membranes coupled to biosynthesis was also observed for phosphatidylethanolamine. When intact mitochondria were incubated with [^{3}H]serine and CDP-diacylglycerol, an appreciable percentage of [^{3}H]phosphatidylserine formed in the outer mitochondrial membrane was decarboxylated by phosphatidylserine decarboxylase in the inner membrane. After 2, 5, 10, and 15 min, the outer and inner mitochondrial membranes were separated and their radioactively labeled phospholipids analyzed. Even after short time intervals, the specific radioactivity of [^{3}H]phosphatidylethanolamine was markedly higher in outer than inner mitochondrial membrane vesicles, indicating that newly synthesized phosphatidylethanolamine was preferentially translocated to the outer mitochondrial membrane without first entering the pool of phosphatidylethanolamine in the inner mitochondrial membrane.

3.4. Phospholipid Transport Proteins

3.4.1. Occurrence, Isolation and Characterization

Phospholipid transfer activity in *S. cerevisiae* cytosol was first reported by Cobon *et al.* (1976). In an assay with ^{32}P-labeled mitochondria and [^{3}H]leucine-labeled microsomes from *S. cerevisiae* or rat liver, protein-catalyzed transfer of phosphatidylcholine, phosphatidylinositol, and, to a lesser extent, phosphatidylethanolamine and phosphatidylserine was demonstrated. With sonicated vesicles composed of [^{3}H]oleate-labeled total *S. cerevisiae* phospholipids as the donor membrane and yeast mitochondria as the acceptor membrane, phospholipid transfer protein(s) present in *S. cerevisiae* cytosol facilitated the transfer of all the phospholipids contained in the donor vesicles. The highest transfer rate was observed for phosphatidic acid, followed by phosphatidylinositol, phosphatidylethanolamine, phosphatidylserine, phosphatidylcholine, and cardiolipin (Daum and Paltauf, 1984). In a more defined system with donor membranes consisting of small unilamellar vesicles composed of either a single phospholipid or 1 : 1 mixtures of phosphatidylcholine and another radioactively labeled phospholipid, transfer of phosphatidic acid and cardiolipin could no longer be observed. Furthermore, the preference for phospholipid classes was different (see below) from that found with the complex system mentioned above.

Chavant and Kader (1982) reported the presence of a phospholipid transfer protein in the soluble protein fraction from the filamentous fungus *Mucor mucedo*. Phosphatidylcholine transfer activity was attributed to a 20-kDa protein.

Purification to homogeneity was reported only for phospholipid transfer proteins from *S. cerevisiae*. Bozzato and Tinker (1987) isolated a phospholipid transfer protein (protein I) by using ammonium sulfate precipitation and dye-ligand and anion exchange chromatography. The purification factor was 530, with 13.3% yield. The molecular weight was 33,700, and the pI was 6.3. Upon storage at 4°C for 2 months, a second protein (protein II) was formed which had the same molecular weight but a lower isoelectric point (6.1). The protease inhibitor phenylmethylsulfonyl fluoride almost completely inhibited the conversion of protein I to protein II, suggesting that protein II is the product of proteolytic modification of protein I. Substrate specificities were identical for both proteins. Phosphatidylcholine was the most rapidly transferred phospholipid, followed by phosphatidylinositol, phosphatidylethanolamine, and phosphatidylserine; the relative rates were 3, 2, 1, and 0.3 when donor (unilamellar vesicles) and acceptor (multilamellar vesicles) membranes contained 40 : 30 : 20 : 10 mol% phosphatidylcholine, phosphatidylethanolamine, phosphatidylinositol, and phosphatidylserine, respectively. By kinetic analysis, the mechanism of protein-catalyzed transfer of phosphatidylcholine was shown to conform to a ping-pong Bi-

Bi model with excess substrate inhibition; a similar model had previously been used to describe the kinetics of phospholipid transfer by a bovine liver phosphatidylcholine transfer protein (Van den Besselaar *et al.*, 1975) and a bovine brain phosphatidylinositol transfer protein (Helmkamp *et al.*, 1976).

Isolation and characterization of an *S. cerevisiae* phosphatidylinositol/phosphatidylcholine transfer protein was reported by Daum and Paltauf (1984) (see Section 3.4.2). Yeast cytosol contains at least one additional phospholipid transfer protein, which catalyzes preferentially the transfer of phosphatidylserine and phosphatidylethanolamine but not phosphatidylcholine and phosphatidylinositol (G. Lafer *et al.*, unpublished data).

3.4.2. Phosphatidylinositol/Phosphatidylcholine-Specific Transfer Protein from *S. cerevisiae*

By a combination of conventional separation techniques including Sephadex, DEAE-Sephacel, and phenyl-Sepharose chromatography, a phospholipid transfer protein was isolated from *S. cerevisiae* cytosol (Daum and Paltauf, 1984; G. Szolderits *et al.*, 1989); it differs in several respects from that described by Bozzato and Tinker (1987). The protein was purified approximately 2700-fold, with a yield of 15%. It has a molecular mass of 35-kDa and a pI of 5.2 as determined by isoelectric focusing. During chromatofocusing it is eluted at pH 4.6. Chromatofocusing was used in the original purification scheme but was then replaced by phenyl-Sepharose chromatography because of the instability of the protein at pH values lower than 5 (Daum *et al.*, 1986b). The protein loses 50% of its activity after 10 min at approximately 50°C. In contrast to the protein preparation described by Bozzato and Tinker (1987), it can be kept at 4°C for more than a year with no significant loss of activity and without apparent alteration of electrophoretic properties. The protein resists digestion by trypsin, chymotrypsin, or papain but is degraded by pronase (Daum *et al.*, 1986b).

When fluorescently labeled phosphatidylcholine and phosphatidylinositol were used, complex formation of these phospholipids with the *S. cerevisiae* phospholipid transfer protein was observed. Accordingly, the protein specifically catalyzes transfer of these two phospholipid classes, similar to the phosphatidylinositol/phosphatidylcholine transfer protein from bovine brain (Helmkamp *et al.*, 1974). The two proteins share further similarities (Szolderits *et al.*, 1989): they have the same size and similar isoelectric points (5.2 for the *S. cerevisiae* protein and 5.5 for the brain protein) and are both inhibited by negatively charged glycerophospholipids. On the other hand, sulfhydryl-group-specific reagents, which inhibit the activity of the brain phospholipid transfer protein (DiCorleto *et al.*, 1979), have no effect on the *S. cerevisiae* protein.

Membrane lipid composition has distinct effects on the *S. cerevisiae* phospholipid transfer protein isolated by Daum and Paltauf (1984) compared with the effects on the protein described by Bozzato and Tinker (1987). Donor and acceptor membranes in the standard transfer assay used by Bozzato *et al.* (1987) contained 30 mol% negatively charged phospholipids (phosphatidylinositol and phosphatidylserine). With phosphatidylcholine alone, transfer rates were reduced to 5%, compared with the rates under standard conditions. Furthermore, this protein transferred phosphatidylethanolamine at rather high rates. In contrast, the pI 5.2. protein studied in our laboratory (Szolderits *et al.*, 1989) was inhibited by 10% phosphatidylserine in donor or acceptor small unilamellar vesicles; phosphatidylethanolamine transfer could not be detected under conditions (membrane lipid composition) similar to those used by Bozzato *et al.* (1987). The reason for the striking difference in the apparent properties of the two proteins is not clear. Most likely differences in assay conditions resulted in modifications of substrate specificity and susceptibility to inhibition by negatively charged lipids. Bozzato *et al.* (1987) used unilamellar vesicles as donor membranes and multilamellar vesicles as acceptor membranes (5 μM and 1 mM, respectively), whereas we used either unilamellar vesicles as donor membranes and *S. cerevisiae* mitochondria as acceptor membranes or, more conveniently, unilamellar vesicles consisting of pyrenedecanoyl-labeled phospholipids as donor membranes and unilamellar phospholipid vesicles as acceptor membranes. With both systems, essentially the same results were obtained. Inhibition by negatively charged phospholipids was most pronounced with phosphatidylserine, but was also observed with phosphatidylglycerol, phosphatidic acid, and cardiolipin, the inhibitory effect decreasing in this order.

The size of vesicles prepared by sonication or ethanol injection could be varied between 20 and 60 nm without any effect on the transfer rate. The fluidity state of the acceptor membrane had a pronounced effect on the transfer efficiency. The phosphatidylcholine transfer rate was markedly reduced when dipalmitoyl phosphatidylcholine was used as the acceptor phospholipid instead of egg yolk phosphatidylcholine (consisting of approximately 40% palmitoyl oleoyl phosphatidylcholine). The addition of ergosterol to acceptor membrane vesicles led to a reduction in the phosphatidylcholine transfer rate to 70% of the control rate (sterol-free vesicles) at 10 mol% ergosterol; further increase of the sterol content to 50 mol% resulted in a 50% decrease of the transfer rate, most probably due to the membrane rigidifying effect of the sterol.

3.5. Physiological Implications

The ubiquity of phosphatidylinositol/phosphatidylcholine transfer proteins in eukaryotic cells indicates that these proteins play a fundamental role in phospholipid transport and metabolism. In accordance with this assumption are recent results obtained by Dowhan and coworkers (W. Dowhan, personal communication) who produced, by gene disruption, a *pit*-mutation which was lethal. This finding demonstrates for the first time that the phosphatidylinositol/ phosphatidylcholine transfer protein is essential for cell viability, although it does not reveal the physiological function of this protein. Therefore, its specific role in cellular processes is purely speculative. It has been suggested that phosphatidylinositol/phosphatidylcholine transfer proteins might be involved in the net transport of phosphatidylinositol to membranes, e.g., the plasma membrane, where stimulated phosphatidylinositol turnover occurs. Regulation of phosphatidylinositol transport by cellular levels of phosphatidylinositol-4-phosphate or phosphatidylinositol-4,5-bisphosphate has been suggested on the basis of experiments with artificial phospholipid vesicles (Van Paridon *et al.*, 1988). In *S. cerevisiae,* as in higher eukaryotes, phosphatidylinositol is essential for cellular growth. The existence of the phosphatidylinositol cycle in *S. cerevisiae* and its involvement in cell proliferation have recently be demonstrated (Kaibuchi *et al.*, 1986; Uno *et al.*, 1988). In addition, the *S. cerevisiae* plasma membrane contains phospholipase B (Witt *et al.*, 1982, 1984), which deacylates phosphatidylinositol and phosphatidylcholine to glycerophosphoinositol and glycerophosphocholine, respectively, which are secreted into the culture medium. Any of these processes requires a constant supply of phosphatidylinositol; the latter also requires a constant supply of phosphatidylcholine. As discussed in Section 3.2, secretory vesicles do not make a significant contribution to the transport of phospholipids to the *S. cerevisiae* plasma membrane; this finding supports the notion that phospholipid transfer proteins are the vehicles involved in phospholipid transport to the cell surface.

Yeast mitochondria are unable to synthesize phosphatidylcholine by the *de novo* methylation pathway. Therefore, growth of cells under derepressed (low-glucose) conditions, where mitochondria are fully developed, requires efficient transport of phosphatidylcholine from the endoplasmic reticulum to mitochondrial membranes. It was indeed found that growth on glycerol, ethanol, or lactate (Bozzato and Tinker, 1987; Daum and Paltauf, 1984) resulted in higher cellular levels of phospholipid transfer activity than growth on glucose. A clear correlation between growth phase and transfer activity was also demonstrated (Bozatto and Tinker, 1987; Daum *et al.*, 1986b): Phospholipid transfer activity is highest during the exponential phase of growth and decreases during entering the stationary phase. However, during further incubation of cells, levels of phospholipid transfer protein were not significantly reduced, indicating that this protein has a very long half-life, at least in stationary cells. Direct evidence for any of the proposed functions of the phosphatidylinositol/phosphatidylcholine transfer protein in *S. cerevisiae* has yet to be provided.

4. FUTURE PERSPECTIVES

Several approaches can be envisaged to study the physiological role of phospholipid transfer proteins in microorganisms. Correlation of cellular phospholipid transfer activity with induced membrane biogenesis by varying the growth conditions can provide the basis for more detailed analyses. Application of specific antibodies directed against phospholipid transfer proteins, e.g., by electroporation, would provide a tool to inactivate phospholipid transfer proteins *in situ*. *S. cerevisiae* is especially suited for genetic manipulation. With the availability of the PIT gene the function of the phosphatidylinositol/phosphatidylcholine transfer protein can be studied in mutants which express the protein under the control of a regulated promoter. Similar experiments should allow to study the role of the phosphatidylserine/phosphatidylethanolamine transfer protein of yeast, and perhaps of other lipid binding or transfer proteins still to be discovered. Such genetic and biochemical analyses of mutants should eventually reveal the true physiological function of phospholipid transfer proteins in eukaryotic cells.

ACKNOWLEDGMENTS. Financial support of our work by the Fonds zur Förderung der wissenschaftlichen Forschung in Österreich (projects 5906 and 6958) is gratefully acknowledged.

5. REFERENCES

Atkinson, K., 1983, Membrane function in yeast lipid synthesis mutants, in *Biosynthesis and Function of Plant Lipids* (W. W. Thomson, J. B. Madd, and M. Gibbs, eds.), pp. 229–249, American Society of Plant Physiology Monograph, Waverly Press, Baltimore.

Backer, J. A., and Dawidowicz, E. A., 1987, Reconstitution of a phospholipid flippase from rat liver microsomes, *Nature* **327**:341–343.

Blok, M. C., Wirtz, K. W. A., and Scherphof, G. L., 1971, Exchange of phospholipids between microsomes and inner and outer mitochondrial membranes of rat liver, *Biochim. Biophys. Acta* **233**:61–75.

Bogdanov, M. V., Kulaev, I. S., and Nesmeyanova, M. A., 1984, Study on lipid-protein interaction during protein transfer through bacterial membranes. 1. Coordinated inhibition of the exoprotein synthesis, secretion and phospholipid metabolism by a protonophore, *Biol. Membr.* **1**:495–502.

Bogdanov, M. V., Tsfasman, I. M., and Nesmeyanova, M. A., 1985, Study of lipid–protein interaction during protein transfer across bacterial membranes. 2. Phospholipids in the translocation site of the alkaline phosphatase across Escherichia coli membrane, *Biol. Membr.* **2**:623–629.

Bodganov, M. V., Fedoseeva, G. N., and Nesmeyanova, M. A., 1986, A study of lipid–protein interaction during protein transfer across bacterial membranes. 3. The turnover of phospholipids and the content of membrane derived oligosaccharides in biosynthesis and secretion of alkaline phosphatase in Escherichia coli, *Biol. Membr.* **3**:1241–1249.

Bozzato, R. P., and Tinker, D. O., 1987, Purification and properties of two phospholipid transfer proteins from yeast, *Biochem. Cell Biol.* **65**:195–202.

Bozzato, R. P., Woolley, D., and Tinker, D. O., 1987, Catalytic properties of the yeast phospholipid transfer protein, *Biochem. Cell Biol.* **65**:203–209.

Cain, B. D., Deal, C. D., Fraley, R. T., and Kaplan, S., 1981, In vivo intermembrane transfer of phospholipids in the photosynthetic bacterium *Rhodopseudomonas sphaeroides, J. Bacteriol.* **145:**1154–1166.

Cain, B. D., Donohue, T. J., Shepherd, W. D., and Kaplan, S., 1984, Localization of phospholipid biosynthetic enzyme activities in cell-free fractions from Rhodopseudomonas sphaeroides, *J. Biol. Chem.* **259:**942–948.

Chavant, L., and Kader, J.-C., 1982, The presence of phospholipid transfer proteins in filamentous fungi, in *Biochemistry and Metabolism of Plant Lipids* (J. F. G. M. Wintermans, and P. J. C. Kuiper, eds.), pp. 125–128, Elsevier Biomedical Press, Amsterdam.

Cobon, G. S., Crowfoot, P. D., and Linnane, A. W., 1974, Biogenesis of mitochondria. Phospholipid synthesis in vitro by yeast mitochondrial and microsomal fraction, *Biochem. J.* **144:**265–275.

Cobon, G. S., Crowfoot, P. D., Murphy, M., and Linnane, A. W., 1976, Exchange of phospholipids between mitochondria and microsomes in vitro stimulated by yeast cell cytosol, *Biochim. Biophys. Acta* **441:**255–259.

Cohen, L. K., Lueking, D. R., and Kaplan, S., 1979, Intermembrane phospholipid transfer mediated by cell-free extracts of Rhodopseudomonas sphaeroides, *J. Biol. Chem.* **254:**721–728.

Crain, R. C., and Zilversmit, D. B., 1980, Two nonspecific phospholipid exchange proteins from beef liver. 1. Purification and characterization, *Biochemistry* **19:**1433–1439.

Daum, G., and Paltauf, F., 1984, Phospholipid transfer in yeast. Isolation and partial characterization of a phopsholipid transfer protein from yeast cytosol, *Biochim. Biophys. Acta* **794:**385–391.

Daum, G., Böhni, P. C., and Schatz, G., 1982, Import of proteins into mitochondria. Cytochrome b_2 and cytochrome c peroxidase are located in the intermembrane space of yeast mitochondria, *J. Biol. Chem.* **257:**13028–13033.

Daum, G., Heidorn, E., and Paltauf, F., 1986a, Intracellular transfer of phospholipids in the yeast, Saccharomyces cerevisiae, *Biochim. Biophys. Acta* **878:**93–101.

Daum, G., Schwelberger, H. G., and Paltauf, F., 1986b, Stability and effects of culture conditions on the cellular content and the activity of a phosphatidylcholine/phosphatidylinositol transfer protein from Saccharomyces cerevisiae, *Biochim. Biophys. Acta* **879:**240–246.

Devaux, P. F., 1988, Phospholipid flippases, *FEBS Lett.* **234:**8–12.

DiCorleto, P. E., Warach, J. B., and Zilversmit, D. B., 1979, Purification and characterization of two phospholipid exchange proteins from bovine heart, *J. Biol. Chem.* **254:**7795–7802.

Dowhan, W., Wickner, W. T., and Kennedy, E. P., 1974, Purification and properties of phosphatidylserine decarboxylase from Escherichia coli, *J. Biol. Chem.* **249:**3079–3084.

Eilers, M., Oppliger, W., and Schatz, G., 1987, Both ATP and an energized inner membrane are required to import a purified precursor protein into mitochondria, *EMBO J.* **6:**1073–1077.

Helmkamp, G. M., Jr., 1986, Phospholipid transfer proteins: Mechanisms of action, *J. Bioenerg. Biomembr.* **18:**71–91.

Helmkamp, G. M., Jr., Harvey, M. S., Wirtz, K. W. A., and van Deenen, L. L. M., 1974, Phospholipid exchange between membranes. Purification of bovine brain proteins that preferentially catalyze the transfer of phosphatidylinositol, *J. Biol. Chem.* **249:**6382–6389.

Helmkamp, G. M., Jr., Wirtz, K. W. A., and van Deenen, L. L. M., 1976, Phosphatidylinositol exchange protein. Effects of membrane structure on activity and evidence for a ping-pong mechanism, *Arch. Biochem. Biophys.* **174:**592–602.

Holcomb, C. L., Hansen, W. J., Etcheverry, T., and Schekman, R., 1988, Secretory vesicles externalize the major plasma membrane ATPase in yeast, *J. Cell Biol.* **106:**641–648.

Jones, N. C., and Osborn, M. J., 1977a, Interaction of Salmonella typhimurium with phospholipid vesicles. Incorporation of exogenous lipids into intact cells, *J. Biol. Chem.* **252:**7398–7404.

Jones, N. C., and Osborn, M. J., 1977b, Translocation of phospholipids between the outer and inner membrane of Salmonella typhimurium, *J. Biol. Chem.* **252:**7405–7412.

Jungalwala, F. B., and Dawson, R. M. C., 1970, Phospholipid synthesis and exchange in isolated liver cells, *Biochem. J.* **117**:481–490.

Kaibuchi, K., Miyajima, A., Arai, K.-I., and Matsumoto, K., 1986, Possible involvement of RAS-encoded proteins in glucose-induced inositolphospholipid turnover in Saccharomyces cerevisiae, *Proc. Natl. Acad. Sci. U.S.A.* **83**:8172–8176.

Kuchler, K., Daum, G., and Paltauf, F., 1986, Subcellular and submitochondrial localization of phospholipid-synthesizing enzymes in *Saccharomyces cerevisiae, J. Bacteriol.* **165**:901–910.

Langley, K. E., Hawrot, E., and Kennedy, E. P., 1982, Membrane assembly: Movement of phosphatidylserine between the cytoplasmic and the outer membranes of *Escherichia coli, J. Bacteriol.* **152**:1033–1041.

Lemaresquier, H., Bureau, G., Mazliak, P., and Kader, J.-C., 1982, Transfer of phospholipids between mesosomes and protoplasts from Bacillus subtilis, *Int. J. Biochem.* **14**:71–74.

Lloyd, G. M., 1987, Synthesis and translocation of lipids in the cell envelope membranes of Micrococcus cryophilus: A comparison of polar and neutral lipids, *Microbios* **52**:29–37.

Makarow, M., 1988, Secretion of invertase in mitotic yeast cells, *EMBO J.* **7**:1475–1482.

Pfanner, N., and Neupert, W., 1986, Transport of F_1 ATPase subunit β into mitochondria depends on both a membrane potential and nucleoside triphosphates, *FEBS Lett.* **209**:152–156.

Raetz, C. R. H., and Kennedy, E. P., 1972, The association of phosphatidylserine synthase with ribosomes in extracts of Escherichia coli, *J. Biol. Chem.* **247**:2008–2014.

Riezman, H., Hay, R., Gasser, S., Daum, G., Schneider, G., Witte, C., and Schatz, G., 1983, The outer membrane of yeast mitochondria: Isolation of outside-out sealed vesicles, *EMBO J.* **2**:1105–1111.

Rothman, J. E., and Kennedy, E. P., 1977, Rapid transmembrane movement of newly synthesized phospholipids during membrane assembly, *Proc. Natl. Acad. Sci. U.S.A.* **74**:1821–1825.

Schekman, R., 1982, The secretory pathway in yeast, *Trends Biochem. Sci.* **7**:243–246.

Schleyer, M., and Neupert, W., 1985, Transport of proteins into mitochondria: Translocational intermediates spanning contact sites between outer and inner membrane, *Cell* **43**:339–350.

Schwaiger, M., Herzog, V., and Neupert, W., 1987, Characterization of translocation contact sites involved in the import of mitochondrial proteins, *J. Cell Biol.* **105**:235–246.

Simbeni, R., Paltauf, F., and Daum, G., 1990, Intramitochondrial transfer of phospholipids in the yeast, Saccharomyces cerevisiae, *J. Biol. Chem.* **265**:281–285.

Sperka-Gottlieb, C. D. M., Hermetter, A., Paltauf, F., and Daum, G., 1988, Lipid topology and physical properties of the outer mitochondrial membrane of the yeast, Saccharomyces cerevisiae, *Biochim. Biophys. Acta,* **946**:227–234.

Szolderits, G., Hermetter, A., Paltauf, F., and Daum, G., 1989, Membrane properties modulate the activity of a phosphatidylinositol transfer protein from the yeast, Saccharomyces cerevisiae, *Biochim. Biophys. Acta,* **986**:301–309.

Tai, S.-P., and Kaplan, S., 1984, Purification and properties of a phospholipid transfer protein from Rhodopseudomonas sphaeroides, *J. Biol. Chem.* **259**:12178–12183.

Tai, S.-P., and Kaplan, S., 1985a, Intracellular localization of phospholipid transfer activity in *Rhodopseudomonas sphaeroides* and a possible role in membrane biogenesis, *J. Bacteriol.* **164**:181–186.

Tai, S.-P., and Kaplan, S., 1985b, Phospholipid transfer in microorganisms, *Chem. Phys. Lipids* **38**:41–50.

Tai, S.-P., Hoger, J. H., and Kaplan, S., 1986, Phospholipid transfer activity in synchronous populations of Rhodobacter sphaeroides, *Biochim. Biophys. Acta* **859**:198–208.

Uno, I., Fukami, K., Kato, H., Tanekawa, T., and Ishikawa, T., 1988, Essential role for phosphatidylinositol 4,5-bisphosphate in yeast cell proliferation, *Nature* **333**:188–190.

Van den Besselaar, A. M. H. P., Helmkamp, G. M., Jr., and Wirtz, K. W. A., 1975, Kinetic model of the protein mediated phosphatidylcholine exchange between single bilayer liposomes, *Biochemistry* **14**:1852–1858.

Van Paridon, P. A., Gadella, T. W. J., Jr., and Wirtz, K. W. A., 1988, The effect of phosphoinositides and phosphatidic acid on the phosphatidylinositol transfer protein from bovine brain: A kinetic study, *Biochim. Biophys. Acta* **943**:76–86.

Voelker, D. R., 1985, Disruption of phosphatidylserine translocation to the mitochondrion in baby hamster kidney cells, *J. Biol. Chem.* **260**:14671–14676.

Walworth, N. C., and Novick, P. J., 1987, Purification and characterization of constitutive secretory vesicles from yeast, *J. Cell Biol.* **105**:163–174.

Witt, W., Brüller, H.-J., Falkner, G., and Fuhrmann, G. F., 1982, Purification and properties of a phospholipid acyl hydrolase from plasma membranes of Saccharomyces cerevisiae, *Biochim. Biophys. Acta* **711**:403–410.

Witt, W., Schweingruber, M. E., and Mertsching, A., 1984, Phospholipase B from the plasma membrane of Saccharomyces cerevisiae. Separation of two forms with different carbohydrate content, *Biochim. Biophys. Acta* **795**:108–116.

Yaffe, M. P., and Kennedy, E. P., 1983, Intracellular phospholipid movement and the role of phospholipid transfer proteins in animal cells, *Biochemistry* **22**:1497–1507.

Chapter 10

Intracellular Trafficking of Sterols

J. T. Billheimer and M. P. Reinhart

1. INTRODUCTION

Sterols are almost universally found as major membrane components of eukaryotic cells but are absent among prokaryotes. Whereas prokaryotic cells contain only a plasma membrane, eukaryotes have evolved an elaborate network of membrane-bounded organelles, or "membranelles," in addition to the plasma membrane. Interestingly, the majority of cellular free cholesterol resides in the plasma membrane, with cholesterol-to-phospholipid (C/P) ratios several-fold higher than those of intracellular membranes. Thus, the lipid aspect of intracellular membranes of eukaryotes biochemically resembles the prokaryotic plasma membrane.

Although the marked sterol enrichment of plasma membranes suggests that this membrane might be the site of sterol synthesis, efforts aimed at localizing the site of sterolgenesis have failed to support this. Rather, sterols seem to be

Abbreviations used in this chapter: ACAT, acyl-CoA-cholesterol acyltransferase; AMG, aminoglutethimide; CHO, Chinese hamster ovary; CBP, cholesterol-binding protein; CoA, coenzyme A; C/P, cholesterol-to-phospholipid ratio; DEAE, diethylaminoethyl; EM, electron microscope; ER, endoplasmic reticulum; HDL, high-density lipoprotein; HMGR, 2-hydroxy-2-methyl-glutaryl-CoA reductase; LDL, low-density lipoprotein, RER, rough endoplasmic reticulum; S_{105}, 105,000 × *g* supernatant; SCP, sterol carrier proteins, SCP_1, sterol carrier protein 1; SCP_2, sterol carrier protein 2; SER, smooth endoplasmic reticulum; SPF, supernatant protein factor.

J. T. Billheimer E. I. du Pont de Nemours & Co., Medical Products Department, Experimental Station, Wilmington, Delaware 19880-0400 **M. P. Reinhart** USDA-ARS-ERRC, Wyndmoor, Pennsylvania 19118.

generated in intracellular membranes of much lower sterol content. Implied in these findings is the existence of a trafficking mechanism capable of transporting newly synthesized sterol to target membranes, particularly the plasma membrane, against a large concentration gradient.

In eukaryotes, secretory and plasma membrane proteins (in addition to proteins of the various membranelles) are also synthesized on internal membranes and trafficked to target destinations. Over the past 2 decades a general understanding of the processes of protein targeting and trafficking has been obtained (Verner and Schatz, 1988). Unfortunately, our knowledge of intracellular lipid transport, and in particular, of cholesterol, has not kept pace with the work on protein targeting and secretion. The mechanisms of sterol trafficking are not well understood, owing to the experimental difficulties involved in such study. Three basic models have previously been described (Green, 1983): (1) spontaneous transfer via the aqueous phase (aqueous diffusion), (2) vesicular transfer, and (3) transfer via carrier proteins. As in the case of membrane and secretory proteins, it will be necessary to define the target membranes, the site of synthesis, and the timing of delivery of sterols to fully define the route of intracellular sterol transfer.

Complicating the study of sterol trafficking mechanisms is the fact that many cell types utilize sterols for purposes other than membrane biogenesis. Examples are steroid hormone production in steroidogenic tissues, bile acid production, and lipoprotein generation by hepatocytes and other cell types. Thus, many interlacing pathways of sterol movement exist in cells, and there is no *a priori* reason to believe that all are mediated by the same mechanism. Many of the known cell routes of cholesterol acquisition and movement are summarized in Figure 1.

In this review, we will discuss the evidence which suggests that trafficking occurs: the unequal distribution of sterols among membranes and the intracellular site of sterolgenesis. Further, we will examine the details that are known regarding the delivery of sterols to intracellular targets, focusing mainly on the plasma membrane and the mitochondrion. Possible mechanisms for the mediation of sterol movement will be discussed, with particular attention to the role of the so-called sterol carrier proteins (SCPs).

2. SUBCELLULAR STEROL DISTRIBUTION

Because of the difficulties encountered in preparing subcellular organelles of sufficient purity, measurement of the sterol content of individual organelles has been a particular challenge, but has been approached by using a variety of strategies: (1) conventional subcellular fractionation coupled with subsequent lipid extraction and quantitative analysis, (2) rapid plasma membrane isolation

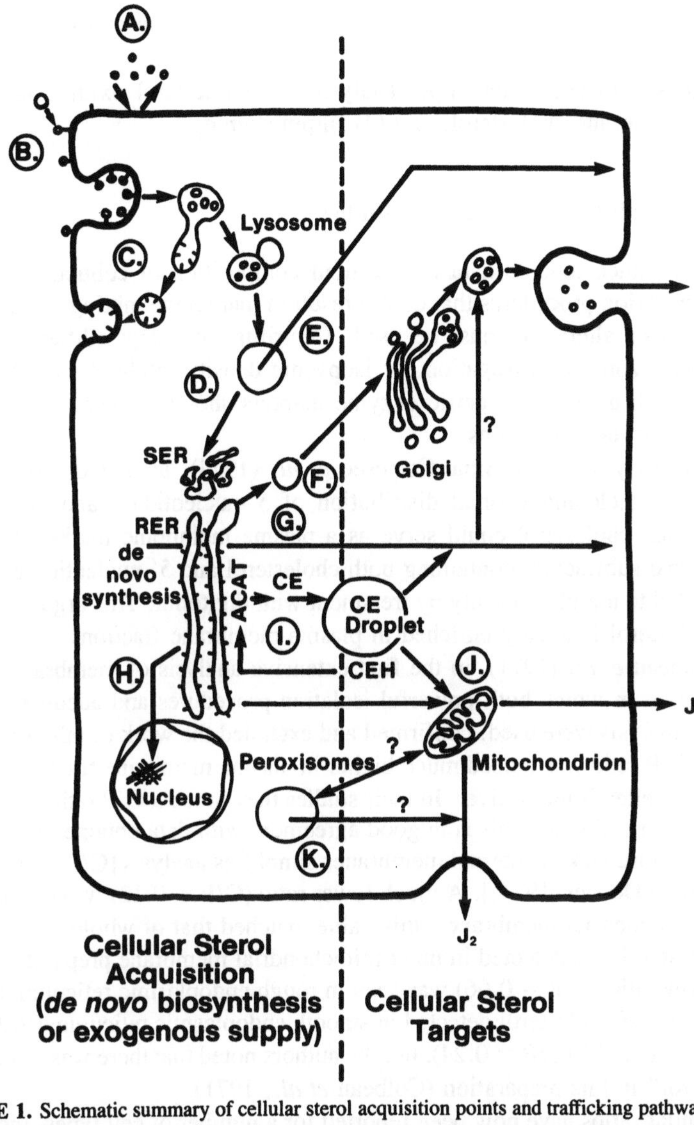

FIGURE 1. Schematic summary of cellular sterol acquisition points and trafficking pathways. Cholesterol can be acquired by or lost from the plasma membrane by interaction with circulating lipoproteins without uptake (A). It is also accumulated through receptor-mediated endocytosis (B). Cholesterol in the endosomal membrane can be recirculated to the plasma membrane (C). Lysosomes fuse with the remaining endocytosed material, and free cholesterol is released and moved to the ER (D) or directly to the plasma membrane (E). At the ER, *de novo* biosynthesis proceeds and the sterol thus generated might move through the exocytic pathway (F) or by a specific route to the plasma membrane (G). Excess sterol can be converted to regulatory oxysterols and transported to the nucleus (H). It can also be converted by ACAT to cholesteryl ester for storage (I). This ester is returned to the free cholesterol by cholesterol ester hydrolase as part of a cholesteryl ester cycle. In certain tissues, cholesterol is diverted to mitochondria (J), where steroid hormone (J1) or bile acid (J2) production commences. Peroxisomes (K) might also generate sterols and, either alone or in interaction with mitochondria, bile acids.

techniques with subsequent lipid analysis, (3) cholesterol oxidase tagging of plasma membrane cholesterol, and (4) filipin staining.

2.1. Subcellular Fractionation Studies

To produce reliable data on the lipid composition of cellular organelles, homogenization procedures that do not result in membrane mixing (by exchange or fusion) or alteration must be used. Centrifugation steps (differential centrifugation, zonal centrifugation, or isopycnic density gradient centrifugation) coupled with assays of specific enzyme markers should be used to verify the identity of isolated fractions.

The early work of Amar-Costesec *et al.* (1974), among others, demonstrated the nucleomicrosomal distribution of 5′-nucleotidase and cholesterol, arguing that cholesterol could serve as a plasma membrane marker. Only the microsomal subfraction containing both cholesterol and 5′-nucleotidase activity was shifted to a higher density by treatment with digitonin. This argues strongly that cholesterol is greatly enriched in plasma membrane fractions.

Colbeau *et al.* (1971), in the first extensive analysis of membranelle lipid composition in which both powerful isolation procedures and adequate marker enzyme analysis were used, confirmed and extended the work of others showing that the C/P ratio was indeed much higher in plasma membrane fractions than in other organelles from rat liver. In their studies they noted a C/P ratio of 0.76 for the plasma membrane. This is in good agreement with data obtained with erythrocytes, whose lack of internal membranes simplifies analysis [C/P = 0.82 (Van Deenen and DeGier, 1974)]. A much lower ratio (C/P = 0.12) was obtained for outer mitochondrial membranes; this value matched that of whole microsomes. No cholesterol was detected in inner mitochondrial membrane preparations, and only a low ratio (C/P = 0.06) was seen in rough endoplasmic reticulum (RER). The amount of cholesterol detected in smooth endoplasmic reticulum (SER) was larger than in RER (C/P = 0.24), but the authors noted that there was "nonstructural sterol" in this preparation (Colbeau *et al.*, 1971).

Similar ratios have now been reported for a number of cell types, including animal (Comte *et al.*, 1976; Schroeder *et al.*, 1976), plant (Sinensky and Strobel, 1976; Hodges *et al.*, 1972), and lower eukaryote (Ulsamer *et al.*, 1971) cells.

2.2. Rapid Plasma Membrane Isolation Techniques

Gotlib and Searls (1980) developed a technique to rapidly isolate plasma membranes from cells in suspension culture by using DEAE-Sephadex beads. The plasma membranes of the cells adhere strongly to the beads. Application of shear force ruptures the cells and releases the contents, but plasma membranes remain in large part associated with the beads. These are easily recovered for lipid analysis. This technique was used by DeGrella and Simoni (1982) in the estimation of sterol content of whole cell versus plasma membranes in Chinese hamster ovary (CHO) cells. In good agreement with the cell fractionation studies above, CHO cell plasma membranes yielded a C/P ratio of 0.93, a marked enrichment over whole cells (C/P = 0.4). Unfortunately, the technique is not applicable to other cell membranelles.

2.3. Enzymatic Labeling Of Plasma Membrane Cholesterol

Another approach to the analysis of the subcellular distribution of sterol is to selectively tag the sterol of a specific organelle. A useful enzyme for such tagging is cholesterol oxidase, which is found in a number of bacteria and oxidizes cholesterol in the presence of oxygen to cholest-4-en-3-one. Gottlieb (1977) showed that the cholesterol of the erythrocyte membrane was not subject to oxidation by the enzyme cholesterol oxidase. If, however, the cells were made leaky, resealed with internalized oxidase, or made inside out, all of the cholesterol was quickly converted to cholestenone. These studies have been extended by Lange *et al.* (1980), who found that enrichment of the membrane with exogenous cholesterol or treatment with chlorpromazine or very-low-ionic-strength iso-osmotic buffer makes erythrocyte cholesterol totally available to cholesterol oxidase attack.

After the demonstration that the oxidase does not gain access to the erythrocyte interior, the technique was applied to the analysis of cholesterol distribution in human fibroblasts, CHO cells, and rat liver hepatocytes (Lange and Ramos, 1983). Respectively, 92, 94, and 80% of the cellular cholesterol in these cells was oxidized. These values were taken to reflect the percentage of total cell cholesterol restricted to the plasma membrane. Dawidowicz (1987) recently pointed out, however, that the use of the plasma membrane cholesterol content suggested by Lange and Ramos (1983) coupled with the plasma membrane C/P ratios reported above would require that 27–40% (hepatocyte) or 30–45% (fibroblast) of the cellular phospholipid also be in the plasma membrane. These amounts are larger than those which have been reported (Pagano and Longmuir, 1983; but see also Blouin *et al.*, 1977).

In contrast to these studies, Thurnhofer *et al.* (1986) concluded that plasma membrane cholesterol becomes a substrate for cholesterol oxidase only under conditions where plasma membranes are disrupted (detergents, phospholipase C, etc.). It was concluded that cholesterol oxidase cannot be used as a probe for cholesterol distribution in brush border membranes until more is known regarding the interaction of this enzyme with membranes. Similarly, Brasaemle *et al.* (1988) recently demonstrated that magnesium ions must be present in the incubation buffer to stabilize the membrane during oxidation by cholesterol oxidase.

2.4. Filipin Staining

The polyene antibiotic filipin binds to unesterified cholesterol in membranes, causing the formation of morphologically distinct pits and can therefore be used to approach the question of sterol distribution microscopically. Samples, which must equilibrate with filipin for up to several hours, are analyzed by electron microscopy, where the frequency of appearance of pits in the membranes correlates with the concentration of cholesterol. This technique has been used to demonstrate the presence of cholesterol in a number of cell membranes, but the technique is not without limitations. Vesicles that contain cholesterol but are clathrin coated do not stain characteristically with filipin until the clathrin coat is removed. The cholesterol content of endocytic vesicles was confirmed to resemble that of plasma membrane in the elegant density shift experiments of Helmy *et al.* (1986). The cholesterol content of other membranes may be similarly shielded from filipin staining. Other sterol-specific staining procedures would be invaluable for further analysis of subcellular sterol distribution.

Although much remains to be accomplished in the measurement of sterols in cellular membranes, it is safe to conclude that sterols are severalfold more abundant in plasma membranes than in most other cell membranes. The general pattern of sterol concentrations observed follows the pattern: plasma membrane > *trans*-Golgi > cis-Golgi > SER ≫ RER and thus forms a gradient reflecting the route followed by newly synthesized secretory and plasma membrane proteins.

3. STEROL ACQUISITION

Demonstration of widely variant sterol compositions of various intracellular membranes does not establish the existence of intracellular sterol trafficking mechanisms. It is possible that each membrane contains the synthetic capacity to fill its own sterol needs. Thus, it is necessary to firmly establish the intracellular site of sterolgenesis.

Chesterton (1968) originally proposed that the endoplasmic reticulum (ER) is the site of cholesterol biosynthesis in rat liver. He injected [2-^{14}C]mevalonic acid into rats, which were sacrificed after 2, 10, or 30 min. Newly synthesized sterol and squalene were associated largely with the microsomal fraction (both granular and agranular components), although the presence of both squalene and cholesterol esters in the cytosolic fraction was noted.

With this experiment, the ER as the site of sterol synthesis became the accepted dogma. Lange and Steck (1985), however, correctly pointed out that membranes other than ER (plasma membrane, Golgi, and perhaps others) are found in the microsomal fraction (Thines-Sempoux *et al.*, 1969). The use of sophisticated double-label techniques coupled with density fractionation, digitonin, and/or cholesterol oxidase treatment of human fibroblasts, provided the first evidence that the conversion of lanosterol to cholesterol might not be carried to completion in the ER membranes. They described a membrane that contained newly synthesized cholesterol but was not the plasma membrane (it lacks 5′-nucleotidase activity), Golgi (separable following digitonin treatment), or SER [since it did not codistribute with the bulk of 2-hydroxy-2-methyl-glutaryl-coenzyme A reductase (HMGR) activity].

We have reevaluated the subcellular location of sterol biosynthesis in rat liver by localizing the cholesterolgenic enzymes (rather than newly synthesized products), reasoning that synthesis can occur only in the vicinity of the requisite enzymes (Reinhart *et al.*, 1987). Since it has been previously demonstrated that the bulk of activity of these enzymes is recovered in the microsomal elements (Gaylor and Delwiche, 1973), our studies were restricted to these. We found that not only do individual enzymes of lanosterol-to-cholesterol conversion codistribute with RER and SER, but also all of the enzymes are present there, as evidenced by their ability to generate cholesterol from mevalonic acid in the presence of cytosol. The RER localization was confirmed by EDTA treatment, which removes ribosomes and causes the RER membranes [stripped RER (sRER)] to codistribute with SER in density gradients.

Thus, the analysis of the subcellular distribution of newly synthesized cholesterol, as well as intermediates, suggests that they are topographically separate from the enzymes that mediate synthesis. To arrive at a model that accommodates both of these conclusions, a reexamination of the initial assumptions is warranted. Several sets of data exist which demonstrate that one of the most widely held assumptions, that the sterol intermediates beyond farnesyl pyrophosphate are membrane bound, might not necessarily be correct. Loud and Bucher (1958), for instance, identified a pool of squalene in rat liver cytoplasm. Chesterton (1968) also noted that some newly synthesized squalene is recovered in the S_{105} fraction of a rat liver homogenate. Bloch and co-workers (Tchen and Bloch, 1957; Tai and Bloch, 1972; Saat and Bloch, 1976; Ferguson and Bloch, 1977; Fuks-Holmberg and Bloch, 1983), as well as others (Gavey and Scallen, 1978), have demonstrated that a protein isolable with the soluble fraction is necessary for the conversion of squalene to lanosterol. This protein, known variously as supernatant protein factor (SPF) and sterol carrier protein 1 (SCP_1) also requires phospholipid for activity, suggesting that it might act as a lipoprotein-type structure that interacts with the ER membrane to facilitate sterol synthesis. Thus, at least one, and possibly many hydrophobic sterol intermediates might transfer in ping-pong-type fashion between these locations during the biosynthetic process. This idea is summarized in Figure 2.

Recent data from both our own and Dr. Lange's laboratories can be interpreted as supporting evidence for this model. Lange and Muraski (1988)

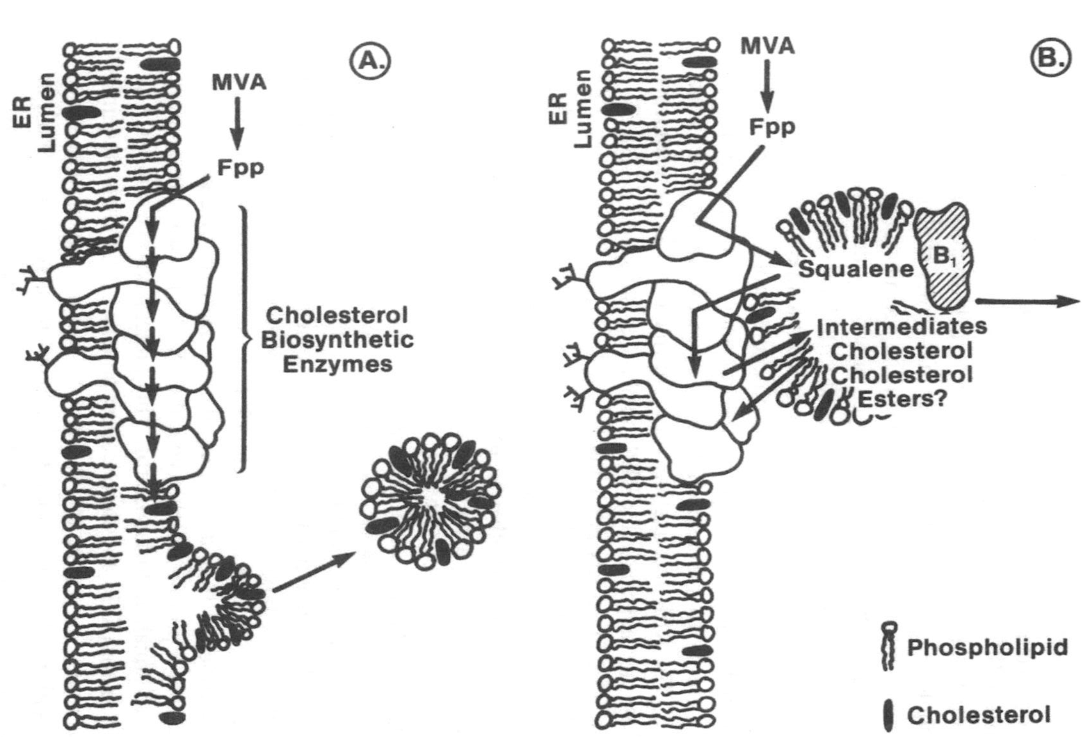

FIGURE 2. Models of release of sterol and intermediates from the biosynthetic site. Recent data suggest that sterol intermediates and cholesterol can be resolved from biosynthetic enzymes on density gradients. This might occur as a natural budding process (A) or might be driven by cell homogenization procedures. Panel B shows a cytoplasmic lipoprotein that interacts with the membrane to orchestrate cholesterol synthesis. Intermediates could (possibly by obligation) ping-pong between these compartments during the course of conversion. Such a phenomenon might explain the presence of squalene in liver cell cytoplasm and would suggest possible roles for proteins such as SPF and SCP_2. The hypothetical protein (B_1) could act as a ligand for organellar receptors to aid in the intracellular delivery of sterol, much like the function of serum apolipoproteins.

demonstrated the topographic heterogeneity of sterol synthesis. Using inhibitors of specific steps of sterol biosynthesis, they demonstrated that newly synthesized squalene, lanosterol, and cholesterol can be resolved by density gradient analysis. Interestingly, 800 × *g* supernatants were used in this work, so that a cytoplasmic rather than membrane component cannot be ruled out. We have recently observed that following *in vitro* mevalonate incorporation assays with highly purified stripped RER (with S_{105}), the biosynthetic enzymes can be resolved from a substantial portion of the newly synthesized cholesterol and intermediates (Reinhart, unpublished data). Thus, synthesis might be a collaborative process between an enzyme containing membrane (ER) and a soluble but membrane-associated lipoproteinlike particle. Such lipoproteinlike particles have been found in the cytoplasm of diverse eukaryotes (Reinhart *et al.*, 1989).

Recent immunohistochemical data suggest the presence of HMGR in rat liver peroxisomes (Keller *et al.*, 1985). Cell fractionation studies have confirmed HMGR activity in isolated peroxisomes (Keller, *et al.*, 1986). These observations have now been extended to show the synthesis of cholesterol from mevalonate by highly purified peroxisomes (Thompson *et al.*, 1987; Appelkvist, 1987). The physiological importance of peroxisomal sterolgenesis remains to be determined.

Although many details of cholesterol biosynthesis remain to be discerned, we believe that the available data support the ER as the major site of sterol biosynthesis.

In addition to *de novo* sterol biosynthesis, many cells obtain sterol and steryl esters from circulating lipoproteins via receptor-mediated endocytosis. This field has been reviewed (Brown and Goldstein, 1986) and will not be discussed here. Release of sterols from lysosomes is discussed below.

4. MOVEMENT OF STEROLS TO THE PLASMA MEMBRANE

Sterol trafficking patterns within the cell appear complex (Figure 1), and there is no *a priori* reason to assume only one operative transport mechanism. The major force driving intracellular sterol transport appears to be the maintenance and growth of the plasma membrane. Thus, the major trafficking route in most cell types is from the sites of sterol acquisition (including both *de novo* biosynthesis and endocytosis) to the plasma membrane. Elaborate regulatory mechanisms have evolved to balance sterol availability against demand. The presence of sterol in excess of cell demands for growth and maintenance causes rapid readjustment of sterol biosynthetic rates, primarily by reducing the levels of HMGR activity at several levels (transcriptional, posttranslational, and protein degradation). The receptors for circulating cholesterol [low-density lipoprotein (LDL)] are also down regulated, resulting in decreased uptake. Upon renewed demand, both systems are again up regulated. Cells also have the ability to buffer against excess free cholesterol by diverting the excess as cholesterol ester to lipid droplets. This is accomplished by acyl-CoA-cholesterol acyltransferase (ACAT), an enzyme of the ER. Lipid droplet cholesterol ester is in rapid equilibrium with free cholesterol via the enzyme cholesterol ester hydrolase; thus, although cholesteryl ester droplets are considered storage compartments, dynamic turnover is ongoing. This is known as the cholesterol ester cycle (McGookey and Anderson, 1983).

The pathway from the site of sterol generation to the plasma membrane could be a default pathway, as has been demonstrated for secretory proteins (Wieland *et al.*, 1987). Cells that store cholesterol ester droplets or use cholesterol as a precursor for other products (bile acids, steroid hormones, etc.) could obtain sterol for these functions by diverting it from the default pathway.

Little is known about the mechanism responsible for sterol delivery to the plasma membrane. The process might be mediated by (1) monomolecular diffusion, (2) vesicular transport along either an exocytic pathway or a specific vesicular pathway, or (3) specific sterol carriers including lipoproteinlike particles or specific sterol carrier proteins (Figure 3). It is important to note that delivery to other organelles might require unique trafficking modes. In addition, it must be realized that even a single route (e.g., ER to plasma membrane) might require consecutive action of several trafficking mechanisms.

4.1. Monomolecular Diffusion

Cholesterol moves rapidly, relative to phospholipids, not only across bilayers (Lange *et al.*, 1981; Brasaemle *et al.*, 1988) but between membranes as well (Bloj and Zilversmit, 1977b). It has been demonstrated that cholesterol partitions into model phospholipid membranes in the order sphingomyelin > phosphatidylcholine > phosphatidylethanolamine. Although this could explain qualitatively the enhanced localization of sterol in the plasma membrane, Wattenberg and Silbert (1983) have demonstrated that partitioning alone could not account for the quantitative sterol patterns observed in biomembranes. Also, the half time for cholesterol exchange between membranes, about 2 hr (Phillips *et al.*, 1987), is slow relative to the appearance of newly synthesized cholesterol in the plasma membrane (see Section 4.2). Thus, it is unlikely that monomolecular diffusion mediates sterol trafficking in cells.

4.2. Vesicular Trafficking

Unfortunately, many of the techniques used to discern the process of protein trafficking are not applicable to the study of the lipid phase of the membrane. Sterols undergo no postsynthetic processing which would allow label incorpora-

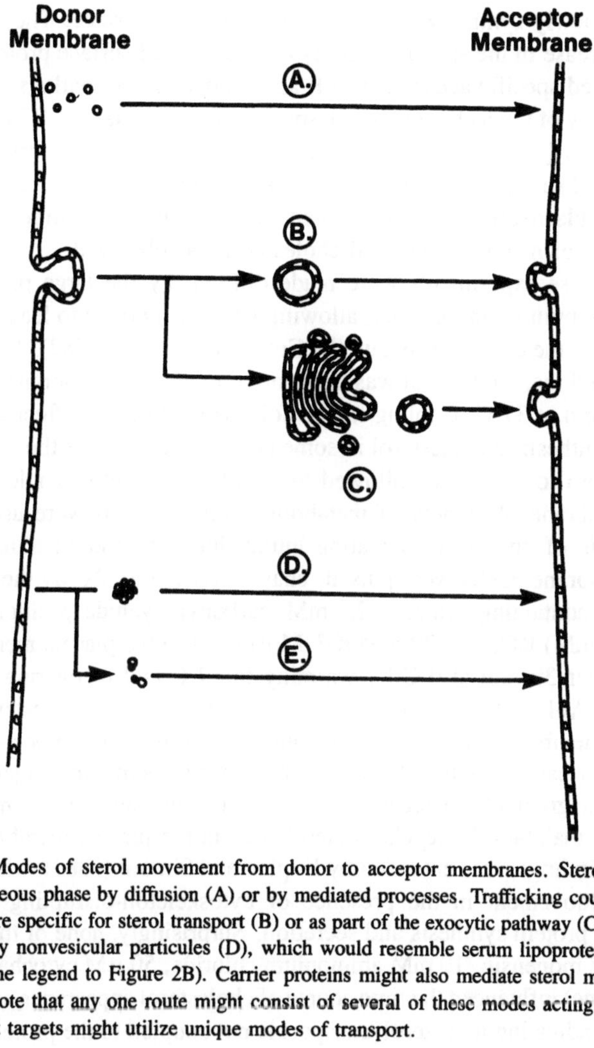

FIGURE 3. Modes of sterol movement from donor to acceptor membranes. Sterol might move across the aqueous phase by diffusion (A) or by mediated processes. Trafficking could proceed via vesicles that are specific for sterol transport (B) or as part of the exocytic pathway (C). It could also be mediated by nonvesicular particules (D), which would resemble serum lipoprotein particles (as described in the legend to Figure 2B). Carrier proteins might also mediate sterol movement. It is important to note that any one route might consist of several of these modes acting consecutively. Also, different targets might utilize unique modes of transport.

tion in specific compartments. In addition, exchange of lipids between membranes during fractionation or fixation has prevented microscopic tracing of newly synthesized sterols. Two techniques, however, have been used to assess the movement of newly synthesized sterol to the plasma membrane: (1) the DEAE bead technique, which allows rapid plasma membrane harvesting (Gotlib and Searls, 1980), and (2) the cholesterol oxidase technique for tagging plasma membrane sterol (Gottlieb, 1977).

DeGrella and Simoni (1982) used the DEAE bead assay to monitor the delivery of newly synthesized cholesterol to the cell surface. They observed a ninefold increase in the specific activity of ^{125}I-labeled surface proteins coupled with decreased specific activities of ER and mitochondrial markers, demonstrating the validity of the technique for plasma membrane isolation. When cells were incubated with [^{3}H]acetate at 37°C, label appeared in cholesterol with only a short lag (1–2 min), after which time incorporation was linear for at least 1 hr. Recovery of plasma membranes revealed a longer lag (10–12 min) before linear accumulation of newly synthesized cholesterol was observed.

Pulse-chase experiments were made difficult by the large pools of intermediates that built up in the cells, allowing label to continue to leak into cholesterol well after the chase was begun (DeGrella and Simoni, 1982). The investigators concluded that cholesterol was transported to plasma membranes with a half time of 10 min at 37°C. Cooling to 0°C following the chase effectively trapped the newly synthesized cholesterol in some compartment before the plasma membrane; a return to 37°C was followed by rapid movement of cholesterol to the plasma membrane. A variety of metabolic energy poisons were used to determine the role of energy in mediating intracellular cholesterol transport to the plasma membrane. Cells were pulsed, chased, cooled to 0°C by the addition of cold buffer containing either 0.25 mM carbonyl cyanide *p*-chlorophenylhydrazone, (2 mM) KCN, or 20 mM KF. Movement to the plasma membrane was measured after 30 min at 0°C by warming to 37°C and recovering the plasma membranes. Although each inhibitor reduced the amount of cholesterol reaching the plasma membrane, none completely abolished trafficking. None of the metabolic poisons caused sterol already in the plasma membrane (against a large concentration gradient) to leave the membrane, indicating that energy is not necessary to maintain the levels of sterol seen in the plasma membrane.

These findings were extended by Kaplan and Simoni (1985), who examined cholesterol movement in the presence of cytoskeleton-disrupting agents and inhibitors of protein synthesis and secretion. Surprisingly, none of the inhibitors used (10 μM monensin, 10 mM ammonium chloride, 50 μM cytochalasin, or 50 μM colchicine) influenced the time course of cholesterol appearance at the plasma membrane, indicating that sterol transport is not coupled to the protein secretory pathway. The authors provided evidence (density gradient analysis) that after leaving the site of synthesis (ER), but before entering the plasma membrane, cholesterol is found in a low-density, lipid-rich fraction. It was speculated that this represents a lipid-rich vesicle intermediate that delivers sterol to the plasma membrane.

Others have also provided evidence that cholesterol is transported to the plasma membrane by vesicular means. Using density gradient techniques coupled with the previously described cholesterol oxidase procedure, Lange and Steck (1985) demonstrated that cholesterol is found in a low-density membrane that is not accessible to cholesterol oxidase prior to fusion with the plasma membrane. Although the kinetics of appearance of cholesterol at the cell surface were first order, as in the work of DeGrella and Simoni (1982), the half time of delivery was longer: about 1 hr when using continuous labeling techniques versus about 2 hr in pulse-chase studies. These times cannot rule out delivery to the membrane by diffusion. The significantly lower rate of transfer observed by Lange and Matthies (1984) relative to that reported by DeGrella and Simoni (1982) might be due to the use of confluent cells in the former case.

The nature of the intermediate vesicles observed by both Kaplan and Simoni (1985) and Lange and Matthies (1984) and Lange and Steck (1985) is not known. One possibility is that it is the Golgi compartment. This would suggest that, in the same way that plasma membrane proteins flow from ER to Golgi to plasma membrane, the lipid aspect of the membrane follows a similar route. Arguing against this possibility are the kinetic data obtained by DeGrella and Simoni (1982). In agreement with others, they found that newly synthesized glycoproteins (using labeled mannose) arrived at the plasma membrane with a lag of about 20 min, whereas cholesterol arrived in only 10 min. Thus, cholesterol moves to the plasma membrane by a faster route. Wieland *et al.* (1987), however, have recently demonstrated that bulk flow (as measured by using the glycosylatable tripeptide Asn-H-Ser/Thr) procedes faster than the flow of surface proteins, with the appearance of the proteins lagging by 10 min rather than 20 min. Since this matches the rate of sterol appearance, more data are necessary to distinguish whether cholesterol moves from ER to Golgi to plasma membrane with nascent plasma membrane proteins.

The lack of complete inhibition of sterol transport to the plasma membrane by monensin argues strongly that cholesterol bypasses the Golgi en route to the plasma membrane. Special lipid-rich vesicles could transport the newly synthesized sterol. Such vesicles have already been found in *Dictyostelium discoideum* (De Silva and Siu, 1981), where they are believed to transport phospholipids to the plasma membrane in the initial stages of starvation (development). The electron microscopy (EM) data showing these vesicles, however, clearly illustrate lipid droplets as well as membranes in this fraction. Therefore, further characterization of these membranes is warranted.

4.3. Lipoprotein-Like Particles

That lipid droplets might act as sterol-transporting bodies has not, to our knowledge, been seriously considered before. A recent study, however, has demonstrated that HDL-LDL-sized lipid particles containing a small number of specific proteins are found in the cytoplasm of eukaryotes as diverse as rat, *Tetrahymena*, and *Zea mays* (Reinhart *et al.*, 1989). Such a possibility suggests that a system of organellar receptors for these particles might exist that could mediate specific lipid delivery to intracellular structures. Obviously this is speculation, but initial data warrant further experimentation.

5. STEROL TRAFFICKING TO AND FROM OTHER ORGANELLES

Certain cells utilize large amounts of cholesterol for purposes other than membrane maintenance. Examples are hepatocytes, which actively incorporate cholesterol into nascent lipoproteins prior to secretion and which generate bile acids by using cholesterol as a substrate. Steroidogenic cells also require stores of cholesterol for steroidogenesis when induced to do so. For this purpose, such cells are often able to form relatively large and numerous cholesterol ester droplets. These cells are able to divert cholesterol from what appears to be a default pathway and reroute it to the secretory apparatus and to the mitochondrian. Macrophages may also accumulate large droplets of cholesterol esters, and in the environment of the artery wall, such buildup leads to atherosclerotic plaque progression. The ability to reroute this cholesterol to the cell surface for availability of retrotransport would be beneficial in the treatment of this disease.

The sparing effects of exogenous sterols on the biosynthetic demands of the cell suggest that endocytically acquired sterol is properly trafficked to the plasma membrane. Exogenously derived steryl esters can also be utilized for membrane growth and maintenance since lysosomes contain acid steryl ester hydrolases. Currently, it is known that exogenously derived cholesterol can be transported to the ER, since a portion of it becomes esterified by the ER-specific enzyme ACAT. It is becoming apparent, however, that esterification is not a prerequisite for utilization and that a direct route to the plasma membrane might exist (W. Johnson, personal communication).

Recently, type C Niemann-Pick disease has been characterized as a disease in which free cholesterol accumulates in cells of the affected individual (Liscum and Faust, 1987; Pentchev *et al.*, 1987). This cholesterol, however, is slow to activate the normal cholesterol homeostatic mechanisms (i.e., HMGR and LDL receptor down regulation). In addition, ACAT activity is not stimulated. It is important to note that normal regulation can still be mediated after exposure of cells to mevalonate or 25-hydroxycholesterol (Liscum and Faust, 1987). These observations suggest that cholesterol is not trafficked properly to various cell organelles. Fractionation and, more recently, filipin staining of cells have confirmed that the free cholesterol accumulates in the lysosomes (Blanchette-Mackie *et al.*, 1988). Filipin staining also revealed that the Golgi, particularly the maturing cisternae and condensing vacuoles, accumulated sterol within 2 hr of LDL loading, whereas normal cells did not show such loading until 1 day. These observations indicate that a sterol transport defect plays a fundamental role in the cellular pathology of Niemann-Pick type C disease. They also indicate that at least some trafficking does occur, since cholesterol moves relatively quickly from the lysosomes to the Golgi. This process could occur as part of a recycling of endocytosed material. Further analysis of this genetic disorder should provide valuable insight into the intracellular machinery of sterol movement.

Many steroidogenic tissues contain numerous large droplets of cholesteryl ester, which serve as a supply of precursor material for hormone production. Initial steps in steroidogenesis (sterol side chain cleavage) are carried out on the inner mitochondrial membrane by cytochrome P-450$_{SCC}$. Following hormonal stimulation (e.g., ACTH in adrenals, luteinizing hormone in ovaries, etc.), cholesterol ester hydrolysis provides free cholesterol, which is rapidly (1.25 μg/mg of mitochondria per min; Crivello and Jefcoate, 1980) and specifically transported to the mitochondria for conversion. This conversion can be inhibited in several ways, including the direct inhibition of the cytochrome P-450$_{SCC}$ by aminoglutethimide (AMG). Other inhibitors, however, have been implicated in blocking the transport of cholesterol to the inner mitochondrial membrane. Hall *et al.* (1979) demonstrated that anti-actin antibodies introduced into Y1 cells prevented side chain cleavage without inhibiting the cytochrome P-450$_{SCC}$ per se. Other inhibitors of microfilaments (cytochalasin B) and of microtubules (vinblastine) can also prevent steroidogenesis in adrenal cells under some conditions (Crivello and Jefcoate, 1980) by blocking sterol transport to the mitochondria. Protein synthesis inhibitors such as cycloheximide can also prevent side chain cleavage without directly influencing cytochrome P-450$_{SCC}$, although sterol builds up in the mitochondria. It has been hypothesized that a cycloheximide activator peptide (Pederson and Brownie, 1983) is necessary to transport sterol to cytochrome P-450$_{SCC}$. Thus, a cytoplasmic and an intramitochondrial transport system might have to act sequentially to present sterol to the inner mitochondrial membrane. It has also been suggested that cholesterol sulfate might regulate the entry of intramitochondrial cholesterol into the steroidogenic pool, suggesting a reason for the high content of cholesterol sulfate in the adrenal mitochondria (Lambeth *et al.*, 1987).

Interestingly, Ma-10 Leydig tumor cells preferentially store free cholesterol instead of cholesterol ester and use this for progesterone production in response to human chorionic gonadotrophin and cyclic AMP (cAMP) analogues. Freeman (1987) has demonstrated that up to 30% of the free cholesterol in these cells resides in the mitochondria. When the cells are stimulated to produce progesterone, mitochondrial free cholesterol is depleted. By using the cholesterol oxidase technique of plasma membrane tagging, it was demonstrated that plasma membrane cholesterol becomes internalized to serve as additional substrate for steroidogenesis (Freeman, 1987). AMG blocked both side chain cleavage and plasma membrane cholesterol internalization. Plasma membrane cholesterol also can be transported directly to the ER, since it can be a substrate for ACAT under some conditions (Slotte *et al.*, 1984; Slotte and Bierman, 1987).

Although it is apparent that much remains to be characterized about sterol trafficking at the cellular level, it has been known for some time that a group of proteins (SCPs) can mediate sterol movement *in vitro* if not *in vivo*. These proteins are the focus of the remainder of this review.

6. CHOLESTEROL (STEROL) CARRIER PROTEINS

If a protein is responsible for the intracellular transport of cholesterol among organelles, it might be expected to have the following properties. It should have a high affinity for cholesterol, as well as some specificity for cholesterol as opposed to other lipids. It should be water soluble, since it must carry cholesterol through an aqueous milieu, but it should also have some hydrophobic character, since it must presumably interact with various membranes to pick up as well as deliver cholesterol. This concept has some precedent, since steroidal hormones are known to be translocated to the nucleus by cytosolic proteins that bind various steroids with high affinity and selectivity (O'Malley and Schrader, 1976).

The general procedure being used by investigators to isolate cholesterol-binding proteins (CBP) is to dry a small amount of radioactive cholesterol in a siliconized tube and incubate it with cell cytosol or to add the cholesterol in an organic solvent to the cytosol. The protein-bound cholesterol can then be separated from free cholesterol, usually by gel filtration. By this procedure, several CBPs have been isolated from various tissues (Table 1). After the initial characterization of cholesterol binding, additional information is available only for the pancreatic protein. Sziegoleit (1982) has demonstrated that the pancreatic CBP belongs to the elastase family of proteases and, therefore, the cholesterol-binding property is most probably coincidental. This emphasizes a primary problem in trying to isolate a noncatalytic protein that binds lipids; a number of enzymes, especially those that interact with membranes, may be expected to have a hydrophobic pocket that could bind cholesterol or other lipids. A similar case has been observed among proteins that bind fatty acids; ligandin, which was studied initially for its ability to bind fatty acids and acyl-CoAs, has been found to be identical to glutathione-*S*-transferase (Habig *et al.*, 1974). The ability of a protein to bind cholesterol *in vitro* does not in itself demonstrate an involvement in lipid metabolism, let alone cholesterol transport.

A second set of proteins called sterol carrier proteins that might be involved in intracellular transfer was initially isolated because of their possible involvement in cholesterol synthesis. Investigations from several laboratories demonstrated that both microsomal membranes and the cytosolic fraction from rat liver were required for the biosynthesis of cholesterol (Bucher and McGarrahan, 1956; Tchen and Bloch, 1957). Several cytosolic proteins were isolated which, when added to an *in vitro* assay, resulted in increased activity of one or more enzymic

Table I
Intracellular Sterol-Binding (Carrier) Proteins

Protein name	Molecular weight	Reference
Cholesterol-binding proteins (liver)	26,500	Erickson *et al.* (1978)
Cholesterol-binding protein (liver)	60,000	Ohta *et al.* (1982)
Cholesterol-binding protein (adrenal)	70,000	LeFevre *et al.* (1978)
Cholesterol-binding protein (pancreas)	28,000	Sziegoleit (1982)
Cholesterol-binding protein (dental cyst acid)	410,000	Higuchi *et al.* (1981)
Sterol carrier protein (fatty-acid-binding protein, Z protein)	14,000	Ritter and Dempsey (1971)
Sterol carrier protein$_1$ (supernatant protein factor)	47,000	Scallen *et al.* (1971)
Sterol carrier protein$_2$ (nonspecific lipid transfer protein)	14,000	Scallen *et al.* (1974)
Sterol carrier protein$_3$	10,000	Scallen *et al.* (1975)
Oxysterol-binding protein	236,000	Kandutsch *et al.* (1977)
Cholesterol ester transfer protein	200,000	Wetteraw and Zilversmit (1984)

steps in the synthesis of cholesterol from squalene. Since experimental results were consistent with a noncatalytic protein(s) which binds precursor sterols and makes them available for sterol-synthesizing enzymes, it was proposed that they be called sterol carrier proteins (SCPs) (Scallen *et al.*, 1971). Four different SCPs have been isolated from rat liver cytosol: squalene and sterol carrier protein (SCP) and sterol carrier proteins 1 through 3 (SCP1, SCP2, and SCP3); the similarity of nomenclature has led to much confusion in the scientific field as to the various properties each possesses. This confusion has been heightened by the fact that as structural data became available, it was discovered that some of the SCPs were identical to proteins isolated by other investigators, because they exhibited various properties which are unrelated to sterol binding (see below).

SCP was originally thought to play a general role as a vehicle for the transport of cholesterol and precursor sterols (Ritter and Dempsey, 1971). Crude preparations of the enzyme were shown to bind various sterols and to increase the *in vitro* activity of cholesterol biosynthetic enzymes. Billheimer and Gaylor (1980) purified a protein that stimulates 4-methyl sterol oxidase activity. The protein had an amino acid composition and ligand-binding properties similar to SCP and also to fatty-acid-binding protein. Fatty-acid-binding protein exhibits an affinity for a wide variety of ligands, including fatty acids, organic anions, and bile pigments, but does not bind cholesterol (Warner and Neims, 1975; Scallen *et al.*, 1985). In the last step of purification, a basic protein was removed which was capable of cholesterol transfer and was later shown to be identical to SCP2

(Trzaskos and Gaylor, 1983). It has since been confirmed that SCP and fatty-acid-binding protein are identical (Gordon *et al.*, 1983). Together, the data suggest that SCP does not bind cholesterol; however its possible involvement in cholesterol synthesis has not been entirely ruled out (Dempsey *et al.*, 1985).

SCP1 was originally isolated as one of three SCPs named according to which step(s) in cholesterol synthesis were affected (Scallen *et al.*, 1975). SCP1 increases the conversion of squalene to lanosterol but not the conversion of lanosterol to cholesterol (Srikantaiah *et al.*, 1976). A protein called supernatant protein factor (SPF), isolated by Ferguson and Bloch (1977), has been shown to be identical to SCP1 (Noland *et al.*, 1980). SCP1 increases squalene epoxidase activity in intact microsomes but not after the enzyme has been solubilized (Ono and Bloch, 1975). SCP1 is equally effective whether squalene is generated *in situ* or added to the assay medium. The exact mechanism of action is not known, but SCP1 does not appear to bind sterols; in this respect SCP1 is a misnomer.

SCP2 was originally named because of its possible involvement in the microsomal conversion of lanosterol to cholesterol (Scallen *et al.*, 1975), but has since been implicated in the intracellular transfer of cholesterol and the metabolism of cholesterol to steroidal hormones (Chanderbhan *et al.*, 1983). The characterization of SCP2 has demonstrated that it is distinct from SCP, but it is identical to nonspecific lipid transfer protein, which was isolated because of its ability to transfer phospholipids between membranes (Bloj and Zilversmit, 1977a). Section 7 is dedicated to reviewing the properties of SCP2, especially in regard to its possible involvement in the intracellular transport of sterols.

Two additional proteins have been isolated which may be involved in the transfer of cholesterol metabolites. A triglyceride and cholesteryl ester transfer protein has been partially purified from liver (Wetterau and Zilversmit, 1984, 1986). Protein activity was found in the liver and intestine, whereas negligible activity was found in the brain, heart, kidneys, or plasma. It has been suggested that the protein may play a role in the incorporation of cholesterol esters into nascent lipoproteins.

Oxygenated derivatives of cholesterol are potent suppressors of cholesterol biosynthesis in mammalian cells (Kandutsch *et al.*, 1978). The oxysterol 25-hydroxycholesterol has been shown to inhibit the activity of HMGR and the synthesis of mRNAs for the enzyme and to suppress the expression of a chimeric gene consisting of the 5′ end of the reductase gene and the coding region for chloramphenicol acetyltransferase (Kandutsch *et al.*, 1978; Luskey *et al.*, 1983; Osborne *et al.*, 1985). This regulation by oxysterols appears to involve a specific, high-affinity binding of the oxysterols to a cytosolic protein (Kandutsch *et al.*, 1977; Taylor and Kandutsch, 1987), which suggests that an oxysteroid-binding protein may have a role in regulation of transcription similar to that of the steroid hormone-binding proteins. There is a linear correlation between the relative *in vitro* binding affinities of more than 50 sterols and their activities as repressors of cholesterol synthesis in cultured cells (Taylor *et al.*, 1984, 1986). A procedure for photoaffinity labeling of the oxysterol-binding protein has been reported recently, which should facilitate the purification of the protein (Taylor *et al.*, 1988).

7. STEROL CARRIER PROTEIN2

Among the proteins initially isolated for their ability to bind (transfer) cholesterol, the one most often suggested to be involved in intracellular cholesterol transport, and therefore the one most extensively studied, has been SCP2. For simplicity, since SCP2 and nonspecific lipid transfer protein are known to be the same protein, only the term SCP2 is used in this section.

7.1. Purification and Physical Properties

SCP2 has been purified from the cytosol of rat liver (Bloj and Zilversmit, 1977; Noland *et al.*, 1980; Trzaskos and Gaylor, 1983), bovine liver (Crain and Zilversmit, 1980), goat liver (Basu *et al.*, 1988), human liver (Van Amerongen *et al.*, 1987), and rat ovary (Tanaka *et al.*, 1984). It is a basic, heat-stable, 12,500-Da protein that is somewhat unusual in that it contains none of the amino acids histidine, arginine, or tyrosine. The amino acid sequence is known for the protein isolated from rat and bovine liver; >90% homology exists between the bovine and rat SCP2 sequences (Westerman and Wirtz, 1985; Pastuszyn *et al.*, 1987; Morris *et al.*, 1988). Recent data suggest that SCP2 may be synthesized as a higher-molecular-mass (14,400-Da) precursor protein (Trzeciak *et al.*, 1987). Steinschneider *et al.* (1989) have demonstrated that SCP2 is phosphorylated *in vitro* by protein kinase C, but not by protein kinase A, calmodulin protein kinase II, or myosin light-chain kinase.

The tissue levels as well as the subcellular localization of SCP2 have been examined by using antibodies prepared against pure SCP2, with the results varying depending on the investigator and method used. One problem is that all SCP2 antibodies obtained so far also cross-react with a 55,000-Da protein, which makes the direct use of immunological techniques difficult (Teerlink *et al.*, 1984; Trzeciak *et al.*, 1987; Tzuneoka *et al.*, 1988). The relationship, if any, between the 55,000-Da protein and SCP2 is not known. Teerlink *et al.* (1984) overcame this problem by heat treating preparations prior to analysis, which removes the 55,000-Da protein. They determined the concentration of SCP2 present in the supernatant fraction of various tissues and found that it was present in highest quantity in the liver (0.78 μg/mg), followed by the intestine (0.46 μg/mg) and then all the other tissues (ca. 0.1 μg/mg). Since this initial analysis, it has been recognized that appreciable amounts of SCP2 are present in subcellular organelles. A more recent determination of SCP2 in rat liver and adrenal gland revealed that the two tissues contain similar amounts of total SCP2, but whereas two-thirds of the liver SCP2 was found in the supernatant fraction of the cell, only 19% of the adrenal SCP2 was cytosolic (Van Amerongen *et al.*, 1989). This demonstrates that analysis of only the supernatant fraction can lead to an underestimation of the cellular content of SCP2. The subcellular location of SCP2 also appeared to vary within some tissues. The SCP2 level in the mitochondria of luteal cells was increased threefold in estradiol-treated rats (McLean *et al.*, 1989), and in Leydig cells, lutropin treatment caused a twofold increase in the SCP2 level in the supernatant fraction (Van Noort *et al.*, 1986). The increase in the level in Leydig cells occurred within 2 min of hormonal stimulation and was independent of protein synthesis (Van Noort *et al.*, 1988).

Several investigators have shown by immunohistochemical staining that a peroxisomal protein reacted with SCP2 antibody (Van der Krift *et al.*, 1985; Tsuneoka *et al.*, 1988; Keller *et al.*, 1989). Upon cell fractionation and analysis of the peroxisomal fraction by immunoblot, Van der Krift *et al.* (1985) detected only the 55,000-Da protein, whereas the more recent investigations detected both SCP2 and the 55,000-Da protein. The differing results may be due to the fragile nature of peroxisomes; if they are broken during tissue homogenization, SCP2 might leak out into the soluble fraction. The likelihood of a peroxisomal location of SCP2 was strengthened by the fact that its C terminus contains what is thought to be a peroxisomal targeting signal (Morris *et al.*, 1988). Also, the SCP2 concentration is extremely low in the livers of infants exhibiting Zellweger's syndrome, a disease that affects peroxisomal assembly (Van Amerongen *et al.*, 1987). The exact amount of SCP2 present in peroxisomes has been difficult to estimate because of the cross-reacting 55,000-Da protein and the fragile nature of the organelle, but Tsuneoka *et al.* (1988) estimated that at least 50% of SCP2 is peroxisomal. A peroxisomal location of SCP2 would argue against its being involved in cholesterol transport, since this organelle has not been thought to play a major role in either cholesterol synthesis or metabolism.

7.2. Activity

SCP2 has been shown *in vitro* to increase the transfer of cholesterol from lipid droplets and liposomes to mitochondria (Noland *et al.*, 1980; Chanderbhan *et al.*, 1982; Trzaskos and Gaylor, 1983) and to increase the rate of exchange of cholesterol, lanosterol, and vitamin D_3 between liposomes and mitochondria (Billheimer, unpublished results). It has been demonstrated that SCP2 also increases the activity of several membrane-bound enzymes involved in the metabolism of cholesterol, including the synthesis of cholesterol from lanosterol (Noland *et al.*, 1980), the esterification of cholesterol by acyl-CoA:cholesterol acyltransferase (Gavey *et al.*, 1981), and the oxidation of cholesterol by side-chain cleavage enzyme or cholesterol-7α-hydroxylase (Chanderbhan *et al.*, 1982; Seltman *et al.*, 1985). The observed increase in enzymatic activity would appear to be due to the transfer of sterol from liposomes (vesicles) not containing enzyme to organelles containing the enzyme. SCP2 did not increase the activity of purified side-chain cleavage enzyme which had been reconstituted in a liposomal system (Vahouny *et al.*, 1983), 7-hydroxylase or 12-hydroxylase activity in a reconstituted system (Lidstrom-Olsson and Wikvall, 1986), or cholesterol esterification when the enzyme has been saturated with cholesterol added in Triton-WR1339 (J. T. Billheimer, unpublished results).

In addition to sterols, SCP2 has been shown to facilitate the *in vitro* transfer of the major phosphtidylglycerols, sphingomyelin, neutral glycosphingolipids, and gangliosides, but not cholesteryl ester and triglyceride, between natural and artificial membranes (Bloj and Zilversmit, 1977a, 1981; Crain and Zilversmit, 1980). The data suggested that SCP2 is nonspecific, facilitating the transfer of phospholipids as well as a number of different sterols. However, since the intracellular transport of cholesterol would necessitate the interaction between a putative carrier protein and lipid bilayers, it might be expected that SCP2 would interact with phospholipids.

Attempts to extend *in vitro* observations to the intracellular transport of cholesterol *in vivo* have, for the most part, gone unrewarded. In general, investigators have attempted to correlate the cellular content of SCP2 with the rate of cholesterol metabolism (e.g., bile acid synthesis). The level of SCP2 in the cytosol of hepatomas was found to be only 10% of the level of control livers (Crain *et al.*, 1983). However, the ability of Morris hepatoma to form cholesteryl ester is similar to that of control liver (van Heusden *et al.*, 1985), as is the synthesis of cholesterol from mevalonate (van Heusden *et al.*, 1985). Also, there was no correlation between hepatic SCP2 levels and bile acid synthesis over a 10-fold range of bile acid output (Geelen *et al.*, 1987). Although these studies have not substantiated the involvement of SCP2, they have not excluded it; it is not immediately evident that the cellular amount of SCP2 would vary even if the delivery of cholesterol is the rate-limiting step in cholesteryl ester and bile acid synthesis, which at present is not known. The best evidence for the involvement of SCP2 has been in steroidogenic cells, where it is generally accepted that the transport of cholesterol from intracellular stores to the side-chain cleavage enzyme is the rate-limiting step in steroid hormone synthesis. Trzeciak *et al.* (1987) demonstrated a threefold increase in SCP2 synthesis in adrenocortical cells treated with ACTH; moreover, the fusion of the adrenocortical cells with liposome-containing anti-SCP2 antibody decreased ACTH-stimulated steroidogenesis by 45% (Chanderbhan *et al.*, 1986).

7.3. Mechanism of Action

The term SCP implies a mechanism of action, i.e., binding of sterol and transporting it through an aqueous milieu. Unfortunately, the bulk of the data does not support SCP2 binding of sterols or phospholipids, at least not with high affinity. There are four mechanisms which could account for the SCP2-mediated transfer of sterols between membranes: (1) promotion of membrane fusion, (2) an increase in the off-rate of sterols from a membrane, (3) carrying of bound sterol across an aqueous milieu, and (4) facilitation of the close association of two membranes without either fusion or an aqueous space. The fusion of membranes would allow equilibration of all lipid components. Since triglyceride and cholesteryl ester are not transferred in the presence of SCP2 (Crain and Zilversmit, 1980), fusion appears to be ruled out as a mechanism of transfer.

In the absence of SCP2, the exchange of cholesterol between vesicles is consistent with the diffusion of cholesterol across the aqueous layer between the membranes. The rate-limiting step in diffusion is the desorption of the cholesterol from the bilayer (McLean and Phillips, 1981). Sequence data (Pastuszyn *et al.*, 1987) and monolayer studies (Demel *et al.*, 1984) demonstrate that SCP2 is an amphipathic protein. It is possible, therefore, that SCP2 increases the off-rate of cholesterol by disrupting the lipid bilayer. In fact, an increased transfer of cholesterol is observed from liposomes containing more unsaturated phospholipids in both the presence and absence of SCP2 (Bloj and Zilversmit, 1977a). However, SCP2 also increases the transfer of phosphatidylcholine, with the rate of transfer comparable to that of cholesterol, yet the desorption rate of phosphatidylcholine in the absence of SCP2 is only about 1/20 that of cholesterol (McLean and Phillips, 1981; Phillips *et al.*, 1987). Therefore, SCP2 would have to increase the desorption rate of phosphatidylcholine to a much greater extent than that of cholesterol, which is unlikely if SCP2 is affecting the transfer of both lipids in a similar manner.

Purified SCP2 does not contain any bound lipid (Chanderbhan *et al.*, 1982). In addition, experiments in which SCP2 was incubated with liposomes containing either radiolabeled phosphatidylcholine or cholesterol, no radiolabel was found associated with the reisolated SCP2 (Van Amerongen *et al.*, 1985; Billheimer, unpublished results). Indirect evidence for cholesterol binding was demonstrated by Chanderbahn *et al.* (1982). They incubated adrenal lipid droplets labeled with [^{14}C]cholesterol with SCP2 and noted an increase in the level of radiolabeled cholesterol in the infranatant after centrifugation. They suggest that the [^{14}C]cholesterol is bound to SCP2, although no SCP2–cholesterol complex was isolated. The studies suggest that if bound, cholesterol rapidly dissociates from SCP2; SCP2 does not function as a classic binding protein.

It is probable, therefore, that SCP2 facilitates the transport of cholesterol by allowing a close association between donor and acceptor membranes, which promotes the intermembrane exchange of lipids (Figure 4). This mechanism, in which a molecule of SCP2 associates with both the donor and acceptor membrane and a dimerization of SCP2 causes the close association of the two bilayers, was first proposed by Van Amerongen *et al.* (1985). One molecule of SCP2 with two bilayer-binding domains is also possible. In either model, the

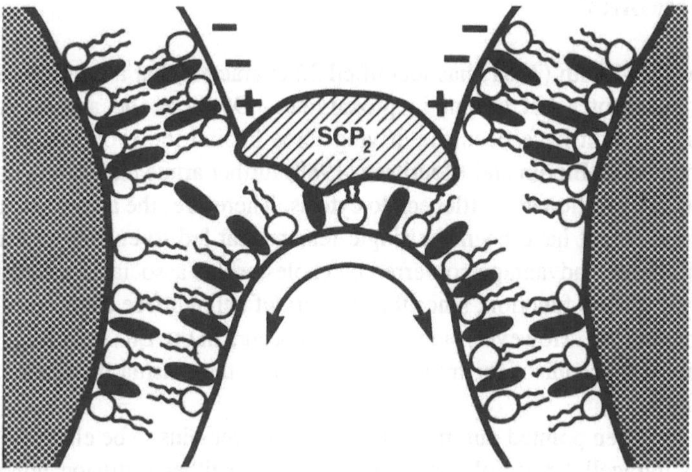

FIGURE 4. Possible mode of action of SCP_2 in facilitating the transfer of lipids between membranes.

negatively charged phospholipids would aid the association of the positively charged SCP2 with the bilayer.

7.4. Physiological Role

The physiological role of SCP2 remains an enigma. Much of the data concerning the properties of SCP2 are contradictory, and so its tissue concentration and subcellular location are uncertain. The ability of SCP2 to facilitate the *in vitro* transfer of cholesterol is not in dispute. However, at least two other proteins, ligandin (glutathione-*S*-transferase) and CBP (elastase), which were initially isolated because of their lipid-binding properties, were shown upon further analysis not to be involved in lipid metabolism (Sziegoleit, 1982; Habig *et al.*, 1974). The above data suggest that SCP2 probably does not act as a classic binding protein which carries cholesterol between membranes. However, if cholesterol is transported in a vesicle, SCP2 may be important in loading and/or unloading the cholesterol. Definitive evidence on the involvement of SCP2 in the intracellular trafficking of cholesterol will come only when it is shown that in an intact cell, affecting the amount or activity of SCP2 affects sterol transport accordingly. Recently, a cDNA of SCP2 was isolated which should aid in such investigations (Moncecchi *et al.*, 1987). Insertion of a construct of DNA coding for antisense message of SCP2 might produce an SCP2-negative cell. The ability of a cell to transport cholesterol to the plasma membrane or other membranelles in the absence of SCP2 could then be tested.

8. SUMMARY

Cavalier-Smith (1981) has identified 22 characters that are universally present in eukaryotes but absent in prokaryotes. Of these, he argues that one, exocytosis, might have been the driving force behind the evolution of modern eukaryotic cells. Bloom and Mouritsen (1988) further argue that sterols may have removed an evolutionary bottleneck to cytosis. Therefore, the advent of sterols in membranes might have been the single feature that led to eukaryote evolution. The evolutionary advantage conferred by cholesterol is associated primarily with plasma membrane function, since the majority of cellular free cholesterol resides in that membrane. However, sterol synthesis occurs in the ER; therefore, the cell must have a mechanism for transporting sterol to the plasma membrane and its regulation.

As has been pointed out in this review, much remains to be elucidated in the study of intracellular sterol trafficking. To date, neither diffusion nor vesicle-mediated transport can be fully confirmed or ruled out. Microtubule and microfilament involvement appears important in some routes (e.g., mitochondria) but not in others. In addition, trafficking roles of cytoplasmic lipoproteinlike particles have not been addressed. Finally, although some "sterol carrier proteins" demonstrate the ability to mediate intervesicular transfer of cholesterol *in vitro,* the true physiological role of these proteins remains obscure.

Future research in this field awaits the refinement of available techniques. Particularly valuable would be cytochemical methods for detection of sterol at the ultrastructural level. Possibly, direct microscopic visualization of radiolabeled components in cells represents the necessary approach. Purification of elements carrying newly synthesized sterols would allow the proteins mediating transport to be identified. Continued analysis of mutants defective in transport, such as in type C Niemann-Pick disease, will shed light on this complex problem.

The importance of extracellular trafficking of cholesterol owing to its involvement in the progression of atherosclerosis, has been emphasized in recent years. Little emphasis has been placed on intracellular trafficking of sterol; however, it can be argued that such transport also plays a major role in atherosclerosis, possibly by fueling retrotransport of cholesterol to the liver and secretion in the bile. Therefore, we hope this review will serve to stimulate research interest in this area.

9. REFERENCES

Amar-Costesec, A., Wibo, M., Thines-Sempoux, D., Beaufay, H., and Berthet, J., 1974, Analytical study of microsomes and isolated subcellular membranes from rat liver. IV. Biochemical,

physical, and morphological modifications of microsomal components induced by digitonin, EDTA, and pyrophosphate, *J. Cell Biol.* **62:**717–745.
Appelkvist, E. L., 1987, In vitro labeling of peroxisomal cholesterol with radioactive precursors, *Biosci. Rep.* **7:**853–858.
Basu, J., Kundu, M., Bhattacharya, U., Mazumder, C., and Chakrabarti, P., 1988, Purification and characterisation of a non-specific lipid transfer protein from goat liver, *Biochim. Biophys. Acta* **959:**134–142.
Billheimer, J. T., and Gaylor, J. L., 1980, Cytosolic modulators of activities of microsomal enzymes of cholesterol biosynthesis, *J. Biol. Chem.* **255:**8128–8135.
Blanchette-Mackie, E. J., Dwyer, N. K., Amende, L. M., Kruth, M. S., Butter, J. D., Sokol, J., Comly, M. E., Vanier, M. T., August, J. T., Brady, R. O., and Pentcher, P. G., 1988, Type-C Nieman-Pick disease: Low density lipoprotein uptake is associated with premature cholesterol accumulation in the Golgi complex and excessive cholesterol storage in lysosomes, *Proc. Natl. Acad. Sci. U.S.A.* **85:**8022–8026.
Bloj, B., and Zilversmit, D. B., 1977a, Rat liver proteins capable of transferring phosphatidylethanolamine-purification and transfer activity for other phospholipids and cholesterol, *J. Biol. Chem.* **252:**1613–1619.
Bloj, B., and Zilversmit, D. B., 1977b, Complete exchangeability of cholesterol in phosphatidylcholine/cholesterol vesicles of different degrees of unsaturation, *Biochemistry* **16:**3943–3948.
Bloj, B., and Zilversmit, D. B., 1981, Accelerated transfer of neutral glycosphingolipids and ganglioside G by a purified lipid transfer protein, *J. Biol. Chem.* **256:**5988–5991.
Bloom, M., and Mouritsen, O. G., 1988, The evolution of membranes, *Can. J. Chem.* **66:**706–712.
Blouin, A., Bolender, R. P., and Weibel, E. R., 1977, Distribution of organelles and membranes between hepatocytes and nonhepatocytes in the rat liver parenchyma, *J. Cell Biol.* **72:**441–455.
Brasaemle, D. L., Robertson, A. D., and Attie, A. D., 1988, Transbilayer movement of cholesterol in the human erythrocyte membrane, *J. Lipid Res.* **29:**481–489.
Brown, M. S., and Goldstein, J. L., 1986, A receptor-mediated pathway for cholesterol homeostasis, *Science* **232:**34–47.
Bucher, N. L. R., and McGarrahan, K., 1956, The biosynthesis of cholesterol from acetate-1-C^{14} by cellular fractions of rat liver, *J. Biol. Chem.* **222:**1–15.
Cavalier-Smith, T., 1981, In: Molecular and cellular aspects of microbial evolution. Society for General Microbiology, ltd., Symposium 32, Carlie, M. J., Collins, J. F., and Mosely, B. E. B., eds., Cambridge University Press, Cambridge, England, pp. 33–84.
Chanderbhan, R., Noland, B. J., Scallen, T. J., and Vahouny, G. V., 1982, sterol carrier protein$_2$, delivery of cholesterol from adrenal lipid droplets to mitochondria for pregnenolone synthesis, *J. Biol. Chem.* **257:**8928–8934.
Chanderbhan, R., Tanaka, T., Strauss, J. F., Irwin, D., Noland, B. J., Scallen, T. J., and Vahouny, G. V., 1983, Evidence for sterol carrier protein$_2$-like activity in hepatic, adrenal and ovarian cytosol, *Biochem. Biophys. Res. Commun.* **117:**702–709.
Chanderbhan, R. F., Kharroubi, A. T., Noland, B. J., Scallen, T. J., and Vahouny, G. V., 1986, Sterol carrier protein$_2$: Further evidence for its role in adrenal steroidogenesis, *Endocrine Res.* **12:**351–370.
Chesterton, C. J., 1968, Distribution of cholesterol precursors and other lipids among rat liver intracellular structures: Evidence for the endoplasmic reticulum as the site of cholesterol and cholesterol ester synthesis, *J. Biol. Chem.* **243:**1147–1151.
Colbeau, A., Nachbaur, J., and Vignais, P. M., 1971, Enzymic characterization and lipid composition of rat liver subcellular membranes, *Biochim. Biophys. Acta* **249:**462–492.
Comte, J., Maisterrena, B., and Gautheron, D. C., 1976, Lipid composition and protein profiles of outer and inner membranes from pig heart mitochondria—comparison with microsomes, *Biochim. Biophys. Acta* **419:**271–284.

Crain, R. C., and Zilversmit, D. B., 1980, Two nonspecific phospholipid exchange proteins from beef liver. 1. Purification and characterization, *Biochemistry* **19:**1433–1439.

Crain, R. C., Clark, R. W., and Harvey, B. E., 1983, Role of lipid transfer proteins in the abnormal lipid content of morris hepatoma mitochondria and microsomes, *Cancer Res.* **43:**3197–3203.

Crivello, J. F., and Jefcoate, C. R., 1980, Intracellular movement of cholesterol in rat adrenal cells. Kinetics and effects of inhibitors, *J. Biol. Chem.* **255:**8144–8151.

Dawidowicz, E. A., 1987, Dynamics of membrane lipid metabolism and turnover, *Annu. Rev. Biochem.* **56:**43–61.

DeGrella, R. F., and Simoni, R. D., 1982, Intracellular transport of cholesterol to the plasma membrane, *J. Biol. Chem.* **257:**14256–14262.

Demel, R. A., Louwers, H., Jackson, R. L., and Wirtz, K. W. A., 1984, Protein-mediated lipid transfer between monolayers and bilayers, *Colloids Surf.* **10:**301–311.

Dempsey, M. E., Hargis, P. S., McGuire, D. M., McMahon, A., Olson, C. D., Salati, L. M., Clarke, S. D., and Towle, H. C., 1985, Role of sterol carrier protein in cholesterol metabolism, *Chem. Phys. Lipids* **38:**223–237.

DeSilva, N. S., and Siu, C.-H., 1981, Vesicle-mediated transfer of phospholipids to plasma membrane during cell aggregation of Dicytostelium discoideum, *J. Biol. Chem.* **256:**5845–5850.

Erickson, S. K., Meyer, D. J., and Gould, R. G., 1978, Purification and characterization of a new cholesterol binding protein from rat liver cytosol, *J. Biol. Chem.* **253:**1817–1826.

Ferguson, J. B., and Bloch, K., 1977, Purification and properties of a soluble protein activator of rat liver squalene epoxidase, *J. Biol. Chem.* **252:**5381–5385.

Freeman, D. A., 1987, Cyclic AMP mediated modification of cholesterol traffic in leydig tumor cells, *J. Biol. Chem.* **262:**13061–13068.

Fuks-Holmberg, D., and Bloch, K., 1983, Intermembrane transfer of squalene promoted by supernatant protein factor, *J. Lipid Res.* **24:**402–408.

Gavey, K. L., and Scallen, T. J., 1978, Studies on the conversion of enzymatically generated microsome-bound squalene to sterol, *J. Biol. Chem.* **253:**5476–5483.

Gavey, K. L., Noland, B. J., and Scallen, T. J., 1981, The participation of sterol carrier protein$_2$ in the conversion of cholesterol to cholesterol ester by rat liver microsomes, *J. Biol. Chem.* **6:**2993–2999.

Gaylor, J. L., and Delwiche, C. V., 1973, Investigation of the multienzymic system of microsomal cholesterol biosynthesis, *Ann. N.Y. Acad. Sci.* **212:**122–138.

Geelen, M. J. H., Beynen, A. C., and Wirtz, W. A., 1987, Cholesterol metabolism and sterol carrier protein$_2$ (non-specific lipid transfer protein), *Int. J. Biochem.* **19:**619–623.

Gordon, J. I., Alpers, D. H., Ockner, R. K., and Strauss, A. W., 1983, The nucleotide sequence of rat liver fatty acid binding protein mRNA, *J. Biol. Chem.* **5:**3356–3363.

Gotlib, L. J., and Searls, D. B., 1980, Plasma membranes isolation of DEAE-sephadex beads, *Biochim. Biophys. Acta* **602:**207–212.

Gottlieb, M. H., 1977, The reactivity of human erythrocyte membrane cholesterol with a cholesterol oxidase, *Biochim. Biophys. Acta* **466:**422–428.

Green, C., 1983, The movement of cholesterol within cells, in Sterol biosynthesis and function, Suckling, K. E., and Baldwin, B. C., eds., *Biochem. Soc. Trans.* **11:**637–639.

Habig, W. H., Pabst, M. J., Fleischner, G., Gatmaitan, Z., Arias, I. M., and Jakoby, W. B., 1974, The identity of glutathione s-transferase B with ligandin, a major binding protein of liver, *Proc. Natl. Acad. Sci. U.S.A.* **71:**3879–3882.

Hall, P. F., Charpponnier, C., Nakamura, M., and Gabbiani, G., 1979, The role of microfilaments in the response of adrenal tumor cells to adrenocorticotropic hormone, *J. Biol. Chem.* **254:**9080–9084.

Hashimoto, S., Drevon, C. A., Weinstein, D. B., Bernett, J. S., Dayton, S., and Steinberg, D., 1983, Activity of acyl-CoA:cholesterol acyltransferase and 3-hydroxy-3-methylglutaryl-CoA reductase in subfractions of hepatic microsomes enriched with cholesterol, *Biochim. Biophys. Acta* **754:**126–133.

Helmy, S., Porter-Jordan, K., Dawidowicz, E. A., Pilch, P., Schwartz, A. L., and Fine, R. E., 1986, Separation of endocytic from exocytic coated vesicles using a novel cholinesterase mediated density shift technique, *Cell* **44:**497–506.

Higuchi, K., Hayakawa, M., Yoshimitsu, A., and Hirotsugu, I., 1981, Comparative studies on a heat-stable cholesterol-binding protein in dental cyst fluid and serum, *Int. J. Biochem.* **13:**777–782.

Hodges, T. K., Leonard, R. T., Bracker, C. E., and Keenan, T. W., 1972, Purification of an ion stimulated adenosine triphosphatase from plant roots: Association with plasma membranes, *Proc. Natl. Acad. Sci. U.S.A.* **69:**3307–3311.

Kandutsch, A. A., Chen, H. W., and Shown, E. P., 1977, Binding of 25-hydroxycholesterol and cholesterol to different cytosolic proteins, *Proc. Natl. Acad. Sci. U.S.A.* **74:**2500–2503.

Kandutsch, A. A., Chen, H. W., and Heiniger, H. J., 1978, Biological activity of some oxygenated sterols, *Science* **201:**498–501.

Kaplan, M. R., and Simoni, R. D., 1985, Transport of cholesterol from the endoplasmic reticulum to the plasma membrane, *J. Cell Biol.* **101:**446–453.

Keller, G. A., Barton, M. C., Shapiro, D. J., and Singer, S. S., 1985, 3-Hydroxy-3-methylglutaryl-coenzyme A reductase is present in peroxisomes in normal rat liver cells, *Proc. Natl. Acad. Sci. U.S.A.* **82:**770–774.

Keller, G. A., Pazirandeh, M., and Krisans, S., 1986, 3-Hydroxy-3-methylglutaryl coenzyme A reductase localization in rat liver peroxisomes and microsomes of control and cholestyramine-treated animals: Quantitative biochemical and immunoelectron microscopical analyses, *J. Cell Biol.* **103:**875–886.

Keller, G. A., Scallen, T. J., Clarke, D., Maher, P. A., Krisans, S. K., and Singer, S. J., 1989, Subcellular localization of sterol carrier protein$_2$ in rat hepatocytes: Its primary localization to peroxisomes, *J. Cell Biol.* **108:**1353–1361.

Lambeth, J. D., Xu, X. X., and Glover, M., 1987, Cholesterol sulfate inhibits adrenal mitochondrial cholesterol side chain cleavage at a site distinct from cytochrome P-450. Evidence for an intramitochondrial cholesterol translocator, *J. Biol. Chem.* **262:**9181–9188.

Lange, Y., and Matthies, H. J. G., 1984, Transfer of cholesterol from its site of synthesis to the plasma membrane, *J. Biol. Chem.* **259:**14624–14630.

Lange, Y., and Muraski, M. F., 1988, Topographic heterogeneity in cholesterol biosynthesis, *J. Biol. Chem.* **263:**9366–9373.

Lange, Y., and Ramos, B. V., 1983, Analysis of the distribution of cholesterol in the intact cell, *J. Biol. Chem.* **258:**15130–15134.

Lange, Y., and Steck, T. L., 1985, Cholesterol-rich intracellular membranes: A precursor to the plasma membrane, *J. Biol. Chem.* **260:**15592–15597.

Lange, Y., Cutler, M. B., and Steck, T. L., 1980, The effect of cholesterol and other intercalated amphipaths on the contour and stability of the isolated red cell membrane, *J. Biol. Chem.* **255:**9331–9337.

Lange, Y., Dolde, J., and Steck, T. L., 1981, The rate of transmembrane movement of cholesterol in the human erythrocyte, *J. Biol. Chem.* **256:**5321–5323.

Lefevre, A., Morera, A. M., and Saez, J. M., 1978, Adrenal cholesterol-binding protein: Properties and partial purification, *FEBS Lett.* **89:**287–292.

Lidstrom-Olsson, B., and Wikvall, K., 1986, The role of sterol carrier protein$_2$ and other hepatic lipid-binding proteins in bile-acid biosynthesis, *Biochem. J.* **238:**879–884.

Liscum, L., and Faust, J. R., 1987, Low density lipoprotein (LDL)-mediated suppression of cholesterol synthesis and LDL uptake is defective in Nieman-Pick type C fibroblasts, *J. Biol. Chem.* **262:**17002–17608.

Loud, A. V., and Bucher, N. L. R., 1958, The turnover of squalene in relation to the biosynthesis of cholesterol, *J. Biol. Chem.* **233**:37–41.

Luskey, K. L., Faust, J. R., Chin, D. J., Brown, M. S., and Goldstein, J. L., 1983, Amplification of the gene for 3-hydroxy-3-methylglutaryl coenzyme A reductase, but not for the 53-kDa protein, in UT-1 cells, *J. Biol. Chem.* **258**:8462–8469.

McGookey, D. J., and Anderson, R. G. W., 1983, Morphological characterization of the cholesteryl ester cycle in cultured mouse macrophage foam cells, *J. Cell Biol.* **97**:1156–1168.

McLean, L. R., and Phillips, M. C., 1981, Mechanism of cholesterol and phosphatidylcholine exchange of transfer between unilamellar vesicles, *Am. Chem. Soc.* **20**:2893–2900.

McLean, M. P., Puryear, T. K., Khan, I., Azhar, S., Billheimer, J. T., Orly, J., and Gibori, G., 1989, Estradiol regulation of sterol carrier protein-2 independent of cytochrome P450 side chain cleavage expression in the rat corpus luteum, *Endocrinology* **125**:1337–1344.

Moncecchi, D., Keightley, J. A., Simmons, P. C., and Scallen, T. J., 1987, Isolation and nucleotide sequence of mouse liver sterol carrier protein$_2$ cDNA, *Fed. Proc.* **46**:2188, abstract 1534.

Morris, H. R., Larsen, B. S., and Billheimer, J. T., 1988, A mass spectrometric study of the structure of sterol carrier protein SCP_2 from rat liver, *Biochem. Biophys. Res. Commun.* **154**:476–482.

Noland, B. J., Arebalo, R. E., Hansbury, E., and Scallen, T. J., 1980, Purification and properties of sterol carrier protein$_2$, *J. Biol. Chem.* **255**:4282–4289.

Ohta, M., Yamano, T., and Miyake, Y., 1982, Purification and some properties of a new cholesterol binding protein from rat liver cytosol, *Biomed. Res.* **3**:239–243.

O'Malley, B. W., and Schrader, W. T., 1976, The receptors of steroid hormones, *Sci. Am.* **234**:32–43.

Ono, T., and Bloch, K., 1975, Solubilization and partial characterization of rat liver squalene epoxidase, *J. Biol. Chem.* **250**:1571–1579.

Osborne, T. F., Goldstein, J. L., and Brown, M. S., 1985, 5′ end of HMG CoA reductase gene contains sequences responsible for cholesterol-mediated inhibition of transcription, *Cell* **42**:203–212.

Pagano, R. E., and Longmuir, K. J., 1983, Intracellular translocation and metabolism of fluorescent lipid analogues in cultured mammalian cells, *Trends Biochem. Sci.* **8**:157–161.

Pastusyzn, A., Noland, B. J., Bazan, J. F., and Fletterick, R. J., 1987, Primary sequence and structural analysis of sterol carrier protein$_2$ from rat liver: Homology with immunoglobulins, *J. Biol. Chem.* **262**:13219–13227.

Pederson, R. C., and Brownie, A. C., 1983, Cholesterol side-chain cleavage in the rat adrenal cortex: Isolation of a cycloheximide-sensitive activator peptide, *Proc. Natl. Acad. Sci. U.S.A.* **80**:1882–1886.

Pentchev, R. G., Comly, M. E., Kruth, M. S., Tokoro, T., Butler, J., Sokol, J., Filling-Katz, M., Quirk, J. M., Marshall, D. C., Patel, S., Vanier, M. T., and Brady, R. O., 1987, Group C Neimann-Pick disease: Faulty regulation of low-density lipoprotein uptake and cholesterol storage in cultured fibroblasts, *FASEB J.* **1**:40–45.

Phillips, M. C., Johnson, W. J., and Rothblat, G. H., 1987, Mechanisms and consequences of cellular cholesterol exchange and transfer, *Biochim. Biophys. Acta* **906**:223–276.

Rahman, K., and Coleman, R., 1987, Effect of chloroquine on biliary lipid and lysosomal enzyme output in the isolated perfused rat liver at low bile salt output rates, *Biochim. Biophys. Acta* **922**:395–397.

Reinhart, M. P., Billheimer, J. T., Faust, J. R., and Gaylor, J. L., 1987. Subcellular localization of the enzymes of cholesterol biosynthesis and metabolism in rat liver, *J. Biol. Chem.* **262**:9649–9655.

Reinhart, M. P., Billheimer, J. T., and Usher, D., 1989, Cytolipophorins: Lipoprotein-like particles in the cytoplasm of phylogenetically diverse eukaryotes, *J. Cell Biol.* **107**:654a (abstr. no. 3715).

Ritter, M. C., and Dempsey, M. E., 1971, Specificity and role in cholesterol biosynthesis of a squalene and sterol carrier protein, *J. Biol. Chem.* **246:**1536–1547.

Saat, Y. A., and Bloch, K. E., 1976, Effect of a supernatant protein on microsomal squalene epoxidase and 2,3-oxidosqualene-lanosterol cyclase, *J. Biol. Chem.* **251:**5155–5160.

Scallen, T. J., Schuster, M. W., Dhar, A. K., and Skirdlant, H. B., 1971, Enzymic synthesis of cholesterol: Use of a liver acetone powder, *Lipids* **6:**162–164.

Scallen, T. J., Srikantaiah, M. V., Seetharam, B., Hansbury, E., and Gavey, K. L., 1974, Sterol carrier protein hypothesis, *Fed. Proc.* **33:**1733–1746.

Scallen, T. J., Seetharam, B., Srikantaiah, M. V., Hansbury, E., and Lewis, M. K., 1975, Sterol carrier protein hypothesis: requirement for three substrate-specific soluble proteins in liver cholesterol biosynthesis, *Life Sciences* **16:**853–874.

Scallen, T. J., Noland, B. J., Gavey, K. L., Bass, N. M., Ockner, R. K., Chanderbhan, R., and Vahouny, G. V., 1985a, Sterol carrier $protein_2$ and fatty acid-binding protein, *J. Biol. Chem.* **260:**4733–4739.

Scallen, T. J., Pastuszyn, A., Noland, B. J., Chanderbhan, R., Kharroubi, A., and Vahouny, G. V., 1985b, Sterol carrier and lipid transfer proteins, *Chem. Phys. Lipids* **38:**239–261.

Schroeder, F., Perlmutter, J. F., Glaser, M., and Vagelos, P. R., 1976, Isolation and characterization of subcellular membranes with altered phospholipid composition from cultured fibroblasts, *J. Biol. Chem.* **251:**5015–5026.

Seltman, H., Diven, W., Rizk, M., Noland, B. J., Chanderbhan, R., Scallen, T. J., Vahouny, G., and Sanghvi, A., 1985, Regulations of bile-acid synthesis of 7a-hydroxycholesterol, *Biochem. J.* **230:**19–24.

Singer, I. I., Scott, S., Kazazis, D. M., and Huff, J. W., 1988, Lovastatin, an inhibitor of cholesterol synthesis, induces hydroxymethylglutaryl-coenzyme A reductase directly on membranes of expanded smooth endoplasmic reticulum in rat hepatocytes, *Proc. Natl. Acad. Sci. U.S.A.* **85:**5264–5268.

Sinensky, M., and Strobel, G., 1976, Chemical composition of cellular fractions enriched in plasma membranes from sugar cane, *Plant Sci. Lett.* **6:**209–214.

Slotte, J. P., and Bierman, E. L., 1987, Movement of plasma-membrane sterols to the endoplasmic reticulum in cultured cells, *Biochem. J.* **248:**237–242.

Slotte, P. J., Lundberg, B., and Bjorkerud, S., 1984, Intracellular transport and esterification of exchangeable cholesterol in cultured human lung fibroblasts, *Biochim. Biophys. Acta* **793:**423–428.

Srikantaiah, M. V., Hansbury, E., Loughran, E. D., and Scallen, T. J., 1976, Purification and properties of sterol carrier $protein_1$, *J. Biol. Chem.* **251:**5496–5504.

Steinschneider, A., McLean, M. P., Billheimer, J. T., Azhar, S., and Gibori, G., 1989, Protein kinase C catalyzed phopshorylation of sterol carrier $protein_2$, *Endocrinology* **125:**569–571.

Sziegoleit, A., 1982, Purification and characterization of a cholesterol-binding protein from human pancreas, *Biochem. J.* **207:**573–582.

Sziegoleit, A., 1984, A novel proteinase from human pancreas, *Biochem. J.* **219:**735–742.

Tai, M.-H., and Bloch, K., 1972, Squalene epoxidase of rat liver, *J. Biol. Chem.* **247:**3767–3773.

Tanaka, T., Billheimer, J. T., and Strauss, J. F., 1984, Luteinized rat ovaries contain a sterol carrier protein, *Endocrinology* **114:**533–540.

Taylor, F. R., and Kandutsch, A. A., 1987, The oxysterol receptor, in *Sterol/Steroid Hormone Mechanism of Action* (R. Kumar and T. Spelsberg, eds.), pp. 395–407. Martinus Nijhoff, Boston.

Taylor, F. R., Saucier, S. E., Shown, E. P., Parish, E. J., and Kandutsch, A. A., 1984, Correlation between oxysterol binding to a cytosolic binding protein and potency in the repression of hydroxymethylglutaryl coenzyme A reductase, *J. Biol. Chem.* **259:**12382–12387.

Taylor, F. R., Kandutsch, A. A., Gayen, A. K., Nelson, J. A., Nelson, S. S., Phirwa, S., and

Spencer, T. A., 1986, 24,25-Epoxysterol metabolism in cultured mammalian cells and repression of 3-hydroxy-3-methylglutaryl-CoA reductase, *J. Biol. Chem.* **261**:15039–15044.

Taylor, F. A., Kandutsch, A. A., Anzalone, L., Phirwa, S., and Spencer, J. A., 1988, Photo affinity labeling of the oxysterol receptor, *J. Biol. Chem.* **263**:2264–2269.

Tchen, T. T., and Bloch, K., 1957, On the conversion of squalene to lanosterol in vitro, *J. Biol. Chem.* **226**:921–930.

Teerlink, T., Van Der Krift, T. P., Van Heusden, G. P., and Wirtz, K. W. A., 1984, Determination of nonspecific lipid transfer protein in rat tissues and morris hepatomas by enzyme immunoassay, *Biochim. Biophys. Acta* **793**:251–259.

Thines-Sempoux, D., Amar-Costesec, A., Beaufray, H., and Berthet, J., 1969, The association of cholesterol 5′-nucleotidase and alkaline phosphodiesterase with a distinct group of microsomal particles, *J. Cell Biol.* **43**:189–192.

Thompson, S. L., Burrows, R., Laub, R. J., and Krisans, S. K., 1987, Cholesterol synthesis in rat liver perosixomes. Conversion of mevalonic acid to cholesterol, *J. Biol. Chem.* **262**:17420–17425.

Thurnhofer, M., Gains, N., Mutsch, B., and Houser, H., 1986, Cholesterol oxidase as a structural probe of biological membranes: Its application to brush border membrane, *Biochim. Biophys. Acta* **856**:174–181.

Trzaskos, J. M., and Gaylor, J. L., 1983, Cytosolic modulators of activities of microsomal enzymes of cholesterol biosynthesis. Purification and characterization of a non-specific lipid-transfer protein, *Biochim. Biophys. Acta* **751**:52–65.

Trzeciak, W. H., Simpson, E. R., Scallen, T. J., Vahouny, G. V., and Waterman, M. R., 1987, Studies on the synthesis of sterol carrier protein$_2$ in rat adrenocortical cells in monolayer culture, *J. Biol. Chem.* **262**:3713–3717.

Tsuneoka, M., Yamamoto, A., Fujiki, Y., and Tashiro, Y., 1988, Nonspecific lipid transfer protein (sterol carrier protein$_2$) is located in rat liver peroxisomes, *J. Biochem.* **104**:560–564.

Ulsamer, A. G., Wright, P. L., Wetzel, M. G., and Korn, E. D., 1971, Plasma and phagosome membranes of Acanthamoeba castellanii, *J. Cell Biol.* **51**:193–215.

Vahouny, G. V., Chanderbhan, R., Noland, B. J., Irwin, D., Dennis, P., Lambeth, J. D., and Scallen, T. J., 1983, Sterol carrier protein$_2$, *J. Biol. Chem.* **258**:11731–11737.

Van Amerongen, A., Teerlink, T., Van Heusden, G. P. H., and Wirtz, K. W. A., 1985, The nonspecific lipid transfer protein (sterol carrier protein$_2$) from rat and bovine liver, *Chem. Phys. Lipids* **38**:195–204.

Van Amerongen, A., Helms, J. B., van der Krift, T. P., Schutgens, R. B. H., and Wirtz, K. W. A., 1987, Purification of nonspecific lipid transfer protein (sterol carrier protein$_2$) from human liver and its deficiency in livers from patients with cerebro-hepato-renal (Zellweger) syndrome, *Biochim. Biophys. Acta* **919**:149–155.

Van Amerongen, A., van Noort, M., van Beckhoven, J. R. C. M., Rommerts, F. F. G., Orly, J., and Wirtz, K. W. A., 1989, The subcellular distribution of the nonspecific lipid transfer protein (sterol carrier protein$_2$) in rat liver and adrenal gland, *Biochim. Biophys. Acta* **1001**:243–248.

Van Deenen, L. L. M., and deGier, J., 1974, Lipids of the red cell membrane in *The Red Blood Cell*, 2nd ed (Surgenor, D. M., ed.), pp. 147–211, Academic Press, New York.

Van der Krift, T. P., Leunissen, J., Teerlink, T., Van Heusden, G. P. H., Verkleij, A. J., and Wirtz, K. W. A., 1985, Ultrastructural localization of a peroxisomal protein in rat liver using the specific antibody against the non-specific lipid transfer protein (sterol carrier protein$_2$), *Biochim. Biophys. Acta* **812**:387–392.

Van Heusden, G. P. H., Souren, J., Geelen, M. J. H., and Wirtz, K. W. A., 1985, The synthesis and esterification of cholesterol by hepatocytes and H35 hepatoma cells are independent of the level of nonspecific lipid transfer protein, *Biochim. Biophys. Acta* **846**:21–25.

Van Noort, M., Rommerts, F. F. G., Van Amerongen, A., and Wirtz, K. W. A., 1986, Localization and hormonal regulation of the non-specific lipid transfer protein (sterol carrier protein$_2$) in the rat testis, *J. Endocrinol.* **109**:R13–R16.

Van Noort, M., Rommerts, F. F. G., Van Amerongen, A., and Wirtz, K. W. A., 1988a, Intracellular redistribution of SCP_2 in leydig cells after hormonal stimulation may contribute to increased pregnenolone production, *Biochem. Biophys. Res. Commun.* **154:**60–65.

Van Noort, M., Rommerts, F. F. G., Van Amerongen, A., and Wirtz, K. W. A., 1987b, Hormonal regulation of sterol carrier $protein_2$ (nonspecific lipid transfer protein) in rat leydig cells, Ann. N.Y. Acad. Sci. **513:**373–376.

Verner, K., and Schatz, G., 1988, Protein translocation across membranes, *Science* **241:**1307–1313.

Warner, M., and Neims, A. H., 1975, Studies on Z-fraction. I. Isolation and partial characterization of low molecular weight ligand binding protein from rat hepatic cytosol, *J. Physiol. Pharmacol.* **53:**493–500.

Wattenberg, B. W., and Silbert, D. F., 1983, Sterol partitioning among intracellular membranes—testing a model for cellular sterol distribution, *J. Biol. Chem.* **258:**2284–2289.

Westerman, J., and Wirtz, K. W. A., 1985, The primary structure of the nonspecific lipid transfer protein, *Biochem. Biophys. Res. Commun.* **127:**333–338.

Wetterau, J. R., and Zilversmit, D. B., 1984, A triglyceride and cholesteryl ester transfer protein associated with liver microsomes, *J. Biol. Chem.* **259:**10863–10866.

Wetterau, J. R., and Zilversmit, D. B., 1986, Localization of intracellular triacylglycerol and cholesteryl ester transfer activity in rat tissues, *Biochim. Biophys. Acta* **875:**610–617.

Wieland, F. T., Gleason, M. L., Serafini, T. A., and Rothman, J. E., 1987, The rate of bulk flow from the endoplasmic reticulum to the cell surface, *Cell* **50:**289–300.

Chapter 11

Spontaneous Transfer of Lipids between Membranes

Rhoderick E. Brown

1. INTRODUCTION

Lipid transfer between different biological surfaces was first observed many years ago (Hahn and Hevesy, 1939a,b; Hagerman and Gould, 1951). Since those early observations, several different types of lipid transfer have been discerned: (1) lipid transfer mediated by soluble proteins; (2) lipid transfer due to vesicle budding from one membrane followed by fusion with a target membrane; and (3) lipid transfer that occurs spontaneously either by lipid monomer diffusion through the aqueous medium or by transient contact following intervesicular collisions. Although types (1) and (2) have been the subject of several reviews (Spener and Wirtz, 1985; Wirtz *et al.*, 1986; Helmkamp, 1986; Sleight, 1987; Dawidowicz, 1987a), far less attention has been devoted to spontaneous lipid transfer. In the recent reviews in which spontaneous lipid transfer has been

Abbreviations used in this chapter: PC, phosphatidylcholine (1,2-diacyl-*sn*-glycero-3-phosphocholine); LysoPC, 1-*O*-hexadecanoyl-*sn*-glycero-3-phosphocholine; PG, phosphatidylglycerol (1,2-diacyl-*sn*-glycero-3-phospho-1′-*sn*-glycerol); PA, phosphatidic acid; PE, phosphatidylethanolamine; PS, phosphatidylserine; PI, phosphatidylinositol C_{12}-NBD, (7-nitro-2,1,3-benzoxadiazol-4-yl)aminododecanoic acid; GM_1, II^3-*N*-acetylneuraminosyl-gangliotetraosylceramide; asialo-GM_1, gangliotetraosylceramide; NMR, nuclear magnetic resonance spectroscopy; PAF, platelet-activating factor (1-alkyl-2-acetyl-*sn*-glycero-3-phosphocholine).

Rhoderick E. Brown The Hormel Institute, University of Minnesota, Austin, Minnesota 55912.

discussed, the coverage usually has been focused on cholesterol and/or phospholipids (Lange, 1986; Sleight, 1987; Dawidowicz, 1987b; Phillips *et al.*, 1987). This review will be devoted primarily to spontaneous lipid transfer as it occurs between membranes and will include several different lipids.

Before beginning this review, a few words of clarification are in order. Often, the processes of lipid transfer and lipid exchange are used interchangeably. However, strictly speaking, lipid transfer involves the redistribution of lipid from a donor structure to an acceptor structure that is initially devoid of that lipid. Thus, a concentration gradient of lipid often exists in the system. In contrast, lipid exchange is a process occurring between structures that are already at equilibrium with respect to their lipid concentration, and there is no concentration gradient of lipid in the system.

Gaining an understanding of the spontaneous lipid transfer and/or exchange processes has been, and continues to be, important for several reasons. First, the knowledge gained can be used to make the lipid transfer process an experimental tool. Two types of spontaneous lipid transfer have been used successfully as experimental tools. Pagano and co-workers have demonstrated the value of spontaneous interbilayer lipid transfer as a technique for introducing fluorescent lipids into cellular plasma membranes (Struck and Pagano, 1980; Pagano *et al.*, 1981; Pagano and Sleight, 1985). These investigators used fluorescence microscopy to study the movement and processing of fluorescent lipid analogs through different compartments of cultured Chinese hamster ovary cells. The fluorescent phospholipid and sphingolipid analogs were introduced into the plasma membranes of the cells by spontaneous transfer from small unilamellar vesicles composed of the fluorescent lipids. A second type of spontaneous lipid transfer used as an experimental tool is that from micelles to bilayers. This type of lipid transfer is limited to lipids that form micelles in an aqueous environment. Several different investigators have used this type of spontaneous lipid transfer to investigate the biological effects and functions of lysophospholipids (Weltzien, 1979) and of different gangliosides (Ledeen, 1984; Fishman, 1980; Hanai *et al.*, 1988b), Thus, the spontaneous lipid transfer process has been, and continues to be, an important and effective tool for elucidating certain cell biological phenomena.

Studying spontaneous lipid transfer is also important because of its potential role in the biological transport of certain lipids between cells or between subcellular compartments. Although such a role might appear unlikely, it cannot be ruled out *a priori*. The spontaneous interbilayer transfer rates of free fatty acids and lysophosphatidylcholine (lysoPC) are high enough to be of physiological relevance. Other lipids, such as cholesterol and anionic phospholipids, display reasonably high intervesicular transfer rates. The idea that spontaneous lipid transfer can and does occur between subcellular compartments is not new (Stuhne-Sekalec and Stanacev, 1978, 1980; Barańska and Wojtczak, 1984). In fact, a recent study by Barańska and Wojtczak (1988) reveals that a substantial fraction of phosphatidic acid (PA) (20–30%), but not phosphatidylcholine (PC) (2.5%), transfers between carefully washed microsomal and mitochondrial membranes. Thiol-blocking reagents do not inhibit the transfer. This and other experimental evidence has led Barańska and Wojtczak (1988) to suggest that the PA transfer is not mediated by protein and occurs spontaneously. Clearly, however, more work is needed to establish the role that spontaneous lipid transfer might play in lipid transport within cells or during membrane biogenesis.

A final reason for studying the spontaneous transfer and/or exchange of lipids is the structural information that is revealed about lipid bilayers participating in the transfer reaction. The spontaneous transfer rate of a lipid provides clues about its interactions with other lipid molecules in the membrane surface. The ease with which a lipid desorbs from the bilayer surface is indicative of its immediate structural environment. Thus, the kinetic data reflect the lateral arrangements of lipids within the bilayer surface. Theoretical work by Jahnig (1984) and experimental studies by Phillips and co-workers (McLean and Phillips, 1982; Lund-Katz *et al.*, 1988) and by Thompson and co-workers (Brown *et al.*, 1985; Brown and Thompson, 1987; Bar *et al.*, 1987) illustrate how the kinetic exchange parameters reveal information about lipid lateral-phase structure within bilayers. Such information is fundamental to our understanding of membrane structure and function.

2. LIPIDS TRANSFERRING SPONTANEOUSLY

2.1. Diacyl Phospholipids

Phospholipid transfer between erythrocytes and plasma components was first described many years ago (Hahn and Hevesy, 1939a; Sakagami *et al.*, 1965; Soula *et al.*, 1967; Reed *et al.*, 1968). However, in many of these early studies, it is difficult to discern whether the observed phospholipid exchange was truly spontaneous or whether plasma lipid transfer proteins (Tall, 1986) played a role in the observed transfer events.

With the development of various model membrane systems in the late 1960s and early 1970s came interest in the dynamic properties of lipids. In fact, at that time, several different laboratories reported that spontaneous exchange of diacyl phospholipids between model membranes was negligible (Kornberg and McConnell, 1971; Ehnholm and Zilversmit, 1973; Hellings *et al.*, 1974; Papahadjopoulos *et al.*, 1974). In cases in which diacyl phospholipid exchange was observed, other complicating factors existed (Barsukov *et al.*, 1974; Maeda and Ohnishi, 1974). However, in subsequent studies carried out over longer time intervals and/or with different phospholipids, spontaneous intervesicular transfer of diacyl phospholipids was clearly documented (Martin and MacDonald, 1976;

Papahadjopuoulos *et al.*, 1976; Duckwitz-Peterlein *et al.*, 1977; Kremer *et al.*, 1977). Phospholipid transfer was monitored by coincubating vesicles of differing lipid composition and measuring the time-dependent changes in vesicle thermotropic properties either optically or calorimetrically. The details of these studies have been reviewed previously (Lange, 1986).

From these initial studies controversy evolved over the mechanism by which the phospholipid transfer process occurred. The controversy centered on whether phospholipid transfer arose by the diffusion of lipid molecules through the aqueous phase between vesicles (Martin and MacDonald, 1976; Papahadjopoulos *et al.*, 1976; Duckwitz-Peterlein and Moraal, 1978) or by collisional contact occurring between vesicles (Kremer *et al.*, 1977). Kinetic analyses were developed to compare data (Thilo, 1977; Lawaczeck, 1978). Subsequently, different approaches were used to study spontaneous phospholipid transfer kinetics and analyze the mechanism of transfer. Among the techniques used were nuclear magnetic resonance (NMR) spectroscopy (de Kruijff and van Zoelen, 1978; Barsukov *et al.*, 1980), monolayers (Schindler, 1979), free-flow electrophoresis (DeCuyper *et al.*, 1980), radioactive tracers (McLean and Phillips, 1981; Schroit and Madsen, 1983), and fluorescence spectroscopy (Roseman and Thompson, 1980; Nichols and Pagano, 1981, 1982). As a result, much has been learned about the details of diacyl phospholipid intervesicular transfer.

Zwitterionic phospholipids with acyl chains of 16 or more carbon atoms exhibit relatively low intervesicular spontaneous transfer rates. For example, the half time for 1-palmitoyl-2-oleoyl PC transfer between dilute sonicated vesicles (37°C) is about 48 hr (McLean and Phillips, 1981; Jones and Thompson, 1989). However, decreasing the acyl chain length increases the transfer rate (Martin and MacDonald, 1976; Duckwitz-Peterlein *et al.*, 1977). The half time for dimyristoyl PC intervesicular transfer is 1.5–2.0 hr (37°C) (de Kruijff and van Zoelen, 1978; De Cuyper *et al.*, 1983; McLean and Phillips, 1984). Including cholesterol or dicetyl phosphate (to confer negative charge) in the vesicles does not affect dimyristoyl PC exchange rates at 37°C. However, the dimyristoyl PC transfer half time does increase by an order of magnitude if the surrounding lipid matrix is in the gel state (McLean and Phillips, 1984).

Comparison of the exchange rates of dimyristoyl PC, dipalmitoyl PC, and 1-palmitoyl-2-oleoyl PC has allowed McLean and Phillips (1984) to estimate the activation free energies for molecular exchange. The free energies of transfer from self-aggregates to water increase by 2.1 $kJ \cdot mol^{-1}$ per methylene group and are a good predictor of the relative exchange rates of lipid molecules. However, the activation energies are about 30 $kJ \cdot mol^{-1}$ greater than the free energies of transfer. The excess free energy is proposed to be associated with restriction of the phospholipid molecule to the vesicle surface in a transition-state complex (McLean and Phillips, 1984).

Introducing charge into the headgroup region of phospholipids has dramatic effects on the intervesicular spontaneous transfer rate. De Cuyper *et al.* (1983) measured the transfer kinetics of dimyristoyl phosphatidylglycerol (PG) between dimyristoyl PC vesicles (33°C). The transfer half time was 41 min at low ionic strength (15 mM salt; pH 6.0) but increased at higher salt concentrations. Changing the fatty acyl chains to oleoyl or palmitoyl groups had dramatic effects on the observed transfer rates. Similarly, varying the anionic phospholipid headgroup structure affected the transfer rate, and differences in rates could be roughly correlated to headgroup hydration (De Cuyper and Joniau, 1985). Not surprisingly, the surface charge of the bilayer surface also strongly influenced the transfer rates of anionic but not zwitterionic phospholipids (De Cuyper *et al.*, 1984).

A technique that has been particularly valuable for studying spontaneous intervesicular phospholipid transfer is fluorescence spectroscopy. A major advantage of the fluorescence technique is that no separation of donor and acceptor vesicles is required for measurement of the transfer event. For example, Roseman and Thompson (1980) used a PC derivative containing decanoic acid with covalently bound pyrene. Intervesicular transfer of the pyrenyl PC resulted in a concentration-dependent change in the emission spectrum (i.e., a change in ratio of excimer:monomer emission intensity) as the probe moved to nonfluorescent vesicles. Roseman and Thompson (1980) found that under the conditions described, the pyrenyl PC transfer rate was independent of acceptor vesicle concentration. The rate-limiting step for pyrenyl PC transfer was identified as departure from the donor vesicle by a mechanism involving diffusion through the aqueous medium.

Nichols and Pagano (1981, 1982) performed in-depth kinetic analyses of diacyl phospholipid spontaneous transfer by using fluorescent derivatives displaying concentration-dependent changes in emission intensity, as well as probes with spectral properties favoring energy transfer-related quenching when placed in close proximity. Using a PC derivative labeled with (7-nitro-2,1,3-benzoxadiazol-4-yl)aminododecanoic acid (C_{12}-NBD), Nichols and Pagano (1981) reported that diffusion of soluble monomers accurately predicts the kinetics of the intervesicular transfer process. The transfer rate also depends on the fatty acyl chain length and the headgroup composition of the transferring derivative, as well as the phase state of the donor and acceptor bilayers (Nichols and Pagano, 1982).

Low concentrations of bile salts increase the rate of spontaneous intervesicular C_{12}-NBD–PC transfer without disrupting the vesicles (Nichols, 1986). The binding of bile salts to the vesicles appears to alter the dissociation and/or association rate constants for phospholipid monomer–vesicle interaction (Nichols, 1985) and increase the rate of phospholipid transfer via monomer diffusion through the aqueous phase. Interestingly, appropriate concentrations of calcium and phosphate decrease (~10-fold) the transfer rate of C_{12}-NBD-phosphatidylserine from erythrocyte lipid vesicles to dioleoyl PC vesicles (Tanaka and Schroit, 1986). The process is dependent on the presence of cholesterol and phosphatidylethanolamine in the donor membrane and does not appear to be due to calcium-mediated phase separations of PS or to vesicle–vesicle fusion. Under similar conditions, however, the transfer rates of C_{12}-NBD-PC, -PE, -PG, -PA, and -succinoyl ethanolamine are not affected.

Silvius and co-workers (Silvius *et al.*, 1987; Gardam *et al.*, 1989) have used an approach similar to the fluorescence energy transfer technique of Nichols and Pagano (1982) to study the partitioning of exchangeable phospholipids and sphingolipids between different lipid bilayer environments. Phospholipid derivatives with different polar headgroups differed only modestly in their relative affinities for vesicles composed of "hydrogen-bonding" lipids (PE and PS) versus "non-hydrogen-bonding" lipids (PC and PG). Probes with different headgroups also showed modest but reproducible differences in their relative affinities for vesicles composed of PC, PG, and cholesterol (48 : 12 : 40) compared with PC-PG (80 : 20) vesicles without cholesterol (Gardam *et al.*, 1989).

When the C_{12}-NBD fluorophore is attached to the headgroup of dilauryl PE, the time-dependent increase in energy transfer upon coincubating donor and acceptor vesicles is no longer monoexponential, but is better described as the sum of two exponentials (Arvinte and Hildenbrand, 1984). Only the faster of the two exponentials can be attributed to spontaneous intervesicular transfer. The lower rate appears to be due to transbilayer migration (flip-flop) of the C_{12}-NBD–dilauryl PE.

Recently, Nichols (1988) compared the transfer rates of a homologous series of *N*-NBD-PEs differing in saturated acyl chain length from 11 to 16 carbons. Transfer half times between PC vesicles were 200 to 6000 times shorter than those between mixed micelles composed of PC and taurocholate. Interestingly, the transfer kinetics suggested that exchange was due to transient intermicellar collisions, which were not evident between vesicles. However, Jones and Thompson (1989) concluded that significant PC transfer can occur by collision between vesicles if the vesicle concentration is high enough. Similar findings were reported for dipalmitoyl PC transfer from egg PC vesicles to brush border membrane vesicles (Mutsch *et al.*, 1986). These findings are important because, for many years, spontaneous PC transfer was thought to be too slow to have any biological relevance. However, the PC flux generated by intervesicular collisions appears to be considerably higher than the flux due to molecular desorption from vesicles followed by diffusion through the aqueous medium. Thus, under conditions of high intracellular membrane concentrations, spontaneous PC transfer might play a physiological role in membrane biogenesis (Jones and Thompson, 1989) or in lipid uptake by enterocytes (Mutsch *et al.*, 1986).

The biological membrane most often used to study the kinetics of spontaneous diacyl phospholipid transfer is the erythrocyte plasma membrane (Rindlisbacher and Zahler, 1983; Ferrell *et al.*, 1985a,b; Daleke and Huestis,

1985; Fujii and Tamura, 1983, 1984). Transfer of PC, PE, and PS to erythrocytes is easily monitored morphologically, isotopically, or with spin labels (Tanaka and Schroit, 1983; Ferrell *et al.*, 1985a,b; Daleke and Huestis, 1985; Tamura *et al.*, 1986). Highly unsaturated PCs transfer from vesicles to erythrocyte membranes faster than other PCs do (Rindlisbacher and Zahler, 1983). When various saturated PCs are compared, their transfer rates decrease exponentially with increasing acyl chain length (Ferrell *et al.*, 1985b; Tamura *et al.*, 1986). The behavior is consistent with a transfer mechanism involving diffusion of phospholipid monomers through the aqueous phase and confirms other results involving fluorescent PC transfer to model lipoproteins (Massey *et al.*, 1982a,b, 1984). Interestingly, the marked dependence of the lipid transfer rate on the acyl chain length may explain why natural membranes adjust their fluidity by introducing unsaturation into long-chain fatty acids rather than by simply decreasing the acyl chain length. A membrane composed of longer-chain lipids would be expected to maintain its integrity for long periods (Ferrell *et al.*, 1985b).

Diacyl phospholipid derivatives, be they fluorescent, spin-labeled, or photoactivatable, can often be incorporated into biological membranes by the spontaneous transfer route. This makes spontaneous lipid transfer an extremely valuable tool for studying many biological processes. Pagano and associates have shown that certain fluorescent derivatives of phospholipids spontaneously transfer from small phospholipid vesicles to the plasma membranes of Chinese hamster ovary cells at 2°C (Struck and Pagano, 1980; Pagano and Sleight, 1985). Following insertion into the plasma membrane, the phospholipid derivatives can be localized by fluorescence microscopy. Subsequent transport to various cellular organelles is easily observed. Analysis of the fluorescent lipid after its transport to the endoplasmic reticulum and Golgi complex reveals that it has been converted to various metabolites. Important information about phospholipid intracellular transport and metabolism has been obtained by such studies (Pagano and Sleight, 1985).

Many other studies exist that demonstrate the value of spontaneous phospholipid transfer as an experimental tool. For example, much of the recent work with phospholipid flippases has been aided by spontaneous introduction of various diacyl phospholipid probes into membranes (Devaux, 1988; Bishop and Bell, 1988). Also, many studies involving protein–lipid spatial orientations in membranes have benefited by spontaneous incorporation of certain photoactivatable diacyl phospholipids (Brunner *et al.*, 1983; Schroit and Madsen, 1983; Schroit *et al.*, 1987). Thus, it is clear that spontaneous lipid transfer has become an extremely valuable tool. However, the physiological importance of spontaneous diacyl phospholipid transfer remains controversial.

Traditionally, the spontaneous transfer of diacyl phospholipids has been thought to play an insignificant role in membrane biogenesis and maintenance (Dawidowicz, 1987b; Sleight, 1987). The reasons are twofold: (1) spontaneous transfer of long-chain diacyl PCs is relatively slow and (2) spontaneous transfer would lead to randomization of the lipid components of different membranes rather than support differential lipid compositions between organelles. However, these reasons in themselves no longer appear to be sufficient to dismiss a role for spontaneous lipid transfer in certain biological processes. As was pointed out above, both Mutsch *et al.* (1986) and Jones and Thompson (1989) have noted that the spontaneous transfer of long-chain diacyl PCs is enhanced significantly when membrane concentrations are high. Furthermore, the membrane lipid composition appears to influence rather dramatically the distribution of individual lipids, such as cholesterol, between membranes (see Section 2.5). Finally, a few reports indicate that spontaneous phospholipid transfer may be important in certain situations (Stuhne-Sekalec and Stanacev, 1978, 1980; Barańska and Wojtczak, 1984). One recent study by Barańska and Wojtczak (1988) reveals that a substantial fraction of PA (20–30%), but not PC (2.5%), transfers between carefully washed microsomal and mitochondrial membranes and that thiol-blocking reagents do not inhibit the transfer. The initial rate appears to be independent of acceptor membrane concentration (mitochondria). These findings suggest that the process is not mediated by protein but occurs spontaneously by phospholipid diffusion through the aqueous phase rather than by membrane collision. Clearly, however, many questions remain before the role of spontaneous lipid transfer in lipid transport between and/or within cells is understood.

2.2. Monoacyl Phospholipids

Interest in the spontaneous transfer of lysophospholipids and, in particular, lysoPC, first evolved because of the potent cytolytic properties of these lipids [for a review, see Weltzien (1979)]. Traditionally, erythrocytes have been the cells of choice for the study of cytolytic events. These cells are easily isolated in large numbers, and their lysis is easily monitored by hemoglobin release. From the studies of erythrocyte hemolysis, five consecutive steps were identified: (1) adsorption of lysophospholipid to the cell surface; (2) penetration into the membrane matrix; (3) induction of changes in membrane molecular organization; (4) ion permeability changes of the membrane; and (5) osmotic lysis and leak of hemoglobin (Reman *et al.*, 1969).

Implication of spontaneous lipid transfer as the *modus operandi* for steps (1) and (2) comes from several lines of evidence. By comparing micellar size and structure of different lysophospholipid derivatives with binding and lysis kinetics, one concludes that hemolysis rates are dominated by the adsorption rate of lysosphospholipid to cells. In fact, lysolipids showing reduced lysis rates exhibit slow adsorption to cells and tend to form lamellar rather than micellar dispersions in aqueous environments (Weltzien *et al.*, 1976). Lysis kinetics also show a marked correlation with the degree of hydrophobicity of the individual lytic molecule. Increasing the chain length of the hydrophobic region increases the lag time and decreases the rate of hemolysis (Weltzien, 1979; Reman *et al.*, 1969). More importantly, no intracellular contact need occur for lysolipid transfer between cells. When erythrocytes are separated by a dialysis membrane, lysophospholipid transfers rapidly between the two populations of cells (Weltzien, 1979). Thus, under the conditions used, these findings are consistent with the transfer of lysophospholipid by aqueous diffusion of individual molecules from the micelle to the cell surface (Weltzien, 1979).

Spontaneous transfer of lysophospholipids has been used as an experimental tool to study shape changes in erythrocytes. At sublytic concentrations, lysoPC and lysoPS cause erythrocytes to crenate [for review, see Weltzien (1979); Ferrell *et al.* (1985a); and Daleke and Huestis (1985)]. Even a slight increase in the membrane lysolipid content leads to transformation from the normal biconcave disk to an echinocytic form. The original shape may be restored upon removal of the lysoPC or lysoPS by washing the cells with serum (Sato, 1973; Sato and Fujii, 1974) or albumin (Klibansky and de Vries, 1963). More recent studies have focused on elucidating the mechanism by which lysolecithin induces the shape changes (Lange and Slayton, 1982; Ferrell *et al.*, 1985a,b). Apparently, the transferred lysoPC intercalates into the outer monolayer of the erythrocyte membrane and causes area expansion relative to the inner monolayer. To accommodate the expanded outer leaflet, the cells crenate. An area expansion of 1–2% of the total cell surface area appears sufficient to induce shape changes (Lange and Slayton, 1982; Ferrell *et al.*, 1985a). The shape change depends directly on the amount of lysoPC incorporated, irrespective of chain length. However, the chain length appears to govern the rate and maximal extent of lysoPC transfer from the medium into the membrane (Fujii and Tamura, 1983, 1984). Interestingly, the level of membrane cholesterol has a distinctive influence on the incorporation of lysoPC and the accompanying shape change in the erythrocyte (Lange and Slayton, 1982).

In addition to causing gross morphological changes, the spontaneous transfer of lysoPC into human erythrocyte ghost membranes causes protein and lipid immobilization (Golan *et al.*, 1986, 1988). Interestingly, similar changes occur in erythrocyte membranes that are under attack by the parasite *Schistosoma mansoni*. Schistosomula incubated with labeled palmitate release lysoPC into the culture medium (Golan *et al.*, 1986, 1988). These data are consistent with the conclusion that lysoPC is transferred from the parasites to erythrocytes and initiates events in the membrane that ultimately result in erythrocyte lysis. Whether the lysoPC transfer occurs by aqueous monomer diffusion or by sloughing of lysoPC-enriched vesicles or micelles is not clear. Nevertheless, the spontaneous transfer of lysoPC has been an important tool for elucidating the molecular events involved in membrane disruption that result from these parasitic infections.

Interest in the spontaneous transfer of lysoPC and other lysophosphatides continues to grow because of newly recognized biological functions. For example, lysoPC has been implicated as a chemotactic factor for human monocytes (Quinn *et al.*, 1988) and as a regulator of protein kinase C (Oishi *et al.*, 1988). Lysophosphatides, such as lysophosphatidylinositol (lysoPI), are taken up by rat islets of Langerhans and mobilize cellular calcium release (Metz, 1988). LysoPS rapidly incorporates into and translocates across erythrocyte membranes (Daleke and Huestis, 1985).

Model membrane systems have been particularly useful in studying the spontaneous transfer of lysoPC from its micelle to phospholipid vesicles (de Kruijff *et al.*, 1977; Elamrani and Blume, 1982) and between phospholipid vesicles (McLean and Phillips, 1984). Carbon-13 NMR experiments with shift reagents show that lysoPC (15 mol%) taken up by sonicated PC vesicles is localized in the outer monolayer (de Kruijff *et al.*, 1977). Stopped-flow rapid-mixing techniques reveal that 1-palmitoyl-lysoPC (16.6 mol%) incorporates rapidly into dimyristoyl PC or dipalmitoyl PC vesicles (half times between 50 and 500 msec) (Elamrani and Blume, 1982). The incorporation rate is pseudo-first-order and strongly temperature dependent, attaining a maximum value near the vesicle phase transition temperature. From these observations and estimations of the aqueous diffusion rate of lysoPC monomers by using the Smouluchowski equation, Elamrani and Blume (1982) argued that the rate-limiting step is the incorporation of the lysoPC monomer into the vesicle bilayer.

The rate of spontaneous transfer of 1-palmitoyl-lysoPC between egg PC unilamellar vesicles is very high. Using a charged-vesicle assay in which ion exchange minicolumns separate donor and acceptor vesicles, McLean and Phillips (1984) reported that the half time for lysoPC transfer is less than 2 min. This value is consistent with their theoretically predicted value of 21 sec.

Ferrell *et al.* (1985b) reported that lysoPC transfers from sonicated vesicles to human erythrocytes with a half time of less than 7 sec. Recently, Steck *et al.* (1988) measured the spontaneous transfer rate of lysoPC from erythrocytes to ghosts. The reaction proceeded extremely rapidly, with a half time estimated to be 10 sec. The transfer kinetics were insensitive to a fivefold change in the concentration of each reactant, suggesting an aqueous diffusion mechanism.

In general, the higher lysoPC transfer rates (ca. 50-fold) observed for micelle-to-bilayer transfer compared with bilayer-to-bilayer transfer are consistent with predictions discussed by Israelachvili *et al.* (1980); see also Cevc and Marsh (1987).

2.3. Sphingolipids

The ability of sphingolipids to transfer between biological surfaces has been recognized for many years (Yokoyama *et al.*, 1963; Dobiasova and Radin, 1968;

Marcus and Cass, 1969). However, the mechanistic details of sphingolipid transfer have become better understood only within the past decade. Recent spontaneous interbilayer transfer studies have focused on three different classes of sphingolipids: sphingomyelins and ceramides, neutral glycosphingolipids, and gangliosides.

2.3.1. Sphingomyelin and Ceramides

Both tritiated sphingomyelin (Frank *et al.*, 1983) and sphingomyelins with covalently attached pyrene fatty acids (Pownall *et al.*, 1982; Frank *et al.*, 1983) have been used to study the spontaneous transfer kinetics of sphingomyelin. Pyrenyl-sphingomyelin departure from donor surfaces exhibits first-order kinetics. The rate of pyrenyl-sphingomyelin transfer varies greatly depending upon the donor surface organizational state (e.g., micelles versus lipoprotein recombinants versus phospholipid vesicles), although the activation energies (21–25 kcal/mol) remain similar for most of these donors. Structure-breaking solutes (NaSCN and CH_3OH) accelerate the transfer rate, whereas structure-making salts (NaCl and $MgCl_2$) inhibit spontaneous transfer (Pownall *et al.*, 1982). Sphingomyelin transfer from a dimyristoyl PC or dipalmitoyl PC vesicle matrix that is in the gel phase (20°C) is almost 300-fold slower than sphingomyelin transfer between dimyristoyl PC or dipalmitoyl PC vesicles that are in the liquid-crystalline phase (50°C) (Frank *et al.*, 1983). However, within this temperature range (20–50°C), a surprisingly abrupt decrease in the sphingomyelin transfer rate occurs below 30°C, when the vesicle matrices are composed of 1-palmitoyl-2-oleoyl PC. Such behavior is consistent with the existence of a sphingomyelin-enriched gel phase in this system even at low sphingomyelin concentrations (Frank *et al.*, 1983).

In addition to providing important information about the phase structure of sphingolipid–phospholipid mixtures, spontaneous lipid transfer has been a valuable tool in the study of sphingolipid transport and metabolism within cells. Pagano and Martin (1988) have shown that certain fluorescent derivatives of ceramides spontaneously transfer between small phospholipid vesicles with half times of 0.29–4.0 min at 25°C. Rapid spontaneous transfer of these fluorescent ceramides also occurs from bovine serum albumin to the plasma membranes of Chinese hamster ovary cells (Lipsky and Pagano, 1983, 1985; Pagano and Sleight, 1985; Pagano and Martin, 1988). Following insertion into the plasma membrane, the fluorescent derivatives can be localized by fluorescence microscopy. Subsequent transport to various cellular organelles is easily observed. Analysis of the fluorescent lipid following transport to the Golgi complex reveals metabolic conversion to sphingomyelin and glucosylceramide. Invaluable information about sphingolipid intracellular transport and metabolism has been obtained by such studies (Lipsky and Pagano, 1983, 1985; Pagano and Sleight, 1985). Similar approaches have been used to study the sorting of sphingolipids in epithelial cells (van Meer *et al.*, 1987; Simons and van Meer, 1988).

2.3.2. Neutral Glycosphingolipids

In general, the spontaneous transfer rate of neutral glycosphingolipids between liquid-crystalline phospholipid bilayers at physiological temperatures is very slow compared with the transfer rates of cholesterol and phospholipids (Thompson *et al.*, 1986). In most cases, kinetic data were obtained by using tritiated neutral glycosphingolipids, although studies with a few fluorescent glycolipid analogs have also been performed. Correa-Freire *et al.* (1982) first measured the spontaneous transfer time for glucosylceramide (half time > 30 days) by using tritiated and pyrene-labeled derivatives. The initial transfer rate did not change when the concentration of acceptor vesicles varied from 5- to 20-fold excess. Such behavior suggests that under the experimental conditions used, the transfer mechanism is due primarily to diffusion through the aqueous medium rather than to vesicle–vesicle collisions.

The spontaneous interbilayer transfer of one neutral, polyglycosylated ceramide has been examined in detail. Brown *et al.* (1985) noted that tritiated gangliotetraosylceramide (asialo-GM_1) derived from bovine brain gangliosides exhibits biphasic transfer kinetics between the outer monolayers of liquid-crystalline vesicles. About 16% of the asialo-GM_1 was a fast-transferring pool (transfer half time, 42 hr), whereas the remaining 84% of the asialo-GM_1 accounted for a slowly transferring pool (transfer half time, 24 days). The identification of these two rates and their pools is most simply explained by an in-plane microdomain model. Asialo-GM_1 molecules may coexist as both monomers and glycolipid-enriched microdomains with the phospholipid matrix. The departure of the monomers is characterized by the fast pool and its associated rate constant. The transfer of asialo-GM_1 from the domain-like clusters is characterized by the slow pool and its associated rate constant (Brown *et al.*, 1985).

Recently, similar kinetic behavior was observed for certain types of cerebrosides by Jones and Thompson (1988). Measurement of the spontaneous interbilayer transfer rate of bovine brain galactosylceramides between 1-palmitoyl-2-oleoyl PC bilayers at 45°C also revealed biphasic kinetics. Both phrenosin (*N*-α-hydroxyacylgalactosylsphingosine) and kerasin (*N*-acylgalactosylsphingosine) exhibited biphasic kinetics. Approximately 20% of the lipid transfers with a half time of 12 hr, and the remaining 80% transfers with a half time of about 100 days. The results are consistent with the presumed phase structure of bilayers composed of these lipids in 1-palmitoyl-2-oleoyl PC (Thompson and Tillack, 1985; Curatolo, 1986).

2.3.3. Gangliosides

Gangliosides differ from many sphingolipids because of their ability to form micelles in excess water. However, when present as minor components of membranes, as is the case in mammalian plasma membranes, gangliosides do not grossly alter the lamellar structure of phospholipid bilayers. If the ganglioside concentration in PC rises too high, the bilayer structure is no longer maintained (Thompson and Tillack, 1985; Thompson and Brown, 1988). Rather, mixed micelles of ganglioside and phopsholipid form. Because of this phase behavior, spontaneous transfer studies of gangliosides can be grouped into two categories: (1) ganglioside interbilayer transfer and (2) ganglioside transfer from their micelles into bilayers.

Ganglioside interbilayer transfer kinetics have been studied by using both fluorescent and tritiated derivatives (Masserini and Freire, 1987; Brown and Thompson, 1987). In both of these studies, the ganglioside studied was II^3-*N*-acetylneuraminosyl-gangliotetraosylceramide (GM_1). Masserini and Freire (1987) studied the rate dependence for pyrenyldecanoyl-GM_1 transfer from dipalmitoyl PC donors and dimyristoyl PC acceptors as a function of temperature. The GM_1 transfer rate is markedly influenced by the phase state of the lipid matrix, since transfer was not detected below 23°C, remained slow between 23 and 42°C, and increased rapidly above 42°C. The addition of excess calcium (20 mM) revealed only a slight reduction in the transfer rate at 54°C, suggesting that no strong association exists between this divalent cation and GM_1. The initial departure of tritiated GM_1 from either sonicated dipalmitoyl PC or 1-palmitoyl-2-oleoyl PC unilamellar vesicles exhibits single-exponential decay kinetics (Brown and Thompson, 1987). Interestingly, this single-exponential kinetic behavior is in sharp contrast to the biphasic kinetics displayed by asialo-GM_1, which differs from GM_1 only by the lack of sialic acid. These findings are consistent with the conclusion that GM_1 is molecularly dispersed at low concentrations within liquid-crystalline phospholipid bilayers. Slightly higher transfer rates are observed at 45°C when dipalmitoyl PC is the donor matrix rather than 1-palmitoyl-2-oleoyl PC (transfer half times, 40 and 55 hr, respectively). Others factors affecting the GM_1 interbilayer transfer rate include the initial GM_1 concentration in donor vesicles and the GM_1 transbilayer distribution in donor vesicles (Brown and Thompson, 1987).

Spontaneous ganglioside transfer between biological membranes has been studied using fluorescent gangliosides. Masserini and Freire (1987) measured the transfer kinetics of pyrenyldecanoyl-GM_1 between synaptic plasma membranes and from dipalmitoyl PC vesicles to synaptic membranes as a function of temperature. Spontaneous GM_1 transfer rates between synaptic membranes increased monotonically with temperatures up to 40°C and then began to decrease. When dipalmitoyl PC donors were used, GM_1 transfer to synaptic membrane acceptors was slower than that observed with dimyristoyl PC acceptors. This difference was attributed to the lower fluidity of synaptic membranes.

The transfer of different gangliosides and ganglioside analogues from their micelles to model membranes has been examined in different laboratories (Felgner *et al.*, 1981, 1983; Masserini and Freire, 1987; Thompson and Brown, 1988). When micelles composed of bovine brain trisialoganglioside are incubated with dipalmitoyl PC unilamellar vesicles at 46°C for 1 hr, ganglioside becomes associated with the phospholipid vesicles (Felgner *et al.*, 1981). The ganglioside association could be demonstrated by molecular-sieve chromatography and by changes in the thermotropic behavior of the dipalmitoyl PC vesicles. Neuraminidase treatment indicated that all of the added ganglioside was present in the outer monolayer of the vesicles. The added ganglioside did not disrupt vesicle permeability (Felgner *et al.*, 1981). Such a ganglioside-phospholipid vesicle system serves as a valuable model system for the plasma membrane because many gangliosides appear to be located primarily in the outer surface of cellular plasma membranes.

Closer examination of the kinetics of ganglioside transfer to phospholipid vesicles reveals that although both the ganglioside and phospholipid components are in transit in this system, the rate of transfer of the ganglioside from micelles to vesicles is much higher than the transfer of phospholipid to micelles (Felgner *et al.*, 1983). As a result, up to 15 mol% of the ganglioside can be incorporated into the phospholipid vesicles. Thus, if the initial micelle–phospholipid system is below 15 mol%, the micelles quickly disappear from the system as the ganglioside incorporates into the outer surface of the vesicles. This incorporation process has also been studied by using ganglioside-like molecules containing an electron spin resonance probe covalently linked to the acyl chain (Kanda *et al.*, 1982a,b). Masserini and Freire (1987) have also examined this process by using a pyrene-labeled ganglioside.

Two interesting aspects of the micelle-to-vesicle transfer process emerge from these studies. First, under the reported conditions, it is clear that the process is rate-limited by the rate at which ganglioside molecules leave the micelle at low system ganglioside concentrations. This was also the case for many other systems in which the spontaneous transfer of lipids was examined (Dawidowicz, 1987b). However, at system concentrations above about 15 mol%, the external bilayer surface of the vesicle appears to saturate. Then, the overall transfer rate becomes limited by a much slower process occurring in the acceptor vesicle bilayer (Felgner *et al.*, 1983). Second, the activation energy for ganglioside transfer at low concentrations has an unusually high temperature coefficient (15–25 kcal · mol^{-1}). The origin of this large temperature dependence probably reflects temperature-dependent changes in micelle structure. Such changes are detectable by other physical techniques (Hinz *et al.*, 1981; Cantù *et al.*, 1986).

Studies of ganglioside insertion into model membranes have been extremely useful in understanding how exogenously added gangliosides associate with complex membranes such as cellular plasma membranes. Reports describing the molecular nature of ganglioside association with cellular plasma membranes have been controversial, to say the least. Central to this controversy is whether exogenous gangliosides become inserted and assume the same orientation in the lipid bilayer of cellular plasma membranes as do endogenous gangliosides. From their studies, Wiegandt and co-workers (Keenan *et al.*, 1974, 1976; Callies *et al.*, 1977; Radsak *et al.*, 1982) proposed that the bulk of exogenously added gangliosides attach to trypsin-sensitive membrane components and do not integrate into the plasma membrane in the same manner as endogenous gangliosides do. Factors such as serum reportedly removed part of the exogenous gangliosides from the surface of chicken and canine erythrocytes and chicken and mouse fibroblasts. Also, the cells did not appear to internalize or metabolize exogenously added gangliosides (Keenan *et al.*, 1976; O'Keefe and Cuatrecasas, 1977). In contrast, Fishman and co-workers (Moss *et al.*, 1976; Fishman *et al.*, 1976, 1980, 1983; Spiegel *et al.*, 1984) found that exogenous gangliosides do become functionally inserted into the outer surface of cultured cells and are both internalized and metabolized by the cells. GM_1-deficient cells incorporated exogenous GM_1, internalized it, and metabolized it to both higher and lower homologs (Fishman *et al.*, 1983). Other researchers have noted that exogenously added gangliosides modify the growth behavior of certain cells in culture (Laine and Hakomori, 1973; Bremer *et al.*, 1984, 1986; Nakajima *et al.*, 1986; Hanai *et al.*, 1988a,b).

There are several plausible explanations for the differences in the way exogenous gangliosides are taken up, internalized, and metabolized by cells. First, different cell types may vary in their ability to incorporate and metabolize exogenous gangliosides. Incubation conditions, such as the presence of serum, reportedly interfere with ganglioside incorporation (Fishman *et al.*, 1978; Sonderfeld *et al.*, 1985). Second, the metabolic state and/or cell density may also be important (Spiegel *et al.*, 1984; Facci *et al.*, 1984). Finally, the physical state of the gangliosides will probably influence the final outcome. From the studies with model membranes described in preceding paragraphs (Section 2.3.3) it is likely that the monomeric form of the ganglioside is the one that inserts itself into the lipid bilayer. Indeed, the fraction of spin-labeled ganglioside analogue associated with mouse fibroblasts, but not released by trypsin, appears to be anchored in the hydrophobic region of the plasma membrane (Schwarzmann *et al.*, 1983). The trypsin releasable fraction appears to be due to micelles that adsorb to the cellular surface.

Understanding the spontaneous transfer kinetics of other glycosphingolipid derivatives to biological membranes may be helpful in deciphering the molecular details of their biological activities. Taketomi *et al.* (1970, 1976) prepared certain lysosphingolipids and studied their chemical and hemolytic properties. Hemolysis was caused by the sponteneous incorporation of the lysosphingolipids into the erythrocyte membrane. More recently, sphingosine and specific lysosphingolipids have been identified as potent but reversible inhibitors of protein kinase C (Hannun *et al.*, 1986; Hannun and Bell, 1987, 1989). Lysosphingolipids also affect cellular growth by modulating the tyrosine phosphorylation of epidermal growth factor receptor (Hanai *et al.*, 1988a,b). However, to date, the dynamic properties of these important glycosphingolipid derivatives have received minimal attention.

2.4. Free Fatty Acids

Long-chain fatty acids serve as cellular substrates for energy production as well as essential metabolites (e.g., arachidonic acid) in the generation of prostaglandins and leukotrienes. There is considerable interest, therefore, in the mechanism(s) by which fatty acids are taken up by, transported within, and released from cells (Scow and Blanchette-Mackie, 1985; Sleight, 1987; Figard *et al.*, 1986; Stoll and Spector, 1987). In keeping with the central theme of this chapter, the following discussion will focus primarily on spontaneous transfer of fatty acids as it occurs between lipid membranes and/or micelles. However, it should be emphasized that fatty acid transfer from other biological surfaces (e.g., lipoproteins and albumin) to membranes can also occur spontaneously (Noy *et al.*, 1986; Hamilton and Cistola, 1986; Cooper *et al.*, 1987).

From a physical standpoint, there is no absolute need for fatty acid transfer to be facilitated by specific receptors (Pownall *et al.*, 1983; Storch and Kleinfeld, 1986; Noy *et al.*, 1986; Cooper *et al.*, 1987). Molecular diffusion through the aqueous media cannot be rejected as a mechanism for fatty acid transport in tissues (Scow and Blanchette-Mackie, 1985) simply on the basis of long-chain fatty acid aqueous solubility measurements in the absence of other membranes and proteins. Abundant evidence documenting long-chain fatty acid spontaneous transfer between membranes and from micelles to membranes exists in the literature. It is hoped that a discussion of this literature will help eliminate misconceptions about the possible mechanisms by which spontaneous fatty acid transfer can occur.

Free fatty acids do transfer spontaneously between model membranes. Sengupta *et al.* (1976) measured pyrenyldecanoic acid (10 carbons) transfer rates between dipalmitoyl PC vesicles using stopped-flow techniques. Between 20 and 39°C, the rate constant varied little ($k = 0.2\ \text{sec}^{-1}$), but it increased nearly fivefold between 39 and 50°C. Including 5 mol% cholesterol in the dipalmitoyl PC vesicles reduced the transfer rate by 50% at 50°C. Kremer and Wiersema (1977) noted that above the phase transition temperature, myristic acid exchange between dipalmitoyl PC vesicles is complete within a few seconds. Pownall and co-workers studied fatty acid interbilayer transfer in detail (Doody *et al.*, 1980; Pownall *et al.*, 1983; Wolkowicz *et al.*, 1984). Pyrenylnonanoic acid (9 carbons) transfer is a first-order process showing no rate dependence on donor or acceptor vesicle concentrations or on the chemical composition of acceptors. A high ionic strength (4 M NaCl) reduces the transfer rate 25- to 60-fold depending on temperature and pH. A large decrease in the transfer rate occurs at low pH owing to protonation of the acid. The transfer rate constants for the protonated and ionized forms of the fatty acid at 34°C are 0.17 and 9.8 sec^{-1}, respectively. After allowing for ionization suppression known to occur for probes in membranes, calculations show that less than 1% of the fatty acid transferring at pH 2.8 is ionized, whereas less than 1% of the transferring species is protonated at pH 7.4 (Doody *et al.*, 1980). Taken together, these observations strongly suggest that the mechanism of transfer is by monomer diffusion through the aqueous phase and that the rate-limiting step is dissociation from the donor vesicle into the aqueous phase. Thermodynamic analysis of the activated state and of fatty acid partitioning indicates that formation of an aqueous intermediate in the interfacial region is the rate-limiting step. The implications, as pointed out by Doody *et al.* (1980), are that fatty acid transfer kinetics are not dictated by the partition equilibrium alone, but are also influenced by the solvation properties of interfacial water.

In studies of the spontaneous interbilayer transfer of pyrenyldodecanoic acid (12 carbons), Wolkowicz *et al.* (1984) focused on the physical properties of the bilayer matrix that affect transfer kinetics. Parameters such as bilayer curvature, fluidity, and composition all influenced the fatty acid transfer rate. An increase in vesicle size significantly decreases the rate of transfer. Similarly, transfer rates decrease by changing from a liquid-crystalline matrix to a gel matrix. Increasing the negative surface charge of the matrix by incorporating PS affects the transfer rate in a biphasic manner. With low concentrations of negative charge, the transfer increases and reaches a maximum around 10 to 15 mol% PS. Further increases in the PS concentration cause a steady decline in the transfer rate and eventually become inhibitory compared with pure PC vesicles. Such studies demonstrate how phospholipid curvature, fluidity, and composition can play important roles in controlling fatty acid interbilayer transfer.

More recently, Storch and Kleinfeld (1986) have studied the spontaneous interbilayer transfer of long-chain free fatty acids (16 and 18 carbons) containing covalently attached fluorescent anthroyloxy groups. The transfer of these long-chain fatty acids is slower than that of short-chain fatty acids. More importantly, the long-chain fatty acid transfer is best described as the sum of two exponential rates. This biexponential behavior, as pointed out by Storch and Kleinfeld (1986), may arise from transbilayer movement (flip-flop) and the departure of fatty acid from the donor vesicle outer monolayer. In contrast, short-chain fatty acid transfer is apparently monoexponential (Sengupta *et al.*, 1976; Doody *et al.*, 1980).

Increasing the carbon length of the fatty acids to 16 or 18 carbons slows the spontaneous transfer rate by 1–2 orders of magnitude compared with that for fatty acids composed of 9, 10, or 12 carbons (see above). However, the average rate of the 16-carbon derivative is twice as fast as the 18-carbon derivative. Introducing a single double bond in the middle of the 18-carbon fatty acid doubles its average transfer rate (1.11 versus 2.5 min^{-1}). Similar correlations between transfer rate and acyl chain length and unsaturation have been observed for phospholipids (Martin and MacDonald, 1976; Massey *et al.*, 1982a,b, 1984; Cevc and Marsh, 1987) and other alkanes and alcohols (Pownall *et al.*, 1983). Varying the fatty acid concentration between 0.5 and 20 mol% in egg PC vesicles has no effect on the transfer rate, but varying the anthroyloxy attachment site from the second to the C-12 of stearic acid shifts the average rate constant from 0.56 to 2.5 min^{-1}. Thus, properties of both the membrane bilayer and the fatty acid itself may contribute to regulation of the transfer process.

Fatty acid uptake by mammalian cell preparations is often carried out in the presence of albumin because such conditions mimic the physiological environment found in serum (Spector *et al.*, 1965; Noy *et al.*, 1986; Cooper *et al.*, 1987). To simplify the kinetic analysis associated with fatty acid multiple binding to serum albumin, DeGrella and Light (1980a) studied fatty acid uptake and metabolism in heart myocytes in the absence of albumin. The study was performed at various concentrations well below the critical micelle concentrations and toxic levels of all fatty acids. Kinetic data for fatty acid uptake and metabolism consisted of a saturable and nonsaturable component. By studying how various metabolic reagents, fatty acid homologs, and albumin inhibited the kinetics, DeGrella and Light (1980b) proposed a simple diffusion mechanism for fatty acid uptake. The nonsaturable component represents fatty acid diffusion to the membrane, whereas the saturable component is due to a metabolic reaction such as fatty acid activation.

Spontaneous fatty acid transfer from bile salt micelles to enterocytes has been studied in detail because of the relevance to intestinal digestion and absorption of fats (Carey *et al.*, 1983; Börgstrom *et al.*, 1985; Tso, 1985). Theoretical calculations involving fatty acid mass distribution between the aqueous and micellar compartments show a good correlation between calculated aqueous monomer concentration and measured fatty acid uptake rates (Wilson *et al.*, 1971; Westergaard and Dietschy, 1976). Bile salt micelles apparently act as efficient, large-capacity vehicles for moving fatty acids and other lipids across unstirred water layers adjacent to enterocytes (Wilson *et al.*, 1971; Westergaard and Dietschy, 1976; Thompson and Dietschy, 1981). Once the fatty acids are positioned close to the plasma membrane, uptake is thought to occur by diffusion of lipid molecules through the aqueous phase (Hoffman, 1970; Hoffman and Yeoh, 1971; Thompson and Dietschy, 1981).

2.5. Cholesterol

Spontaneous cholesterol transfer involving biological membranes as well as lipoproteins has been reviewed in detail recently by Dawidowicz (1987b) and by Phillips *et al.* (1987). Also, discussions of cholesterol transfer are included in reviews by Sleight (1987) and Lange (1986). Rather than duplicating these efforts, I will focus on work that has appeared since these reviews.

Many recent studies involving cholesterol exchange have focused on how bilayer composition influences the distribution of sterols between model membranes (Bar *et al.*, 1986, 1987; Yeagle and Young, 1986; Lund-Katz *et al.*, 1988; Thomas and Poznansky, 1988a,b). Bar *et al.* (1986) reported that 80% of the cholesterol present in sonicated, 1-palmitoyl-2-oleoyl PC vesicles exchanged monoexponentially ($k = 0.0117\ \text{min}^{-1}$ at 37°C) while the remaining 20% of the cholesterol did not exchange within 8 hr. The size of the nonexchangeable cholesterol pool varied according to the lipid composition of the vesicles and the temperature. Meanwhile, Yeagle and Young (1986) also noted that cholesterol distribution between sonicated donor vesicles and large unilamellar acceptor vesicles is markedly influenced by vesicle composition. Both the lipid hydrocarbon chain and the headgroup composition affected the cholesterol distribution. Although neither PS nor PI altered cholesterol distribution, both sphingomyelin and high concentrations of PE slowed intervesicular transfer rates and inhibited equilibration of the cholesterol.

In subsequent studies, various laboratories continued to probe how bilayer composition affects cholesterol exchange rates (Bar *et al.*, 1987; Lund-Katz *et al.*, 1988; Thomas and Poznansky, 1988a,b). All reports agreed that vesicle composition plays an important role in regulating cholesterol exchange and that sphingomyelin, in particular, slows cholesterol intervesicular exchange rates. As a result of such findings, both Bar *et al.* (1987) and Lund-Katz *et al.* (1988) performed experiments designed to define the molecular interactions responsible for the different cholesterol exchange rates observed for vesicles of differing lipid composition. Bar *et al.* (1987) measured cholesterol exchange parameters from donor vesicles composed of different binary phospholipid mixtures and compared the values with those obtained from the mole fraction weighted averages of individual phospholipids. Deviations from simple averaging of the kinetic parameters were assumed to reflect mixing nonideality of the two phospholipids, along with a concomitant preferential interaction of cholesterol for one of the phospholipids. Accordingly, when the binary phospholipid mixtures contain sphingomyelin, cholesterol preferentially interacts with the sphingomyelin relative to other selected PCs.

Lund-Katz *et al.* (1988) performed intervesicular exchange measurements as well as NMR and monolayer experiments designed to evaluate directly the physical state of cholesterol and its interactions with phospholipid. However, the NMR relaxation parameters for [4-^{13}C]cholesterol revealed no differences in the molecular interactions of cholesterol with different PCs compared with sphingomyelin. In contrast, surface pressure–molecular area isotherms for mixed monolayers of cholesterol and different PCs or sphingomyelins revealed that sphingomyelin lateral packing density is greater than that of PC with the same acyl chain saturation and length. A larger cholesterol condensing effect is observed in sphingomyelin than in PC. Such an observation indicates a greater van der Waals interaction by sphingomyelin than by PC.

Bhuvaneswaran and Mitropoulos (1986) studied how liposomal phospholipid composition affects cholesterol transfer between microsomal and liposomal vesicles. Interestingly, the data indicate that vesicles composed of bovine brain sphingomyelin accept the sterol from microsomal vesicles at the same rate as do vesicles composed of egg yolk PC. However, sphingomyelin vesicles give up the sterol at a much lower rate than the PC vesicles do. This observation is consistent with the positive correlation between cholesterol and sphingomyelin localization found in biological membranes as well as the differential lipid compositions between various organelles.

Sanyal and co-workers have recently examined the kinetics of cholesterol transfer between lipid vesicles and monkey small intestine brush border membranes (Sadana *et al.*, 1986; Sanyal *et al.*, 1987). Cholesterol transfer displayed first-order kinetics, although the transfer rate was three times faster from brush border membrane vesicles to sonicated phospholipid vesicles than in the opposite direction. The rate of cholesterol transfer was not affected when the acceptor vesicle concentration was increased more than fivefold. Such findings suggest that the rate-limiting step is the desorption of cholesterol from the donor surface and that the transfer mechanism involves diffusion of cholesterol monomers through the aqueous phase. In contrast, Mutsch *et al.* (1986) have argued that lipid exchange between phospholipid vesicles and brush border membrane vesicles occurs by a collisional contact mechanism. These investigators showed that the exchange rates for spin-labeled cholestane, radiolabeled cholesteryl oleate, and spin-labeled PCs were nearly identical. Overall, the exchange was a second-order reaction, with rate constants being directly proportional to the brush border membrane vesicle concentration. To explain their results, Mutsch *et al.* (1986) proposed that a collisional contact mechanism would dominate the kinetic picture if high levels of donor and acceptor vesicles are present, whereas an aqueous diffusional mechanism would predominate at lower concentrations of donor and acceptor vesicles.

Recently, a two-step pathway invoking transient collisional contacts has been proposed to explain how cholesterol exchanges between biological membranes (Steck *et al.*, 1988). These investigators argued that the aqueous diffusion mechanism of cholesterol transfer does not adequately describe transfer between biological membranes. Thus, they proposed a different model. Accordingly, the initial step is a first-order process involving a unimolecular activation event within the donor membrane surface. In the second step, the activated cholesterol transfers as a result of transient collisions between the donor and acceptors (second-order process).

2.6. Other Lipids

Studies devoted to the spontaneous transfer kinetics of other lipids are sparse. Of the few that exist, most are devoted to transfer of diglycerides, tocopherols, and prostaglandins to lipoproteins (see, e.g., Charlton *et al.*, 1978; Charlton and Smith, 1982; Massey, 1984; Muzya *et al.*, 1987). However, a few studies involving intermembrane lipid transfer have been performed.

Lipid metabolites, such as fatty acyl-coenzyme A and fatty acyl-carnitine, exhibit very high spontaneous intermembrane transfer rates. Wolkowicz *et al.* (1984) examined the transfer kinetics of pyrenyldodecanoyl-coenzyme A, pyrenyldodecanoylcarnitine, and pyrenyldodecanoic acid. The more water-soluble carnitine and coenzyme A esters transferred 3.5 to 5 times more rapidly than the free acid. On the other hand, Pownall *et al.* (1983) compared the fatty acid transfer rates and activation energies with those of the corresponding methyl ester, alcohol, and alkane derivatives. For any given derivative, the transfer rate increases as the temperature rises and decreases as the hydrocarbon chain lengthens. The activation energy for transfer increases with the chain length, but at a given chain length, the activation energy increases in the following order: ester (pH 2.8) < acid (pH 9.0) < alcohol < acid (pH 2.8) < ester (pH 7.4). In contrast, the transfer rate generally decreases in the same order. Thus, the transfer rate is related to both the hydrophobicity (chain length) and the hydrophilicity (polar or nonpolar functional group) of the transferred species. Comparison of the activation free energy and the transfer free energy from lipid to water are consistent with transfer via the aqueous phase.

Other bioactive lipids, such as platelet-activating factor, have been studied. Lumb *et al.* (1983) examined the spontaneous transfer of 1-alkyl-2-acetyl-*sn*-glycero-3-phosphocholine [platelet-activating factor (PAF)] between phospholipid vesicles. Two transferable pools were identified in donor vesicles in which the PAF was colyophilized with phospholipid prior to sonication. One pool transferred almost instantaneously, whereas the second pool transferred much more slowly. The slower-transferring pool had access to a soluble protein fraction that accelerates PAF intermembrane transfer. Dialysis of the donor vesicles for up to 300 hr against buffer alone resulted in a 2% release of [^{3}H]PAF. However, when liposomes were included in the dialysis buffer, a 20-fold increase of label in the dialysate appeared.

The extremely low spontaneous transfer rates of cholesteryl esters, cholesteryl ethers, cholesteryl esters, and triglycerides between phospholipid vesicles has made these lipids popular "nontransferable markers" in the study of lipid transfer proteins. In fact, [^{14}C]cholesteryl oleate was used as a nontransferable marker in a study of the transfer kinetics of glycosphingolipids between sonicated phospholipid vesicles (Brown *et al.*, 1985; Brown and Thompson, 1987). Only 5–10% of the cholesteryl oleate transferred from small dipalmitoyl PC donor vesicles (25 nm in diameter) to large dipalmitoyl PC acceptor vesicles (65 nm in diameter) after 550 hr at 45°C. In contrast, Mutsch *et al.* (1986) observed much faster transfer of cholesteryl oleate from sonicated egg PC vesicles to intestinal brush border vesicles. After 30 hr at 20°C, nearly total equilibration of cholesteryl oleate occurred. A transfer mechanism involving lipid exchange by collisional contact appeared consistent with the kinetics reported by Mutsch *et al.* (1986). It is difficult to distinguish whether the high cholesteryl transfer rates involving intestinal brush border vesicles are related to the presence of proteins or to the high vesicle concentrations necessary to induce transfer by intervesicular collisional contacts. The latter explanation appears more likely.

3. CONCLUDING REMARKS

Since investigators began studying spontaneous lipid transfer events, controversy has persisted about the mechanism by which lipid molecules move between membranes. In the late 1970s and early 1980s, the majority of evidence favored a transfer mechanism involving the diffusion of lipid monomers through the aqueous medium between membranes rather than by transient collisional contacts between membranes. However, more recently, evidence has appeared from different laboratories implicating a collisional-contact mechanism for spontaneous lipid transfer when membrane concentrations are high. This observation is important because the lipid flux accompanying intermembrane collisional contacts appears to be significantly greater than that observed when individual lipid monomers diffuse through the aqueous phase between membranes. The higher flux rates reinforce the possibility that spontaneous lipid transfer plays a role in the biological transport of lipid or in membrane biogenesis. Furthermore, if vesicles were involved in lipid trafficking and/or membrane biogenesis, the targeting of lipids to needed cellular sites might be aided by proteins associated with the vesicles. In contrast, it is difficult to envision how individual lipid monomers diffusing freely through the aqueous medium might be targeted to specific sites within cells.

Regardless of whether biological lipid transport or membrane biogenesis involves spontaneous lipid transfer, the information presented in this review attests to the importance of the process. Spontaneous lipid transfer can and continues to be a valuable experimental tool for studying the biological effects of different lipids in cells, as well as a way to obtain information about the physical organization of components of membranes.

ACKNOWLEDGMENTS. I would like to extend special thanks to Thomas E. Thompson for fostering my interest in spontaneous lipid transfer processes and to acknowledge Julie Knutson for editorial assistance and the Hormel Foundation and U.S. Public Health Service Program Project Grant HL08214 for support.

4. REFERENCES

Arvinte, T., and Hildenbrand, K., 1984, N-NBD-L-α-dilauroylphosphatidylethanolamine: A new fluorescent probe to study spontaneous lipid transfer, *Biochim. Biophys. Acta* **775:**86–94.

Bar, L. K., Barenholz, Y., and Thompson, T. E., 1986, Fraction of cholesterol undergoing spontaneous exchange between small unilamellar phosphatidylcholine vesicles, *Biochemistry* **25:**6701–6705.

Bar, L. K., Barenholz, Y., and Thompson, T. E., 1987, Dependence on phospholipid composition of the fraction of cholesterol undergoing spontaneous exchange between small unilamellar vesicles, *Biochemistry* **26:**5460–5465.

Barañska, J., and Wojtczak, L., 1984, Transfer of phosphatidic acid between microsomal and mitochondrial outer and inner membranes, *Biochim. Biophys. Acta* **773:**23–31.

Barañska, J., and Wojtczak, L., 1988, Non-protein-mediated transfer of phosphatidic acid between microsomal and mitochondrial membranes, *Arch. Biochem. Biophys.* **260:**301–308.

Barsukov, L. I., Shapiro, Y. E., Viktorov, V. I., Bystorv, V. F., and Bergelson, L. D., 1974, Study of intervesicular phospholipid exchange by NMR, *Biochem. Biophys. Res. Commun.* **60:**196–203.

Barsukov, L. I., Victorov, A. V., Vasilenko, I. A., Evstigneeva, R. P., and Bergelson, L. D., 1980, Investigation of the inside–outside distribution, intermembrane exchange and transbilayer movement of phospholipids in sonicated vesicles by shift reagent NMR, *Biochim. Biophys. Acta* **598:**153–168.

Bhuvaneswaran, C., and Mitropoulos, K. A., 1986, Effect of liposomal phospholipid composition on cholesterol transfer between microsomal and liposomal vesicles, *Biochem. J.* **238:**647–652.

Bishop, W. R., and Bell, R. M., 1988, Assembly of phospholipids into cellular membranes: Biosynthesis, transmembrane movement, and intracellular translocation, *Annu. Rev. Cell Biol.* **4:**579–610.

Börgstrom, B., Barrowman, J. A., and Lindstrom, M., 1985, Roles of bile acids in intestinal lipid digestion and absorption, in *Sterols and Bile Acids* (H. Danielson and J. Sjovall, eds.), Vol. 12, pp. 405–425, Elsevier Science Publishing, New York.

Bremer, E. G., Hakomori, S., Bowen-Pope, D. F., Raines, E., and Ross, R., 1984, Ganglioside-mediated modulation of cell growth, growth factor binding, and receptor phosphorylation, *J. Biol. Chem.* **259:**6818–6825.

Bremer, E. G., Schlessinger, J., and Hakomori, S., 1986, Ganglioside-mediated modulation of cell growth, *J. Biol. Chem.* **261:**2434–2440.

Brown, R. E., and Thompson, T. E., 1987, Spontaneous transfer of ganglioside GM_1 between phospholipid vesicles, *Biochemistry* **26:**5454–5460.

Brown, R. E., Sugàr, I. P., and Thompson, T. E., 1985, Spontaneous transfer of gangliotetraosylceramide between phospholipid vesicles, *Biochemistry* **24:**4082–4091.

Brunner, J., Spiess, M., Aggeler, R., Huber, P., and Semenza, G., 1983, Hydrophobic labeling of a single leaflet of the human erythrocyte membrane, *Biochemistry* **22**:3812–3820.

Callies, R., Schwarzmann, G., Radsak, K., Siegert, R., and Wiegandt, H., 1977, Characterization of the cellular binding of exogenous gangliosides, *Eur. J. Biochem.* **80**:425–432.

Cantù, L., Corti, M., Sonnino, S., and Tettamanti, G., 1986, Light scattering measurements on gangliosides: Dependence of micellar properties on molecular structure and temperature, *Chem. Phys. Lipids* **41**:315–328.

Carey, M. C., Small, D. M., and Bliss, C. M., 1983, Lipid digestion and absorption, *Annu. Rev. Physiol.* **45**:651–677.

Cevc, G., and Marsh, D., 1987, Phospholipid bilayers: Physical principles and models, in *Cell Biology: A Series of Mongraphs* (E. E. Bittar, ed.), Vol. 5, pp. 51–55, Wiley-Interscience Press, New York.

Charlton, S. C., and Smith, L. C., 1982, Kinetics of transfer of pyrene and rac-1-oleoyl-2-[4-(3-pyrenyl)butanoyl]glycerol between human plasma lipoproteins, *Biochemistry* **21**:4023–4030.

Charlton, S. C., Hong, K.-Y., and Smith, L. C., 1978, Kinetics of rac-1-oleoyl-2-[4-(3-pyrenyl)butanoyl]glycerol transfer between high density lipoproteins, *Biochemistry* **17**:3304–3309.

Cooper, R., Noy, N., and Zakim, D., 1987, A physical-chemical model for cellular uptake of fatty acids: Prediction of intracellular pool sizes, *Biochemistry* **26**:5890–5896.

Correa-Freire, M. C., Barenholz, Y., and Thompson, T. E., 1982, Glucocerebroside transfer between phosphatidylcholine bilayers, *Biochemistry* **21**:1244–1248.

Curatolo, W., 1986, The interactions of 1-palmitoyl-2-oleoylphosphatidylcholine and bovine brain cerebroside, *Biochim. Biophys. Acta* **861**:373–376.

Daleke, D. L., and Huestis, W. H., 1985, Incorporation and translocation of aminophospholipids in human erythrocytes, *Biochemistry* **24**:5406–5416.

Dawidowicz, E. A., 1987a, Dynamics of membrane lipid metabolism and turnover, *Annu. Rev. Biochem.* **56**:43–61.

Dawidowicz, E. A., 1987b, Lipid exchange: Transmembrane movement, spontaneous movement, and protein-mediated transfer of lipids and cholesterol, *Curr. Top. Membr. Transp.* **29**:175–202.

De Cuyper, M., and Joniau, M., 1985, Spontaneous intervesicular transfer of anionic phospholipids differing in the nature of their polar headgroup, *Biochim. Biophys. Acta* **814**:374–380.

De Cuyper, M., Joniau, M., and Dangreau, H., 1980, Spontaneous phospholipid transfer between artificial vesicles followed by free-flow electrophoresis, *Biochem. Biophys. Res. Commun.* **95**:1224–1230.

De Cuyper, M., Joniau, M., and Dangreau, H., 1983, Intervesicular phospholipid transfer: A free-flow electrophoresis study, *Biochemistry* **22**:415–420.

De Cuyper, M., Joniau, M., Engberts, J. B., and Sudholter, E. J., 1984, Exchangeability of phospholipids between anionic, zwitterionic, and cationic membranes, *Colloids Surf.* **10**:313–319.

DeGrella, R. F., and Light, R. J., 1980a, Uptake and metabolism of fatty acids by dispersed adult rat heart myocytes. I. Kinetics of homologous fatty acids, *J. Biol. Chem.* **255**:9731–9738.

DeGrella, R. F., and Light, R. J., 1980b, Uptake and metabolism of fatty acids by dispersed adult rat heart myocytes. II. Inhibition by albumin and fatty acid homologues, and the effect of temperature and metabolic reagents, *J. Biol. Chem.* **255**:9739–9745.

de Kruijff, B., and van Zoelen, E. J. J., 1978, Effect of the phase transition on the transbilayer movement of dimyristoylphosphatidylcholine in unilamellar vesicles, *Biochim. Biophys. Acta* **511**:105–115.

de Kruijff, B., van den Besselaar, A., and van Deenen, L., 1977, Outside-inside distribution and translocation of lysophosphatidylcholine in phosphatidylcholine vesicles as determined by ^{13}C-NMR using (N-$^{13}CH_3$)-enriched lipids, *Biochim. Biophys. Acta* **465**:443–453.

Devaux, P. F., 1988, Phospholipid flippases, *FEBS Lett.* **234:**8–12.

Dobiasova, M., and Radin, N. S., 1968, Uptake of cerebroside, cholesterol, and lecithin by brain myelin and mitochondria, *Lipids* **3:**439–448.

Doody, M. C., Pownall, H. J., Kao, Y. J., and Smith, L. C., 1980, Mechanism and kinetics of transfer of a fluorescent fatty acid between single-walled phosphatidylcholine vesicles, *Biochemistry* **19:**108–116.

Duckwitz-Peterlein, G., and Moraal, H., 1978, Transport of lipids through water as exchange mechanism between two liposomes populations, *Biophys. Struct. Mech.* **4:**315–326.

Duckwitz-Peterlein, G., Eilenberger, G., and Overath, P., 1977, Phospholipid exchange between bilayer membranes, *Biochim. Biophys. Acta* **469:**311–325.

Ehnholm, C., and Zilversmit, D. B., 1973, Exchange of various phospholipids and of cholesterol between liposomes in the presence of highly purified phospholipid exchange protein, *J. Biol. Chem.* **248:**1719–1724.

Elamrani, K., and Blume, A., 1982, Incorporation kinetics of lysolecithin into lecithin vesicles. Kinetics of lysolecithin-induced vesicle fusion, *Biochemistry* **21:**521–526.

Facci, L., Leon, A., Toffano, G., Sonnino, S., Ghidoni, R., and Tettamanti, G., 1984, Promotion of neuritogenesis in mouse neuroblastoma cells by exogenous gangliosides. Relationship between the effect and the cell association of ganglioside GM_1, *J. Neurochem.* **42:**299–305.

Felgner, P. L., Freire, E., Barenholz, Y., and Thompson, T. E., 1981, Asymmetric incorporation of trisialoganglioside into dipalmitoylphosphatidylcholine vesicles, *Biochemistry* **20:**2168–2172.

Felgner, P. L., Thompson, T. E., Barenholz, Y., and Lichtenberg, D., 1983, Kinetics of transfer of gangliosides from their micelles to dipalmitoylphosphatidylcholine vesicles, *Biochemistry* **22:**1670–1674.

Ferrell, J. E., Lee, K.-J., and Huestis, W. H., 1985a, Membrane bilayer balance and erythrocyte shape: A quantitative assessment, *Biochemistry* **24:**2849–2857.

Ferrell, J. E., Lee, K.-J., and Huestis, W. H., 1985b, Lipid transfer between phosphatidylcholine vesicles and human erythrocytes: Exponential decrease in rate with increasing acyl chain length, *Biochemistry* **24:**2857–2864.

Figard, P. H., Hejlik, D. P., Kaduce, T. L., Stoll, L. L., and Spector, A. A., 1986, Free fatty acid release from endothelial cells, *J. Lipid Res.* **27:**771–780.

Fishman, P. H., 1980, Mechanism of action of cholera toxin: Studies on the lag period, *J. Membr. Biol.* **54:**61–72.

Fishman, P. H., Moss, J., and Vaughan, M., 1976, Uptake and metabolism of gangliosides in transformed mouse fibroblasts. Relationship of ganglioside structure to choleragen response, *J. Biol. Chem.* **251:**4490–4494.

Fishman, P. H., Bradley, R. M., Moss, J., and Manganiello, V. C., 1978, Effect of serum on ganglioside uptake and choleragen responsiveness of transformed fibroblasts, *J. Lipid Res.* **19:**77–81.

Fishman, P. H., Pacuszka, T., Hom, B., and Moss, J., 1980, Modification of ganglioside GM_1: Effect of lipid moiety on choleragen action, *J. Biol. Chem.* **255:**7657–7664.

Fishman, P. H., Bradley, R. M., Hom, B. E., and Moss, J., 1983, Uptake and metabolism of exogenous gangliosides by cultured cells: Effect of choleragen on the turnover of GM_1, *J. Lipid Res.* **24:**1002–1011.

Frank, A., Barenholz, Y., Lichtenberg, D., and Thompson, T. E., 1983, Spontaneous transfer of sphingomyelin between phospholipid bilayers, *Biochemistry* **22:**5647–5651.

Fujii, T., and Tamura, A., 1983, Dynamic behavior of amphiphilic lipids to penetrate into membrane of intact erythrocytes and to induce change in the cell shape, *Biomed. Biochim. Acta* **42:**81–85.

Fujii, T., and Tamura, A., 1984, Shape change of human erythrocytes induced by phosphatidylcholine and lysophosphatidylcholine species with various acyl chain lengths, *Cell Biochem. Funct.* **2:**171–176.

Gardam, M. A., Itovitch, J. J., and Silvius, J. R., 1989, Partitioning of exchangeable fluorescent phospholipids and sphingolipids between different lipid bilayer environments, *Biochemistry* **28**:884–893.

Golan, D. E., Brown, C. S., Cianci, C. M. L., Furlong, S. T., and Caulfield, J. P., 1986, Schistosomula of *Schistosoma mansoni* use lysophosphatidylcholine to lyse adherent human red blood cells and immobilize red cell membrane components, *J. Cell Biol.* **103**:819–828.

Golan, D. E., Furlong, S. T., Brown, C. S., and Caulfield, J. P., 1988, Monopalmitoylphosphatidylcholine incorporation into human erythrocyte ghost membranes causes protein and lipid immobilization and cholesterol depletion, *Biochemistry* **27**:2661–2667.

Hagerman, J. S., and Gould, R. G., 1951, The *in vitro* interchange of cholesterol between plasma and red cells, *Proc. Soc. Exp. Biol. Med.* **78**:329–332.

Hahn, L., and Hevesy, G., 1939a, Interaction between the phosphatides of the plasma and the corpuscles, *Nature* **144**:72.

Hahn, L., and Hevesy, G., 1939b, Phosphatide exchange between plasma and organs, *Nature* **144**:204–205.

Hamilton, J. A., and Cistola, D. P., 1986, Transfer of oleic acid between albumin and phospholipid vesicles, *Proc. Natl. Acad. Sci. U.S.A.* **83**:82–86.

Hanai, N., Dohi, T., Nores, G. A., and Hakomori, S., 1988a, A novel ganglioside, de-N-acetyl-GM_1 ($II^3NeuNH_2LacCer$), acting as a strong promoter for epidermal growth factor receptor kinase and as a stimulator for cell growth, *J. Biol. Chem.* **263**:6296–6301.

Hanai, N., Nores, G. A., MacLeod, C., Torres-Mendez, C.-R., and Hakomori, S., 1988b, Ganglioside-mediated modulation of cell growth, *J. Biol. Chem.* **263**:10915–10921.

Hannun, Y. A., and Bell, R. M., 1987, Lysophingolipids inhibit protein kinase C: Implications for the sphingolipidoses, *Science* **235**:670–674.

Hannun, Y. A., and Bell, R. M., 1989, Functions of sphingolipids and sphingolipid breakdown products in cellular regulation, Science **243**:500–507.

Hannun, Y. A., Loomis, C. R., Merrill, A. H., Jr., and Bell, R. M., 1986, Sphingosine inhibition of protein kinase C activity and of phorbol dibutyrate binding *in vitro* and in human platelets, *J. Biol. Chem.* **261**:12604–12609.

Hellings, J. A., Kamp, H. H., Wirtz, K. W. A., and van Deenen, L. L. M., 1974, Transfer of phosphatidylcholine between liposomes, *Eur. J. Biochem.* **47**:601–605.

Helmkamp, G. M., Jr., 1986, Phospholipid transfer proteins: Mechanism of action, *J. Bioenerg. Biomembr.* **18**:71–91.

Hinz, H.-J., Korner, O., and Nicolau, C., 1981, Influence of gangliosides GM_1 and GD_{1a} on structural and thermotropic properties of sonicated small 1,2-dipalmitoyl-L-α-phosphatidylcholine vesicles, *Biochim. Biophys. Acta* **643**:557–571.

Hoffman, N. E., 1970, The relationship between uptake *in vitro* of oleic acid and micellar solubilization, *Biochim. Biophys. Acta* **196**:193–203.

Hoffman, N. E., and Yeoh, V. J., 1971, The relationship between concentration and uptake by rat small intestine, *in vitro*, from two micellar solutes, *Biochim. Biophys. Acta* **233**:49–52.

Israelachvili, J. N., Marcelja, S., and Horn, R. G., 1980, Physical principles of membrane organization, *Q. Rev. Biophys.* **13**:121–200.

Jahnig, F., 1984, Lipid exchange between membranes, *Biophys. J.* **46**:687–694.

Jones, J. D., and Thompson, T. E., 1988, Spontaneous interbilayer transfer of galactosyl ceramides, *Biophys. J.* **53**:119a.

Jones, J. D., and Thompson, T. E., 1989, Spontaneous phosphatidylcholine transfer by collision between vesicles at high lipid concentration, *Biochemistry* **28**:129–134.

Kanda, S., Inoue, K., Nojima, S., Utsumi, H., and Wiegandt, H., 1982a, Incorporation of spin-labeled ganglioside analogues into cell and liposomal membranes, *J. Biochem.* **91**:1707–1718.

Kanda, S., Inoue, K., Nojima, S., Utsumi, H., and Wiegandt, H., 1982b, Incorporation of a ganglioside and a spin-labeled ganglioside analogue into cell and liposomal membranes, *J. Biochem.* **91**:2095–2098.

Keenan, T. W., Franke, W. W., and Wiegandt, H., 1974, Ganglioside accumulation by transformed murine fibroblasts (3T3) cells and canine erythrocytes, *Hoppe-Seyler's Z. Physiol. Chem.* **355:**1543–1558.

Keenan, T. W., Schmid, E., Franke, W. W., and Wiegandt, H., 1976, Exogenous glycosphingolipids suppress growth rate of transformed and untransformed 3T3 mouse cells, *Exp. Cell Res.* **92:**259–270.

Klibansky, C., and de Vries, A., 1963, Quantitative study of erythrocyte–lysolecithin interaction, *Biochim. Biophys. Acta* **70:**176–187.

Kornberg, R. D., and McConnell, H. M., 1971, Lateral diffusion of phospholipids in a vesicle membrane, *Proc. Natl. Acad. Sci. U.S.A.* **78:**2564–2568.

Kremer, J. M. H., and Wiersema, P. H., 1977, Exchange and aggregation in dispersions of dimyristoyl phosphatidylcholine vesicles containing myristic acid, *Biochim. Biophys. Acta* **471:**348–360.

Kremer, J. M., Kops-Werkhoven, M. M., Pathmamanoharan, C., Gijzeman, O. L., and Wiersema, P. H., 1977, Phase diagrams and the kinetics of phospholipid exchange for vesicles of different composition and radius, *Biochim. Biophys. Acta* **471:**177–188.

Laine, R. A., and Hakomori, S., 1973, Incorporation of exogenous glycosphingolipids in plasma membranes of cultured hamster cells and concurrent change of growth behavior, *Biochem. Biophys. Res. Commun.* **54:**1039–1045.

Lange, Y., 1986, Molecular dynamics of lipids, in *Handbook of Lipid Research: The Physical Chemistry of Lipids* (D. M. Small, ed.), pp. 538–549. Plenum Press, New York.

Lange, Y., and Slayton, J. M., 1982, Interaction of cholesterol and lysophosphatidylcholine in determining red cell shape, *J. Lipid Res.* **23:**1121–1127.

Lawaczeck, R., 1978, Intervesicular lipid transfer and direct fusion of phospholipid vesicles: A comparison on a kinetic basis, *J. Colloid Interface Sci.* **66:**247–256.

Ledeen, R. W., 1984, Biology of gangliosides: Neuritogenic and neuronotrophic properties, *J. Neurosci. Res.* **12:**147–159.

Lipsky, N., and Pagano, R. E., 1983, Sphingolipid metabolism in cultured fibroblasts: Microscopic and biochemical studies employing a fluorescent ceramide analogue, *Proc. Natl. Acad. Sci. U.S.A.* **80:**2608–2612.

Lipsky, N., and Pagano, R. E., 1985, Intracellular translocation of fluorescent sphingolipids in cultured fibroblasts: Endogenously synthesized sphingomyelin and glucocerebroside analogues pass through the Golgi apparatus en route to the plasma membrane, *J. Cell Biol.* **100:**27–34.

Lumb, R. H., Pool, G. L., Bubacz, D. G., Blank, M. L., and Snyder, F., 1983, Spontaneous and protein-catalyzed transfer of 1-alkyl-2-acetyl-*sn*-glycero-3-phosphocholine (platelet-activating factor) between phospholipid bilayers, *Biochim. Biophys. Acta* **750:**217–222.

Lund-Katz, S., Laboda, H. M., McLean, L. R., and Phillips, M. C., 1988, Influence of molecular packing and phospholipid type on rates of cholesterol exchange, *Biochemistry* **27:**3416–3423.

Maeda, T., and Ohnishi, S., 1974, Membrane fusion: Transfer of phospholipid molecules between phospholipid bilayer membranes, *Biochem. Biophys. Res. Commun.* **60:**1509–1516.

Marcus, D. M., and Cass, L. E., 1969, Glycosphingolipids with Lewis blood group activity: Uptake by human erythrocytes, *Science* **164:**553–555.

Martin, F. J., and MacDonald, R. C., 1976, Phospholipid exchange between bilayer membrane vesicles, *Biochemistry* **15:**321–327.

Masserini, M., and Freire, E., 1987, Kinetics of ganglioside transfer between liposomal and synaptosomal membranes, *Biochemistry* **26:**237–242.

Massey, J. B., 1984, Kinetics of transfer of α-tocopherol between model and native plasma lipoproteins, *Biochim. Biophys. Acta* **793:**387–392.

Massey, J. B., Gotto, A. M., Jr., and Pownall, H. J., 1982a, Kinetics and mechanism of the spontaneous transfer of fluorescent phosphatidylcholines between apolipoprotein-phospholipid recombinants, *Biochemistry* **21:**3630–3636.

Massey, J. B., Gotto, A. M., Jr., and Pownall, H. J., 1982b, Kinetics and mechanism of the spontaneous transfer of fluorescent phosphatidylcholines between apolipoprotein-phospholipid recombinants: Effect of the polar head group, *J. Biol. Chem.* **257**:5444–5448.

Massey, J. B., Hickson, D., She, H. S., Sparrow, J. T., Via, D. P., Gotto, A. M., and Pownall, H. J., 1984, Measurement and prediction of the rates of spontaneous transfer of phospholipids between plasma lipoproteins, *Biochim. Biophys. Acta* **794**:274–280.

McLean, L. R., and Phillips, M. C., 1981, Mechanism of cholesterol and phosphatidylcholine exchange or transfer between unilamellar vesicles, *Biochemistry* **20**:2893–2900.

McLean, L. R., and Phillips, M. C., 1982, Cholesterol desorption from clusters of phosphatidylcholine and cholesterol in unilamellar vesicle bilayers during lipid transfer or exchange, *Biochemistry* **21**:4053–4059.

McLean, L. R., and Phillips, M. C., 1984, Kinetics of phosphatidylcholine and lysophosphatidylcholine exchange between unilamellar vesicles, *Biochemistry* **23**:4624–4630.

Metz, S. A., 1988, Mobilization of cellular Ca^{2+} by lysophospholipids in rat islets of Langerhans, *Biochim. Biophys. Acta* **968**:239–252.

Moss, J., Fishman, P. H., Manganiello, V. C., Vaughan, M., and Brady, R. O., 1976, Functional incorporation of ganglioside into intact cells: Induction of choleragen responsiveness, *Proc. Natl. Acad. Sci. U.S.A.* **73**:1034–1037.

Mutsch, B., Gains, N., and Hauser, H., 1986, Interaction of intestinal brush border membrane vesicles with small unilamellar phospholipid vesicles. Exchange of lipids between membranes is mediated by collisional contact, *Biochemistry* **25**:2134–2140.

Muzya, G. I., Korobkova, E. N., Golovanova, N. K., and Bergelson, L. D., 1987, Influence of prostaglandins on the lipid transfer between human high density and low density lipoproteins, *FEBS Lett.* **220**:371–375.

Nakajima, J., Tsuji, S., and Nagai, Y., 1986, Bioactive gangliosides: Analysis of functional structures of the tetrasialoganglioside GQ_{1b} which promotes neurite outgrowth, *Biochim. Biophys. Acta* **876**:65–71.

Nichols, J. W., 1985, Thermodynamics and kinetics of phospholipid monomer-vesicle interaction, *Biochemistry* **24**:6390–6398.

Nichols, J. W., 1986, Low concentrations of bile salts increase the rate of spontaneous phospholipid transfer between vesicles, *Biochemistry* **25**:4596–4601.

Nichols, J. W., 1988, Phospholipid transfer between phosphatidylcholine-taurocholate mixed micelles, *Biochemistry* **27**:3925–3931.

Nichols, J. W., and Pagano, R. E., 1981, Kinetics of soluble lipid monomer diffusion between vesicles, *Biochemistry* **20**:2783–2789.

Nichols, J. W., and Pagano, R. E., 1982, Use of resonance energy transfer to study the kinetics of amphiphile transfer between vesicles, *Biochemistry* **21**:1720–1726.

Noy, N., Donnelly, T. M., and Zakim, D., 1986, Physical-chemical model for the entry of water-soluble components into cells. Studies of fatty acid uptake by the liver, *Biochemistry* **22**:2013–2021.

O'Keefe, E., and Cuatrecasas, P., 1977, Persistence of exogenous, inserted ganglioside GM_1 on the cell surface of cultured cells, *Life Sci.* **21**:1649–1654.

Oishi, K., Raynor, R. L., Charp, P. A., and Kuo, J. F., 1988, Regulation of protein kinase C by lysophospholipids, *J. Biol. Chem.* **263**:6865–6871.

Pagano, R. E., and Martin, O. C., 1988, A Series of fluorescent *N*-acylsphingosines: Synthesis, physical properties, and studies in cultured cells, *Biochemistry* **27**:4439–4445.

Pagano, R. E., and Sleight, R. G., 1985, Defining lipid transport pathways in animal cells, *Science* **229**:1051–1057.

Pagano, R. E., Martin, O. C., Schroit, A. J., and Struck, D. K., 1981, Formation of asymmetric phospholipid membranes via spontaneous transfer of fluorescent lipid analogues between vesicle populations, *Biochemistry* **20**:4920–4927.

Papahadjopoulos, D., Poste, G., Schaeffer, B. E., and Vail, W. J., 1974, Membrane fusion and molecular segregation in phospholipid vesicles, *Biochim. Biophys. Acta* **352:**10–28.

Papahadjopoulos, D., Hui, S., Vail, W. J., and Poste, G., 1976, Studies on membrane fusion: Interactions of pure phospholipid membranes and the effect of myristic acid, lysolecithin, proteins, and dimethylsulfoxide, *Biochim. Biophys. Acta* **448:**245–264.

Phillips, M. C., Johnson, W. J., and Rothblat, G. H., 1987, Mechanisms and consequences of cellular cholesterol exchange and transfer, *Biochim. Biophys. Acta* **906:**223–276.

Pownall, H. J., Hickson, D., Gotto, A. M., and Massey, J. B., 1982, Kinetics of spontaneous and plasma-stimulated sphingomyelin transfer, *Biochim. Biophys. Acta* **712:**169–176.

Pownall, H. J., Hickson, D. L., and Smith, L. C., 1983, Transport of biological lipophiles: Effect of lipophile structure, *J. Am. Chem. Soc.* **105:**2440–2445.

Quinn, M. T., Parthasarathy, S., and Steinberg, D., 1988, Lysophosphatidylcholine: A chemotactic factor for human monocytes and its potential role in atherogenesis, *Proc. Natl. Acad. Sci. U.S.A.* **85:**2805–2809.

Radsak, K., Schwarzmann, G., and Wiegandt, H., 1982, Studies on the cell association of exogenously added sialo-glycolipids, Hoppe-Seyler's *Z. Physiol. Chem.* **363:**263–272.

Reed, C. F., Murphy, M., and Roberts, G., 1968, Phospholipid exchange between plasma and erythrocytes in man and the dog, *J. Clin. Invest.* **47:**749–760.

Reman, F. C., Demel, R. A., de Gier, J., van Deenen, L. L. M., Eibl, H., and Westphal, O., 1969, Studies on the lysis of red cells and bimolecular lipid leaflets by synthetic lysolecithins, lecithins, and structural analogs, *Chem. Phys. Lipids* **3:**221–233.

Rindlisbacher, B., and Zahler, P., 1983, Interaction of phosphatidylcholine liposomes and plasma lipoproteins with sheep erythrocyte membranes, *Biochim. Biophys. Acta* **732:**485–491.

Roseman, M. A., and Thompson, T. E., 1980, Mechanism of the spontaneous transfer of phospholipids between bilayers, *Biochemistry* **19:**439–444.

Sadana, T., Sanyal, S. N., Majumdar, S., Dhall, K., and Chakravarti, R. N., 1986, Spontaneous cholesterol movement between lipid vesicles and monkey small intestinal brush border membrane, *Biochem. Cell Biol.* **64:**575–582.

Sakagami, T., Minari, O., and Orii, T., 1965, Behavior of plasma lipoproteins during exchange of phospholipids between plasma and erythrocytes, *Biochim. Biophys. Acta* **98:**111–116.

Sanyal, S., Sadana, T., and Majumdar, S., 1987, Kinetics of cholesterol movement between liposomes and intestinal brush border membrane vesicles, *Indian J. Exp. Biol.* **25:**674–680.

Sato, T., 1973, Variability in the lysolecithin content of human erythrocyte membranes, *Chem. Pharm. Bull.* **21:**176–183.

Sato, T., and Fujii, T., 1974, Changes in shape and osmotic resistance of human erythrocytes resulting from changes in the lysolecithin content of the membranes, *Chem. Pharm. Bull.* **22:**152–156.

Schindler, H., 1979, Exchange and interactions between lipid layers at the surface of a liposome solution, *Biochim. Biophys. Acta* **555:**316–336.

Schroit, A. J., and Madsen, J., 1983, Synthesis and properties of radioiodinated phospholipid analogs that spontaneously undergo vesicle-vesicle and vesicle-cell transfer, *Biochemistry* **22:**3617–3623.

Schroit, A. J., Madsen, J., and Ruoho, A. E., 1987, Radioiodinated, photoactivatable phosphatidylcholine and phosphatidylserine: Transfer properties and differential photoreactive interaction with human erythrocyte membrane proteins, *Biochemistry* **26:**1812–1819.

Schwarzmann, G., Hoffmann-Bleihauer, P., Schubert, J., Sandhoff, K., and Marsh, D., 1983, Incorporation of ganglioside analogues into fibroblast cell membranes, *Biochemistry* **22:**5041–5048.

Scow, R. O., and Blanchette-Mackie, E. J., 1985, Why fatty acids flow in cell membranes, *Prog. Lipid Res.* **24:**197–241.

Sengupta, P., Sackmann, E., Kuhnle, W., and Scholz, H. P., 1976, An optical study of the exchange kinetics of membrane bound molecules, *Biochim. Biophys. Acta* **436**:869–878.

Silvius, J. R., Leventis, R., Brown, P. M., and Zuckermann, M., 1987, Novel fluorescent phospholipids for assays of lipid mixing between membranes, *Biochemistry* **26**:4279–4287.

Simons, K., and van Meer, G., 1988, Lipid sorting in epithelial cells, *Biochemistry* **27**:6197–6202.

Sleight, R. G., 1987, Intracellular lipid transport in eukaryotes, *Annu. Rev. Physiol.* **49**:193–208.

Sonderfeld, S., Conzelmann, E., Schwarzmann, G., Burg, J., Hinrichs, U., and Sandhoff, K., 1985, Incorporation and metabolism of ganglioside G_{M2} in skin fibroblasts from normal and G_{M2} gangliosidosis subjects, *Eur. J. Biochem.* **149**:247–255.

Soula, G., Valdiguie, P., and Douste-Blazy, L., 1967, Phospholipid metabolism in rabbit erythrocytes. II. *In vitro* exchange of ^{32}P-labeled phospholipids between red blood cells and plasma, *Bull. Soc. Chim. Biol.* **49**:1317–1330.

Spector, A. A., Steinberg, D., and Tanaka, A., 1965, Uptake of free fatty acids by Ehrlich ascites tumor cells, *J. Biol. Chem.* **240**:1032–1041.

Spener, F., and Wirtz, K. W. A., (eds.), 1985, Intracellular lipid binding and transfer proteins, *Chem. Phys. Lipids* **38**:1–261.

Spiegel, S., Schlessinger, J., and Fishman, P. H., 1984, Incorporation of fluorescent gangliosides into human fibroblasts: Mobility, fate, and interaction with fibronectin, *J. Cell Biol.* **99**:699–704.

Steck, T. L., Kezdy, F. J., and Lange, Y., 1988, An activation-collision mechanism for cholesterol transfer between membranes, *J. Biol. Chem.* **263**:13023–13031.

Stoll, L. L., and Spector, A. A., 1987, Lipid transfer between endothelial and smooth muscle cells in co-culture, *J. Cell. Physiol.* **133**:103–110.

Storch, J., and Kleinfeld, A. M., 1986, Transfer of long-chain fluorescent free fatty acids between unilamellar vesicles, *Biochemistry* **25**:1717–1726.

Struck, D. K., and Pagano, R. E., 1980, Insertion of fluorescent phospholipids into the plasma membrane of a mammalian cells, *J. Biol. Chem.* **255**:5404–5410.

Stuhne-Sekalec, L., and Stanacev, N. Z., 1978, Cytosol protein-independent translocation of isomeric spin-labelled radioactive lipids from isolated guinea pig liver microsomal to mitochondrial membranes, *Can. J. Biochem.* **56**:407–413.

Stuhne-Sekalec, L., and Stanacev, N. Z., 1980, Mitochondrial importation of lipids and liponucleotides from microsomes independent of and facilitated by purified cytosol proteins, *Can. J. Biochem.* **58**:1082–1090.

Taketomi, T., and Kawamura, N., 1970, Preparation of lysohematoside (neuraminyl-galactosylglucosylsphingosine) from hematoside of equine erythrocyte and its chemical and hemolytic properties, *J. Biochem.* **68**:475–485.

Taketomi, T., Kawamura, N., Hara, A., and Murakami, S., 1976, Comparative studies on chemical, hemolytic and diffusion-in-gel precipitation properties of various lysophingolipids, *Biochim. Biophys. Acta* **424**:106–113.

Tall, A., 1986, Plasma lipid transfer proteins, *J. Lipid Res.* **27**:361–367.

Tamura, A., Yoshikawa, K., Fujii, T., Ohki, K., Nozawa, Y., and Sumida, Y., 1986, Effect of fatty acyl chain length of phosphatidylcholine on their transfer from liposomes to erythrocytes and transverse diffusion in the membranes inferred by TEMPO-phosphatidylcholine spin probes, *Biochim. Biophys. Acta* **855**:250–256.

Tanaka, Y., and Schroit, A. J., 1983, Insertion of fluorescent phosphatidylserine into the plasma membrane of red blood cells, *J. Biol. Chem.* **258**:11335–11343.

Tanaka, Y., and Schroit, A. J., 1986, Calcium/phosphate-induced immobilization of fluorescent phosphatidylserine in synthetic membranes: Inhibition of lipid transfer between vesicles, *Biochemistry* **25**:2141–2148.

Thilo, L., 1977, Kinetics of phospholipid exchange between bilayer membranes, *Biochim. Biophys. Acta* **469:**326–334.

Thomas, P. D., and Poznansky, M. J., 1988a, Effect of surface curvature on the rate of cholesterol transfer between lipid vesicles, *Biochem. J.* **254:**155–160.

Thomas, P. D., and Poznansky, M. J., 1988b, Cholesterol transfer between lipid vesicles. Effects of phospholipids and gangliosides, *Biochem. J.* **251:**55–61.

Thompson, A. B. R., and Dietschy, J. M., 1981, Intestinal lipid absorption: Major extracellular and intracellular events, in *Physiology of the Gastrointestinal Tract* (L. R. Johnson, ed.), pp. 1147–1220, Raven Press, New York.

Thompson, T. E., and Brown, R. E., 1988, Biophysical properties of gangliosides, in *New Trends in Ganglioside Research: Neurochemical and Neuroregenerative Aspects* (R. W. Ledeen, E. L. Hogan, G. Tettamanti, A. J. Yates, and R. K. Yu, eds.), Fidia Research Series, Vol. 14, pp. 65–78, Liviana Press, Padova, Italy.

Thompson, T. E., and Tillack, T. W., 1985, Organization of glycosphingolipids in bilayers and plasma membranes of mammalian cells, *Annu. Rev. Biophys. Biophys. Chem.* **14:**361–386.

Thompson, T. E., Barenholz, Y., Brown, R. E., Correa-Freire, M., Young, W. W., and Tillack, T. W., 1986, Molecular organization of glycosphingolipids in phosphatidylcholine bilayers and biological membranes, in *Enzymes of Lipid Metabolism II* (L. Freysz, H. Dreyfus, R. Massarelli, and S. Gatt, eds.), pp. 387–396, Plenum Press, New York.

Tso, P., 1985, Gastrointestinal digestion and absorption of lipid, *Adv. Lipid Res.* **21:**143–186.

van Meer, G., Stelzer, E. H. K., Wijnaendts-van-Resandt, R. W., and Simons, K., 1987, Sorting of sphingolipids in epithelial (Madin-Darby canine kidney) cells, *J. Cell Biol.* **105:**1624–1635.

Weltzien, H. U., 1979, Cytolytic and membrane-perturbing properties of lysosphophatidylcholine, *Biochim. Biophys. Acta* **559:**259–287.

Weltzien, H. U., Arnold, B., and Kalkoff, H. G., 1976, Quantitative studies on lysolecithin mediated hemolysis. Benzylated lysolecithin as a probe to study effects of temperature and red cell species on the hemolytic reaction, *Biochim. Biophys. Acta* **455:**56–65.

Westergaard, H., and Dietschy, J. M., 1976, The mechanism whereby bile salt micelles increase the rate of fatty acid and cholesterol uptake into the intestinal mucosal cell, *J. Clin. Invest.* **58:**97–108.

Wilson, F. A., Sallee, V. L., and Dietschy, J. M., 1971, Unstirred water layers in intestine: Rate determinant of fatty acid absorption from micellar solutions, *Science* **174:**1031–1033.

Wirtz, K. W. A., op den Kamp, J. A. F., and Roelofsen, B., 1986, Phosphatidylcholine transfer protein: Properties and applications in membrane research, in *Progress in Protein-Lipid Interactions* (A. Watts and S. de Pont, eds.), Vol. 2, pp. 221–265, Elsevier Science Publishing, New York.

Wolkowicz, P. E., Pownall, H. J., Pauly, D. F., and McMillin-Wood, J. B., 1984, Pyrenedodecanoylcarnitine and pyrenedodecanoyl coenzyme A: Kinetics and thermodynamics of their intermembrane transfer, *Biochemistry* **23:**6426–6432.

Yeagle, P. L., and Young, J. E., 1986, Factors contributing to the distribution of cholesterol among phospholipid vesicles, *J. Biol. Chem.* **261:**8175–8181.

Yokoyama, M., Trams, E. G., and Brady, R. O., 1963, Immunological studies with gangliosides, *J. Immunol.* **90:**372–380.

Chapter 12

Extra- and Intracellular Transport of Retinoids

Ulf Eriksson

1. INTRODUCTION

The essential role of vitamin A in the normal development and maintenance of physiological functions in higher organisms is well known. The major effects of its deficiency are seen in various epithelia, in the reproductive organs, and in the eyes, but submicroscopic lesions may occur in other tissues. Although the observed effects of vitamin A deficiency are dramatic, the precise molecular function of the vitamin in nonocular tissues is unknown. This is in contrast to the detailed knowledge of the function of vitamin A in the visual process (Wald, 1968).

This chapter will focus on the transport of vitamin A compounds in extra- and intracellular compartments. In particular, the current knowledge of specialized vitamin A-binding proteins and their roles in the metabolism of the

Abbreviations used in this chapter: ACAT, acyl-CoA: cholesterol acyltransferase; ARAT, acyl-CoA: retinol acyltransferase; BiP, immunoglobulin heavy-chain-binding protein; CRABP, cellular retinoic acid-binding protein CRBP, cellular retinol-binding protein; CRBP II, cellular retinol-binding protein type II; FGI, murine fibroblast growth inhibitor; IRBP, interstitial retinol-binding protein; MGDI, bovine mammary-derived growth inhibitor; RAR, retinoic acid receptor; RBP, retinol-binding protein; RPE cells; retinal pigment epithelial cells.

Ulf Eriksson Ludwig Institute for Cancer Research, Stockholm Branch, S-104 01 Stockholm, Sweden.

vitamin will be reviewed. The recent identification of a class of nuclear receptors for retinoic acid (vitamin A acid) will also be discussed in the context of the physiology of the extra- and intracellular vitamin A-binding proteins. In this chapter the term "retinoid" will be used as a collective name for compounds with structural and/or functional characteristics similar to those of vitamin A.

1.1. Structure of Vitamin A Compounds

The chemical structures of the major physiological forms of vitamin A, all-*trans*-retinol, all-*trans*-retinal, all-*trans*-retinoic acid, and the provitamin β-carotene, are shown in Figure 1. Although the vitamin A compounds are very similar in structure, their biological functions are rather diverse. Retinol (vitamin A alcohol) is the major form of the vitamin, and under normal conditions it is stored in significant amounts in the body. Most cells obtain retinol as their major source of vitamin A. Retinal (vitamin A aldehyde) is the chromophore of the visual pigments, and its *cis-trans* isomerization in response to light is the trigger for a series of events that eventually lead to visual perception. Retinoic acid (vitamin A acid) occurs more sparsely, is likely to be generated by oxidation of retinol, and is rapidly metabolized. In intact animals retinoic acid cannot substitute for all the biological effects of retinol (Dowling and Wald, 1960) but in several *in vitro* systems it appears to be significantly more potent.

A schematic illustration of the absorption of vitamin A from the diet, its storage in and mobilization from the liver, its secretion into plasma, and finally its uptake and metabolism in vitamin A-requiring cells is outlined in Figure 2.

1.2. Retinoid-Binding Proteins

One common property of retinoids is their hydrophobic character, which renders them almost insoluble in aqueous surroundings. To circumvent this problem, several vitamin A-binding proteins have evolved. Apart from solubilizing the hydrophobic compounds, a protein-mediated metabolism allows several possibilities for detailed regulation. Furthermore, the binding to specific carrier proteins abolishes the well-known cytotoxic effects of free retinoids. The well-characterized extra- and intracellular retinoid-binding proteins are listed in Table I.

2. INTESTINAL UPTAKE AND LYMPHATIC TRANSPORT OF RETINOL

Retinol can be obtained from the provitamins, the carotenes, or in the diet from long-chain retinyl esters present in animal tissues. Both retinyl esters and carotenes are lipophilic compounds, and a normal intestinal adsorption of fat is

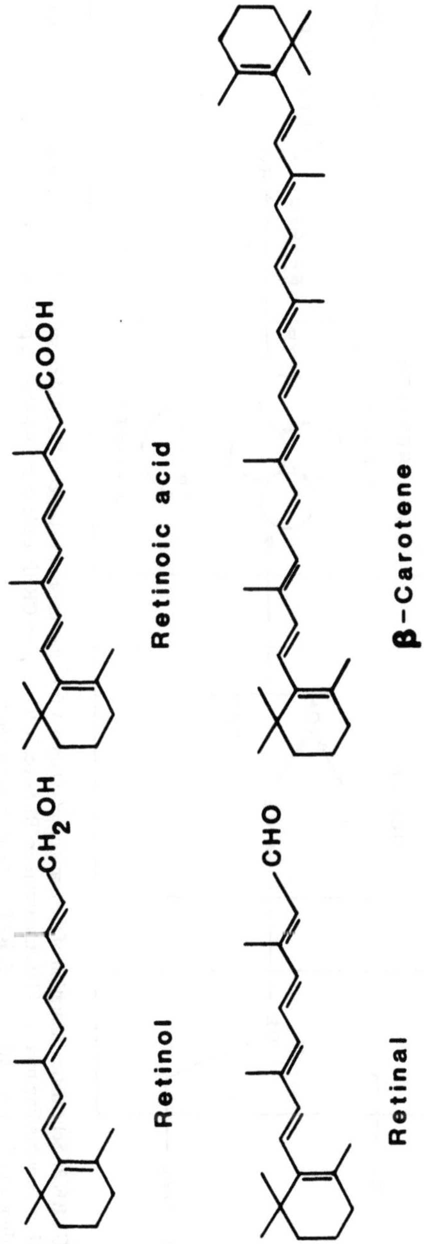

FIGURE 1. Schematic structures of retinol, retinal, retinoic acid and the provitamin β-carotene.

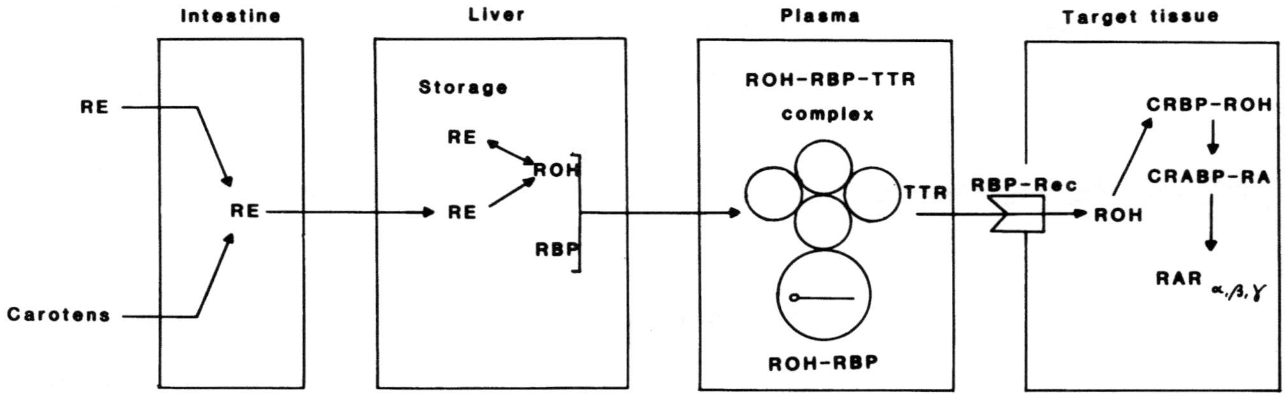

FIGURE 2. Schematic illustration of the transport of vitamin A from the intestine to target cells. (Abbreviations: RE, retinyl esters; ROH, retinol; RBP, retinol-binding protein; TTR, transthyretin; RBP-rec, RBP-receptor; CRBP, cellular retinol-binding protein; CRABP, cellular retinoic acid-binding protein; RA, retinoic acid; RAR, retinoic acid receptor.)

Table I
Soluble Retinoid-Binding Proteins[a]

Name	Mol. wt.	Number of amino acids	Endogenous ligand
Retinol-binding protein (RBP)	21,000	183 (182)	All-*trans* retinol
Cellular retinol-binding protein (CRBP)	15,700	134	All-*trans* retinol
Cellular retinol-binding protein type II (CRBPII)	15,600	134	All-*trans* retinol
Cellular retinoic acid-binding protein (CRABP)	15,500	136	All-*trans* retinoic acid
Cellular retinal-binding protein (CRA1BP)	36,400	316	11-*cis* retinal and 11-*cis* retinol
Interphotoreceptor or interstitial retinol-binding protein (IRBP)	133,000	1230	All-*trans* and 11-*cis* retinol

[a]References are given in the text.

necessary to adsorb the vitamin from the diet (Goodman and Blaner, 1984; Wolf, 1980).

The carotenes, mainly β-carotene, are absorbed by enterocytes of the proximal part of the intestine and cleaved at the central double bond, thus generating two molecules of retinal. Subsequent reduction of retinal yields two molecules of retinol. The cleavage reaction is carried out by the enzyme β-carotene-15,15′-dioxygenase in a reaction requiring molecular oxygen. A second enzyme, retinal reductase, converts the formed retinal to retinol. The dietary retinyl esters are hydrolyzed in the intestinal lumen by the pancreatic nonspecific lipase (Erlandson and Borgström, 1968; Lombardo and Guy, 1980). The formed retinol is subsequently almost quantitatively absorbed by the enterocytes (Hollander, 1981).

Following absorption, retinol, either newly absorbed from the intestinal lumen or obtained from β-carotene, is esterified with long-chain fatty acids (Goodman *et al.*, 1966; Huang and Goodman, 1965). The esters are mainly palmitate and stearate esters, with retinyl oleate and linoleate esters being less abundant.

The metabolism of vitamin A in enterocytes may involve a specific intracellular binding protein. This protein, named cellular retinol-binding protein type II (CRBPII), is expressed at high levels in enterocytes of the proximal part of the intestine (Crow and Ong, 1985; Ong, 1984). CRBPII makes up more than 1% of the total cytosolic protein of the jejunal mucosa and is as abundant as several of the fatty acid-binding proteins. The precise molecular role of CRBPII is not fully established, but several lines of evidence support the notion that it is involved in adsorption and esterification of retinol. CRBPII binds both retinol and retinal with high affinity (MacDonald and Ong, 1987). Since retinal is formed from the cleavage of β-carotene, CRBPII is likely to complex with newly generated retinal and to act as a substrate carrier in the subsequent reduction to retinol (Ong *et al.*, 1987). Furthermore, CRBPII appears to be the physiological carrier of retinol in the esterification reactions that occur in the intestinal mucosa (Ong *et al.*, 1987).

The enzyme responsible for the esterification of retinol in the enterocytes was for some time believed to be the microsomal enzyme acyl-coenzyme A (CoA):retinol acyltransferase (ARAT). This enzyme is similar to but clearly distinct from the acyl-CoA cholesterol acyl transferase (ACAT) that is responsible for the synthesis of several other esters formed in enterocytes (Helgerud *et al.*, 1982, 1983; Norum *et al.*, 1983; Ross *et al.*, 1984). The ARAT studies involved the use of free retinol and isolated microsomes to study the formation of retinyl esters. The lack of selective incorporation of the appropriate fatty acids into the retinyl esters, compared with the situation *in vivo* (Helgerud *et al.*, 1982), could not be satisfactorily explained. When retinol was introduced bound to CRBPII, the relative amounts of retinyl esters formed from the endogenous acyl donors closely resembled those observed in chylomicrons and lymph from intact animals. Furthermore, the protein-bound retinol was not available for the acyl-CoA-dependent reaction, suggesting that the esterification by ARAT is not a physiological reaction (Ong *et al.*, 1987). The presence of an acyl-CoA-independent esterification process for retinol in the liver and in retinal pigment epithelial cells of the eyes has recently been verified (Ong *et al.*, 1988; Saari and Bredberg, 1988).

The generated retinyl esters are incorporated into chylomicrons, together with triacylglycerols, cholesteryl esters, and phospholipids, and transported via the lymphatic system into the circulation (Goodman *et al.*, 1966). The triacylglycerols are rapidly hydrolyzed in extrahepatic tissues by the enzyme lipoprotein lipase, and after an exchange of some of the chylomicron constituents with lipoproteins and cell membranes, chylomicron remnants are formed. Retinyl esters seem not to participate in the latter exchange reactions (Blomhoff *et al.*, 1984a) and are consequently almost quantitatively taken up by the liver.

3. HEPATIC UPTAKE, STORAGE, AND METABOLISM OF VITAMIN A

3.1. Hepatic Uptake of Vitamin A by Chylomicron Remnants

Chylomicron remnants are taken up from plasma by the parenchymal liver cells. The uptake mechanism involves receptor-mediated endocytosis, and several studies have shown that high-affinity cell surface receptors for apolipoprotein E are responsible for the accumulation of chylomicron remnants (Brown *et al.*, 1981). Following internalization, the remnant particles are transferred to lysosomes and degraded by lysosomal hydrolases.

The fate of the newly absorbed retinyl esters is not known in detail. Studies by Blomhoff *et al.* (1985a) demonstrated that [^{3}H]retinyl esters initially are located in endosomes. The authors concluded that the intracellular transport of the radiolabeled retinyl esters and of ^{125}I-labeled asialofetuin differed following transit in the endosomes. Although the asialofetuin was transferred to lysosomes and degraded, the radiolabeled retinoid was transferred to the endoplasmic reticulum. It has not been established whether the esters are hydrolyzed during the intracellular transport, and the possible involvement of retinoid-binding proteins is not known. However, the transfer to the endoplasmic reticulum may indicate that at least a fraction of the esters are hydrolyzed and that generated retinol may subsequently become complexed with retinol-binding protein (RBP). RBP is abundant in the endoplasmic reticulum, and binding of retinol to the presynthesized RBP is a prerequisite for its secretion into the bloodstream.

3.2. Storage and Intrahepatic Metabolism of Retinol

Following the initial uptake of chylomicron remnants by the hepatocytes, retinol is transferred to perisinusoidal stellate cells of the liver for storage (Blomhoff *et al.*, 1982, 1984b). Injection of chylomicron remnants labeled with radioactive retinyl esters results in a rapid accumulation of radioactivity in the hepatocytes. During the next 2–4 hr the radioactivity is redistributed in the liver, and after more than 4 hr up to 80% of the total radioactivity recovered in the liver is found in the stellate cells. Endothelial cells and Kupffer cells accumulated only minute amounts of the labeled retinyl esters at all time points measured.

In well-nourished animals 80% or more of the total hepatic store of vitamin A is localized in the stellate cells. A majority of the vitamin A stored in this cell type is esterified to long-chain fatty acids, mainly palmitate. The remaining hepatic store of vitamin A is found in the hepatocytes. In this cell type, approximately one-third occurs as retinol, most probably bound to the CRBP, while the rest is esterified as in the stellate cells. The unique localization of retinol in different liver cell populations is in agreement with fluorescence microscopy data. These analyses clearly demonstrate that the stellate cells contain considerably higher concentrations of the autofluorescent vitamin A than does any other hepatic cell type.

The unique localization of retinol in the liver is reflected in the distribution of CRBP and of the enzymes involved in esterification and hydrolysis of retinol and retinyl esters, respectively (Blaner *et al.*, 1985; Blomhoff *et al.*, 1985b). The hepatocytes account for 75–90% of the total liver content of CRBP and of the two enzymes, but the concentrations of CRBP and the two enzymes are significantly higher in stellate cells.

The transfer of retinol to the stellate cells seems to be rather specific, since neither cholesterol nor vitamin D_3, both normal constituents of chylomicron remnants, is transferred to these cells (Blomhoff *et al.*, 1982; Dueland *et al.*, 1983). Experiments with retinyl esters labeled with radioactive fatty acids or a nonhydrolyzable retinyl ether analogue have shown that the retinyl esters must be hydrolyzed prior to transfer to the stellate cells (Blaner *et al.*, 1987; Blomhoff *et al.*, 1988).

The routing of retinol in the hepatocytes is dependent on the vitamin A status. Under vitamin A-deficient conditions, only a minor portion of the retinol is directed to stellate cells. Instead, a major fraction of the absorbed retinol is secreted from the hepatocytes as a complex with RBP (Blomhoff *et al.*, 1982).

3.3. Intercellular Transfer of Retinol in the Liver

Several mechanisms may be considered for the intercellular traffic of retinol between hepatocytes and stellate cells. The transfer mechanism may involve retinoid-binding proteins expressed in liver or involve a transfer of retinol or retinyl esters bound to lipoproteins or lipid particles. Two of the possible mechanisms involving retinoid-binding proteins are illustrated in Figure 3. Any hypothesis about a possible transfer mechanism has to consider the fact that retinol stored in the stellate cells must be transferred back to the hepatocytes to be mobilized by RBP. This can be inferred since RBP is synthesized only in hepatocytes of the liver (Yamada *et al.*, 1987).

The intercellular transfer of retinol may be mediated by RBP. By using RBP labeled *in vivo* with radiolabeled retinol, it was concluded that retinol secreted into plasma in association with RBP is efficiently recycled back to the liver (Blomhoff *et al.*, 1985c). When separating the different liver cell types, it became evident that the retinol had initially accumulated in the hepatocytes. However, it was subsequently transferred to the stellate cells for storage, although 50–60 hr was required before 70% of the labeled retinol was found in those cells. The time required for an efficient transfer of retinol to stellate cells, when administrating the retinol as chylomicron remnants, is significantly shorter (80% of the labeled retinol is transferred with 4 hr). Thus, based on the kinetic data, the possible role of RBP as the shuttle between hepatocytes and stellate cells appears less likely. Hendriks *et al.* (1987) have come to the same general conclusion.

Blomhoff and co-workers have presented data arguing for a possible role of RBP in the transfer process (Blomhoff *et al.*, 1985c, 1988; Gjöen *et al.*, 1986). Injection of RBP derivatized with ^{125}I-labeled tyramine cellobiose was used to analyze the tissue accumulation of RBP. Most of the radioactivity was recovered in the liver at all times. The distribution of the labeled RBP was determined in

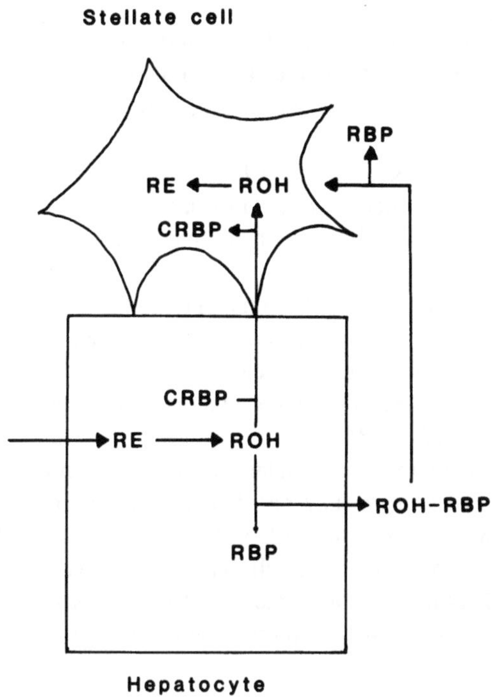

FIGURE 3. Hypothetical mechanisms for the transfer of retinol from hepatocytes to stellate cells of the liver. (Abbreviations: see legend of Fig. 2.)

various liver cell fractions. It was found that equal amounts of radioactivity were recovered in hepatocytes and in stellate cells when expressed per gram of total liver. This result suggested that these two cell types of the liver possessed a specific mechanism for the accumulation of RBP, presumably specific RBP receptors. However, it should be emphasized that the authors did not demonstrate a specific and saturable mechanism for the accumulation. In a recent paper the same authors demonstrated that antibodies to RBP efficiently inhibited the transfer of retinol from hepatocytes to stellate cells in *in situ* perfused rat liver (Blomhoff *et al.*, 1988). Although the latter sets of experiments argue for a direct role of RBP in the transfer process, it cannot be ruled out that the observed effect is indirect. As shown by Blomhoff *et al.* (1982), accumulated retinol is selectively secreted in association with RBP and is not transferred to stellate cells in retinol-deficient rat liver. Interfering with the normal recirculation of retinol back to the hepatocytes with antibodies to RBP may mimic the situation of the deficient liver. Consequently, the inhibitory effect of the antibodies may be indirect and related to the selective routing of retinol in hepatocytes. Information on the cellular localization of specific RBP receptors in liver is needed and should definitely shed light on the role of RBP in the intercellular transfer of retinol.

A role for CRBP as an intercellular transport vehicle for retinol is, at present, unlikely. Immunohistochemical data show that the level of CRBP in the stellate cells, in contrast to most other cell types, is regulated by the retinol status (Eriksson *et al.*, 1984; Kato *et al.*, 1984). Since the stellate cells express the mRNA for CRBP (Nordlinder *et al.*, 1989), it appears unlikely that CRBP is shuttled between the two cell types of the liver as a consequence of the retinol status.

A direct transfer of vitamin A from cell to cell is unlikely, since no connections between hepatocytes and stellate cells have been demonstrated (Wake, 1980). However, the pinocytotic vesicles demonstrated in areas of close proximity between hepatocytes and stellate cells may offer a possible mechanism for the transfer (Mak *et al.*, 1984).

By analogy with the intercellular transfer of retinol in the eye, a third possible transfer mechanism may be offered by an extracellular liver specific retinol-binding protein other than RBP. The intercellular traffic of retinol between the retinal pigment epithelial cells and the neuroretina of the eye is carried out by interstitial retinol-binding protein (IRBP) (Lai *et al.*, 1982; Liou *et al.*, 1982). It has not been established whether the liver is able to synthesize a similar protein.

4. THE PLASMA RETINOL-BINDING PROTEIN

4.1. Primary Structure and Biosynthesis of RBP

The complete primary structure of human RBP was established more than a decade ago (Rask *et al.*, 1979). The amino acid sequence encompasses 182 amino acids and has a molecular weight of close to 21,000 (Figure 4). Following the isolation of cDNA clones for both rat and human RBP, it became evident that the newly synthesized protein has 183 amino acids (Sundelin *et al.*, 1985a; Colantuoni *et al.*, 1983). The difference, an extra leucine in the carboxyl terminus of the protein, may be the result of a posttranslational event or a sequencing error. RBP is not glycosylated, and it has not been fully established whether other posttranslational modifications occur. The six cysteine residues in the protein form three disulfide bonds (between residues 4 and 160, 70 and 174, and 120 and 129; Rask *et al.*, 1987).

Translation of RBP mRNA results in a polypeptide precursor with a molecular weight of 24,000. Similar to other secreted proteins, its signal sequence is cleaved cotranslationally, resulting in the mature polypeptide of 21,000 (Soprano *et al.*, 1981).

```
Residues              10        20        30        40        50
                       *         *         *         *         *

Human     ERDCRVSSFRVKENFDKARFSGTWYAMAKKDPEGLFLQDNIVAEFSVDEN

Rabbit    --------------------A----------------------------T

Rat       ----------------------L---I--------------IV------K

                      60        70        80        90       100
                       *         *         *         *         *

Human     GQMSATAKGRVRLLNNWDVCADMVGTFTDTEDPAKFKMKYWGVASFLQKG

Rabbit    -H----------------------------------------------R-

Rat       --------------S--E--------------------------------

                     110       120       130       140       150
                       *         *         *         *         *

Human     NDDHWIVDTDYDTYAVQYSCRLLNLDGTCADSYSFVFSRDPNGLPPEAQK

Rabbit    ------I------F----------F----------------H----DV--

Rat       ---------------L------Q---------------------T--TRR

                     160       170       180
                       *         *         *

Human     IVRQRQEELCLARQYRLIVHNGYCDGRSERNLL

Rabbit    L----------S-------------DK-V----

Rat       -----------E----W-E-----QSRPS--S
```

FIGURE 4. Primary structures of human, rabbit, and rat RBPs. Positions in rabbit and rat RBP exhibiting amino acid differences to human RBP are indicated.

In plasma, RBP occurs as a 1 : 1 complex with the tetrameric thyroxine-binding protein transthyretin (previously termed prealbumin; Peterson, 1971). This protein complex circulates in plasma and delivers its retinol to specific cell surface receptors on vitamin A-requiring cells. As a result, the interaction between transthyretin and RBP is diminished and RBP eventually undergoes glomerular filtration and degradation in the kidneys (Vahlquist *et al.*, 1973).

RBP is evolutionarily highly conserved, and the identity between human, rabbit, and rat RBPs is close to 90%. (Sundelin *et al.*, 1985a). The amino acid substitutions are not equally distributed along the polypeptide chains. Residues 145–151 and 175–179 have accumulated several mutations, whereas residues 1–20, 69–98, and 126–141 are totally invariant. A detailed analysis of the structure–function relations of RBP in the context of its three-dimensional structure is discussed below.

4.2. Exon-Intron Organization of the RBP Gene

Isolation of the complete gene encoding rat RBP revealed that the gene spans 6.9 kb and contains six exons (Laurent *et al.*, 1985). The first exon and part of the second encodes the 5′ untranslated region. In addition to an untranslated sequence, the second exon encodes the complete signal sequence (18 amino acids) and the first 19 residues of the mature protein. The third exon encodes residues 20–65; the fourth, residues 66–100; the fifth, residues 101–171; and the sixth, the remaining residues (172–183) and the complete 3′ untranslated region. The introns vary in size from 78 bp (intron 2) to 4.4 kb (intron 4) and account for close to 90% of the gene. The organization of the human RBP gene is very similar to that of the rat gene (D'Onofrio *et al.*, 1985).

4.3. Tertiary Structure of RBP and a Comparison with the RBP Gene Organization

The tertiary structure of human RBP containing retinol was established in 1984 (Newcomer *et al.*, 1984). RBP consists of a single globular domain made up of an N-terminal coil (residues 1–19), a β-barrel core (residues 20–146), an α-helix (residues 147–160), and a carboxy-terminal coil (residues 161–182), as shown in Figure 5. The eight antiparallel β-strands have the form a flattened cone. When viewed from the flat side of the cone, the strands appears as two orthogonal β-sheets with the retinol molecule buried in between. The barrel is closed at the surface at one end by a salt link between residues K-17 and D-79 and in the interior of the protein by five phenylalanine rings. The retinol molecule lies along the barrel axis, with the β-ionone ring innermost and the isoprene tail stretching toward the surface of the protein. The retinol-binding pocket is built up of hydrophobic or uncharged residues. The three-turn α-helix is formed from residues 147–160 and ends with C-160 making a disulfide bridge to C-4 of the N-terminal coil.

As pointed out earlier, RBP is highly conserved during evolution (Sundelin *et al.*, 1985a). One explanation for this observation is that RBP participates in several molecular interactions, e.g., interactions with retinol, with transthyretin in plasma, and with a cell surface receptor upon delivering retinol to requiring

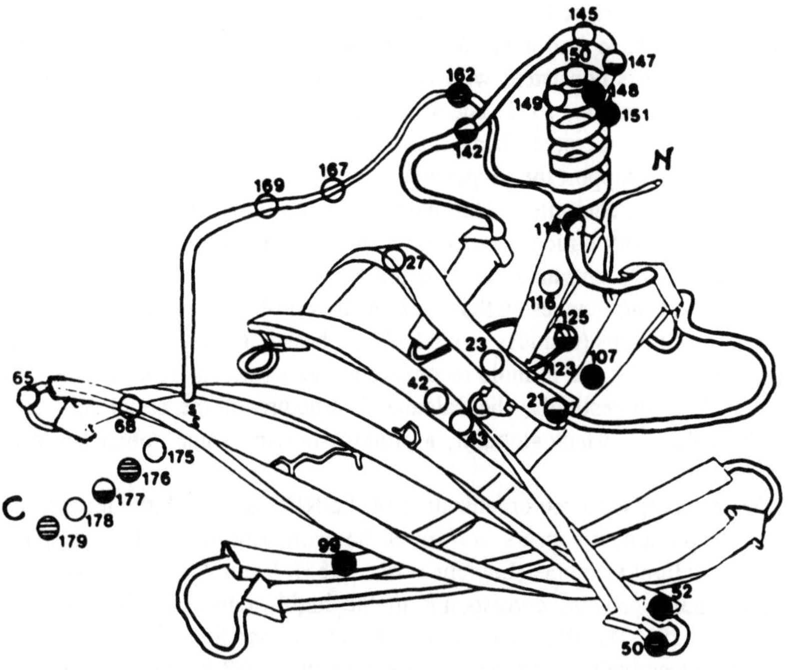

FIGURE 5. Three dimensional structure of human RBP with bound retinol. Positions exhibiting amino acid differences between human, rabbit, and rat RBP have been marked as follows: [open circle], human and rabbit alike; [half-filled circle], human and rat alike; [filled circle], rabbit and rat alike; [striped circle] all three RBPs differ. Sections of β-sheets are marked as ribbon arrows, pointing in the direction of the carboxyl terminus of the protein. Reproduced from Sundelin *et al.* (1985a) with permission of the publisher.

cells. Molecular interactions tend, in general, to conserve the structure of proteins. Mutations in one protein may have to be compensated for by a simultaneous and complementary substitution in the other protein to preserve biological function. The retinol-binding site in RBP is known at the atomic level, but the corresponding information on the transthyretin and receptor-binding sites is lacking. The N-terminal coil, the C-terminal of the α-helix, and the loop region around residue 80 are invariant in human, rabbit, and rat RBPs, suggesting that these parts of the molecule are good candidates for the transthyretin-binding site (Newcomer *et al.*, 1984). Several of the other loops connecting the β-strands in RBP are highly conserved (residues 58–66) or invariant (residues 29–35 and 91–98) among the three species of RBP. These regions may be involved in recognition of and binding to the cell surface receptor.

No information is currently available on the structure of RBP in the absence of retinol. Newcomer *et al.* (1984) proposed that removal of retinol from RBP most probably triggered a dramatic structural change in the protein. Such changes would have profound effects on the biological properties of the molecule. This view is consistent with the decreased affinity of the apoprotein for transthyretin (Peterson, 1971) and with the inability of newly synthesized apoprotein to become secreted (Ronne *et al.*, 1983). Recent studies using molecular dynamics do not support the suggestion of Newcomer *et al.* (1984) (Åqvist *et al.*, 1986). When this computer technique was used to characterize atomic fluctuations and secondary structure motions, major conformational changes were observed only in the entrance loops of the β-barrel and in the α-helical region. Åqvist *et al.* (1986) suggested that the transthyretin-binding site is located in one of the entrance loops (residues around W-91), based on these calculations and on tryptophan labeling and retinoid reconstitution experiments (Horwitz and Heller, 1973, 1974). Interestingly, the calculated structure of the apoprotein closely resembles that of β-lactoglobulin, a protein showing some homology to RBP (see Section 4.4).

When comparing the organization of the RBP gene with the tertiary structure of the protein, it appears as if all of the exon regions closely correspond to tertiary structural elements (Laurent *et al.*, 1985). Exon 2, the first exon encoding the mature protein, corresponds to the highly conserved N-terminal coil (residues 1–19). The third and the fourth exons, encoding residues 20–65 and 66–100, respectively, correspond closely to two βββ-building blocks which terminate in or close to a loop. The fifth exon encodes several structural units, e.g., an additional βββ-unit (residues 101–146), the α-helix (residues 147–160), and part of the C-terminal coil (residues 161–171). The sixth exon encodes the highly variable residues of the C terminus of the protein (residues 172–182).

4.4. RBP Is a Member of a Protein Family—the Lipocalins

The primary structure of human RBP, determined in 1979, did not reveal significant homologies to any other sequenced protein at that time (Rask *et al.*, 1979). However, during the last few years, several homologous proteins have been identified. The protein family consists, so far, of eight members apart from RBP, and the tertiary structures of several of the proteins are known. The proteins belonging to this family are listed in Table II. It was recently proposed that the members of this protein family should be named lipocalins (Pervaiz and Brew, 1987).

Sequence analyses show that all members of the protein superfamily display 25–30% identical residues in pairwise comparisons, with one notable exception. Purpurin, a protein isolated from chicken retina, displays 50% homology to RBP (Berman *et al.*, 1987).

Apart from RBP, the tertiary structures are known for β-lactoglobulin (Pa-

Table II
Proteins Belonging to the Lipocalin Family

Protein	Function	Reference[a]
RBP	Plasma transport of retinol	
Purpurin	Adhesion and survival protein for chicken neurons	1
β-Lactoglobulin	Abundant milk protein	2, 3
Apolipoprotein D	Binding and transport of cholesteryl esters in plasma	4
protein HC	Thought to bind porphyrin	5
α_2-Microglobulin	Thought to bind and transport pheromones	6
Androgen-dependent protein	Epididymal secretory protein involved in sperm maturation	7
Protein BG	Abundant protein of the olfactory epithelium with a suggested role in olfaction	8
Insecticyanin (or bilin-binding protein)	Binding of dyes involved in insect camouflage	9, 10

[a]References: 1, Berman *et al.* (1987); 2, Godovac-Zimmerman *et al.* (1985); 3, Sawyer *et al.* (1985); 4, Drayna *et al.* (1986); 5, Lopez *et al.* (1981); 6, Unterman *et al.* (1981); 7, Brooks *et al.* (1986); 8, Lee *et al.* (187); 9, Huber *et al.* (1987); 10, Riley *et al.* (1984). The references refer to articles demonstrating primary structure similarities to the lipocalin family.

piz *et al.*, 1986) and the insecticyanins (Holden *et al.*, 1987; Huber *et al.*, 1987). The overall levels of sequence homology seem marginal, but the three-dimensional structures of the three proteins are remarkably similar. Like RBP, β-lactoglobulin and the insecticyanins consist of a β-barrel formed by two orthogonal β-sheets and an α-helix. The ligands are bound within the central cavity of the β-barrel. The use of this structural framework offers an excellent ability to bind, protect, and transport a variety of small hydrophobic compounds in aqueous surroundings.

The *in vivo* ligands for all lipocalins are not known at present, but it appears as if both β-lactoglobulin and purpurin are able to bind retinol (Fugate and Song, 1980; Schubert *et al.*, 1986). The physiological significance of this is not understood. The striking homology of purpurin to RBP and its unique expression in the interstitial space of the eye (Berman *et al.*, 1987) are consistent with a role as a tissue-specific retinol-binding and transporting protein. Its functional relation to IRBP (Lai *et al.*, 1982; Liou *et al.*, 1982) awaits further analyses.

4.5. Ligand-Dependent Secretion of RBP

It has long been recognized that an adequate level of retinol is necessary to maintain normal levels of RBP in plasma (Peterson *et al.*, 1973; Smith *et al.*, 1975). Concomitant with the decreased plasma levels of RBP, the concentration of RBP in the liver was found to be elevated. Replenishing retinol-deficient animals with retinol resulted in a rapid secretion of RBP from the hepatocytes. By using intact animals, it was shown that retinol promotes the secretion of presynthesized RBP, since treatment with cycloheximide, an inhibitor of protein synthesis, or inhibitors of transcription did not affect the retinol-induced secretion. The *in vivo* studies were later confirmed by using freshly isolated hepatocytes (Ronne *et al.*, 1983).

The precise mechanism by which retinol promotes the secretion of RBP in retinol-deficient hepatocytes remains elusive. Studies on the subcellular localization of RBP in normal and deficient cells have demonstrated that the presence of retinol is a prerequisite for RBP to leave the endoplasmic reticulum (Harrison *et al.*, 1980; Rask *et al.*, 1983). The ultrastructural localizations of RBP in normal and deficient cells have not been determined. Consequently, it is not known whether the block in secretion is actually the exit from the endoplasmic reticulum or the entry to the Golgi complex.

Several factors may be responsible for retention of RBP in retinol-deficient hepatocytes. The conformation of the newly synthesized apo-RBP may be incompatible with putative receptors (Fries *et al.*, 1984; Lodish *et al.*, 1983) whose function might be to shuttle proteins between the endoplasmic reticulum and the *cis*-Golgi compartment. Alternatively, apo-RBP may be associated with resident proteins of the endoplasmic reticulum, similar to the association of improperly folded and/or assembled proteins to the immunoglobulin heavy-chain-binding protein, BiP (Hendershoot *et al.*, 1987; Kassenbrock *et al.*, 1988).

5. UPTAKE OF VITAMIN A BY VITAMIN A-REQUIRING CELLS

Several mechanisms may be involved in the uptake of retinoids by target cells. During embryogenesis, free diffusion of retinoic acid is believed to be of importance in establishing retinoic acid gradients which might be involved in determining anterior–posterior specification during limb morphogenesis (Thaller and Eichele, 1987). During hypervitaminosis A, lipoproteins may be involved in the transport of retinoids to cells (Mallia *et al.*, 1975). However, under normal conditions plasma RBP is the major vitamin A-transporting vehicle.

The delivery of retinol from RBP to cells should be highly discriminatory, since only some cells are dependent on vitamin A for proper function. To generate the specific uptake, it appears reasonable to assume that vitamin A-requiring cells express a specific cell surface receptor for RBP. RBP receptors have been identified on bovine retinal pigment epithelial cells (Heller, 1975; Bok and Heller, 1976), on epithelial cells of monkey small intestine (Rask and Peterson, 1976), and on interstitial cells of rat testis (McGuire *et al.*, 1981). However, the structural characteristics of the receptor and its mode of interaction with RBP have largely remained elusive. Recently, Sivaprasadaro and Findlay (1988) presented some details on the binding properties of a membrane receptor for RBP in placenta.

Using membrane fractions derived from bovine retinal pigment epithelial cells (RPE cells), we have been able to characterize the binding of ^{125}I-labeled RBP to a membrane receptor. Cross-linking techniques were also used to determine some of the structural features of the membrane receptor of RBP. In this section some of our results will be briefly summarized (C. Båvik, P. A. Peterson, and U. Eriksson, unpublished data).

The RBP-binding activity on the RPE membranes was specific for RBP, inasmuch as proteins other than RBP did not compete for binding. Notable is that addition of transthyretin did not affect the binding of RBP. This observation suggests that the binding sites for the receptor and for transthyretin on the RBP molecule are distinct. Proteinase K digestions of the RPE membranes prior to the binding assay completely abolished the binding, thus verifying the protein nature of the receptor structure. Maximal binding was achieved at neutral pH and was reduced by more than 90% at pH 5. The retinol content of RBP did not influence the binding characteristics, since apo-RBP and holo-RBP showed almost identical binding profiles. Equilibrium binding experiments indicated a single class of binding sites, with a binding constant of $1.3 \times 10^{7}\ M^{-1}$. Considering that the plasma concentration of RBP is close to 2 μM, the binding constant appears relevant. Binding-kinetics experiments demonstrated that the association of RBP to the receptor was rapid and reached saturation within 15 min. The membrane receptor for RBP was abundant in the RPE membranes. This observation is in accordance with the extensive use of vitamin A in thc visual process.

Cross-linking of radiolabeled RBP to RPE membranes by using an UV-activated cross-linker revealed a major molecular complex of 86 kDa on sodium dodecyl sulfate-polyacrylamide gel electrophoresis. Endoglycosidase digestions demonstrated that the complex contained carbohydrate residues, and the results indicate that the receptor is probably present on the cell surface of RPE cells. Physicochemical characterization demonstrated that the cross-linked complex is part of a high-molecular-weight complex. The properties of this complex have not been determined in great detail, but the results indicate that the RBP receptor may be composed of several noncovalently associated protein subunits. Further studies of the RBP receptor, including its structural determination, are aimed at understanding the basis for the interaction with RBP and the possible involvement of the receptor structure in the transmembrane transport of retinol to intracellular acceptors, presumably CRBP.

6. THE CELLULAR RETINOID-BINDING PROTEINS

6.1. Molecular Characterization of the Cellular Retinoid-Binding Proteins CRBP, CRBP(II), AND CRABP

The cellular retinoid-binding proteins are small acidic proteins with molecular weights of approximately 15,000–16,000, and they have several physicochemical properties in common. Each protein has one binding site for its endogenous ligand, and apart from CRBP(II), the binding sites are exquisitely discriminatory. The discriminatory effect is largely restricted toward the polar group of the side chain of the retinoids, whereas the binding ability is less sensitive to stereoisomers and changes in the cyclohexenyl ring. CRBP binds retinoids with a free hydroxyl group, and the endogenous ligand has been identified as all-*trans* retinol (Saari *et al.*, 1982). CRABP, in contrast, requires a carboxyl group in the side chain of the ligand, and its *in vivo* ligand is identified as all-*trans* retinoic acid (Saari *et al.*, 1982). The ligand specificity of CRBP(II) is not absolute, since both retinol and retinal can bind (MacDonald *et al.*, 1987). However, the *in vivo* ligand appears to be all-*trans* retinol (Ong, 1984).

The complete amino acid sequence of rat liver CRBP has been determined (Sundelin *et al.*, 1985b). The polypeptide chain is 134 amino acids long, and the calculated molecular weight is 15,700. The amino acid sequence of human CRBP, as deduced from a near full-length cDNA clone, shows 96% identity to the rat CRBP sequence (Colantuoni *et al.*, 1985). The replacements are confined to the terminal portions, leaving residues 13–120 totally invariant. The strong evolutionary conservation of the CRBP molecule suggests that the central portion of CRBP is under strong selective pressure. Apart from binding of retinol, CRBP may be involved in several other molecular interactions that conserve the molecule during evolution. Although CRBP and RBP have the same *in vivo* ligand, retinol, no obvious sequence homologies can be observed when the entire sequences are compared. CRBP has two regions of internal sequence homology (residues 1–53 and 54–106), in analogy with the pseudosymmetry displayed by plasma RBP. These similarities may indicate that CRBP and RBP share significant structural features. This suggestion is supported by the recent determination of the tertiary structure the myelin protein P2 (Jones *et al.*, 1988), a protein showing significant homologies to CRBP and CRABP.

The amino acid sequence of CRBPII encompasses 134 amino acids and shows striking homologies to CRBP (Figure 6) (Li *et al.*, 1986). Identical resi-

```
Residues                   10        20        30        40        50        60        70
                            *         *         *         *         *         *         *
CRABP   bovine   --PNFAGTWKMRSSENFDELLKALGVNAMLRKVAVAAASKPHVEIRQDGDQFYIKTSTTVRTTEIN-FKVG
CRBP    rat        PVD N Y    L N    E Y R  D  VA    I NLL--   DK  V     HMI R LS F NYIMD- Q
CRBPII  rat      MTKDQN    E E N   EGYM    DIDFAT   I  RL--TQTKI V     N KT   NS F NYDLD- T
P2      bovine   -SNK-L     LV        YM      GLAT  LGNL --    R I SKK   IIT R ESPFKN    S-  L
aP2     murine   MCDA V     LV       DYM EV   GFAT     GMA--   NMI SVN   LVT RSES FKN    S-  L
H-FABP  rat      TEK  V     LVD K    DYM S    GFAT Q  SMT--   TTI EKN   TI G  HS FKN    SN QL
I-FABP  rat      --MA D     VYRN  YEKFMEKM I VVK   LGAHD--NLKLT T E NK TV E SNF NIDVV- EL
L-FABP  rat      --M    S KYQVQ QE   EPFM  M LPED IQKGKDI-- GVS  VHE KKVKLTITYGSKVIHNE- TL
                 --M    S KYQVQ QE   EPFM  M LPED IQKGKDI-- GVS  VHE KKVKLTITYGSKVIHNE- TL

                /_______exon 1________/_______________________________exon 2______

                          80        90       100       110       120        130       140
                           *         *         *         *         *          *         *
CRABP   bovine   EGFEEETV--DGRKCRSLPTWENENKIHCTQTLLEGDGPKTYWTRELANDELILTFGADDVVCTRIYVRE-
CRBP    rat      KE    DLTGI D    MTTVS DG-D LQ V KGE---KEGRG  QWIEG    HLEMR EGVT  KQVFKKVH
CRBPII  rat      VE D H KGI     NVKT V     G- TLV V KGE---KENRG KQWVEG K Y ELTCG Q  RQVFKKK-
P2      bovine   QE    T A-- N   TK TV  LAR-GSLNQV KWN---  NE TIK K VDGKMVVECKMK          EKV-
aP2     murine   VE D I A-- D   VK II  LDG-GALVQV KVD---  KS TIK KRDG K VVECVMKG TS   V ERA-
H-FABP  rat      VEDD V A-- D   VK VV  LDG-G LVHV KVD---  QE TL     SDGK     LTHGN   S   T EK A
I-FABP  rat      VD AYSLA--   TELTGTL  M G-   LVGKFKRVDN- KELIAV   ISGN   IQ YTYEG EAK   FKK -
L-FABP  rat       EC L   M--T  E VKAVVKM  GD   MVT FKG-------IKSVT FNG TITN MTLG I YK VSK I-

                 _______________/_____________exon 3_____________/_____exon 4_____/
```

FIGURE 6. Sequence aligment of cellular retinoid-binding proteins and the related fatty acid-binding proteins. A blank space indicates identity to the CRABP sequence. Optimal alignment was obtained by introducing gaps in some sequences (-). The exon/intron boundaries of the corresponding genes are indicated below the sequences (/).

dues are found in 75 of 133 comparable residues. The identical residues are not randomly distributed. Thus, the peptide stretches corresponding to residues 3–25, 55–70, and 95–118 display the highest degree of homology.

The complete amino acid sequence of bovine CRABP has also been determined (Sundelin *et al.*, 1985c). The protein is composed of 136 amino acids, and the molecular weight was calculated to 15,500. Like CRBP, CRABP is highly conserved during evolution. Murine CRABP is identical to the bovine counterpart (Vaessen *et al.*, 1989) and to the amino terminus of rat CRABP (Eriksson *et al.*, 1981).

6.2. The Cellular Retinoid-Binding Proteins Are Members of a Protein Family

Once the amino acid sequences of CRBP and CRABP were available, it became clear that they were highly homologous and belonged to a family of small cytosolic proteins with several other members. Eight members of this protein family are well characterized (Table III), and cDNA clones are available. A recently identified retinoic acid-binding protein, named CRABPII, also belongs to this group of proteins (Bailey and Siu, 1988). In addition, based on amino acid sequence similarities and antigenic characteristics, two growth inhibitors [a bovine mammary-derived growth inhibitor (MGDI) and a murine fibro-

Table III
Cellular Retinoid-Binding Proteins and Related Proteins and Their *In Vivo* Ligands

Protein	Ligand	Reference[a]
CRBP	Retinol	1
CRBPII	Retinol	2
CRABP	Retinoic acid	3
Myelin protein P2	Oleic acid	4
Adipocyte P2	Presumably fatty acids	5
Fatty acid-binding proteins from		
Heart	C_{16}–C_{20} fatty acids	6
Intestine	C_{16}–C_{20} fatty acids	7
Liver	C_{16}–C_{20} fatty acids	8

[a]References: 1, Sundelin *et al.* (1985b); 2, Li *et al.* (1986); 3, Sundelin *et al.* (1985c); 4, Kitamura *et al.* (1980); 5, Bernlohr *et al.* (1984); 6, Sacchettini *et al.* (1986); 7, Alpers *et al.* (1984); 8, Takahashi *et al.* (1982). The references refer to articles demonstrating primary structure similarities to this protein family.

blast growth inhibitor (FGI)] are related to this protein family (Böhmer *et al.*, 1987a,b).

Statistical analyses of the amino acid sequences indicate that the intracellular retinoid-binding proteins are more closely related to the myelin protein P2 and the adipocyte P2 than to the heart, liver, and intestinal fatty acid-binding proteins (Table IV). In fact, the retinoid-binding proteins are as related to each other as they are to the two P2 proteins, with the exception of CRBP and CRBPII. The myelin protein P2 can bind both retinol and retinoic acid *in vitro* (Uyemura *et al.*, 1984), but the *in vivo* ligand was tentatively identified as oleic acid (Jones *et al.*, 1988). The ability to bind a fatty acid and the close similarities between the two P2 proteins suggest that both P2 proteins are fatty acid-binding proteins.

6.3. Organization and Chromosomal Localization of the Genes Encoding the Cellular Retinoid-Binding Proteins

The organization of the genes encoding CRBP, CRBPII, and CRABP have been determined (Demmer *et al.*, 1987; Nilsson *et al.*, 1988a; Shubeita *et al.*, 1987). Each protein is encoded by four exons, and comparable introns vary considerably in size. The most striking feature of the genomic organization of these genes is the identical positioning of introns relative to exons. Optimal

Table IV
Statistical Analyses of the Relatedness between Cellular Retinoid-Binding Proteins and Fatty Acid-Binding Proteins Using the RELATE Program

						FABP		
Protein	CRABP	CRBP	CRBPII	P2	aP2	Heart	Liver	Intestine
CRABP	—	9.7	9.3	10.0	8.0	7.7	4.7	6.1
CRBP	9.7	—	16.5	9.2	8.4	7.3	2.6	4.6
CRBPII	9.3	16.5	—	8.2	10.4	11.0	5.2	4.5
P2	10.0	9.2	8.2	—	27.1	16.2	4.2	5.6
aP2	8.0	8.4	10.4	27.1	—	19.0	5.2	7.4
FABP								
Heart	7.7	7.3	11.0	16.2	19.0	—	5.7	3.7
Liver	4.7	2.6	5.2	4.2	5.2	5.7	—	4.1
Intestine	6.1	4.6	4.5	5.6	7.4	3.7	4.1	—

[a]The program RELATE was used in conjunction with the mutation data matrix [250 PAMs (Dayhoff, 1978)] to compare the sequences of bovine CRABP (Sundelin *et al.*, 1985c), rat CRBP (Sundelin *et al.*, 1985b), rat CRBPII (Li *et al.*, 1986), bovine myelin protein P2 (Kitamura *et al.*, 1980), murine adipocyte P2 (Bernlohr *et al.*, 1984), rat heart FABP (Sacchettini *et al*, 1986), rat liver FABP (Takahashi *et al.*, 1982), and rat intestine FABP (Alpers *et al*, 1984). The number of random runs per comparison was 100, and the segment comparison scores were generated by using a span of 11 residues.

alignment of the protein sequences (Figure 6) clearly demonstrates the conserved structure of these genes and indicates the exon/intron boundaries in the corresponding genes. The genomic organization of several of the fatty acid-bindig proteins shows the same unique conserved structure (Hunt *et al.*, 1986; Sweetser *et al.*, 1986). This fact may indicate that all members of this protein family are derived from an ancestral gene. Over time, gene duplication and subsequent diversification have created proteins with different ligand specificities and biological functions.

A common feature of some structurally related genes, which have arisen by gene duplications, is that they appear to be closely linked in the genome. Like these gene families, the cellular retinoid-binding proteins are closely linked. The three genes are located on chromosome 3 in humans (Demmer *et al.*, 1987; Nilsson *et al.*, 1988a,b) and chromosome 9 in mice (Demmer *et al.*, 1987; Vaessen *et al.*, 1989).

6.4. Tertiary Structure of the Cellular Retinoid-Binding Proteins

CRBP and apo-CRBPII have been crystallized, but their three-dimensional structures have not yet been solved (Newcomer *et al.*, 1981; Sacchettini *et al.*, 1987). However, the three-dimensional structure of the myelin protein P2 was recently solved (Jones *et al.*, 1988). The close similarities in the primary structures of the cellular retinoid-binding proteins and the fatty acid-binding proteins (including the two P2 proteins), together with the conserved structure of the corresponding genes, make it likely that these proteins will turn out to be very similar. Consequently, it appears relevant to generalize the structural information of the myelin protein P2 to include all members of the protein family.

The P2 protein consists of a flattened globular domain built up of two orthogonal β-sheets (Figure 7). The sheets are generated from 10 antiparallel β-strands that form a β-barrel, and the ligand is sandwiched inbetween the two β-sheets. Unlike RBP, the structural units in the P2 molecule do not show any obvious correlations with the exons of the P2 gene. However, some similarity to the structure of RBP is shown in the architecture of the β-barrel. The main differences are that RBP consists of eight β-strands and the loop connecting strands 1 and 2 in P2 consists of two short α-helices instead of the short reverse turn found in RBP (Newcomer *et al.*, 1984).

The ligand identified in P2 is a fatty acid (presumably oleic acid). The most striking feature is that the polar end group of the ligand is buried deeply in the β-barrel. This is in contrast to the situation in RBP, in which the polar end group of retinol is located close to the surface of the protein (Newcomer *et al.*, 1984). The carboxyl group of the fatty acid interacts with two internal arginines (R106 and R126) in the binding pocket. If retinoids interact with the cellular binding pro-

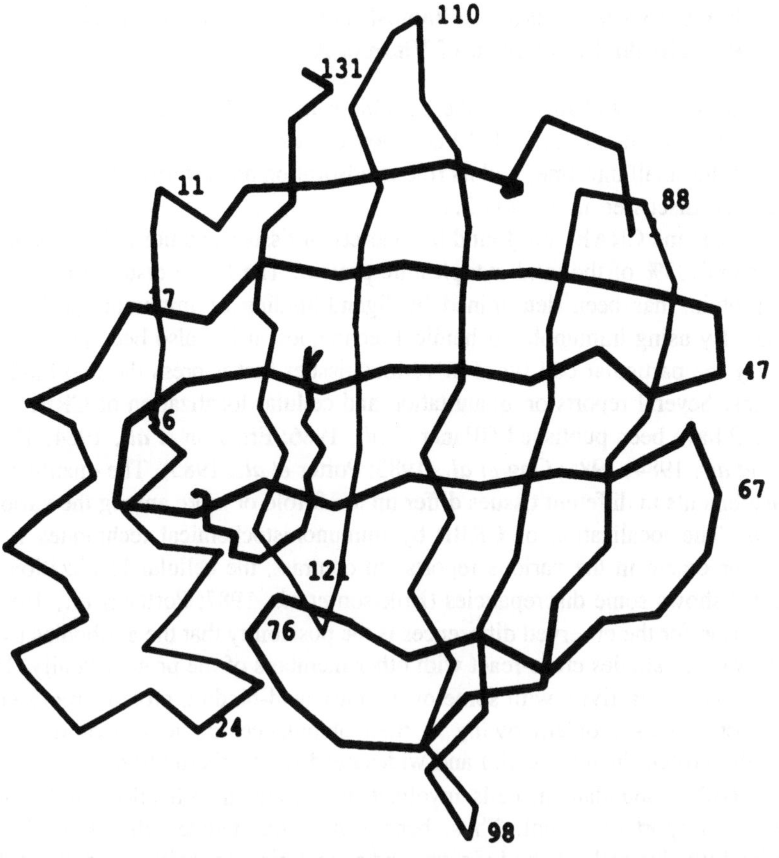

FIGURE 7. Three-dimensional structure of the myelin protein P2, a protein closely related to the cellular retinoid-binding proteins. The bound fatty acid is sandwiched between the two β-sheets. C_α-bonds closer to the viewer are drawn more thickly to enhance the three-dimensional effect. Reproduced from Jones *et al.* (1988) with permission of the publisher.

teins in a similar manner, the observed ligand specificities may be explained by the interactions between the polar end groups and polar or charged amino acid residues in the binding pocket of the proteins. Interestingly, the two arginine residues are totally conserved within the protein family, with the exception of CRBP and CRBPII, which both have glutamines in that position. Jones *et al.* (1988) speculated that these substitutions may account for the observed ligand specificity of the two CRBPs. However, this hypothesis does not fully take into account the fact that whereas CRBP exclusively binds retinoids with a hydroxyl end group, CRBPII is also able to bind retinal (MacDonald and Ong, 1987).

6.5. Expression of Cellular Retinoid-Binding Proteins and Their Roles in the Metabolism of Vitamin A

This section will focus on the possible roles of CRBP and CRABP in the metabolism of vitamin A. CRBPII is almost exclusively expressed in the enterocytes of the small intestine, and its role in adsorption and esterification of retinol have been discussed in Section 2.1.

CRBP and CRABP are found in a variety of tissues and normally constitute less than 0.05% of the total cytoplasmic protein. The tissue distribution of the two proteins has been determined by ligand-binding or immunological techniques. By using immunohistochemical techniques, it has also been possible to identify the particular cell types in various tissues that express the two binding proteins. Several reports on quantitation and cellular localization of CRBP and CRABP have been published (Blaner *et al.*, 1986; Eriksson *et al.*, 1984, 1987; Kato *et al.*, 1984, 1985; Ong *et al.*, 1983; Porter *et al.*, 1985). The quantitative measurements in different tissues differ up to 10-fold or more among the various reports. The localization of CRBP by immunohistochemical techniques is in good agreement in the various reports. In contrast, the cellular localization of CRABP shows some discrepancies (Eriksson *et al.*, 1987; Porter *et al.*, 1985). One reason for the observed differences is the possibility that the antibodies used in the various studies cross-react with other members of the protein family. The possible cross-reactivity with some of the fatty acid-binding proteins represents the most serious problem owing to their abundance (up to 8% of the total cytosolic protein in some cells) and widespread tissue distribution.

CRBP is abundant in cells involved in storage, mobilization, and transcellular transport of retinol. Thus, hepatocytes and stellate cells of the liver, proximal tubulus cells of the kidneys, and retinal pigment epithelial cells of the eyes, contain high levels of CRBP (Bok *et al.*, 1984; Eriksson *et al.*, 1984; Kato *et al.*, 1984). In these cells CRBP may have a role in the intracellular transport of retinol. A dual role of CRBP is likely, since the protein also is found in several other cell types, with no obvious function in trancellular transport of retinol. In the latter types of cells, CRBP may act as an intracellular acceptor of retinol following the receptor-mediated uptake from RBP.

The physiological role of CRABP is less obvious. Its ligand is not stored in the body or transported in plasma by a specific transport protein (Smith *et al.*, 1973). In view of the structural features shared by CRABP, CRBP, and the intracellular fatty acid-binding proteins, it is conceivable that CRABP also exhibits an acceptor and transport function. In contrast to CRBP, CRABP is found almost exclusively in vitamin A-sensitive tissues, e.g., various epithelia and reproductive organs. Since those cells obtain their vitamin A as retinol delivered via RBP, retinoic acid is likely to be generated intracellularly from retinol (Emerick *et al.*, 1967). The mechanism that regulates the conversion to retinoic acid is at present obscure.

The cellular expression of CRBP and CRABP is not regulated by the availability of their ligands (Blaner *et al.*, 1986; Eriksson *et al.*, 1984, 1987; Kato *et al.*, 1985). The exceptions are the expression of CRBP in stellate cells of the liver (Kato *et al.*, 1984) and of CRBP in lung tissue (Sherman *et al.*, 1987). A direct regulatory role of these proteins in the metabolism of vitamin A is consequently rather unlikely. Instead, the cellular retinoid-binding proteins may serve as buffer proteins and substrate carriers and render their ligands easily accessible in an aqueous milieu.

7. RETINOID UPTAKE AND TRANSPORT IN THE EYE

The molecular function of vitamin A in vision has been clearly established (Wald, 1968). However, the precise molecular mechanisms by which the vitamin is transferred from the extracellular RBP to its destination as the chromophore of the opsins are largely unknown. The transport of vitamin A from the blood across the retinal pigment epithelium and the interstitial space to the neuroretina of the eye involves several retinoid-binding proteins. Some of these proteins are involved in uptake and intracellular transport of vitamin A in a variety of tissues, whereas expression of other proteins are restricted to the eye.

The uptake of retinol from RBP to ocular tissues is mediated by RBP receptors specifically expressed in the basolateral membrane of the retinal pigment epithelial cells (Bok and Heller, 1976). The subsequent transfer of retinol across the retinal pigment epithelium is likely to be mediated by CRBP, which is abundant in the cytosol of these cells (Bok *et al.*, 1984). CRBP may also be involved in the esterification of retinol that occurs in the pigment epithelium (Saari and Bredberg, 1988). The process by which retinol is transferred across the apical membrane of the retinal pigment epithelial cells to the interstitial space is unknown. However, it appears that retinol eventually ends up in association with the interstitial retinol-binding protein, IRBP, localized in the interstitial space of the retina (Lai *et al.*, 1982; Liou *et al.*, 1982, 1989). IRBP is an abundant glycoprotein synthesized by the neuroretina, and *in vivo* this protein binds a variety of hydrophobic compounds including all-*trans* and 11-*cis* retinol. This protein may act as a shuttle vehicle for the transfer of retinol between the neuroretina and the retinal pigment epithelium. This conclusion is strengthened by the observation that the endogenous retinoids associated with IRBP vary between dark-adapted and bleached eyes (Adler and Martin, 1982; Liou *et al.*, 1982).

Recently, an additional retinol-binding protein was found to be expressed in the interstitial space of the eyes, namely, purpurin (Berman *et al.*, 1987). This protein displays extensive homology with RBP and binds retinol *in vitro*. Although purpurin was detected and isolated of the basis of its ability to promote attachment and growth of neurons, the striking homology to RBP suggests a physiological role as a retinol-binding and transport protein. The different roles of IRBP and purpurin in the interstitial transfer of retinoids between the retinal pigment epithelium and the neuroretina remain to be established.

As well as IRBP, purpurin, and the opsins, CRBP, CRABP, and the eye-specific cellular retinal-binding protein (CRA1BP; Crabb *et al.*, 1988) are expressed in the neuroretina. It has not been possible to determine the exact molecular role of most of these proteins in the retinoid metabolism of the eyes. In fact, the presence of CRABP may indicate that vitamin A can exert a role in the eyes other than vision. This can be inferred from the observation that retinoic acid, the ligand of CRABP, cannot be converted to retinol or retinal *in vivo* and hence can not serve as a precursor to the chromophore of the opsins, retinal (Dowling and Wald, 1960).

8. NUCLEAR RECEPTORS FOR RETINOIC ACID

The recognized effects of vitamin A on epithelial differentiation and growth led to suggestions that the vitamin was involved in the regulation of gene expression. By analogy with the action of steroid hormones, it was believed that the vitamin bound to cytoplasmic receptors and eventually was translocated to the nucleus. The discovery of two cytoplasmic vitamin A-binding proteins (CRBP and CRABP) displaying high affinity for the two most biologically active vitamin A derivatives, retinol and retinoic acid, substantiated the hypothesis (Bashor *et al.*, 1973; Ong and Chytil, 1975; Sani and Hill, 1974). The ability of CRBP to deliver its ligand to chromatin structures provided further evidence (Liau *et al.*, 1981; Takase *et al.*, 1979). The structural determinations of the cellular retinoid-binding proteins pointed out a different function for these proteins and indicated that other ligand-binding proteins might be responsible for the observed effects on gene expression. However, such proteins have remained elusive until recently. Two independent laboratories reported the existence of a retinoic acid-binding nuclear protein, RAR_{α}, related in structure to the steroid-binding nuclear receptors (Giguere *et al.*, 1988; Petkovich *et al.*, 1987). Later, two additional nuclear receptors for retinoic acid (RAR_{β} and RAR_{γ}) were identified (Brand *et al.*, 1988; Benbrook *et al.*, 1988; Zelent *et al.*, 1989). All three types of nuclear receptors bind retinoic acid with high affinity and are expressed in a wide range of tissues. Like steroid receptors, the retinoic acid receptors are ligand-inducible transacting factors; e.g., in the presence of ligand these receptors regulate the activity of a specific set of cellular genes [for a recent review of steroid receptors, see Evans (1988)]. The genes regulated by the retinoic acid receptors are not known at present. Identification of such genes is of primary interest and should, for the first time, give some insight into the nonocular function of vitamin A.

Considering that the nuclear retinoic acid receptors mediate the molecular function of vitamin A, it appears reasonable to assume that they are expressed in tissues and cells requiring vitamin A for proper function. Furthermore, the cellular retinoid-binding proteins should presumably be expressed in most cells containing the nuclear receptors. Measured by mRNA levels, RAR_{α} is most abundant in brain tissue, with only low, if any, expression in most other tissues (Giguere *et al.*, 1987). RAR_{β}, in contrast, has a wide tissue distribution and is abundant in various epithelia and in the reproductive organs (Benbrook *et al.*, 1988). In adult tissues, RAR_{γ} is restricted largely to the skin (Zelent *et al.*, 1989).

The mechanisms regulating the activity of the retinoic acid receptors, e.g., their ligand saturation, are obscure. The physiological significance of retinoic acid in blood is not known. Although retinoic acid is found at approximatively 10^{-8} M (De Leenheer *et al.*, 1982), it is conceivable that the concentration of free retinoic acid is several orders of magnitude lower as a result of the strong interaction with serum albumin (Smith *et al.*, 1973). The plasma level of retinol bound to RBP is significantly higher (2×10^{-6} M) and is almost constant. One possible site of regulation is the intracellular conversion from retinol to retinoic acid (Emerick *et al.*, 1967). This process is not well characterized, and the possible factors affecting it are unknown. It is conceivable that the intracellular retinoid-binding proteins are important in the metabolism of retinol and retinoic acid. Although both proteins are constitutively expressed at significantly higher concentrations than the nuclear receptors, they may be part of a cellular machinery that regulates the concentration of retinoic acid available to the nuclear receptors.

9. CONCLUSIONS

Studies on vitamin A have identified an essential role of retinoids on epithelial growth and differentiation. The transport routes for the vitamin from the diet to the nucleus of a target cell are well characterized. In particular, the structures of several extra- and intracellular retinoid-binding proteins are known. The forthcoming characterization of the RBP receptor should shed light on the mechanism by which cells accumulate retinol and how retinol is transferred across the plasma membrane.

The recent discovery of nuclear retinoic acid receptors indicate that retinoic acid is a hormone. Identification of genes whose activity is regulated by this compound is of primary interest. The isolation and characterization of such genes may indicate that retinoids exert more widespread functions in cellular physiology than previously thought.

10. REFERENCES

Adler, A. J., and Martin, K. J., 1982, Retinol-binding proteins in bovine interphotoreceptor matrix, *Biochem. Biophys. Res. Commun.* **108:**1601–1608.

Alpers, D. H., Strauss, A. W., Ockner, R. K., Bass, N. M., and Gordon, J. I., 1984, Cloning of a cDNA encoding rat intestinal fatty acid-binding protein, *Proc. Natl. Acad. Sci. U.S.A.* **81:**313–317.

Åqvist, J., Sandblom, P., Jones, T. A., Newcomer, M. E., van Gunsteren, W. F., and Tapia, O., 1986, Molecular dynamics simulations of the holo and apo forms of retinol binding protein, *J. Mol. Biol.* **192:**593–604.

Bailey, J. S., and Siu, C.-H., 1988, Purification and partial characterization of a novel binding protein for retinoic acid from neonatal rat, *J. Biol. Chem.* **263:**9326–9332.

Bashor, M. M., Toft, D. O., and Chytil, F., 1973, *In vitro* binding of retinol to rat-tissue components, *Proc. Natl. Acad. Sci. U.S.A.* **70:**3483–3487.

Benbrook, D., Lernhardt, E., and Pfahl, M., 1988, A new retinoic acid receptor identified from a heptocellular carcinoma, *Nature* **333:**669–672.

Berman, P., Gray, P., Chen, E., Keyser, K., Ehrlich, D., Karten, H., LaCorbiere, Esch, F., and Schubert, D., 1987, Sequence analysis, cellular localization and expression of a neuroretina adhesion and cell survival molecule, *Cell* **51:**135–142.

Bernlohr, D. A., Anjus, C. W., Lane, M. D., Bolanowski, M. A., and Kelly, T. J., Jr., 1984, Expression of specific mRNAs during adipose differentiation: Identification of an mRNA encoding a homologue of myelin P2 protein, *Proc. Natl. Acad. Sci. U.S.A.* **81:**5468–5472.

Blaner, W. S., Hendriks, H. F. J., Brouwer, A., de Leeuw, A. M., Knook, D., and Goodman, D. S., 1985, Retinoids, retinoid-binding proteins, and retinyl palmitate hydrolase distribution in different types of rat liver cells, *J. Lipid Res.* **26:**1241–1251.

Blaner, W. S., Das, K., Mertz, J. R., Das, S. R., and Goodman, D. S., 1986, Effects of dietary retinoic acid on cellular retinol- and retinoic acid binding protein levels in various rat tissues, *J. Nutr.* **27:**1084–1088.

Blaner, W. S., Dixon, J. L., Moriwaki, H., Martino, R. A., Stein, Y., and Goodman, D. S., 1987, Studies on the *in vivo* transfer of retinoids from parenchymal to stellate cells in rat liver, *Eur. J. Biochem.* **164:**301–307.

Blomhoff, R., Helgerud, P., Rasmussen, M., Berg, T., and Norum, K. R., 1982, In vivo uptake of chylomicron [^{3}H] retinyl esters by rat liver: Evidence for retinol transfer from parenchymal to non parenchymal cells, *Proc. Natl. Acad. Sci. U.S.A.* **79:**7326–7330.

Blomhoff, R., Helgerud, P., Dueland, S., Berg, T., Pederson, J. I., Norum, K. R., and Drevon, C. A., 1984a, Lymphatic absorption and transport of retinol and vitamin D-3 from rat intestine, *Biochim. Biophys. Acta* **772:**109–116.

Blomhoff, R., Holte, K., Naess, L., and Berg, T., 1984b, Newly administrated [^{3}H] retinol is transferred from hepatocytes to stellate cells in liver for storage, *Exp. Cell Res.* **150:**186–193.

Blomhoff, R., Eskild, W., Kindberg, G. M., Prydz, K., and Berg, T., 1985a, Intracellular transport of endocytosed chylomicron [^{3}H] retinyl ester in rat liver parenchymal cells: Evidence for translocation of a [^{3}H] retinoid from endosomes to endoplasmic reticulum, *J. Biol. Chem.* **260:**13566–13570.

Blomhoff, R., Rasmussen, M., Nilsson, A., Norum, K. R., Blaner, W. S., Kato, M., Mertz, J. R., Goodman, D. S., Eriksson, U., and Peterson, P. A., 1985b, Hepatic retinol metabolism:

Distribution of retinoids, enzymes, and binding proteins in isolated rat liver cells, *J. Biol. Chem.* **260:**13560–13565.

Blomhoff, R., Norum, K. R., and Berg, T., 1985c, Hepatic uptake of [^{3}H]retinol bound to the serum retinol binding protein involves both parenchymal and perisinusiodal stellate cells, *J. Biol. Chem.* **260:**13571–13575.

Blomhoff, R., Berg, T., and Norum, K. R., 1988, Transfer of retinol from parenchymal to stellate cells in liver is mediated by retinol-binding protein, *Proc. Natl. Acad. Sci. U.S.A.* **85:**3455–3458.

Böhmer, F.-D., Kraft, R., Otto, A., Wernstedt, C., Hellman, U., Kurtz, A., Muller, T., Rohde, K., Etzold, G., Lehmann, W., Langen, P., Heldin, C.-H., and Grosse, R., 1987a, Identification of a polypeptide growth inhibitor from bovine mammary gland (MGDI)—sequence homology to fatty acid and retinoid-binding proteins, *J. Biol. Chem.* **262:**15137–15143.

Böhmer, F.-D., Sun, Q., Pepperle, M., Muller, T., Eriksson, U., Wang, J., and Grosse, R., 1987b, Antibodies against mammary derived growth inhibitor (MGDI) react with a fibroblast growth inhibitor and with heart fatty acid binding protein, *Biochem. Biophys. Res. Commun.* **148:**1425–1431.

Bok, D., and Heller, J., 1976, Transport of retinol from the blood to the retina: An autoradiographic study of the pigment epithelial cell surface receptor for plasma retinol-binding protein, *Exp. Eye Res.* **22:**395–402.

Bok, D., Ong, D. E., and Chytil, F., 1984, Immunocytochemical localization of cellular retinol-binding protein in the rat retina, *Ophthalmol. Visual Sci.* **25:**877–883.

Brand, N., Petkovich, M., Krust, A., Chambon, P., de The, H., Marchio, A., Tiollais, P., and Dejean, A., 1988, Identification of a second human retinoic acid receptor, *Nature* **332:**850–853.

Brooks, D. E., Means, A. R., Wright, E. J., Singh, S. P., and Triver, K. K. J., 1986, Molecular cloning of the cDNA for two major androgen-dependent secretory proteins of 18.5 kilodaltons synthesized by the rat epididymis, *J. Biol. Chem.* **261:**4956–4961.

Brown, M. S., Korvanen, P. T., and Goldstein, J. L., 1981, Regulation of plasma cholesterol by lipoprotein receptors, *Science,* **212:**628–635.

Colantuoni, V., Romano, V., Bensi, G., Santoro, C., Constanzo, F., Raugei, G., and Cortese, R., 1983, Cloning and sequencing of a full-length cDNA coding for human retinol-binding protein, *Nucleic Acids Res.* **11:**7769–7776.

Colantuoni, V., Cortese, R., Nilsson, M., Lundvall, J., Båvik, C.-O., Eriksson, U., Peterson, P. A., and Sundelin, J., 1985, Cloning and sequencing of a full length cDNA corresponding to human cellular retinol-binding protein, *Biochem. Biophys. Res. Commun.* **130:**431–439.

Crabb, J., Johnson, C. M., Carr, S. A., Armes, L. G., and Saari, J. C., 1988, The complete primary structure of the cellular retinaldehyde-binding protein from bovine retina, *J. Biol. Chem.* **263:**18678–18687.

Crow, J. A., and Ong, D. E., 1985, Cell-specific immunohistochemical localization of a cellular retinol-binding protein (type two) in the small intestine of rat, *Proc. Natl. Acad. Sci. U.S.A.* **82:**4707–4711.

Dayhoff, M. O., 1978, in *Atlas of Protein Sequence and Structure* (Dayhoff, M. O., ed.) Vol. 5, Suppl. 3, National Biomedical Research Foundation, Washington, D.C.

De Leenheer, A. P., Lambert, W. E., Claeys, I., 1982, All-trans-retinoic acid: Measurement of reference values in human serum by high performance liquid chromatography, *J. Lipid Res.* **23:**1362–1367.

Demmer, L. A., Birkenmeier, E. H., Sweetser, D. A., Levin, M. S., Zollman, S., Sparkes, R. S., Mohandas, T., Luisis, A. J., and Gordon, J., 1987, The cellular retinol binding protein II gene: Sequence analysis of the rat gene, chromosomal localization in mice and humans, and documentation of its close linkage to the cellular retinol binding protein gene, *J. Biol. Biol.* **262:**2458–2467.

D'Onofrio, C., Colantuoni, V., and Cortese, R., 1985, Structure and cell-specific expression of a cloned human retinol binding protein gene: The 5'-flanking region contains hepatoma specific transcriptional signals, *EMBO J.* **4**:1981–1989.

Dowling, J. E., and Wald, G., 1960, The biological function of vitamin A acid, *Proc. Natl. Acad. Sci. U.S.A.* **46**:587–609.

Drayna, D., Fielding, C., McLean, J., Bear, B., Castro, G., Chen, E., Constock, L., Henzel, W., Kohr, W., Rhee, L., Wion, K., and Lawn, R., 1986, Cloning and expression of human apolipoprotein D cDNA, *J. Biol. Chem.* **261**:16535–16539.

Dueland, S., Helgerud, P., Pedersen, J. I., Berg, T., and Drevon, C. A., 1983, Plasma clearence, transfer and distribution of vitamin D_3 from intestinal lymph, *Am. J. Physiol.* **245**:E326–E331.

Emerick, R. J., Zile, M., and DeLuca, H., 1967, Formation of retinoic acid from retinol in the rat, *Biochem. J.* **102**:606–611.

Eriksson, U., Sundelin, J., Rask, L., and Peterson, P. A., 1981, The NH_2-terminal amino acid sequence of cellular retinoic acid-binding protein from rat testis, *FEBS Lett.* **135**:70–72.

Eriksson, U., Das, K., Busch, C., Nordlinder, H., Rask, L., Sundelin, J., Sällström, J., and Peterson, P. A., 1984, Cellular retinol-binding protein: Quantitation and distribution, *J. Biol. Chem.* **259**:13464–13470.

Eriksson, U., Hansson, E., Nordlinder, H., Busch, C., Sundelin, J., and Peterson, P. A., 1987, Quantitation and tissue localization of the cellular retinoic acid-binding protein, *J. Cell. Physiol.* **133**:482–490.

Erlandson, C., and Borgström, B., 1968, The identity of vitamin A esterase activity of rat pancreatic juice, *Biochim. Biophys. Acta* **167**:629–631.

Evans, R. M., 1988, The steroid and thyroid hormone receptor superfamily, *Science* **240**:889–895.

Fries, E., Gustavsson, L., and Peterson, P. A., 1984, Four secretory proteins synthesized by hepatocytes are transported from endoplasmic reticulum to Golgi complex at different rates *EMBO J.* **3**:147–152.

Fugate, R. D., and Song, P.-S., 1980, Spectroscopic characterization of β-lactoglobulin-retinol complex, *Biochim. Biophys. Acta* **625**:28–42.

Giguere, V., Ong, E. S., Segui, P., and Evans, R. M., 1988, Identification of a receptor for the morphogen retinoic acid, *Nature* **330**:624–629.

Gjöen, T., Bjerkelund, T., Kiil Blomhoff, H., Norum, K. R., Berg, T., and Blomhoff, R., 1986, Liver takes up retinol-binding proteins from plasma, *J. Biol. Chem.* **262**:10926–10930.

Godovac-Zimmerman, J., Conti, A., Liberatori, J., and Braunitzer, G., 1985, Homology between the primary structure of β-lactoglobulin and human retinol-binding protein: Evidence for a similar biological function, *Biol. Chem. Hoppe-Seyler* **366**:431–434.

Goodman, D. S., and Blaner, W. S., 1984, Biosynthesis, adsorption and hepatic metabolism of retinol, in *The Retinoids* (M. B. Sporn, A. B. Roberts, and D. S. Goodman, eds.), Vol. 2, pp. 1–39, Academic Press, New York.

Goodman, D. S., Blomstrand, R., Werner, B., Huang, H. S., and Shiratori, T., 1966, The intestinal absorption and metabolism of vitamin A and β-carotene in man, *J. Clin. Invest.* **45**:1615–1623.

Harrison, E. H., Smith, J. E., and Goodman, D. S., 1980, Effects of vitamin A deficiency on the levels and distribution of retinol-binding protein and marker enzymes in homogenates and Golgi-rich fractions of rat liver, *Biochim. Biophys. Acta* **628**:489–497.

Helgerud, P., Peterson, L. B., and Norum, K. R., 1982, Acyl CoA:retinol acyltransferase in rat small intestine: Its activity and some properties of the enzymatic reaction, *J. Lipid Res.* **23**:609–618.

Helgerud, P., Peterson, L. B., and Norum, K. R., 1983, Retinol esterification by microsomes from the mucosa of human small intestine: Evidence for acyl-coenzyme A retinol acyltransferase activity, *J. Clin. Invest.* **71**:747–753.

Heller, J., 1975, Interactions of plasma retinol-binding protein with its receptor. Specific binding of bovine and human retinol-binding protein to pigment epithelium cells from bovine eyes, *J. Biol. Chem.* **250**:3613–3619.

Hendershot, L., Bole, D., and Kearney, J. F., 1987, The role of immunoglobulin heavy chain binding protein in immunoglobulin transport, *Immunol. Today* **8**:111–114.

Hendriks, H. F. J., Brouwer, A., and Knook, D. L., 1987, The role of hepatic fat-storing (stellate) cells in retinoid metabolism, *Hepatology* **7**:1368–1371.

Holden, H. M., Rypniewski, W. R., Law, J. H., and Rayment, I., 1987, The molecular structure of insecticyanin from the tobacco hornworm *Manduca sexta* L. at 2.6 Å resolution, *EMBO J.* **6**:1565–1570.

Hollander, D., 1981, Intestinal absorption of vitamins A, E, D, and E., *J. Lab. Clin. Med.* **97**:449–462.

Horwitz, J., and Heller, J., 1973, Interactions of all-trans-, 9,-11-, and 13-cis retinal, all-trans retinyl acetate, and retinoic acid with human retinol-binding protein and prealbumin, *J. Biol. Chem.* **248**:6317–6324.

Horwitz, J., and Heller, J., 1974, Modification of tryptophan residues in retinol-binding protein and prealbumin with 2-hydroxy-5-nitrobenzyl bromide: Effects of the modification on the protein retinol and protein-protein interaction, *J. Biol. Chem.* **249**:7181–7185.

Huang, H. S., and Goodman, D. S., 1965, Vitamin A and carotinoids. I. Intestinal adsorption and metabolism of ^{14}C-labelled vitamin A alcohol and β-carotene in the rat, *J. Biol. Chem.* **240**:2839–2844.

Huber, R., Schneider, M., Mayr, I., Muller, R., Deutzmann, R., Suter, F., Zuber, H., Falk, H., and Kayser, H., 1987, Molecular structure of the bilin-binding protein (BBP) from *Pieris brassicae* after refinement at 2.0 Å resolution, *J. Mol. Biol.* **198**:499–513.

Hunt, C. R., Ro, J. H.-S., Dobson, D. E., Min, H. Y., and Spiegelman, B. M., 1986, Adipocyte P2 gene: Developemental expression and homology of 5′-flanking sequences among fat cell-specific genes, *Proc. Natl. Acad. Sci. U.S.A.* **83**:3786–3790.

Jones, T. A., Bergfors, T., Sedzik, J., and Unge, T., 1988, The three-dimensional structure of P2 myelin protein, *EMBO J.* **7**:1597–1604.

Kassenbrock, C. K., Garcia, P. D., Walter, P., and Kelly, R. B., 1988, Heavy-chain binding protein recognizes aberrant polypeptides translocated in vitro, *Nature* **333**:90–94.

Kato, M., Kato, K., and Goodman, D. S., 1984, Immunocytochemical studies on the localization of plasma and of cellular retinol-binding proteins and of transthyretin (prealbumin) in rat liver and kidney, *J. Cell Biol.* **98**:1698–1704.

Kato, M., Blaner, W. S., Mertz, J. R., Das, K., Kato, K., and Goodman, D. S., 1985, Influence of retinoid status on cellular retinol and cellular retinoic acid-binding protein concentrations in various rat tissues, *J. Biol. Chem.* **260**:4832–4838.

Kitamura, K., Suzuki, M., Suzuki, A., and Uyemura, K., 1980, The complete amino acid sequence of the P2 protein in bovine peripheral nerve myelin, *FEBS Lett.* **115**:27–30.

Lai, Y. L., Wiggert, B., Liu, Y. P., and Chader, G. J., 1982, Interphotoreceptor retinol-binding proteins: Possible transport vehicles between compartments of the retina, *Nature* **298**:848–849.

Laurent, B. C., Nilsson, M. H. L., Båvik, C.-O., Jones, T. A., Sundelin, J., and Peterson, P. A., 1985, Characterization of the rat retinol-binding protein gene and its comparison to the three-dimensional structure of the protein, *J. Biol. Chem.* **260**:11476–11480.

Lee, H. K., Wells, R. G., and Reed, B. R., 1987, Isolation of an olfactory cDNA: Similarity to retinol-binding protein suggests a role in olfaction, *Science* **235**:1053–1056.

Li, E., Demmer, L. A., Sweefser, D. A., Ong, D. E., and Gordon, J. I., 1986, Rat cellular retinol-binding protein II: Use of a cloned cDNA to define its primary structure, tissue specific expression and development regulation, *Proc. Natl. Acad. Sci. U.S.A.* **83**:5779–5783.

Liau, G., Ong, D. E., and Chytil, F., 1981, Interaction of the retinol-cellular retinol-binding protein complex with isolated nuclei and nuclear components, *J. Cell Biol.* **91**:63–68.

Liou, G. I., Bridges, C. D. B., Fong, S. L., Alvarez, R. A., and Gonzalez-Fernandez, F., 1982, Vitamin A transport between retina and pigment epithelium—an interstitial protein carrying endogenous retinol (interstitial retinol-binding protein), *Vision Res.* **22:**1457–1467.

Liou, G. I., Ma, D.-P., Yang, Y.-W., Geng, L., Zhu, C., and Baehr, W., 1989, Human interstitial retinoid-binding protein. Gene structure and primary sequence, *J. Biol. Chem.* **264:**8200–8206.

Lodish, H. F., Kong, N., Snider, M., and Strous, G. J. A. M., 1983, Hepatoma secretory proteins migrate from rough endoplasmic reticulum to Golgi at characteristic rates, *Nature* **304:**80–83.

Lombardo, D., and Guy, O., 1980, Studies on the substrate specificity of the carboxyl ester hydrolase from human pancreatic juice: II. Action on cholesterol esters and lipid-soluble vitamin esters, *Biochim. Biophys. Acta* **611:**147–155.

Lopez, C., Grubb, A., Soriani, F., and Mendez, E., 1981, The complete amino acid sequence of human complex-forming glycoprotein heterogenous in charge (protein HC), *Biochem. Biophys. Res. Commun.* **103:**919–925.

MacDonald, P. N., and Ong, D. E., 1987, Binding specificities of cellular retinol-binding protein and cellular retinol-binding protein, type II, *J. Biol. Chem.* **262:**10550–10556.

Mak, K. M., Leo, M. A., and Lieber, C. S., 1984, Alcoholic liver injury in baboons: Transformation of lipocytes to transitional cells, *Gastroenterology* **87:**188–200.

Mallia, A. K., Smith, J. E., and Goodman, D. S., 1975, Metabolism of retinol-binding protein and vitamin A during hypervitaminosis A in the rat, *J. Lipid Res.* **16:**180–188.

McGuire, B. W., Orgebin-Crist, M.-C., and Chytil, F., 1981, Autoradiographic localization of serum retinol-binding protein in rat testis, *Endocrinology* **108:**658–667.

Newcomer, M. E., Liljas, A., Erikssson, U., Sundelin, J., Rask, L., and Peterson, P. A., 1981, Crystallization of and preliminary X-ray data for an intracellular vitamin A-binding protein from rat liver, *J. Biol. Chem.* **256:**8162–8163.

Newcomer, M., Jones, T. A., Åqvist, J., Sundelin, J., Eriksson, U., Rask, L., and Peterson, P. A., 1984, The three-dimensional structure of retinol-binding protein, *EMBO J.* **3:**1451–1454.

Nilsson, M. H. L., Spurr, N. K., Lundwall, J., Rask, L., and Peterson, P. A., 1988a, Human CRBP gene organization and chromosomal location, *Eur. J. Biochem.* **173:**38–44.

Nilsson, M. H. L., Spurr, N. K., Sakena, P., Busch, C., Nordlinder, H., Peterson, P. A., and Sundelin, J., 1988b, Isolation and characterization of a cDNA clone corresponding to bovine cellular retinoic acid-binding protein and chromosomal localization of the corresponding human gene, *Eur. J. Biochem.* **173:**45–51.

Nordlinder, H., Busch, C., Eriksson, U., Gustafsson, S., Mantovaara-Jönsson, T., Nilsson, M., Pertoft, H., and Sakena, P., 1989, Percoll purified subpopulations of stellate cells studied by immunocytochemistry and in situ hybridization, in *Cells of the Hepatic Sinusoid* (E. Wisse, D. Knook, and K. Decker, eds.), Vol. 2, pp. 490–492, The Kupffer Cell Foundation, Rijswijk, The Netherlands.

Norum, K. R., Helgerud, P., Peterson, L. B., Groot, P. H. E., and De Jonge, H. R., 1983, Influence of diets on acyl-CoA:cholesterol acyltransferase and on acyl-CoA:retinol acyltransferase in villous and crypt cells from rat small intestinal mucosa and in the liver, *Biochim. Biophys. Acta* **751:**153–161.

Ong, D. E., 1984, A novel retinol-binding protein from rat: Purification and partial characterization, *J. Biol. Chem.* **259:**1476–1482.

Ong, D. E., and Chytil, F., 1975, Retinoic acid-binding protein in rat tissue, *J. Biol. Chem.* **250:**6113–6116.

Ong, D. E., Crow, J. A., and Chytil, F., 1983, Radioimmunochemical determination of cellular retinol and cellular retinoic acid-binding proteins in cytosols of rat tissues, *J. Biol. Chem.* **257:**13385–13389.

Ong, D. E., Kakkad, B., and MacDonald, P., 1987, Acyl-CoA-independent esterification of retinol bound to cellular retinol-binding protein (type II) by microsomes from rat small intestine, *J. Biol. Chem.* **262:**2729–2736.

Ong, D. E., MacDonald, P. N., and Gubitosi, A. M., 1988, Esterification of retinol in rat liver. Possible participation by cellular retinol-binding protein and cellular retinol-binding protein II, *J. Biol. Chem.* **263:**5789–5796.

Papiz, M. Z., Sawyer, L., Eliopoulos, E. E., North, A. C. T., Findlay, J. B. C., Sivaprasadarao, R., Jones, T. A., Newcomer, M. E., and Kraulis, P. J., 1986, The structure of β-lactoglobulin and its similarity to plasma retinol-binding protein, *Nature* **324:**383–385.

Pervaiz, S., and Brew, K., 1987, Homology and structure-function correlations between α_1-acid glycoprotein and serum retinol-binding protein and its relatives, *FASEB J.* **1:**209–214.

Peterson, P. A., 1971, Studies on the interaction between prealbumin, retinol-binding protein and vitamin A, *J. Biol. Chem.* **246:**44–49.

Peterson, P. A., Rask, L., Ostberg, L., Anderson, L., Kamwendo, F., and Pertoft, H., 1973, Studies on the transport and cellular distribution of vitamin A in normal and vitamin A-deficient rats with special reference to the vitamin A-binding plasma protein, *J. Biol. Chem.* **248:**4009–4022.

Petkovich, M., Brand, N. J., Krust, A., and Chambon, P., 1987, A human retinoic acid receptor which belongs to the family of nuclear receptors, *Nature* **330:**444–450.

Porter, S. B., Ong, D. E., Chytil, F., and Orgebin-Crist, M. C., 1985, Localization of cellular retinol-binding protein and cellular retinoic acid-binding protein in rat testis and epidymis, *J. Androl.* **6:**197–212.

Rask, L., and Peterson, P. A., 1976, In vitro uptake of vitamin A from retinol-binding plasma protein to mucosal epithelial cells of the monkeys small intestine, *J. Biol. Chem.* **251:**6360–6366.

Rask, L., Anundi, H., and Peterson, P. A., 1979, The primary structure of the human retinol-binding protein, *FEBS Lett.* **104:**55–58.

Rask, L., Valterson, C., Anundi, H., Kvist, S., Eriksson, U., Dallner, G., and Peterson, P. A., 1983, Subcellular localization in normal and vitamin A-deficient rat liver of vitamin A serum transport proteins, albumin, ceruloplasmin and class 1 major histocompatibility antigens, *Exp. Cell Res.* **143:**91–102.

Rask, L., Anundi, H., Fohlman, J., and Peterson, P. A., 1987, The complete amino acid sequence of human retinol-binding protein, *Upsala J. Med. Sci.* **92:**115–146.

Riley, C. T., Barbeau, B. K., Keim, P. S., Kezdy, F. J., Heinriksson, R. L., and Law, J. H., 1984, The covalent structure of insecticyanin, a blue biliprotein from the hemolymph of the tobacco hornworm, *Macata sexta* L., *J. Biol. Chem.* **259:**13159–13165.

Ronne, H., Ocklind, K., Wiman, K., Rask, L., Öbrink, B., and Peterson, P. A., 1983, Ligand-dependent regulation of intracellular transport: Effect of vitamin A on the secretion of the retinol-binding protein, *J. Cell Biol.* **96:**907–910.

Ross, A. C., Go, K. J., Heider, J. G., and Rothblat, G. H., 1984, Selective inhibition of acyl coenzyme A:cholesterol acyltransferase by compound 58-035, *J. Biol. Chem.* **259:**815–819.

Saari, J., and Bredberg, D. L., 1988, CoA- and non-CoA-dependent retinol esterification in retinal pigment epithelim, *J. Biol. Chem.* **263:**8084–8090.

Saari, J., Bredberg, D. L., and Garwin, G. G., 1982, Identification of the endogenous retinoids associated with three cellular retinoid-binding proteins from bovine retina and retinal pigment epithelim, *J. Biol. Chem.* **257:**13329–13333.

Sacchettini, J. C., Said, B., Schultz, J., and Gordon, J. I., 1986, Rat heart fatty acid-binding protein is highly homologous to the murine adipocyte 422 protein and the P2 protein of peripheral nerve myelin, *J. Biol. Chem.* **261:**8218–8223.

Sacchettini, J. C., Stockhausen, D., Li, E., Banaszak, L. J., and Gordon, J. I., 1987, Crystallization of rat cellular retinol binding protein II: Preliminary X-ray data obtained from the apoprotein expressed in *Escherichia coli, J. Biol. Chem.* **262:**15756–15758.

Sani, B. P., and Hill, D. L., 1974, Retinoic acid: A binding protein in chicken embryo metatarsal skin, *Biochem. Biophys. Res. Commun.* **61:**1276–1282.

Sawyer, L., Papiz, M. Z., North, A. C. T., and Eliopoulos, S. E., 1985, Structure and function of bovine β-lactoglobulin, *Biochem. Soc. Trans.* **13**:265–266.

Schubert, D., LaCorbiere, M., and Esch, F., 1986, A chick neural retina adhesion and survival molecule is a retinol-binding protein, *J. Cell Biol.* **102**:2295–2301.

Sherman, D. R., Lloyd, R. S., and Chytil, F., 1987, Rat cellular retinol-binding protein: cDNA sequence and rapid retinol-dependent accumulation of mRNA, *Proc. Natl. Acad. Sci. U.S.A.* **84**:3209–3213.

Shubeita, H. E., Sambrook, J. F., and McCormick, A. M., 1987, Molecular cloning and analysis of functional cDNA and genomic clones encoding bovine cellular retinoic acid-binding protein, *Proc. Natl. Acad. Sci. U.S.A.* **84**:5645–5649.

Sivaprasadarao, A., and Findlay, J. B. C., 1988, The interaction of retinol-binding protein with its plasma-membrane receptor, *Biochem. J.* **255**:561–569.

Smith, J. E., Milch, P. O., Muto, Y., and Goodman, D. S., 1973, The plasma transport of retinoic acid in the rat, *Biochem. J.* **132**:821–827.

Smith, J. E., Muto, Y., and Goodman, D. S., 1975, Tissue distribution and subcellular localization of retinol-binding protein in normal and vitamin A-deficient rats, *J. Lipid Res.* **16**:318–323.

Soprano, D. R., Pickett, C. B., Smith, J. E., and Goodman, D. S., 1981, Biosynthesis of plasma retinol-binding protein as a larger molecular weight precursor, *J. Biol. Chem.* **256**:8256–8258.

Sundelin, J., Laurent, B. C., Anundi, H., Trägårdh, L., Larhammar, D., Björck, L., Eriksson, U., Åkerström, B., Jones, A., Newcomer, M., Peterson, P. A., and Rask, L., 1985a, Amino acid sequence homologies between rabbit, rat, and human serum retinol-binding proteins, *J. Biol. Chem.* **260**:6472–6480.

Sundelin, J., Anundi, H., Tragardh, L., Eriksson, U., Lind, P., Ronne, H., Peterson, P. A., and Rask, L., 1985b, The primary structure of rat liver cellular retinol-binding protein, *J. Biol. Chem.* **260**:6488–6493.

Sundelin, J., Das, S., Eriksson, U., Rask, L., and Peterson, P. A., 1985c, The primary structure of bovine cellular retinoic acid-binding protein, *J. Biol. Chem.* **260**:6494–6499.

Sweetser, D. A., Lowe, J. B., and Gordon, J. I., 1986, The nucleotide sequence of the rat fatty acid-binding protein gene, *J. Biol. Chem.* **261**:5553–5561.

Takahashi, K., Odani, S., and Ono, R., 1982, The primary structure of rat liver z-protein, *FEBS Lett.* **115**:27–30.

Takase, S., Ong, D. E., and Chytil, F., 1979, Cellular retinol-binding protein allows specific interaction of retinol with the nucleus *in vitro Proc. Natl. Acad. Sci. U.S.A.* **76**:2204–2208.

Thaller, C., and Eichele, G., 1987, Identification and spatial distribution of retinoids in the developing chick limb bud, *Nature* **327**:625–626.

Unterman, R. D., Lynch, K. R., Nakhasi, H. L., Dolan, K. P., Hamilton, J. W., Cohn, D. V., and Fiegelson, P., 1981, Cloning and sequencing of several $\alpha_2\mu$-globulin cDNAs, *Proc. Natl. Acad. Sci. U.S.A.* **78**:3478–3482.

Uyemura, K., Yoshimura, K., Suzuki, M., and Kitamura, K., 1984, Lipid-binding activities of the P2 protein in peripheral nerve myelin, *Neurochem. Res.* **9**:1509–1514.

Vaessen, M.-J., Kootwijk, E., Mummery, C., Hilkens, J., Bootsma, D., and van Kessel, A. G., 1989, Prefrential expression of cellular retinoic acid-binding protein in a subpopulation of neural cells in the developing mouse embryo, *Differentiation* **40**:99–105.

Vahlquist, A., Peterson, P. A., and Wibell, L., 1973, Metabolism of the vitamin A transporting protein complex, *Eur. J. Clin. Invest.* **3**:352–362.

Wake, K., 1980, Perisinuosidal stellate cells (fat-storing, interstitial cells, lipocytes), their related structure in and around the liver sinusoids, and vitamin A storing cells in extrahepatic organs, *Int. Rev. Cytol.* **66**:303–353.

Wald, G., 1968, Molecular basis of visual excitations, *Science* **162**:230–239.

Wolf, G., 1980, Vitamin A, in *Human Nutrition* (R. B. Alfin-Slater and D. Kritchevsky, eds.), Vol. 3B, pp. 97–203, Plenum Press, New York.

Yamada, M., Blaner, W. S., Soprano, D. R., Dixon, J. L., Keldbye, H. M., and Goodman, D. S., 1987, Biochemical characteristics of isolated rat liver stellate cells, *Hepatology* **7**:1224–1229.
Zelent, A., Krust, A., Petkovich, M., Kastnar, P., and Chambon, P., 1989, Cloning of murine α and β retinoic acid receptors and a novel receptor γ predominantly expressed in skin, *Nature* **339**:714–717.

Index

MIX
Papier aus verantwortungsvollen Quellen
Paper from responsible sources
FSC® C105338
FSC
www.fsc.org

If you have any concerns about our products,
you can contact us on
ProductSafety@springernature.com

In case Publisher is established outside the EU,
the EU authorized representative is:
Springer Nature Customer Service Center GmbH
Europaplatz 3, 69115 Heidelberg, Germany

Printed by Libri Plureos GmbH
in Hamburg, Germany